Biorefining of Renewable Diesel

From Production to Market

Green Chemistry Series

For a list of titles in this series see: rsc.li/green-chemistry

How to obtain future titles on publication:
A standing order plan is available for this series. A standing order will bring delivery of each new volume immediately on publication.

For further information please contact:
Book Sales Department, Royal Society of Chemistry, Thomas Graham House, Science Park, Milton Road, Cambridge, CB4 0WF, UK
Telephone: +44 (0)1223 420066, Fax: +44 (0)1223 420247
Email: booksales@rsc.org
Visit our website at books.rsc.org

Biorefining of Renewable Diesel
From Production to Market

Edited by

Mohammad Aslam
National Institute of Technology Srinagar, India
Email: maslam@nitsri.ac.in

Sanjeev Mishra
Sardar Swaran Singh National Institute of Bioenergy, India
Email: sanjeevmishra@nibe.res.in

and

Jorge Aburto
Mexican Petroleum Institute, Mexico
Email: jaburto@imp.mx

ROYAL SOCIETY OF CHEMISTRY

Green Chemistry Series No. 89

EPUB ISBN: 978-1-83767-562-3
Hardback ISBN: 978-1-83767-262-2
PDF ISBN: 978-1-83767-561-6
PDF ISBN: 978-1-83767-561-6
Hardback ISBN: 978-1-83767-262-2
Print ISSN: 1757-7039
Electronic ISSN: 1757-7047

A catalogue record for this book is available from the British Library

The Royal Society of Chemistry is a charity, registered in England and Wales, Number 207890, and a company incorporated in England by Royal Charter (Registered No. RC000524), registered office: Burlington House, Piccadilly, London W1J 0BA, UK, Telephone: +44 (0)20 7437 8656.

For further information see our website at www.rsc.org

For general enquiries, please contact books@rsc.org

For EU product safety enquiries, please email books@rsc.org or contact Royal Society of Chemistry Worldwide (Germany) GmbH, Römischer Hof, Unter den Linden 10, 10117 Berlin.

Printed in the United Kingdom by CPI Group (UK) Ltd, Croydon, CR0 4YY, UK

Preface

The global energy scenario of the 21st century is clearly envisaged to show a transition of the fuel market and the economy from fossil fuels to carbon-neutral fuels or a decarbonized fuel system. Renewable diesel is a hydrocarbon-type fuel produced from biomass sources such as vegetable oils, animal fats, greases, crop residues, woody biomass and algae. It is composed of straight-chain and branched paraffins in the range of C_{15}–C_{18}. It has the chemical formula of C_nH_{2n+2} and is free of aromatics, oxygen, sulfur, and nitrogen, and can be easily adapted depending on feedstock availability and to meet safety and quality standards of internal combustion (IC) engines. It meets the ASTM D975 specification in the United States and EN 590 in Europe. Unlike biodiesel, which is an ester and has different chemical properties from petroleum diesel, renewable diesel is chemically the same as petroleum diesel, being composed of hydrocarbons. Renewable diesel is currently considered to be an emerging and sustainable green transportation fuel for IC engines due to its efficiency and environmental benefits. Renewable diesel is currently available in international markets. Neste, a Finland-based petroleum-refining company, is the world's top producer of renewable diesel. Other oil and gas companies, such as ENI, TotalEnergies, Exxon Mobil, Diamond, *etc.*, also contribute a significant amount of renewable-diesel production. The U.S. Energy Information Administration (EIA) predicts that the renewable-diesel market supply is expected to grow from approximately 5 million tons per year (Mt/year) in 2018 to approximately 25 Mt/year by 2025 and 30 Mt/year by 2030.

Scientists, researchers and corporate leaders around the world are exploring new dimensions to enhance renewable-diesel production. As of now, renewable diesel has no presence in India while biodiesel has not received adequate policy support. Moreover, a very limited number (two) of books are available specifically on renewable diesel in international markets.

Green Chemistry Series No. 89
Biorefining of Renewable Diesel: From Production to Market
Edited by Mohammad Aslam, Sanjeev Mishra and Jorge Aburto

Published by the Royal Society of Chemistry, www.rsc.org

Keeping in view of significant contribution in the field of green fuels, the editors are highly enthusiastic to edit a specific book on renewable diesel. The editors feel proud to edit and publish such a dedicated book on renewable diesel that can provide an influential foundation for our technocrats, graduate engineers and researchers in the fields of chemical, automobile, and mechanical engineering. Additionally, this book aims to help encourage professionals and entrepreneurs working in the domain of green energy to set up new renewable-diesel refineries and hence contribute towards a sustainable and greener future.

In this book, Chapters 1 and 2 discuss renewable diesel as a sustainable driver for the next generation and its potential for decarbonization of the transport sector. Chapter 3 emphasises the major resources available for renewable-diesel production, including first, second and third generation, as well as its challenges and opportunities. The processing technologies and catalytic systems employed for renewable-diesel production are discussed in Chapters 4 and 5, and the hydrodeoxygenation of triglycerides into renewable diesel and turning lignocellulose biomass into etheric biofuel are covered in Chapters 6 and 7. Biorefining of second-generation liquid biomass and third-generation algal biomass into renewable diesel and high-value products are covered in Chapters 8 and 9, while Chapters 10 and 11 emphasize organic waste as a feedstock to produce renewable diesel and high-value co-products from a renewable diesel-based biorefinery platform. The blending, storage and distribution of renewable diesel from production to market and a real-world scenario of the renewable-diesel industry are considered in Chapters 12 and 13. Important aspects like the sustainability performance by life-cycle and techno-economic assessment of renewable diesel and biorefineries is discussed in Chapter 14. Chapter 15 focuses on integrating renewable-diesel biorefineries with existing petroleum refineries and the scope and challenges of this. Finally, the last chapter of the book, Chapter 16 addresses the current status and future outlook of the biorefining of renewable diesel.

Mohammad Aslam
Sanjeev Mishra
Jorge Aburto

Contents

Green Chemistry Series No. 89
Biorefining of Renewable Diesel: From Production to Market
Edited by Mohammad Aslam, Sanjeev Mishra and Jorge Aburto

Published by the Royal Society of Chemistry, www.rsc.org

CHAPTER 1

Renewable Diesel: A Sustainable Driver for Next Generation

CLAUDIA GUTIÉRREZ-ANTONIO

Facultad de Ingeniería, Universidad Autónoma de Querétaro, Campus Juriquilla, Campus Juriquilla, Avenida de las Ciencias s/n esquina con Villas del Mesón, Juriquilla, Querétaro, 76230, Mexico
Email: claudia.gutierrez@uaq.mx

1.1 Introduction

Nowadays, society faces big challenges such as pollution and climate change; these problems are a consequence of the intensive use of fossil fuels, which boost the growth of industrial production processes aligned to linear economic models. In addition, there is a decrease in oil reserves, a fact that increases the operating costs associated with its production. Moreover, population growth puts special stress on areas such as food and water, which are also related to energy production; this relation is known as the food-energy-water nexus, and it has been widely studied in recent years.

In order to address these challenges, the United Nations proposed the 2030 Agenda, which establishes 17 objectives to guarantee sustainable development.[1] Figure 1.1 shows the 17 Sustainable Development Goals (SDG), which include zero hunger, clean water, responsible production and consumption, quality education, peace, justice, and others. The main purpose of these objectives is the promotion of economic growth along with an improvement in quality of life of the population. In this context, the generation

Green Chemistry Series No. 89
Biorefining of Renewable Diesel: From Production to Market
Edited by Mohammad Aslam, Sanjeev Mishra and Jorge Aburto

Published by the Royal Society of Chemistry, www.rsc.org

Figure 1.1 Sustainable Development Goals proposed by the United Nations. Data from ref. 1.

of affordable and clean energy (SDG 7) is of vital importance, both for survival of society as well as for the development of economic sectors.

In particular, SDG 7 seeks to substantially increase the contribution of renewable energy to the global energy supply, as well as double the rate of improvement in energy efficiency.[2] Renewable energy is that converted from solar radiation, wind movement, water movement, geothermal reservoirs, and biomass. These renewable energy sources can be used to produce electricity and heat; nevertheless, biomass is the only source which also can be used to produce biofuels.

"Biofuel" refers to a fuel made from materials derived from living organisms harvested within the last 10 years.[3] These biofuels, including those in gaseous and solid states, can be burned to produce heat and/or electricity; in addition, liquid biofuels can be used as replacements of fossil fuels in the transport sector. According to the International Energy Agency, the transport sector continues to rely on oil products for nearly 91% of its final energy in 2022.[4] Indeed, motorized transport remains dependent on oil, in particular on combustion engines that run on liquids fuels.[4] In this respect, air, maritime and land transportation are included. For land transport over short distances, the use of electrical vehicles can help to decarbonize this sector. Nevertheless, air, maritime and long-distance land transportation are the most difficult to decarbonize, meaning the development of biofuels offers a unique alternative.

In particular, maritime and long-distance land transportation rely on the use of fossil diesel for operation. For the replacement of fossil diesel, also called petrodiesel, two main biofuels have been proposed: biodiesel and green diesel. According to the ASTM D6751 standard, biodiesel is defined as a fuel comprised of monoalkyl esters of long-chain fatty acids derived from vegetable oils or animal fats, designated B100.[5] On the other hand, green diesel, also known as renewable diesel, is a blend of straight-chain and branched saturated hydrocarbons with 15–18 carbon atoms per molecule (C_{15} to C_{18}).[6] From the previous definitions, it is clear that these two biofuels are completely different regarding their chemical compositions. Due to this, biodiesel can be used in mixtures with fossil diesel, while green diesel can completely replace to its fossil counterpart. Therefore, the use and production of green diesel is the best alternative to replace fossil diesel without adjustments or changes in infrastructure. This also implies a lower environmental impact, contributing to the decarbonization strategy of maritime and long-distance land transportation. Green diesel can be produced from many different raw materials through several conversion pathways. This chapter will provide information related to green diesel, its production processes, and the latest scientific advances regarding its production. In addition, future trends related to the production of green diesel are discussed.

1.2 Renewable Diesel

As mentioned before, green diesel, also known as renewable diesel, is a blend of straight-chain and branched saturated hydrocarbons with 15–18 carbon atoms per molecule (C_{15} to C_{18}).[6] On the other hand, fossil diesel is a hydrocarbon mixture composed of saturated hydrocarbons, monoaromatic hydrocarbons, and polyaromatic hydrocarbons.[7] As we can observe, both fuels consist of hydrocarbons, and therefore green or renewable diesel can be used in combustion engines that operate with fossil diesel. Indeed, renewable diesel must fulfill the same specifications as fossil diesel, which are fully described in the ASTM D975 standard.[8]

According to Szeto and Leung,[9] the properties of density, viscosity and heating value are almost the same between the two fuels. On the other hand, the cetane number is 60% higher in green diesel;[9] the high cetane number value in renewable diesel allows high-speed diesel engines to run efficiently.[10] Moreover, the use of renewable diesel results in soot/particle mass, carbon monoxide emissions and unburnt hydrocarbon emissions that are generally improved.[9] Moreover, green diesel is a low-sulfur fuel (<10 ppm wt) with very low greenhouse-gas emissions.[11]

1.3 Production Processes for Renewable Diesel

Renewable diesel is produced from the catalytic hydroprocessing of triglyceride feedstock and has been used interchangeably with petro-diesel.[12] The catalytic hydroprocessing consists of deoxygenation and isomerization reactions (Figure 1.2). In the hydrodeoxygenation step the main target is the removal of oxygen, which is present in the biomass at almost 167 times the oxygen content in petroleum,[13] to produce linear-chain hydrocarbons. After that, the renewable hydrocarbons are hydroisomerized, so a mixture of normal and iso-paraffins is obtained.[14]

The raw material for renewable-diesel production is triglyceride feedstock, which consists of fatty acids whose composition varies significantly depending on the source and geographical origin.[15] Nevertheless, as can be observed in Figure 1.2, lignocellulose, sugar and starchy feedstocks can also be used to produce renewable diesel; in these cases, additional processing stages are required in order to convert them to a suitable kind of feedstock for hydroprocessing.

Triglyceride feedstocks include vegetable oils, animal fats, and microalgal oils, as well as residual oils; this kind of feedstock contains fatty acids and triglycerides. Residues from forestry, agricultural, and agroindustrial wastes, as well those from pruning, are considered to be lignocellulosic feedstocks; as the name indicates, this type of feedstock is composed of cellulose, hemicellulose, and lignin. Edible biomasses such as sugarcane, beet and corn are considered to be sugar and starchy feedstocks; this category also includes confectionery, edible biomasses in decomposition processes, and expired foods. As can be observed from Figure 1.2, lignocellulosic feedstocks can be conditioned through hydrothermal liquefaction, pyrolysis or

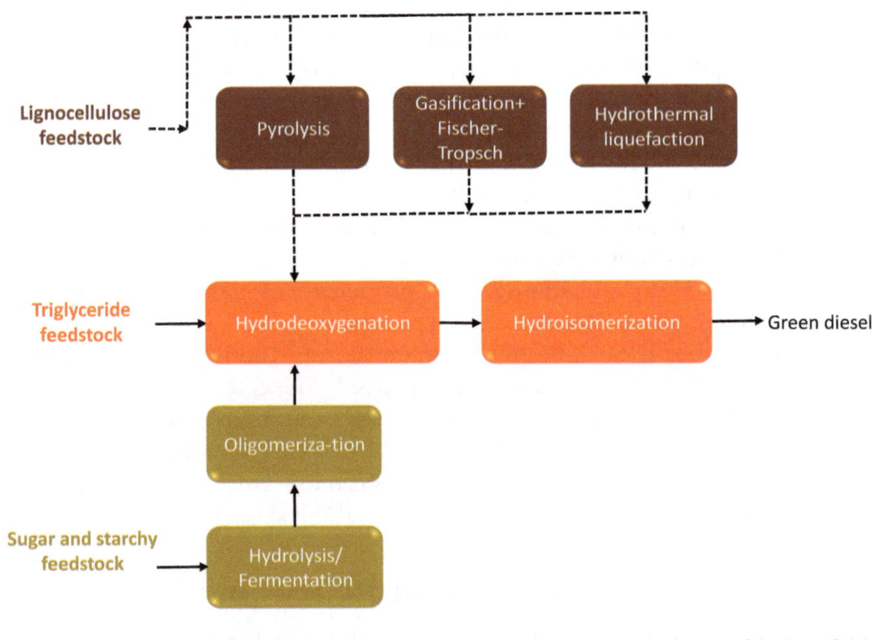

Figure 1.2 Renewable-diesel production processes from different kinds of biomass. Author's own elaboration.[13]

gasification followed by Fischer–Tropsch synthesis, while sugar and starchy feedstocks can be hydrolyzed, fermented, and then oligomerized.

Usually, hydroprocessing operation conditions consider pressures of 30–200 bar, temperatures ranging from 250 °C to 375 °C, and an H_2/oil ratio of around 1000 ml per ml of oil.[16] The most common catalysts include platinum, nickel, molybdenum, and palladium, usually supported on γ-alumina.[16] Regard the yields, the values vary from 35% to 75%. At the moment, Neste is the company with the most plants where renewable diesel is produced, with a total capacity of 1 980 000 tons per year.[12] The aforementioned variables strongly depend on the raw material used; due to this, in the next section a brief review of this aspect will be covered.

1.4 Revision of the State of the Art

Research related to production of renewable diesel has intensified in the last 15 years, as shown in Figure 1.3. We can observe how the number of publications passed 50 in 2009.[17] This tendency is aligned with the introduction of the Renewable Energy Directive of the European Union in 2009.[18] The total number of publications on renewable diesel (2877 documents) contrast with those available for biodiesel (63 940 documents), bioethanol (30 148 documents), and even biogas (77 026 documents).[19–21] This result was unexpected, considering that this biofuel can be directly used in combustion engines that operate with fossil diesel.

In the available literature, the first studies employed edible and non-edible oils, such as those derived from cottonseed,[22] palm,[23] rapeseed[24] and

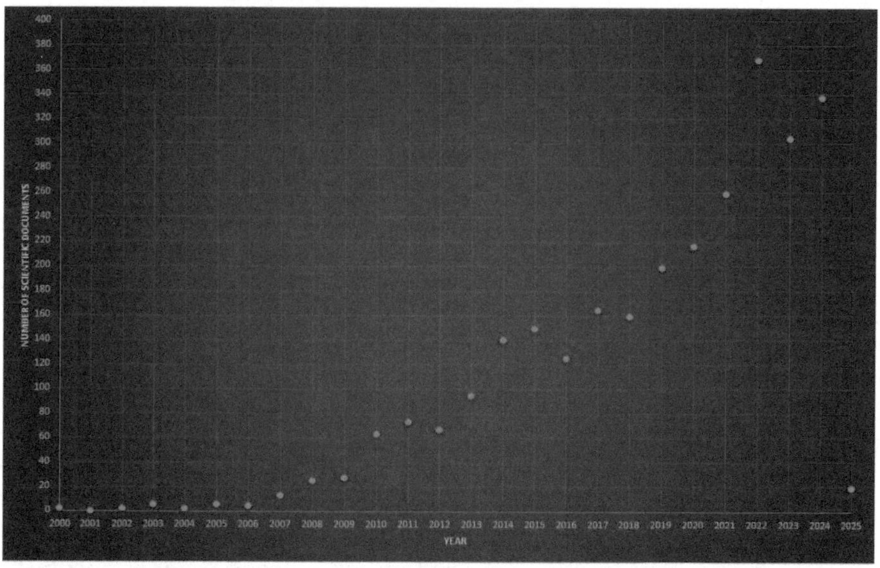

Figure 1.3 Reported renewable-diesel production studies. Author's own elaboration based on data from ref. 17.

microalgae.[25] In 2011 the co-hydroprocessing of palm oil with gasoil was reported.[26] Later, the first studies with oil derived from *Jatropha curcas* for the production of renewable diesel were published[27,28] and Herskowitz *et al.*[29] reported one-step hydroprocessing for the production of green diesel from soybean oil. In the timeline of the use of biomass rich in triglycerides as feedstocks for the production of green diesel, the first studies were focused on the technical feasibility and considered the most direct conversion pathway.

The next research was focused on the use of lignocellulosic materials through pyrolysis processing. Wright *et al.*[30] reported the first work where biomass fast pyrolysis was used for the production of renewable diesel. The considered raw material was corn stover, and, in spite of this, the biomass cost is a significant element in the analyzed scenarios. Later, Frank *et al.*[31] reported the hydrothermal liquefaction of algae for the production of renewable diesel. Moreover, bio-oil from different wood species (beech, pine, and, typically, white spruce and red maple) as well as agroindustrial residues (rice husk) were used as raw materials for the production of renewable diesel.[12]

The next area of work regarded hydrodeoxygenation of other raw materials. Yang *et al.*[32] presented the hydrodeoxygenation of furans over a palladium-based catalyst; they obtained yields between 87% and 94% for alkanes in the jet-fuel range. In the same year, the feasibility of the hydroprocessing of acrylic acid on molybdenum carbide-based catalysts was reported by Rocha *et al.*[33] and Han *et al.*[34] reported the co-hydrotreatment of tire pyrolysis oil and vegetable oil for the production of green diesel, reaching yields of 60.2 wt%. An interesting project is the one proposed by Kumar *et al.*,[35] where they used lignocellulosic biomass as feed for oleaginous yeast and bacteria, and the produced oil is converted to renewable diesel. Also, the co-hydrotreating of non-edible vegetable oils and petrodiesel blends to produce green diesel was reported by Perez-Cisneros *et al.*;[36] they employed reactive distillation, finding that the proposal is feasible from a technical point of view. In these works, it can be noted the interest lay in co-processing with vegetable oils in order to produce renewable diesel. In these studies, the yields were not high, leading to development of new catalysts.

The syntheses of many new catalysts were reported during 2019. New catalysts based on niobium phosphate,[37] nickel supported on palygorskite,[38] platinum–molybdenum supported on zirconia,[39] nickel–molybdenum alumina composite,[40] silver oxide with lanthanum(III) oxide supported on activated carbon,[41] sonochemically synthesized nickel–molybdenum supported on γ-alumina,[42] platinum with tin supported on alumina,[43] and trimetallic sulfided rhenium–nickel–molybdenum supported on γ-alumina[44] were reported.

Later, research on the production of renewable diesel was focused on reducing hydrogen consumption and exploring the use of other atmospheres to perform the hydrodeoxygenation reaction. Krobkrong *et al.*[45] presented the deoxygenation of oleic acid under a nitrogen atmosphere using molybdenum oxide catalysts with promoters (Ni, Co, and Cu); in this case, renewable diesel can be obtained without the use of hydrogen, reaching a conversion of 80%. Gamal *et al.*[46] reported the deoxygenation of palm fatty acid distillate using a bimetallic catalyst CoMo supported on activated carbon under ambient nitrogen; in this case a conversion of 92% was observed. An outstanding work was the use of transition metal oxide (TMO) on zeolite Y that was able to produce green diesel *via* deoxygenation;[47] in that study, the reported conversion is 76%. A similar approach was presented by Baharudin *et al.*,[48] where green diesel is obtained with a conversion of 83% using synthesized NiO–ZnO catalysts without the use of solvents or hydrogen. In the same line, the production of green diesel from waste cooking oil *via* hydrogen-free and solventless deoxygenation over Ni–Co/organosilane-SBA-15 catalyst was reported by Razak *et al.*[49]

Zhao *et al.*[50] reported reduced hydrogen requirements and carbon loss using nickel–gallium intermetallic compounds in deoxygenation of methyl esters to green diesel; the reported conversion was 99%. The deoxygenation of *Jatropha* oil using Ni_y–Ag_x/activated carbon catalysts without hydrogen was reported by Asikin-Mijan *et al.*;[51] in this case, the activated carbon was derived from coconut fiber residues. In the same line, Kaewtrakulchai *et al.*[52] presented the deoxygenation of palm oil using metal phosphides supported on activated biochar derived from palm fiber waste to produce green diesel fuel.

In addition, green diesel has been obtained through one-step deoxygenation, without hydrogen, using a Co–Ag carbon-based catalyst;[53] in that study the conversion was 92% and the selectivity for renewable diesel was 95%. An interesting result was reported by Pimenta *et al.*[54] related to the use of aerogel-based catalysts to produce renewable diesel; the aerogel catalyst produced over 4 times more hydrocarbon than a typical catalyst.

Interesting contributions were also reported in other related fields. For instance, Lorenzi *et al.*[55] reported the use of a high temperature in solid oxide electrolysis cells to generate hydrogen in the hydrotreating of waste oils. El-Sawy *et al.*[56] reported co-hydroprocessing and hydrocracking of a mixture of vacuum gas oil, waste lubricating oil and waste cooking oil to produce green diesel (selectivity of 60.9%) and kerosene (selectivity of 20.24%).

In the last two years, some of the key advances have been focused on the development of new catalysts for hydrogen-free deoxygenation processes that do not quickly deactivate,[57] the use of methanol as a hydrogen donor in deoxygenation reactions,[58] and *in situ* production of hydrogen from green solvents;[59] in addition, the use of a nanocatalyst to improve yields in the hydroprocessing of bio-oil derived from pyrolysis was reported.[60] Moreover, the use of hydrothermal liquefaction with food waste as a raw material for the production of renewable diesel was proposed by Wang *et al.*,[61] and the development of new catalysts based on hydrothermal carbonization of wastes for hydroprocessing[62] and the recovery of lipids from sewage sludge without solvents[63] were reported. An interesting perspective is the re-engineering of microbes to generate advanced biofuels.[64]

1.5 Perspectives

Unlike other biofuels, renewable diesel is an option to be used in the land and maritime transport sectors. The main advantage of this biofuel is the fact that it can be used in the same combustion engines that operate with fossil diesel without modifications. This advantage facilitates its storage, distribution and use, since the infrastructure requirements are the same as for fossil diesel.

It is important to mention that renewable diesel must fulfill the technical, sustainability and economic requirements in order to be commercialized. From the technical point of view, as mentioned before, green diesel must comply with the same physicochemical properties established for fossil diesel; this set of properties are described in the ASTM D975 standard.[8] In spite of renewable diesel being on the market today, it is important to continue exploring new raw materials in order to increase the availability of this biofuel. It is important to mention that in the case of biofuel production the biomass should be available nearby; this will lead to biofuels with lower production costs and lower environmental impact. From a review of the literature, it can be noted that development of novel catalysts that allow green diesel to be obtained without hydrogen and/or solvents or, in other approaches, the *in situ* production of hydrogen are also considered. Regarding the raw materials, the main interest has been residues, especially agroindustrial residues as well as other end-use products such as tires or sludge. The production of renewable diesel can be improved through the application of process-intensification strategies, as well as its generation through biorefinery schemes. Also, it is important to consider production processes than can manage several raw materials or even mixtures of them; this will help to guarantee the operation of the processing facilities, bringing robustness to the supply chain.

From the economic point of view, it is necessary to guarantee optimal production processes as well as its supply chain. The smaller the supply chain, the lower the operating costs and environmental impacts associated with its operation. Moreover, the production process must also be efficient in the use of materials and energy to allow lower production costs. In this respect, an interesting research line is the development of new catalysts that allow hydrodeoxygenation and hydroisomerization reactions to be carried out in the same vessel with lower pressure, temperature, and H_2 requirements. In addition, the application of process-intensification strategies can lead to more compact and efficient production facilities. Stankiewicz and Moulijn[65] defined the concept of process intensification as the development of novel apparatuses and techniques that, compared to those commonly used today, are expected to bring dramatic improvements in manufacturing and processing, substantially decreasing the equipment-size/production-capacity ratio, energy consumption, or waste production, and ultimately resulting in cheaper, more sustainable technologies. In this respect, the proposal of reactive distillation columns, conventional or thermally coupled, is an important area with opportunity to decrease production costs and environmental impacts.

Regarding the sustainability aspect, the Roundtable on Sustainable Biomaterials (RSB) has developed robust and integral standards to guarantee the sustainability of the processing and co-processing of biomass to generate energy and non-energy products.[66] The sustainability standards are developed considering 12 principles, which are aligned to the Sustainable Development Goals (Figure 1.4). From Figure 1.4, it can be observed that measures and impacts are not exclusively focused on the production process or even the supply chain, but also environmental, legal, and social aspects are considered in this standard. In particular, an interesting characteristic is that this standard includes the complete supply chain, which makes it different from similar environmental standards. Regarding sustainability standards for biofuels, there are three:

- RSB Global Fuels Certification
- RSB EU RED Fuel Certification
- RSB CORSIA Certification

Figure 1.4 Principles involved in the sustainability standards of the RSB. Data from ref. 66.

The first two standards are directly applicable to green diesel, depending on the region where the biofuel is going to be produced; the main difference is that the second one guarantees compliance with the European Union (EU) sustainability criteria and traceability requirements for biofuels and bioliquids considered in the Renewable Energy Directive (RED).[66] In the case of the third standard, renewable diesel can be used for the production of sustainable aviation fuel; in that case, its production process must also be included as part of the supply chain.

1.6 Concluding Remarks

Green diesel is the most suitable alternative for land and maritime transport. Renewable diesel is obtained from the hydroprocessing of triglyceride feedstocks. Nevertheless, lignocellulose, sugar and starchy feedstocks can also be used to produce this biofuel; in this case, additional stages are required to convert them to a suitable feedstock for hydroprocessing.

The production of renewable diesel has mainly been studied considering edible, non-edible and residual biomasses. During the last 15 years, most efforts have been focused on the exploration of different raw materials, development of new catalysts that allow hydrogen-free and solventless processes for the production of green diesel. There has been recent interest in the use of hydrothermal liquefaction for the processing of microalgae and sludge.

Renewable diesel is available commercially; nevertheless, research must be intensified in order to decrease the environmental impact while its profitability increases. This will lead to improved production processes which can boost the decarbonization of the maritime and long-distance transport sectors.

Acknowledgements

This research was supported by the Bioproducts Sustainable Production Processes Research Group from the Engineering Faculty of Universidad Autónoma de Querétaro (UAQ).

References

1. UN, United Nations, Sustainable development goals, 2024. Available from: https://www.undp.org/sustainable-development-goals?gclid=EAIaIQobChMItOW56-SHgQMVwg6zAB3nKwyLEAAYAiAAEgJaG_D_BwE (accessed on 11.09.2024).
2. M. Acheampong, F. C. Ertem, B. Kappler and P. Neubauer, In pursuit of Sustainable Development Goal (SDG) number 7: Will biofuels be reliable?, *Renewable Sustainable Energy Rev.*, 2017, 75, 927–937.
3. M. Doble and A. K. Kruthiventi, Alternate Energy Sources, in *Green Chemistry and Engineering*, ed. M. Doble and A. K. Kruthiventi, Academic Press, 2007, ch. 7, pp. 171–192.

4. IEA, International Energy Agency, Transport, 2024. Available from: https://www.iea.org/energy-system/transport (accessed on 11.09.2024).
5. ASTM, American Society of Testing and Materials, ASTM D6751-20a, Standard Specification for Biodiesel Fuel Blend Stock (B100) for Middle Distillate Fuels, 2020. Available from: https://www.astm.org/d6751-20a.html (accessed on 17.10.2024).
6. A. K. Das, S. K. Sahu and A. K. Panda, Current status and prospects of alternate liquid transportation fuels in compression ignition engines: A critical review, *Renewable Sustainable Energy Rev.*, 2022, **161**, 112358.
7. K. J. Hidalgo, A. J. Giachini, M. R. Schneider, A. U. Soriano, M. P. Baessa, L. F. Martins and V. M. Oliveira, Recent advances in bioremediation of biofuel blends, *Int. Biodeterior. Biodegrad.*, 2024, **188**, 105750.
8. ASTM, American Society of Testing and Materials, ASTM D975-21, Standard Specification for Diesel Fuel, 2021. Available from: https://www.astm.org/d0975-21.html (accessed on 28.10.2024).
9. W. Szeto and D. Y. C. Leung, Is hydrotreated vegetable oil a superior substitute for fossil diesel? A comprehensive review on physicochemical properties, engine performance and emissions, *Fuel*, 2022, **327**, 125065.
10. K. K. Jaiswal, C. R. Chowdhury, S. Dutta, I. Banerjee, K. S. Jaiswal, H. M. D. Nisansala, B. Sangmesh and N. M. S. Sirimuthu, Synthesis of renewable diesel as a substitute for fossil fuels, in *Renewable Diesel*, ed. D. Kumar, B. Singh and S. K. Gupta, Elsevier, 2024, ch. 1, pp. 1–31.
11. S. Bezergianni and A. Dimitriadis, Comparison between different types of renewable diesel, *Renewable Sustainable Energy Rev.*, 2013, **21**, 110–116.
12. M. Patel and A. Kumar, Production of renewable diesel through the hydroprocessing of lignocellulosic biomass-derived bio-oil: A review, *Renewable Sustainable Energy Rev.*, 2016, **58**, 1293–1307.
13. C. Gutiérrez-Antonio, A. G. Romero-Izquierdo, F. I. Gómez Castro and S. Hernández, *Production Processes of Renewable Aviation Fuel – Present Technologies and Future Trends*, Elsevier, 2021, ISBN: 9780128197196.
14. M. Krár, S. Kovács, D. Kalló and J. Hancsók, Fuel purpose hydrotreating of sunflower oil on $CoMo/Al_2O_3$ catalyst, *Bioresour. Technol.*, 2010, **101**(23), 9287–9293.
15. S. K. Maity, Opportunities, recent trends and challenges of integrated biorefinery: Part I, *Renewable Sustainable Energy Rev.*, 2015, **43**, 1427–1445.
16. A. Amin, Review of diesel production from renewable resources: Catalysis, process kinetics and technologies, *Ain Shams Eng. J.*, 2019, **10**(4), 821–839.
17. Sciencedirect, Search for "renewable diesel" OR "green diesel", 2024a. Available from: https://www.sciencedirect.com/search?qs=%22renewable%20diesel%22%20OR%20%22green%20diesel%22 (accessed on 29.10.2024).
18. European Commission, Renewable Energy Directive, 2024. Available from: https://energy.ec.europa.eu/topics/renewable-energy/renewable-energy-directive-targets-and-rules/renewable-energy-directive_en (accessed on 29.10.2024).

19. Sciencedirect, Search for "biodiesel", 2024b. Available from: https://www.sciencedirect.com/search?qs=%22biodiesel%22 (accessed on 29.10.2024).
20. Sciencedirect, Search for "bioethanol", 2024c. Available from: https://www.sciencedirect.com/search?qs=%22bioethanol%22 (accessed on 29.10.2024).
21. Sciencedirect, Search for "biogas", 2024d. Available from: https://www.sciencedirect.com/search?qs=%22biogas%22 (accessed on 29.10.2024).
22. I. Sebos, A. Matsoukas, V. Apostolopoulos and N. Papayannakos, Catalytic hydroprocessing of cottonseed oil in petroleum diesel mixtures for production of renewable diesel, *Fuel*, 2009, **88**(1), 145–149.
23. A. Guzman, J. E. Torres, L. P. Prada and M. L. Nuñez, Hydroprocessing of crude palm oil at pilot plant scale, *Catal. Today*, 2010, **156**(1–2), 38–43.
24. P. Šimáček, D. Kubička, G. Šebor and M. Pospíšil, Fuel properties of hydroprocessed rapeseed oil, *Fuel*, 2010, **89**(3), 611–615.
25. N. H. Tran, J. R. Bartlett, G. S. K. Kannangara, A. S. Milev, H. Volk and M. A. Wilson, Catalytic upgrading of biorefinery oil from micro-algae, *Fuel*, 2010, **89**(2), 265–274.
26. C. Templis, A. Vonortas, I. Sebos and N. Papayannakos, Vegetable oil effect on gasoil HDS in their catalytic co-hydroprocessing, *Appl. Catal., B*, 2011, **104**(3–4), 324–329.
27. J. Liu, K. Fan, W. Tian, C. Liu and L. Rong, Hydroprocessing of Jatropha oil over NiMoCe/Al_2O_3 catalyst, *Int. J. Hydrogen Energy*, 2012, **37**(23), 17731–17737.
28. G. Ramya, R. Sudhakar, J. A. I. Joice, R. Ramakrishnan and T. Sivakumar, Liquid hydrocarbon fuels from jatropha oil through catalytic cracking technology using AlMCM-41/ZSM-5 composite catalysts, *Appl. Catal., A*, 2012, **433–434**, 170–178.
29. M. Herskowitz, M. V. Landau, Y. Reizner and D. Berger, A commercially-viable, one-step process for production of green diesel from soybean oil on Pt/SAPO-11, *Fuel*, 2013, **111**, 157–164.
30. M. Wright, J. A. Satrio, R. C. Brown, D. E. Daugaard and D. D. Hsu, Techno-Economic Analysis of Biomass Fast Pyrolysis to Transportation Fuels, Prepared under Task No. BB07.7510, Technical Report NREL/TP-6A20-46586, Contract No. DE-AC36-08GO28308, 2010.
31. E. D. Frank, A. Elgowainy, J. Han and Z. Wang, Life cycle comparison of hydrothermal liquefaction and lipid extraction pathways to renewable diesel from algae, in *Mitigation Adaption Strategies Global Change*, 2013, vol. 18, pp. 137–158.
32. J. Yang, S. Li, L. Zhang, X. Liu, J. Wang, X. Pan, N. Li, A. Wang, Y. Cong, X. Wang and T. Zhang, Hydrodeoxygenation of furans over Pd–FeO_x/SiO_2 catalyst under atmospheric pressure, *Appl. Catal., B*, 2017, **201**, 266–277.
33. A. S. Rocha, L. A. Souza, R. R. Oliveira, A. B. Rocha and V. T. da Silva, Hydrodeoxygenation of acrylic acid using Mo_2C/Al_2O_3, *Appl. Catal., A*, 2017, **531**, 69–78.

34. Y. Han, F. Stankovikj and M. Garcia-Perez, Co-hydrotreatment of tire pyrolysis oil and vegetable oil for the production of transportation fuels, *Fuel Process. Technol.*, 2017, **159**, 328–339.
35. D. Kumar, B. Singh and J. Korstad, Utilization of lignocellulosic biomass by oleaginous yeast and bacteria for production of biodiesel and renewable diesel, *Renewable Sustainable Energy Rev.*, 2017, **73**, 654–671.
36. E. S. Perez-Cisneros, M. Sales-Cruz, R. Lobo-Oehmichen and T. Viveros-García, A reactive distillation process for co-hydrotreating of non-edible vegetable oils and petro-diesel blends to produce green diesel fuel, *Comput. Chem. Eng.*, 2017, **105**, 105–122.
37. C. A. Scaldaferri and V. M. Duarte Pasa, Production of jet fuel and green diesel range biohydrocarbons by hydroprocessing of soybean oil over niobium phosphate catalyst, *Fuel*, 2019, **245**, 458–466.
38. S. Lycourghiotis, E. Kordouli, L. Sygellou, K. Bourikas and C. Kordulis, Nickel catalysts supported on palygorskite for transformation of waste cooking oils into green diesel, *Appl. Catal., B*, 2019, **259**, 118059.
39. S. Janampelli and S. Darbha, Highly efficient Pt–MoO_x/ZrO_2 catalyst for green diesel production, *Catal. Commun.*, 2019, **125**, 70–76.
40. S. R. Yenumala, P. Kumar, S. K. Maity and D. Shee, Production of green diesel from karanja oil (Pongamia pinnata) using mesoporous NiMo–alumina composite catalysts, *Bioresour. Technol. Rep.*, 2019, 7, 100288.
41. G. Abdulkareem-Alsultan, N. Asikin-Mijan, N. Mansir, H. V. Lee, Z. Zainal, A. Islam and Y. H. Taufiq-Yap, Pyro-lytic de-oxygenation of waste cooking oil for green diesel production over Ag_2O_3–La_2O_3/AC nano-catalyst, *J. Anal. Appl. Pyrolysis*, 2019, **137**, 171–184.
42. M. Ameen, M. T. Azizan, A. Ramli, S. Yusup and M. S. Alnarabiji, Catalytic hydrodeoxygenation of rubber seed oil over sonochemically synthesized Ni–Mo/γ-Al_2O_3 catalyst for green diesel production, *Ultrason. Sonochem.*, 2019, **51**, 90–102.
43. P. A. Zharova, A. V. Chistyakov, S. S. Shapovalov, A. A. Pasynskii and M. V. Tsodikov, Original Pt–Sn/Al_2O_3 catalyst for selective hydrodeoxygenation of vegetable oils, *Energy*, 2019, **172**, 18–25.
44. S. Thongkumkoon, W. Kiatkittipong, U. W. Hartley, N. Laosiripojana and P. Daorattanachai, Catalytic activity of trimetallic sulfided Re–Ni–Mo/γ-Al_2O_3 toward deoxygenation of palm feedstocks, *Renewable Energy*, 2019, **140**, 111–123.
45. N. Krobkrong, V. Itthibenchapong, P. Khongpracha and K. Faungnawakij, Deoxygenation of oleic acid under an inert atmosphere using molybdenum oxide-based catalysts, *Energy Convers. Manage.*, 2018, **167**, 1–8.
46. M. S. Gamal, N. Asikin-Mijan, W. N. A. Wan Khalit, M. Arumugam, S. Mohd Izham and Y. H. Taufiq-Yap, Effective catalytic deoxygenation of palm fatty acid distillate for green diesel production under hydrogen-free atmosphere over bimetallic catalyst CoMo supported on activated carbon, *Fuel Process. Technol.*, 2020, **208**, 106519.

47. M.-Y. Choo, L. E. Oi, T. C. Ling, E.-P. Ng, Y.-C. Lin, G. Centi and J. C. Juan, Deoxygenation of triolein to green diesel in the H_2-free condition: Effect of transition metal oxide supported on zeolite Y, *J. Anal. Appl. Pyrolysis*, 2020, **147**, 104797.
48. K. B. Baharudin, N. Abdullah, Y. H. Taufiq-Yap and D. Derawi, Renewable diesel *via* solventless and hydrogen-free catalytic deoxygenation of palm fatty acid distillate, *J. Cleaner Prod.*, 2020, **274**, 122850.
49. N. A. A. Razak, N.-A. Mijan, Y. H. Taufiq-Yap and D. Derawi, Production of green diesel *via* hydrogen-free and solventless deoxygenation reaction of waste cooking oil, *J. Cleaner Prod.*, 2020, **366**, 132971.
50. N. Zhao, Y. Zheng and J. Chen, Remarkably reducing carbon loss and H_2 consumption on Ni–Ga intermetallic compounds in deoxygenation of methyl esters to hydrocarbons, *J. Energy Chem.*, 2020, **41**, 194–208.
51. N. Asikin-Mijan, J. M. Ooi, G. AbdulKareem-Alsultan, H. V. Lee, M. S. Mastuli, N. Mansir, F. A. Alharthi, A. A. Alghamdi and Y. H. Taufiq-Yap, Free-H_2 deoxygenation of Jatropha curcas oil into cleaner diesel-grade biofuel over coconut residue-derived activated carbon catalyst, *J. Cleaner Prod.*, 2020, **249**, 119381.
52. N. Kaewtrakulchai, M. Fuji and A. Eiad-Ua, Catalytic deoxygenation of palm oil over metal phosphides supported on palm fiber waste derived activated biochar for producing green diesel fuel, *RSC Adv.*, 2022, **12**(40), 26051–26069.
53. M. Safa-Gamal, N. Asikin-Mijan, M. Arumugam, W. N. A. Wan Khalit, I. N. Azreena, F. S. Hafez and Y. H. Taufiq-Yap, Catalytic deoxygenation by H_2-free single-step conversion of free fatty acid feedstock over a Co–Ag carbon-based catalyst for green diesel production, *J. Anal. Appl. Pyrolysis*, 2021, **160**, 105334.
54. J. L. C. W. Pimenta, M. de Oliveira Camargo, R. Belo Duarte, O. A. Andreo dos Santos and L. M. de Matos Jorge, Deoxygenation of vegetable oils for the production of renewable diesel: Improved aerogel based catalysts, *Fuel*, 2021, **290**, 119979.
55. G. Lorenzi, P. Baptista, B. Venezia, C. Silva and M. Santarelli, Use of waste vegetable oil for hydrotreated vegetable oil production with high-temperature electrolysis as hydrogen source, *Fuel*, 2020, **278**, 117991.
56. M. S. El-Sawy, S. A. Hanafi, F. Ashour and T. M. Aboul-Fotouh, Co-hydroprocessing and hydrocracking of alternative feed mixture (vacuum gas oil/waste lubricating oil/waste cooking oil) with the aim of producing high quality fuels, *Fuel*, 2020, **269**, 117437.
57. N. Asikin-Mijan, J. C. Juan, Y. H. Taufiq-Yap, H. C. Ong, Y.-C. Lin, G. AbdulKareem-Alsultan and H. V. Lee, Towards sustainable green diesel fuel production: Advancements and opportunities in acid-base catalyzed H_2-free deoxygenation process, *Catal. Commun.*, 2023, **182**, 106741.
58. L. Ai, H. Shi and J. Chen, In-situ deoxygenation of methyl palmitate to hydrocarbons on highly dispersed Ni–Re/TiO_2 catalyst in aqueous phase using methanol as a hydrogen donor, *J. Energy Inst.*, 2023, **109**, 101265.

59. E. Kordouli, S. Lycourghiotis, K. Bourikas, A. Lycourghiotis and C. Kordulis, Renewable diesel synthesis by hydro-processing in green solvents, *Curr. Opin. Green Sustainable Chem.*, 2024, **48**, 100936.
60. H. Shahbeik, A. Shafizadeh, V. K. Gupta, S. S. Lam, H. Rastegari, W. Peng, J. Pan, M. Tabatabaei and M. Aghbashlo, Using nanocatalysts to upgrade pyrolysis bio-oil: A critical review, *J. Cleaner Prod.*, 2023, **413**, 137473.
61. Z. Wang, B. Si, S. Summers and Y. Zhang, Renewable diesel blends production from hydrothermal liquefaction of food waste: A novel approach coupling fractional distillation and emulsification, *Energy Convers. Manage.*, 2024, **314**, 118716.
62. N. Kaewtrakulchai, S. Chanpee, S. Jadsadajerm, S. Wongrerkdee, K. Manatura and A. Eiad-Ua, Co-hydrothermal carbonization of polystyrene waste and maize stover combined with KOH activation to develop nanoporous carbon as catalyst support for catalytic hydrotreating of palm oil, *Carbon Resour. Convers.*, 2024, 7(4), 100231.
63. L. di Bitonto, V. Locaputo and C. Pastore, Solventless recovery of lipids from urban sewage sludge: How to sustainably turn a waste into a valuable source of oleochemicals, *Chem. Eng. J.*, 2024, **494**, 152991.
64. J. Love, Microbial pathways for advanced biofuel production, *Biochem. Soc. Trans.*, 2022, **50**(2), 987–1001.
65. A. I. Stankiewicz and J. A. Moulijn, Process intensification: Transforming chemical engineering, *Chem. Eng. Prog.*, 2000, **96**(1), 22–33.
66. RSB, Roundtable on Sustainable Biomaterials, Certification, 2024. Available from: https://rsb.org/ (accessed on 29.10.2024).

CHAPTER 2

Biorefining of Renewable Diesel: From Production to Market

FRANCISCO IZAIAS DA SILVA AIRES,[a] DAYANA NASCIMENTO DARI,[a] KAIANY MOREIRA DOS SANTOS,[a] VIVIANE DE CASTRO BIZERRA,[a] PATRICK DA SILVA SOUSA,[b] RODRIGO DA SILVA VIEIRA,[a] IURY BARBOSA PEREIRA,[a] FRANCISCO SIMÃO NETO[b] AND C. JOSÉ C. S. DOS SANTOS*[a]

[a] Instituto de Engenharias e Desenvolvimento Sustentável, Universidade da Integração Internacional da Lusofonia Afro-Brasileira, Campus das Auroras, Redenção CEP 62790-000, Ceará, Brazil; [b] Departamento de Engenharia Química, Universidade Federal do Ceará, Campus do Pici, Bloco 709, Fortaleza 60440-554, Ceará, Brazil
*Email: jcs@unilab.edu.br

2.1 Introduction

The challenge of tackling climate change and reducing emissions from fossil fuels has motivated global-scale efforts towards transitioning to renewable energy sources.[1] Among these efforts, the transportation sector stands out as a significant contributor to carbon emissions, demanding urgent measures to mitigate its environmental impact.[2,3] In this context, renewable diesel has emerged as a promising alternative with the potential to play a key role in decarbonizing the transportation sector.[4]

Green Chemistry Series No. 89
Biorefining of Renewable Diesel: From Production to Market
Edited by Mohammad Aslam, Sanjeev Mishra and Jorge Aburto

Published by the Royal Society of Chemistry, www.rsc.org

Renewable diesel, also known as green diesel or hydrotreated vegetable oil (HVO), is derived from renewable feedstocks such as vegetable oils, animal fats, and residual oils through a hydrotreating process.[5,6] Unlike conventional diesel from fossil fuels, renewable diesel is derived from renewable feedstocks.[7] Renewable diesel offers a sustainable solution by utilizing biomass-derived feedstocks.[8,9] This distinction makes renewable diesel an attractive option for reducing emissions of greenhouse gases (GHGs) and improving environmental sustainability.[10]

The potential of renewable diesel lies in its economic viability to serve as a direct substitute for conventional diesel, requiring no modifications to existing diesel engines.[11] This characteristic makes it an attractive option for fleet operators, transportation companies, and consumers seeking to reduce their carbon footprint without compromising performance or convenience.[12,13] Furthermore, renewable diesel offers improved air quality, reduced emissions of pollutants like particulate matter and nitrogen oxides, and decreased dependence on finite fossil resources.[14–16]

In recent years, the adoption of renewable diesel has gained momentum, driven by a combination of regulatory mandates, corporate sustainability initiatives, and technological advancements.[17,18] European governments have continued to lead and implement policies promoting renewable fuels, including diesel, as part of their broader strategies to combat climate change and achieve carbon-neutrality goals.[19] Similarly, companies increasingly recognize clean energy's economic and environmental benefits by incorporating renewable diesel into their operations.[10] This has led to increased investments in production capacity and infrastructure.[20] This study aims to analyze the production and market of renewable diesel, highlighting its relevance as a sustainable alternative to fossil fuels and discussing the technological and political innovations necessary to promote its large-scale adoption.

2.2 History of Transportation Fuels

Figure 2.1 presents a timeline of the evolution of fuels until the end of the 20th century, where room was made for reflection on the future of the energy matrix and the importance of renewable sources. Fossil fuels originated millions of years ago when the remnants of plants, animals, and other organisms were buried beneath sedimentary layers and exposed to extreme temperatures and pressures.[21–23] From this, coal, oil, and natural gas were formed, becoming indispensable energy sources for human development. The first fossil fuel to be utilized extensively was coal, particularly during the Industrial Revolution in the 18th century when it started to power railroads and steam engines, revolutionizing worldwide production and transportation networks.[24]

In the late 19th century, oil gained prominence due to the demand for efficient energy driven by technological advances.[25] In 1859, the first commercially viable oil drilling occurred in Pennsylvania, United States, becoming a landmark of a new energy era.[26] Crude oil was used to obtain a significant number of by-products, such as gasoline, which was initially

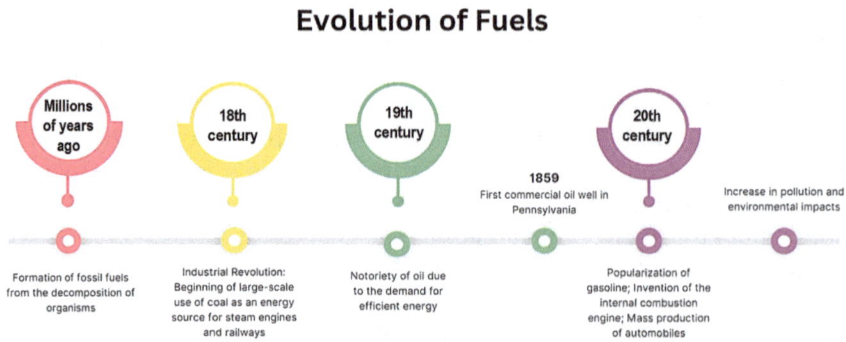

Figure 2.1 Timeline of the evolution of fuels.

considered a residue of the refining process and became the primary energy source due to its use in internal combustion engines.[27–29]

Meanwhile, Rudolf Diesel developed diesel, a petroleum derivative, which emerged as an efficient alternative for industrial engines.[30,31] Due to its energy density and reduced volatility, diesel could be supplied *via* freight trucks, locomotives, and ships.[32,33] Therefore, diesel quickly gained prominence in the transportation sector, although its application and use were initially limited, where power and efficiency were essential.[34,35]

Rapid technological development and the demand for energy-efficient and portable systems drove the transition to the use and commercialization of liquid fuels in the early 20th century.[36,37] A significant event was the invention of the internal combustion engine and its application in automobiles and industrial machinery.[38,39] The use of gasoline in these engine types increased industrial growth and facilitated individual and collective transportation.[40,41] The mass production of automobiles primarily drove this advancement, notably the Ford Model T, which popularized motorized and accessible transportation.[42,43]

Fossil fuels power the chemical industry, electricity production, and transportation, and have been essential to the growth of world economies.[44] However, their extensive usage and heavy exploitation have presented serious problems, including adverse effects on the environment and society.[45] The world's energy landscape has been moulded by reliance on fossil fuels, creating complicated issues of sustainability and climate change while providing the groundwork for advancing contemporary technology.

2.3 Renewable Diesel: A Sustainable Alternative

2.3.1 Definition and Characteristics of Renewable Diesel

"Renewable diesel" refers to diesel-like hydrocarbons produced from vegetable oil. It consists of linear and branched alkanes in the range C_{15} to C_{18}

with the chemical formula C_nH_{2n+2}, and does not contain aromatics, oxygen, sulfur, or ash.[46] It is commonly called second-generation biodiesel, green diesel or hydrotreated vegetable oil.[47,48] The renewable-diesel production process combines oils or fats with hydrogen at high temperatures and pressures. This process removes impurities and produces a chemically similar fuel to fossil diesel but with a reduced environmental impact.[49] Unlike biodiesel, which consists of chemical compounds distinct from traditional diesel, renewable diesel has a chemical composition almost identical to fossil diesel. This allows it to be applied directly in diesel engines without modifications or combinations.[50]

Renewable diesel, produced from renewable sources and with a more environmentally friendly combustion process, produces fewer greenhouse gases, solid particles, and sulfur compounds. This helps to reduce the environmental effect of the transport sector. It has a high resistance to oxidation, which means it does not deteriorate quickly over time.[47] Furthermore, it performs well in low-temperature environments, which makes it suitable for application in cold areas. Renewable diesel can be transported, stored, and distributed using fossil-diesel infrastructure, reducing implementation costs.[51] Because it is produced from renewable raw materials and the reuse of waste, it has a more favourable carbon balance, helping to build a more circular economy.[50]

Diesel from renewable sources has several benefits, such as significantly reducing greenhouse-gas emissions, compatibility with current diesel engines without modifications, better performance, and better durability than other biofuels.[46] Furthermore, it is produced from materials classified as waste, such as animal fats and used oils.[49] However, there are still some restrictions, such as a production cost higher than that of fossil diesel, dependence on renewable resources, and the possible competition with food production, particularly when vegetable oils are used as the raw material.[52,53]

2.3.1.1 *Differences Between Renewable Diesel, Biodiesel and Fossil Diesel*

Biodiesel, diesel from renewable sources, and fossil diesel differ in terms of their origin, production process, chemical composition, impact on the environment, and compatibility with engines.[54] Fossil diesel, derived from petroleum, is produced through fractional distillation in refineries and contains hydrocarbons with varying amounts of sulfide and aromatic compounds.[55] Despite being widely used and compatible with all diesel engines, it is one of the primary sources of greenhouse-gas emissions and a resource that cannot be replaced.[56]

On the other hand, biodiesel and renewable diesel are alternative fuels originating from renewable sources, such as vegetable oils and animal fat. Biodiesel is obtained through a transesterification reaction that generates methyl esters with oxygen in their constitution.[57] Despite reducing pollutant emissions compared to fossil diesel, it requires mixtures (such as B10 or B20) for full compatibility and can present challenges related to oxidative stability and waste generation.[58]

The cost of renewable diesel is still high due to the manufacturing process and the need for renewable raw materials.[51] On the other hand, biodiesel tends to be cheaper. However, its production faces obstacles such as competition with the food chain.[58] Because it is abundant and easily accessible, fossil diesel is cheaper. However, its environmental impact and the scarcity of resources make it unsustainable in the long term. Thus, renewable diesel emerges as a promising option, combining sustainability and efficiency.[57] To facilitate the comparison between the fuels, Table 2.1 summarizes their main characteristics.

2.3.1.2 Renewable-diesel Production Process

Renewable-diesel production occurs primarily through catalytic hydrogenation, also known as water treatment.[59] This procedure transforms renewable raw materials, such as vegetable oils (soy, palm, canola) and fatty waste (such as animal fat and used cooking oil), into high-purity hydrocarbons comparable to fossil diesel.[60,61] The initial phase of the pre-treatment process of the raw material consists of purifying it to eliminate impurities such as water, phosphorus, sulfur, and suspended solids, using techniques such as filtration, centrifugation, and chemical treatment.[62] This purification is

Table 2.1 Comparison of fossil diesel, biodiesel, and renewable diesel.

Characteristic	Fossil diesel	Biodiesel	Renewable diesel
Origin	Petroleum	Vegetable oils and animal fats	Vegetable oils and animal fats
Production process	Fractional distillation	Transesterification	Catalytic hydrogenation
Chemical composition	Hydrocarbons with sulfur and aromatic compounds	Methyl esters (with oxygen)	High-purity hydrocarbons comparable to fossil diesel
Environmental impact	High greenhouse gas emissions, non-renewable	Lower pollutant emissions, but can generate waste	Lower environmental impact, sustainable
Engine compatibility	Fully compatible	Requires blends for full compatibility	Can be used in its pure form (100%) or combined with fossil diesel
Cost	Lower cost	Cheaper than renewable diesel, but with production obstacles	Higher cost due to the manufacturing process and the need for renewable raw materials

essential to ensure the effectiveness of the processing and the quality of the product.[63]

After purifying the raw material, it is subjected to hydrogenation in a reactor, where it is combined with hydrogen under high temperatures (300–450 °C) and high pressures (30–130 bar).[64,65] At this stage, two significant chemical reactions take place: hydrodeoxygenation (HDO), responsible for removing oxygen from the raw material in the form of water, and hydrodecarbonylation or hydrodecarboxylation (HDCX), which converts oxygen into carbon dioxide (CO_2).[65] These reactions completely convert oils and fats into pure hydrocarbons free of oxygen, generating a fuel with chemical characteristics similar to those of petroleum diesel.[64,66]

The product can also be subjected to an optional isomerization process, which reorganizes the hydrocarbon chains to improve characteristics such as fluidity at low temperatures, which makes renewable diesel more practical for application in cold areas.[67] In the last stage, the product is separated and purified. By-products such as propane, CO_2, and water are eliminated, giving renewable diesel as the main fraction. It is ready to be used in its pure form (100%) or combined with fossil diesel to comply with local market standards.[68]

2.3.2 Environmental and Technical Benefits

The transport sector is one of the main beneficiaries of population growth and intense urbanization of cities, responsible for the annual movement of a large volume of passengers and cargo on a global scale.[69] Projections indicate a continued increase in demand for transport, with the expectation that by 2050 more than two-thirds of the growth in energy demand will be directed towards the transport of heavy loads, especially by large vehicles such as trucks.[70] On the other hand, the transport sector is one of the main contributors of direct global emissions of greenhouse gases into the atmosphere, which generate a series of negative impacts on human health and exacerbate climate change on the planet.[1,71] Unfortunately, segments such as road transport, shipping and aviation are projected to remain among the top sources of residual emissions until mid-century, posing a significant challenge to achieving carbon-neutrality goals.[72]

The vehicles used in transportation are powered primarily by fuels, most of which are derived from fossil sources, such as gasoline, ethanol, and regular diesel, among others. Among them, regular petroleum-based diesel has been used for over a century and stands out as a highly efficient fuel for use on highways and in urban areas.[73] Automotive vehicles with diesel engines are preferred over gasoline-powered cars due to their superior fuel economy with better torque, outstanding durability, and excellent energy efficiency.[74] However, the high emission of nitrogen oxides (NO_x), hydrocarbons, and CO_2 negatively affect the environment.[73,75] The main motivation for exploring alternatives to fossil fuels is the deterioration of environmental conditions.[76] For example, in 2022 alone more than a quarter of total greenhouse-gas emissions in the European Union came from the transport sector.[77] Of these emissions, passenger cars and light commercial vehicles contributed 53% of the polluting gases released into the atmosphere.[49]

The exploration of substitutes for common petroleum diesel includes to biofuels, such as FAME biodiesel and renewable diesel, as potential candidates for partial or complete replacement. While biodiesel comprises a base of fatty acid methyl esters (FAME), renewable diesel contains hydrocarbons free of aromatic components.[78] Unlike fossil fuels, they contain lower amounts of toxic chemical compounds and are readily biodegradable when present in the environment.[79] HVO (renewable) diesel is a good option to help solve the environmental problems caused by the burning of fossil fuels and overcome the limitations of FAME biodiesel, such as incompatibility, component corrosion, oxidation and moisture absorption in combustion engines.[70,80] The sources of raw materials used in the HVO-diesel production chain are similar to FAME biodiesel, including vegetable oils, animal fats, residual frying oils, and lignocellulosic biomass.[79] In Brazil, the agricultural sector contributes significantly to the production of green diesel with large quantities of potential raw materials, such as sugarcane bagasse, primarily produced by the sugar and ethanol sector, in addition to agro-industrial residues such as soybean oil and palm oil.[81] The abundant availability of raw material sources and their easy accessibility create a promising opportunity for developing the Brazilian green-diesel industry.[54,81]

The production of HVO diesel is based on the same production goal as FAME biodiesel, namely the reduction of CO_2 emissions from its cleaner combustion compared to petroleum-based diesel. Compared to FAME biodiesel, HVO diesel has some advantages, such as the possibility of prolonged storage due to its properties being similar to petroleum diesel, which contains a mixture of saturated straight-chain and branched-chain hydrocarbons (C_{15} to C_{18}) and is oxygen-free.[82,83] In addition, HVO diesel has excellent performance in cold climates, as its high cetane number facilitates ignition, surpassing biodiesel in this aspect.[78,84] HVO diesel is attractive for commercialization because, in addition to its rich composition, it is compatible with diesel engines and the existing distribution infrastructure, eliminating the need for costly adaptations to automotive vehicles or gas stations.[54] It can be used pure or mixed with petroleum diesel in any proportion without damaging the engine, allowing a gradual and flexible transition to more sustainable alternative fuels.[61]

2.3.2.1 *Reduction of GHG Emissions Compared to Fossil Fuels*

The transportation industry has an active impact on the global energy scenario.[75] This industry makes up 30% of global energy consumption, with road transportation accounting for 60% of the sector's total energy demand, in addition to being directly responsible for almost 75% of greenhouse-gas emissions due to the massive use of fossil fuels.[85,86] Most emissions come from conventional vehicles with internal combustion engines (ICE), which release polluting gases such as CO_2 at 97.4%, followed by hydrofluorocarbons (HFCs) at 2.0%, nitrogen oxides (NO_x) at 0.6%, and methane (CH_4) at 0.1%.[85,86] Despite regulations and international climate agreements, GHG emissions increased from 1.53 billion to 1.88 billion metric tons of CO_2 per year between 1990 and 2019.[85]

Biofuels, or renewable liquid fuels, are essential in transitioning to a more sustainable energy matrix, as they offset the carbon footprint of fossil fuels while simultaneously recovering biomass and urban waste.[87] The use of biofuels such as HVO diesel in internal combustion engines generates low emissions from contaminants associated with paraffinic compounds compared to petroleum-based diesel fuels.[86] The lower distillation range, high cetane number, and the absence of aromatics in HVO diesel are likely responsible for the reduction in the mass of particulate matter and emissions of carbon monoxide (CO) and total hydrocarbons into the atmosphere.[85,86] NO_x emissions are typically slightly reduced when using HVO compared to conventional diesel fuels, although this effect may be ambiguous due to the interaction between ignition delay and the amount and timing of biofuel injection.[86]

Among several production routes to obtain these renewable hydrocarbons, the hydrodeoxygenation of triglycerides and fatty acids (HEFA) is the most well-established technique.[88] However, most production technologies depend on the same input, namely hydrogen, which is generally obtained through steam methane reforming (SMR) from a fossil source and is classified as grey hydrogen due to the associated GHG emissions.[88–92]

2.3.2.2 *Compatibility with Existing Engines (Drop-in Fuel) and Current Infrastructure*

Renewable fuels compatible with existing engines and infrastructure are known as "drop-in fuels" and have no restrictions on blending with petroleum diesel due to their chemical similarity.[81] These fuels are chemically similar to fossil fuels and can partially or fully replace petroleum diesel in current engines and the existing distribution network, such as pipelines, storage tanks, and fuel pumps, without costly adaptations.[87,93]

Drop-in fuels can be directly applied to internal combustion engines in operation.[94] Unlike other biofuels, such as FAME biodiesel, which may require engine modifications due to its particular physical and chemical characteristics, drop-in fuels have properties similar to those of fossil fuels, such as high energy density and oxidative stability.[93,94]

HVO diesel, with a chemical profile almost identical to fossil diesel, is one of these biofuels that can be used in conventional diesel engines without the risk of corrosion, clogging of filters and other parts, or engine degradation.[95] Its compatibility with existing infrastructure and engines will reduce the costs of the energy transition, eliminating the need to replace or modify an entire current fleet and allowing the gradual introduction of low-carbon fuels without interrupting operations in the transportation sector.[96]

2.4 Contribution of Renewable Diesel to Decarbonization

The transport sector strongly influences global warming, air pollution, and climate change, as most vehicles still use internal combustion engines, which run on fossil fuels, the best known being gasoline and diesel.[56] The decarbonization of transport has become necessary. Frequent debates on using more ecological alternatives are being encouraged by government policies in several countries to reduce or eliminate greenhouse-gas emissions, as emphasized by Net Zero by 2050, from the International Energy Agency (IEA) published in 2021.[97,98]

The scenario desired by Net Zero and the IEA is directly related to the transport sector, as the goal is that by 2050 total decarbonization will be achieved in all sectors.[99] In 2024, during COP29, Brazil presented a new target to reduce greenhouse-gas emission rates (IEA, 2021) between 59% and 67% by 2035, becoming the new Nationally Determined Contribution (NDC) presented. California has already defined that by 2035, sales of light-duty vehicles (LDV), such as passenger cars, trucks, and delivery vehicles, will have zero carbon emissions, replacing vehicles with internal combustion engines with cars with electric motors powered by batteries.[100] These ambitious and challenging proposals depend on several variables, such as the high investment rate, requiring consistent and executable policies.[101,102]

Renewable diesel, also called green diesel, biofuel, or hydrotreated vegetable oil, is a type of fuel made from renewable raw materials with characteristics similar to petroleum diesel, and can be produced from vegetable oils, such as soybean or canola oil, animal fats, such as beef tallow, food waste, such as used cooking oils, or, more recently, algae biomass.[54,103] Renewable diesel gives lower emissions of air pollutants, meaning air quality is not aggressively affected in urban or industrial areas.[104] Renewable diesel is superior to biodiesel, as it has a high degree of storage and operational stability, in addition to having superior performance in cold climates, because its chemical composition includes high levels of cetane, which facilitates its ignition when compared to biodiesel, and it can also be used pure or in mixtures with petrodiesel without causing damage to vehicle engines and without drastic changes to performance.[105] However, the costs of producing renewable diesel are considerably higher than those of making biodiesel. When comparing renewable diesel and petroleum diesel, it is possible to see several benefits of renewable diesel, such as non-corrosiveness and high calorific value. These aspects make renewable diesel attractive and promising for reducing greenhouse-gas emissions.[106]

A recent study addresses the potential for decarbonization of road fuels and transmission systems, specifically in the European Union (EU) and the United States (US), evaluating greenhouse-gas emission reductions based on well-to-wheels (WTW) analyses applied to various fuel technologies.[107] Vehicle fuel systems were analyzed, comparing vehicles that use gasoline and diesel in both markets, and it was found that renewable diesel is more efficient in reducing GHGs. However, some difficulties were also highlighted regarding large-scale implementation, such as the availability of raw materials for production, costs, and logistical challenges encountered.[108] It was concluded that transmission systems such as electric powertrains, which are more energy efficient when compared to internal combustion engines, as well as combined vehicle-fuel systems (WTW), present considerable GHG reduction rates, which provides greater understanding of which methodologies are applicable to achieve decarbonization in countries.[107]

About 8% of California's GHG emissions come from the transportation sector. In 2022, the California Air Resources Board set a challenging goal of reducing GHG emissions by 48% by 2030, below 1990 levels (California Air Resources Board, 2022). The work of Hennessy and collaborators (2024) addresses the simulation of strategies to achieve these desired results, especially for the heavy-duty vehicle (HDV) fleet. Accelerated retirement policies and sales mandate policies for zero-emission vehicles were evaluated, and it was found that, even if HDV vehicle sales were zero-emission by 2025, they would still not achieve the desired results by 2045, contrary to the schedule established by the state.[109] The study concludes that California must work on joint policies to achieve carbon neutrality, emphasizing that decarbonization is extremely necessary, since the objective is to decarbonize the entire vehicle fleet, ensuring that the policies of all sectors belonging to the state work together so that there is no overload on other transport. In other words, if a considerable reduction of GHGs in road transport is achieved, there would consequently be an increase in GHG emissions in the aviation or railway sectors. However, the problem would be solved if all transport sectors were entirely decarbonized.[110,111]

In the study by Riazi and collaborators (2020), poultry fat oils and residues, beef tallow, and soybean oil were used for the production of renewable diesel on a laboratory scale, applying life-cycle assessment (LCA) analysis to identify the environmental impacts associated with the raw materials and the process.[112] Life-cycle inventory (LCI) models were used to perform predictions and data collection on renewable diesel's global-warming potential (GWP), and demonstrated a 110% reduction when using renewable diesel from beef fat compared to petroleum-based diesel. At the same time, the GWP is considerably lower than that of biodiesel, which is directly related to the products and treatment used for the life-cycle assessment.[70,113]

The entire transport sector, which also includes road, maritime, and aviation vehicles, is responsible for 7.5 Gt (gigatonnes) of CO_2 emissions, with around 23% of emissions caused by fuel combustion, according to the International Energy Agency (IEA). Although the transport sector is considered complex and challenging for the decarbonization process,[114] the Paris Climate Agreement encourages countries to intensify and strengthen mitigation efforts based on Nationally Determined Contributions (NDCs), with long-term strategies aimed at achieving low greenhouse-gas emissions. To achieve these objectives, it is necessary to establish commitments, especially over the medium and long term, so that the transport sector can achieve significant decarbonization rates, in addition to short-term methodologies to stimulate the implementation of projects that favour the desired objectives and more ambitious plans that emphasize the need for decarbonization in the transport sector, which would have a considerable influence on the scope of sustainable development.[115]

Decarbonization impacts the entire transport sector, especially trucks and buses, as there is a high demand for these vehicles in large urban centres worldwide. There is also this need for maritime and rail transport, as they emit significant amounts of CO_2.[116] The transportation sector is highly dependent on fossil fuels, including in Brazil, despite policies established in 2004 that require biodiesel to be mixed into all diesel-like fuels sold.[117] Decarbonization is promising for reducing carbon emissions and is a sustainable and economically viable alternative in the transport sector. However, it is necessary to use more advanced and accessible technologies so that adequate and large-scale production and application can occur, and the immediate application of agreements and goals designed by countries to achieve the desired objectives through viable, socially responsible, and eco-efficient methodologies is required.

2.5 Regional and Global Perspectives

2.5.1 Countries and Regions: Production and Use Potential

Biofuel production varies significantly around the world and is influenced by various factors. Figure 2.2 provides an overview of the geographical distribution of the primary raw materials used in renewable-diesel production in the Americas, Africa, Asia, and Europe. The availability of raw materials directly limits the production of renewable diesel, and is the main factor that can boost its production.[118] However, other factors are also associated with limitations in production, such as incentive policies, demand for this type of fuel, and even refining infrastructure.[119] In general, all global regions have the potential to produce renewable diesel, but some stand out more than others.

By region, starting in North America, the United States and Canada stand out, as they have numerous oilseeds available for the production of this fuel.[120] Soybeans, corn and canola stand out in these countries, but they also have well-developed and functional infrastructures focused on refining these biofuels.[118] Finally, it is important to note that the United States has strong carbon-reduction policies, and Canada is following suit, encouraging the production and use of renewable diesel fuel.[121]

South America also stands out when it comes to encouraging renewable diesel. Countries such as Brazil, Argentina, and Colombia stand out in this region due to the variety of potential raw materials for this biofuel.[122] These include agro-industrial waste, residual oils, and soybean, palm, and macaw palm (*Acrocomia aculeata*) oils. In this region, Brazil's RenovaBio policy can also be highlighted, which aims to boost the replacement of oil-based fuels with biofuels from sustainable sources.[123]

Figure 2.2 Geographical distribution of the main raw materials, by region, used in the production of renewable diesel.

Europe is a pioneering region producing renewable diesel and has well-developed production and refining technologies.[120] Thus, the production potential for renewable diesel in Europe is high, driving its production in its central countries, such as Germany, France, and the Netherlands. In addition, the European Union is putting incentives and pressure on the industrial sector to significantly reduce greenhouse-gas emissions in the coming years.[124] These standards and conventions intensify the production of biofuels such as renewable diesel.

In Asia, palm oil is the primary raw material for producing renewable diesel, which is responsible for this region's production potential.[125] This industry is driven by the need to reduce emissions in the transportation process and Asia is today the leading consumer of renewable diesel. On the other hand, in Africa, the main driver of renewable-diesel production is agricultural production, which has numerous raw materials with potential for application.[126] However, no large industries in this region are yet focusing on producing renewable diesel, although discussions and incentives are already being developed on this topic.[127]

Finally, the dominant country producing renewable diesel is the United States, which has a robust and well-developed infrastructure for creating and refining renewable diesel.[128] On the other hand, it is also possible to highlight Brazil due to its large amount of potential raw materials for this production, which is an essential factor in boosting renewable diesel in a region.[129]

Indonesia and Germany also stand out in this area. Indonesia has a large production of raw materials for renewable diesel, such as palm oil, and Germany has technology and infrastructure that favours this application.[130] In this sense, China also stands out for its production infrastructure and as an essential biofuel consumer, as it has a large vehicle fleet. Thus, these countries and regions play important roles in transitioning to cleaner and more sustainable energy systems, contributing to reductions in greenhouse-gas emissions, and diversifying energy sources.[131]

2.5.2 Impact on the Supply Chain and Circular Economy

Renewable-diesel production plays a strategic role in transitioning to a low-carbon economy while positively impacting the supply chain and promoting principles of circular economy.[132] Derived from renewable sources such as vegetable oils, animal fats, and agricultural waste, this biofuel aligns with global sustainability and decarbonization goals.[126]

The supply chain for renewable-diesel production is transformed by introducing new dynamics for obtaining and processing raw materials.[131] Instead of relying on oil by-products, it uses biomass, industrial waste, and agricultural by-products, strengthening local supply chains and reducing the need for imports.[119]

Renewable diesel plays a key role in sustainability, transforming and promoting economic development in agricultural and rural areas, creating job opportunities and generating added value.[133] This biofuel also reduces the supply chain's carbon footprint, lowering emissions associated with transportation and processing.[134]

In addition, renewable diesel also influences the circular economy, which is one of the pillars of production, as it takes advantage of residual resources and seeks to close the production cycle.[132] On the other hand, waste cooking oil, discarded animal fats, and agricultural by-products can also be reused as raw materials, avoiding improper disposal and reducing environmental impacts.[124]

In producing renewable diesel, there are numerous benefits for the environment and the economy, such as reducing raw material costs, making renewable diesel competitive compared to fossil fuels.[121] Integrating waste into the production cycle minimizes greenhouse-gas emissions and reduces the environmental impact of improper disposal.[135] There is also a need for technologies for refining and reuse, which is driving the development of innovative and competitive solutions. Regional supply chains are necessary for renewable diesel production and to promote the economic and social development of rural and urban communities.[136]

Using agricultural waste prevents competition with food production for agricultural land, promoting a more ethical and sustainable supply chain.[137] In addition, this approach reduces the environmental impact associated with the disposal of this waste, which often results in greenhouse-gas emissions such as methane, in landfills or burning practices.[130]

Animal fats, often by-products of the meat and fish industries, are another valuable resource in renewable-diesel production.[136] These fats can be converted into biofuels through processes such as transesterification and hydrotreatment of esters and fatty acids.[120] Using animal fats reduces dependence on virgin vegetable oils and uses waste that would otherwise be disposed of inappropriately, contributing to environmental pollution.[121]

Waste oils, such as those from domestic and industrial frying, are widely used as a raw material for renewable diesel.[138] These oils are collected from restaurants, food industries and households, avoiding improper disposal that can contaminate soil and water resources.[139] The collection and reuse of waste oils also promotes a circular economy by transforming common waste into a valuable resource. This process encourages the creation of collection networks and partnerships between different sectors, increasing efficiency and reducing costs.[140]

Thus, using materials that would otherwise be discarded minimizes pollution and environmental impact.[138] Renewable diesel produced from these sources results in reduced greenhouse-gas emissions[141] in addition to promoting the diversification of raw materials, reducing dependence on imported oil and improving energy security.[142] This stimulates the circular economy, generating value at different stages of the supply chain.

2.6 Conclusion

Renewable diesel presents itself as a promising alternative in addressing existing challenges across various sectors, including the transportation sector, which is one of the major contributors to greenhouse-gas emissions. This chapter highlighted the characteristics of renewable diesel, including using renewable raw materials, ranging from vegetable oils to waste. Also, it showed its compatibility with existing diesel engines in the current market. Such information reveals that renewable diesel is essential in the energy transition as it is environmentally friendly and does not require significant technological adjustments, allowing the use of current resources and technologies. Compared to other biofuels, renewable diesel has demonstrated superior efficiency due to its stability, storage, and performance in cold climates, surpassing other biofuels. Additionally, its environmental benefits and production techniques have proven to be promising. Moreover, renewable diesel significantly impacts the circular economy, as it uses waste as raw material, reusing resources and minimizing the environmental impacts of its use.

Thus, renewable diesel is a viable option in the decarbonization process of the transportation sector, contributing to the mitigation of carbon-related issues and directly aiding in meeting established global goals. It reaffirms the importance of environmental care and stands out for its critical role in combating climate change and promoting sustainability and technological innovation in significant nations like Brazil. Despite its promise, renewable diesel must overcome several barriers, including production costs, competition for the same raw materials with other industries, and technical challenges such as collecting, transporting, processing, and storing the necessary raw materials. These hurdles should be overcome with time, but they require encouragement, collaboration, and investment involving researchers, industries, and the government.

Researchers will be responsible for developing techniques, methods, and analyses for more efficient renewable-diesel production. The industry will provide the technical and technological support for this endeavour, and the government will be responsible for providing the fiscal framework and implementing public policies to develop, produce, and use this new resource. These three pillars, working together, can ensure the introduction of renewable diesel into the market, incorporating it into the current national and international energy matrix. That said, integrating renewable diesel into the global energy matrix, aiming for a sustainable and low-carbon transportation sector, requires a collective effort among the state, industry, and research. Using renewable diesel as a viable solution to the energy challenges faced by current society, we seek a clean and sustainable future.

Acknowledgements

This study is supported in part by financial support from the following Brazilian agencies for scientific and technological development: Fundação Cearense de Apoio ao Desenvolvimento Científico e Tecnológico (FUNCAP) (PNE-0112-00048.01.00/16, PS1-0186-00216.01.00/21, PS1-00186-00255.01.00/21), Conselho Nacional de Desenvolvimento Científico e Tecnológico (CNPq) (311062/2019-9, 308452/2022-4), and Coordenação de Aperfeiçoamento de Ensino Superior (CAPES) (Finance Code 001, PROEX 23038.000509/2020-82). The article processing fee for this research was paid by the Coordenação de Aperfeiçoamento de Pessoal de Nível Superior – CAPES (ROR ID: OOxOma614).

References

1. J. Tian, L. Yu, R. Xue, S. Zhuang and Y. Shan, Global Low-Carbon Energy Transition in the Post-COVID-19 Era, *Appl. Energy*, 2022, **307**, 1–11.
2. J. Fan, X. Meng, J. Tian, C. Xing, C. Wang and J. Wood, A Review of Transportation Carbon Emissions Research Using Bibliometric Analyses, *J. Traffic Transp. Eng., Engl. Ed.*, 2023, **10**(5), 878–899.
3. Q.-L. Jing, H.-Z. Liu, W.-Q. Yu and X. He, The Impact of Public Transportation on Carbon Emissions—From the Perspective of Energy Consumption, *Sustainability*, 2022, **14**(10), 1–18.
4. V. J. Reddy, N. P. Hariram, R. Maity, M. F. Ghazali and S. Kumarasamy, Sustainable E-Fuels: Green Hydrogen, Methanol and Ammonia for Carbon-Neutral Transportation, *World Electr. Veh. J.*, 2023, **349**(14), 1–29.
5. G. Di Vito Nolfi, K. Gallucci and L. Rossi, Green Diesel Production by Catalytic Hydrodeoxygenation of Vegetables Oils, *Int. J. Environ. Res. Public Health*, 2021, **18**(24), 1–28.
6. T. J. Hilbers, L. M. J. Sprakel, L. B. J. van den Enk, B. Zaalberg, H. van den Berg and L. G. J. van der Ham, Green Diesel from Hydrotreated

Vegetable Oil Process Design Study, *Chem. Eng. Technol.*, 2015, **38**(4), 651–657.
7. C. Kochaphum, S. H. Gheewala and S. Vinitnantharat, Environmental Comparison of Straight Run Diesel and Cracked Diesel, *J. Cleaner Prod.*, 2012, **37**, 142–146.
8. N. M. Clauser, G. González, C. M. Mendieta, J. Kruyeniski, M. C. Area and M. E. Vallejos, Biomass Waste as Sustainable Raw Material for Energy and Fuels, *Sustainability*, 2021, **13**(2), 1–21.
9. R. Höfer and J. Bigorra, Biomass-Based Green Chemistry: Sustainable Solutions for Modern Economies, *Green Chem. Lett. Rev.*, 2008, **1**(2), 79–97.
10. H. K. Jeswani, A. Chilvers and A. Azapagic, Environmental Sustainability of Biofuels: A Review: Environmental Sustainability of Biofuels, *Proc. R. Soc. A*, 2020, **476**(20), 1–37.
11. O. Ogunkunle and N. A. Ahmed, A Review of Global Current Scenario of Biodiesel Adoption and Combustion in Vehicular Diesel Engines, *Energy Rep.*, 2019, **5**, 1560–1579.
12. S. Aminzadegan, M. Shahriari, F. Mehranfar and B. Abramović, Factors Affecting the Emission of Pollutants in Different Types of Transportation: A Literature Review, *Energy Rep.*, 2022, **8**, 2508–2529.
13. M. Massar, I. Reza, S. M. Rahman, S. M. H. Abdullah, A. Jamal and F. S. Al-Ismail, Impacts of Autonomous Vehicles on Greenhouse Gas Emissions—Positive or Negative?, *Int. J. Environ. Res. Public Health*, 2021, **18**(23), 1–23.
14. L. Lakanen, K. Grönman, S. Väisänen, H. Kasurinen, A. Soininen and R. Soukka, Applying the Handprint Approach to Assess the Air Pollutant Reduction Potential of Paraffinic Renewable Diesel Fuel in the Car Fleet of the City of Helsinki, *J. Cleaner Prod.*, 2021, **290**, 125786.
15. M. L. Melamed, J. Schmale and E. von Schneidemesser, Sustainable Policy—Key Considerations for Air Quality and Climate Change, *Curr. Opin. Environ. Sustainability*, 2016, **23**, 85–91.
16. O. Ogunkunle and N. A. Ahmed, Overview of Biodiesel Combustion in Mitigating the Adverse Impacts of Engine Emissions on the Sustainable Human–Environment Scenario, *Sustainability*, 2021, **10**(13), 1–28.
17. J. Popp, Z. Lakner, M. Harangi-Rákos and M. Fári, The Effect of Bioenergy Expansion: Food, Energy, and Environment, *Renewable Sustainable Energy Rev.*, 2014, **32**, 559–578.
18. S. Turkdogan, Design and Optimization of a Solely Renewable Based Hybrid Energy System for Residential Electrical Load and Fuel Cell Electric Vehicle, *Eng. Sci. Technol., Int. J.*, 2021, **24**(2), 397–404.
19. K. Daszkiewicz, Policy and Regulation of Energy Transition, in *The Geopolitics of the Global Energy Transition*, ed. M. Hafner and S. Tagliapietra, Springer International Publishing, Cham, 2020, pp. 203–226.
20. X. Du, H. Zhang and Y. Han, How Does New Infrastructure Investment Affect Economic Growth Quality? Empirical Evidence from China, *Sustainability*, 2022, **14**(6), 1–30.

21. E. T. Sayed, M. A. Abdelkareem, K. Obaideen, K. Elsaid, T. Wilberforce, H. M. Maghrabie and A. G. Olabi, Progress in Plant-Based Bioelectrochemical Systems and Their Connection with Sustainable Development Goals, *Carbon Resour. Convers.*, 2021, **4**, 169–183.
22. K. Obaideen, M. Nooman AlMallahi, A. H. Alami, M. Ramadan, M. A. Abdelkareem, N. Shehata and A. G. Olabi, On the Contribution of Solar Energy to Sustainable Developments Goals: Case Study on Mohammed Bin Rashid Al Maktoum Solar Park, *Int. J. Thermofluids*, 2021, **12**, 100123.
23. K. Obaideen, M. A. Abdelkareem, T. Wilberforce, K. Elsaid, E. T. Sayed, H. M. Maghrabie and A. G. Olabi, Biogas Role in Achievement of the Sustainable Development Goals: Evaluation, Challenges, and Guidelines, *J. Taiwan Inst. Chem. Eng.*, 2022, **131**, 104207.
24. T. N. Jonell, P. Jones, A. Lucas and S. Naylor, Limited Waterpower Contributed to Rise of Steam Power in British "Cottonopolis", *PNAS Nexus*, 2024, **3**(7), 1–12.
25. M. K. Hubbert, History of Petroleum Geology and Its Bearing Upon Present and Future Exploration, *Am. Assoc. Pet. Geol. Bull.*, 1966, **50**, 2504–2518.
26. S. Wołkowicz, M. Graniczny, K. Wołkowicz and H. Urban, History of the Oil Industry in Poland until 1939, *Geol. Soc., London, Spec. Publ.*, 2017, **442**(1), 401–411.
27. I. Ruble, The U.S. Crude Oil Refining Industry: Recent Developments, Upcoming Challenges and Prospects for Exports, *J. Econ. Asymmetries*, 2019, **20**, e00132.
28. M. K. Al-Sakkaf and S. A. Onaizi, Effects of Emulsification Factors on the Characteristics of Crude Oil Emulsions Stabilized by Chemical and Biosurfactants: A Review, *Fuel*, 2024, **361**, 130604.
29. B. Li, Z. Guo, L. Zheng, E. Shi and B. Qi, A Comprehensive Review of Wax Deposition in Crude Oil Systems: Mechanisms, Influencing Factors, Prediction and Inhibition Techniques, *Fuel*, 2024, **357**, 129676.
30. J. R. Payne and C. R. Phillips, *Petroleum Spills in the Marine Environment*, CRC Press, 2018, DOI: 10.1201/9781351075480.
31. K.-H. Dietsche, History of the Diesel Engine, in *Fundamentals of Automotive and Engine Technology*, Springer Fachmedien Wiesbaden, Wiesbaden, 2014, pp. 8–17.
32. G. Squeo, R. Silletti, G. Napoletano, M. Greco Miani, G. Difonzo, A. Pasqualone and F. Caponio, Characterization and Effect of Refining on the Oil Extracted from Durum Wheat By-Products, *Foods*, 2022, **11**(5), 683.
33. F. Sharifianjazi, A. Esmaeilkhanian, N. Karimi, B. A. Horri, L. Bazli, S. Eskandarinezhad and E. Ahmadi, A Review of Combustion Properties, Performance, and Emission Characteristics of Diesel Engine Fueled with Al_2O_3 Nanoparticle-Containing Biodiesel, *Clean Technol. Environ. Policy*, 2024, **26**(11), 3715–3737.
34. J. Yakhshilikov, M. Cavana and P. Leone, A Review of the Energy System and Transport Sector in Uzbekistan in View of Future Hydrogen Uptake, *Energies*, 2024, **17**(16), 3987.

35. S. Steiner, C. Bisig, A. Petri-Fink and B. Rothen-Rutishauser, Diesel Exhaust: Current Knowledge of Adverse Effects and Underlying Cellular Mechanisms, *Arch. Toxicol.*, 2016, **90**(7), 1541–1553.
36. N. A. A. Qasem and G. A. Q. Abdulrahman, A Recent Comprehensive Review of Fuel Cells: History, Types, and Applications, *Int. J. Energy Res.*, 2024, **2024**(1), 1–36.
37. K. Wu, Q. Lu, Q. Cao, A. S. Dhmees, K. Yang, S. Wang, J. Yu, L. Hu and H. Zhang, Hydrolytic and Pyrolytic Technologies of Pretreatment Lignocellulose for Production of Ethanol Fuels – A Comparative Review, *Ind. Crops Prod.*, 2024, **218**, 118840.
38. S. Valizadeh, B. Valizadeh, M. W. Seo, Y. J. Choi, J. Lee, W.-H. Chen, K.-Y. A. Lin and Y.-K. Park, Recent Advances in Liquid Fuel Production from Plastic Waste *via* Pyrolysis: Emphasis on Polyolefins and Polystyrene, *Environ. Res.*, 2024, **246**, 118154.
39. A. H. Zulkafli, H. Hassan, M. A. Ahmad, A. T. Mohd Din and S. M. Wasli, Co-Pyrolysis of Biomass and Waste Plastics for Production of Chemicals and Liquid Fuel: A Review on the Role of Plastics and Catalyst Types, *Arabian J. Chem.*, 2023, **16**(1), 104389.
40. H. Yang and H. Chen, Biomass Gasification for Synthetic Liquid Fuel Production, in *Gasification for Synthetic Fuel Production*, Elsevier, 2015, pp. 241–275.
41. S. A. Isaacs, M. D. Staples, F. Allroggen, D. S. Mallapragada, C. P. Falter and S. R. H. Barrett, Environmental and Economic Performance of Hybrid Power-to-Liquid and Biomass-to-Liquid Fuel Production in the United States, *Environ. Sci. Technol.*, 2021, **55**(12), 8247–8257.
42. S. Frigo, M. Seggiani, M. Puccini and S. Vitolo, Liquid Fuel Production from Waste Tyre Pyrolysis and Its Utilisation in a Diesel Engine, *Fuel*, 2014, **116**, 399–408.
43. J. Kaur, U. Tyagi, M. Aslam and A. K. Sarma, Green Gasoline as a Commercial Liquid Fuel Throughout Asia, Europe and the USA: A Technical Review, in *Green Gasoline*, Royal Society of Chemistry, 2023, pp. 272–298.
44. J. Li, D. Yu, L. Pan, X. Xu, X. Wang and Y. Wang, Recent Advances in Plastic Waste Pyrolysis for Liquid Fuel Production: Critical Factors and Machine Learning Applications, *Appl. Energy*, 2023, **346**, 121350.
45. J. Liu, M. Lin, P. Han, G. Yao and H. Jiang, Biosynthesis Progress of High-Energy-Density Liquid Fuels Derived from Terpenes, *Microorganisms*, 2024, **12**(4), 706.
46. A. Dimitriadis, I. Natsios, A. Dimaratos, D. Katsaounis, Z. Samaras, S. Bezergianni and K. Lehto, Evaluation of a Hydrotreated Vegetable Oil (HVO) and Effects on Emissions of a Passenger Car Diesel Engine, *Front. Mech. Eng.*, 2018, **4**, 1–19.
47. D. Kumar, B. Singh and J. Korstad, Utilization of Lignocellulosic Biomass by Oleaginous Yeast and Bacteria for Production of Biodiesel and Renewable Diesel, *Renewable Sustainable Energy Rev.*, 2017, **73**, 654–671.

48. G. Knothe, Biodiesel and Renewable Diesel: A Comparison, *Prog. Energy Combust. Sci.*, 2010, **36**(3), 364–373.
49. S. Puricelli, G. Cardellini, S. Casadei, D. Faedo, A. E. M. van den Oever and M. Grosso, A Review on Biofuels for Light-Duty Vehicles in Europe, *Renewable Sustainable Energy Rev.*, 2021, **137**, 110398.
50. T. Manchanda, R. Tyagi and D. K. Sharma, Comparison of Fuel Characteristics of Green (Renewable) Diesel with Biodiesel Obtainable from Algal Oil and Vegetable Oil, *Energy Sources, Part A*, 2018, **40**(1), 54–59.
51. A. Dimitriadis, T. Seljak, R. Vihar, U. Žvar Baškovič, A. Dimaratos, S. Bezergianni, Z. Samaras and T. Katrašnik, Improving PM–NOx Trade-off with Paraffinic Fuels: A Study towards Diesel Engine Optimization with HVO, *Fuel*, 2020, **265**, 116921.
52. P. Zeman, V. Hönig, M. Kotek, J. Táborský, M. Obergruber, J. Mařík, V. Hartová and M. Pechout, Hydrotreated Vegetable Oil as a Fuel from Waste Materials, *Catalysts*, 2019, **9**(4), 337.
53. S. L. Douvartzides, N. D. Charisiou, K. N. Papageridis and M. A. Goula, Green Diesel: Biomass Feedstocks, Production Technologies, Catalytic Research, Fuel Properties and Performance in Compression Ignition Internal Combustion Engines, *Energies*, 2019, **12**(5), 809.
54. A. A. V. Julio, T. S. Milessi, E. A. Ocampo Batlle, E. E. Silva Lora, D. M. Yepes Maya and J. C. Escobar Palacio, Techno-Economic and Environmental Potential of Renewable Diesel as Complementation for Diesel and Biodiesel in Brazil: A Comprehensive Review and Perspectives, *J. Cleaner Prod.*, 2022, **371**, 133431.
55. S. R. Chia, S. Nomanbhay, M. Y. Ong, A. H. B. Shamsuddin, K. W. Chew and P. L. Show, Renewable Diesel as Fossil Fuel Substitution in Malaysia: A Review, *Fuel*, 2022, **314**, 123137.
56. H. Xu, L. Ou, Y. Li, T. R. Hawkins and M. Wang, Life Cycle Greenhouse Gas Emissions of Biodiesel and Renewable Diesel Production in the United States, *Environ. Sci. Technol.*, 2022, **56**(12), 7512–7521.
57. S. B. Glisic, J. M. Pajnik and A. M. Orlović, Process and Techno-Economic Analysis of Green Diesel Production from Waste Vegetable Oil and the Comparison with Ester Type Biodiesel Production, *Appl. Energy*, 2016, **170**, 176–185.
58. S. N. Gebremariam and J. M. Marchetti, Economics of Biodiesel Production: Review, *Energy Convers. Manage.*, 2018, **168**, 74–84.
59. P. Šimáček, D. Kubička, G. Šebor and M. Pospíšil, Fuel Properties of Hydroprocessed Rapeseed Oil, *Fuel*, 2010, **89**(3), 611–615.
60. A. Amin, Review of Diesel Production from Renewable Resources: Catalysis, Process Kinetics and Technologies, *Ain Shams Eng. J.*, 2019, **10**(4), 821–839.
61. M. Ameen, M. T. Azizan, S. Yusup, A. Ramli and M. Yasir, Catalytic Hydrodeoxygenation of Triglycerides: An Approach to Clean Diesel Fuel Production, *Renewable Sustainable Energy Rev.*, 2017, **80**, 1072–1088.
62. J. Gong and F. You, Consequential Life Cycle Optimization: General Conceptual Framework and Application to Algal Renewable Diesel Production, *ACS Sustainable Chem. Eng.*, 2017, **5**(7), 5887–5911.

63. C. A. Scaldaferri and V. M. D. Pasa, Production of Jet Fuel and Green Diesel Range Biohydrocarbons by Hydroprocessing of Soybean Oil over Niobium Phosphate Catalyst, *Fuel*, 2019, **245**, 458–466.
64. V. Larnaudie, M. Bule, K.-Y. San, P. V. Vadlani, J. Mosby, S. Elangovan, M. Karanjikar and S. Spatari, Life Cycle Environmental and Cost Evaluation of Renewable Diesel Production, *Fuel*, 2020, **279**, 118429.
65. M. Krár, S. Kovács, D. Kalló and J. Hancsók, Fuel Purpose Hydrotreating of Sunflower Oil on $CoMo/Al_2O_3$ Catalyst, *Bioresour. Technol.*, 2010, **101**(23), 9287–9293.
66. B. Zhang, J. Wu, C. Yang, Q. Qiu, Q. Yan, R. Li, B. Wang, J. Wu and Y. Ding, Recent Developments in Commercial Processes for Refining Bio-Feedstocks to Renewable Diesel, *Bioenergy Res.*, 2018, **11**(3), 689–702.
67. A. Afshar Taromi and S. Kaliaguine, Green Diesel Production *via* Continuous Hydrotreatment of Triglycerides over Mesostructured γ-Alumina Supported NiMo/CoMo Catalysts, *Fuel Process. Technol.*, 2018, **171**, 20–30.
68. M. Lapuerta, O. Armas, J. J. Hernández and A. Tsolakis, Potential for Reducing Emissions in a Diesel Engine by Fuelling with Conventional Biodiesel and Fischer–Tropsch Diesel, *Fuel*, 2010, **89**(10), 3106–3113.
69. A. Khurshid, K. Khan, S. F. Saleem, J. Cifuentes-Faura and A. Cantemir Calin, Driving towards a Sustainable Future: Transport Sector Innovation, Climate Change and Social Welfare, *J. Cleaner Prod.*, 2023, **427**, 139250.
70. H. Xu, U. Lee and M. Wang, Life-Cycle Energy Use and Greenhouse Gas Emissions of Palm Fatty Acid Distillate Derived Renewable Diesel, *Renewable Sustainable Energy Rev.*, 2020, **134**, 110144.
71. D. Álvarez-Antelo, A. Lauer and I. Capellán-Pérez, Exploring the Potential of a Novel Passenger Transport Model to Study the Decarbonization of the Transport Sector, *Energy*, 2024, **305**, 132313.
72. J. Martin, E. Dimanchev and A. Neumann, Carbon Abatement Costs for Renewable Fuels in Hard-to-Abate Transport Sectors, *Adv. Appl. Energy*, 2023, **12**, 100156.
73. D. Boopathi, S. Thiyagarajan, V. Edwin Geo and S. Madhankumar, Effect of the Second Generation and Third Generation Biofuel Blend on Performance, Emission and Combustion Characteristics of CI Engine, *Int. J. Ambient Energy*, 2020, **41**(7), 767–774.
74. A. Rahman Adib, M. Mizanur Rahman, T. Hassan, M. Ahmed and A. Al Rifat, Novel Biofuel Blends for Diesel Engines: Optimizing Engine Performance and Emissions with C. Cohnii Microalgae Biodiesel and Algae-Derived Renewable Diesel Blends, *Energy Convers. Manage.: X*, 2024, **23**, 100688.
75. M. Filonchyk, M. P. Peterson, L. Zhang, V. Hurynovich and Y. He, Greenhouse Gases Emissions and Global Climate Change: Examining the Influence of CO_2, CH_4, and N_2O, *Sci. Total Environ.*, 2024, **935**, 173359.
76. N. Kumar, R. Koul and R. C. Singh, Comparative Analysis of Ternary Blends of Renewable Diesel, Diesel and Ethanol with Diesel, *Sustainable Energy Technol. Assess.*, 2022, **50**, 101828.

77. A. García, J. Monsalve-Serrano, M. G. Guzmán-Mendoza and E. Iñiguez, Technical Evaluation of Low-Carbon Fuels as a Decarbonization Pathway of the Light-Duty Transport Sector, *Fuel*, 2024, **369**, 131772.
78. S. R. Chia, S. Nomanbhay, M. Y. Ong, A. H. B. Shamsuddin, K. W. Chew and P. L. Show, Renewable Diesel as Fossil Fuel Substitution in Malaysia: A Review, *Fuel*, 2022, **314**, 123137.
79. Z. Yang, K. Shah, C. Pilon-McCullough, R. Faragher, P. Azmi, B. Hollebone, B. Fieldhouse, C. Yang, D. Dey, P. Lambert and V. Beaulac, Characterization of Renewable Diesel, Petroleum Diesel and Renewable Diesel/Biodiesel/Petroleum Diesel Blends, *Renewable Energy*, 2024, **224**, 120151.
80. S. Lucantonio, A. Di Giuliano, L. Rossi and K. Gallucci, Green Diesel Production *via* Deoxygenation Process: A Review, *Energies*, 2023, **16**(2), 844.
81. L. S. dos Santos, H. Gasparetto and N. P. G. Salau, Technological Maturity and Future Perspectives for Green Diesel Production in Brazil, *Cleaner Chem. Eng.*, 2024, **10**, 100127.
82. S. L. Douvartzides, N. D. Charisiou, K. N. Papageridis and M. A. Goula, Green Diesel: Biomass Feedstocks, Production Technologies, Catalytic Research, Fuel Properties and Performance in Compression Ignition Internal Combustion Engines, *Energies*, 2019, **12**(5), 1–41.
83. S. R. Chia, S. Nomanbhay, K. W. Chew, P. L. Show, J. Milano and A. H. Shamsuddin, Indigenous Materials as Catalyst Supports for Renewable Diesel Production in Malaysia, *Energies*, 2022, **15**(8), 1–31.
84. A. A. Stepacheva, L. Z. Nikoshvili, E. M. Sulman and V. G. Matveeva, Hydrodeoxygenation of Stearic Acid for the Production of "Green" Diesel, *Green Process Synth.*, 2014, **3**(6), 441–446.
85. A. M. Kuyumcu, B. Bingül, F. Akar and A. Yıldız, Well-to-Wheel Carbon Footprint and Cost Analysis of Gasoline, Diesel, Hydrogen ICE, Hybrid and Fully Electric City Buses, *Energy*, 2024, **301**, 131685.
86. C. Mata, D. Cárdenas, C. Esarte, J. A. Soriano, A. Gómez, P. Fernández-Yáñez, R. García-Contreras, L. Sánchez, J. I. Nogueira and O. Armas, Performance and Regulated Emissions from a Euro VI-D Hybrid Bus Tested with Fossil and Renewable (Hydrotreated Vegetable Oil) Diesel Fuels under Urban Driving in Bilbao City, Spain, *J. Cleaner Prod.*, 2023, **383**, 135472.
87. L. Tipanluisa, M. V. Prati and M. A. Costagliola, Impact of Diesel/Renewable Fuels Blend on Gaseous and Particle Emissions of a Light-Duty Vehicle under Real Driving Emissions, *Renewable Energy*, 2024, **230**, 120819.
88. G. C. Umenweke, R. B. Pace, E. Santillan-Jimenez and J. A. Okolie, Techno-Economic and Life-Cycle Analyses of Sustainable Aviation Fuel Production *via* Integrated Catalytic Deoxygenation and Hydrothermal Gasification, *Chem. Eng. J.*, 2023, **452**, 139215.
89. M. Hermesmann and T. E. Müller, Green, Turquoise, Blue, or Grey? Environmentally Friendly Hydrogen Production in

Transforming Energy Systems, *Prog. Energy Combust. Sci.*, 2022, **90**, 100996.

90. P. M. Lunardi, P. M. Julio, V. F. Bolson, F. D. Mayer and F. de Castilhos, Comparative Techno-Economic Assessment of Renewable Diesel Production Integrated with Alternative Hydrogen Supply, *Int. J. Hydrogen Energy*, 2024, **89**, 820–835.
91. S. Lucantonio, A. Di Giuliano, L. Rossi and K. Gallucci, Green Diesel Production *via* Deoxygenation Process: A Review, *Energies*, 2023, **16**(2), 1–44.
92. J. Incer-Valverde, A. Korayem, G. Tsatsaronis and T. Morosuk, "Colors" of Hydrogen: Definitions and Carbon Intensity, *Energy Convers. Manage.*, 2023, **291**, 117294.
93. S. Lilonfe, B. Davies, A. F. N. Abdul-Manan, I. Dimitriou and J. McKechnie, A Review of Techno-Economic Analyses and Life Cycle Greenhouse Gas Emissions of Biomass-to-Hydrocarbon "Drop-in" Fuels, *Sustainable Prod. Consumption*, 2024, **47**, 425–444.
94. D. Singh, K. A. Subramanian, R. Bal, S. P. Singh and R. Badola, Combustion and Emission Characteristics of a Light Duty Diesel Engine Fueled with Hydro-Processed Renewable Diesel, *Energy*, 2018, **154**, 498–507.
95. J. Bullermann, N. C. Meyer, A. Krafft and F. Wirz, Comparison of Fuel Properties of Alternative Drop-in Fuels with Standard Marine Diesel and the Effects of Their Blends, *Fuel*, 2024, **357**, 129937.
96. S. Lilonfe, I. Dimitriou, B. Davies, A. F. N. Abdul-Manan and J. McKechnie, Comparative Techno-Economic and Life Cycle Analyses of Synthetic "Drop-in" Fuel Production from UK Wet Biomass, *Chem. Eng. J.*, 2024, **479**, 147516.
97. E. Lindstad, T. Ø. Ask, P. Cariou, G. S. Eskeland and A. Rialland, Wise Use of Renewable Energy in Transport, *Transp. Res. D: Transp. Environ.*, 2023, **119**, 103713.
98. X. Yu, S. LeBlanc, N. Sandhu, L. Wang, M. Wang and M. Zheng, Decarbonization Potential of Future Sustainable Propulsion—A Review of Road Transportation, *Energy Sci. Eng.*, 2024, **12**(2), 438–455.
99. S. Brynolf, J. Hansson, J. E. Anderson, I. R. Skov, T. J. Wallington, M. Grahn, A. D. Korberg, E. Malmgren and M. Taljegård, Review of Electrofuel Feasibility—Prospects for Road, Ocean, and Air Transport, *Prog. Energy*, 2022, **4**(4), 042007.
100. C. Fernández-Dacosta, L. Shen, W. Schakel, A. Ramirez and G. J. Kramer, Potential and Challenges of Low-Carbon Energy Options: Comparative Assessment of Alternative Fuels for the Transport Sector, *Appl. Energy*, 2019, **236**, 590–606.
101. E. Monroe, J. S. Carlson, R. Dhaoui, M. Sarwar, P. T. Benavides, J. Zhu, C. S. McEnally, L. Pfefferle, A. George, N. Sizemore and R. W. Davis, Application of Alkoxyalkanoates (AOAs) as Renewable Diesel Blendstocks from Chemical Coupling of High-Yield Fermentation Products, *Energy Fuels*, 2023, **37**(3), 2091–2099.

102. V. J. Reddy, N. P. Hariram, R. Maity, M. F. Ghazali and S. Kumarasamy, Sustainable Vehicles for Decarbonizing the Transport Sector: A Comparison of Biofuel, Electric, Fuel Cell and Solar-Powered Vehicles, *World Electr. Veh. J.*, 2024, **15**(3), 93.
103. J. Roles, J. Yarnold, K. Hussey and B. Hankamer, Techno-Economic Evaluation of Microalgae High-Density Liquid Fuel Production at 12 International Locations, *Biotechnol. Biofuels*, 2021, **14**(1), 133.
104. L. Fulton, A. W. Schäfer and D. Sperling, Transitions to Deeply Decarbonized Transportation and Energy Systems around the World: Challenges and Solutions, *Transp. Res. D: Transp. Environ.*, 2024, **134**, 104219.
105. C. Sevim and B. Zincir, Biodiesel and Renewable Diesel as a Drop-in Fuel for Decarbonized Maritime Transportation, in *Energy, Environment, and Sustainability*, 2022, pp. 319–345.
106. G. Di Blasio, R. Ianniello and C. Beatrice, Hydrotreated Vegetable Oil as Enabler for High-Efficient and Ultra-Low Emission Vehicles in the View of 2030 Targets, *Fuel*, 2022, **310**, 122206.
107. H. Cai, M. Prussi, L. Ou, M. Wang, M. Yugo, L. Lonza and N. Scarlat, Decarbonization Potential of On-Road Fuels and Powertrains in the European Union and the United States: A Well-to-Wheels Assessment, *Sustainable Energy Fuels*, 2022, **6**(19), 4398–4417.
108. M. Prussi, U. Lee, M. Wang, R. Malina, H. Valin, F. Taheripour, C. Velarde, M. D. Staples, L. Lonza and J. I. Hileman, CORSIA: The First Internationally Adopted Approach to Calculate Life-Cycle GHG Emissions for Aviation Fuels, *Renewable Sustainable Energy Rev.*, 2021, **150**, 111398.
109. E. M. Hennessy, M. Singh, S. Saltzer and I. M. L. Azevedo, Pathways to Zero Emissions in California's Heavy-Duty Transportation Sector, *Environ. Res.: Infrastruct. Sustainability*, 2024, **4**(3), 035001.
110. K. Na, S. Biswas, W. Robertson, K. Sahay, R. Okamoto, A. Mitchell and S. Lemieux, Impact of Biodiesel and Renewable Diesel on Emissions of Regulated Pollutants and Greenhouse Gases on a 2000 Heavy Duty Diesel Truck, *Atmos. Environ.*, 2015, **107**, 307–314.
111. D. Ogunkoya, W. L. Roberts, T. Fang and N. Thapaliya, Investigation of the Effects of Renewable Diesel Fuels on Engine Performance, Combustion, and Emissions, *Fuel*, 2015, **140**, 541–554.
112. B. Riazi, J. M. Mosby, B. Millet and S. Spatari, Renewable Diesel from Oils and Animal Fat Waste: Implications of Feedstock, Technology, Co-Products and ILUC on Life Cycle GWP, *Resour., Conserv. Recycl.*, 2020, **161**, 104944.
113. V. Uusitalo, S. Väisänen, J. Havukainen, M. Havukainen, R. Soukka and M. Luoranen, Carbon Footprint of Renewable Diesel from Palm Oil, Jatropha Oil and Rapeseed Oil, *Renewable Energy*, 2014, **69**, 103–113.
114. E. Shafiei, B. Davidsdottir, J. Leaver, H. Stefansson and E. I. Asgeirsson, Energy, Economic, and Mitigation Cost Implications of Transition toward a Carbon-Neutral Transport Sector: A Simulation-Based Comparison between Hydrogen and Electricity, *J. Cleaner Prod.*, 2017, **141**, 237–247.

115. S. Gota, C. Huizenga, K. Peet, N. Medimorec and S. Bakker, Decarbonising Transport to Achieve Paris Agreement Targets, *Energy Effic.*, 2019, **12**(2), 363–386.
116. D. Oke, J. B. Dunn and T. R. Hawkins, The Contribution of Biomass and Waste Resources to Decarbonizing Transportation and Related Energy and Environmental Effects, *Sustainable Energy Fuels*, 2022, **6**(3), 721–735.
117. A. A. V. Julio, T. S. Milessi, D. M. Y. Maya, E. S. Lora and J. C. E. Palacio, Assessment of the Sustainability and Economic Potential of Hydrotreated Vegetable Oils to Complement Diesel and Biodiesel Blends in Brazil, *Biofuels, Bioprod. Biorefin.*, 2023, **17**(2), 312–323.
118. S. Tamrat, V. R. Ancha, R. Gopal, R. B. Nallamothu and Y. Seifu, Emission and Performance Analysis of Diesel Engine Running with CeO_2 Nanoparticle Additive Blended into Castor Oil Biodiesel as a Substitute Fuel, *Sci. Rep.*, 2024, **14**(1), 1–12.
119. G. Verasoundarapandian, N. N. Zakaria, N. A. Shaharuddin, K. A. Khalil, N. A. Puasa, A. A. Azmi, C. Gomez-Fuentes, A. Zulkharnain, C. Y. Wong, M. F. Rahman and S. A. Ahmad, Coco Peat as Agricultural Waste Sorbent for Sustainable Diesel-Filter System, *Plants*, 2021, **10**(11), 1–21.
120. M. Thillainayagam, K. Venkatesan, D. Rana, S. Subramani, B. Sethuramasamyraja and R. K. Babu, Diesel Reformulation Using Bio-Derived Propanol to Control Toxic Emissions from a Light-Duty Agricultural Diesel Engine, *Environ. Sci. Pollut. Res. Int.*, 2017, **24**(20), 16725–16734.
121. K. Lahti-Leikas, E. Niemistö, H. Talvenmäki, N. Saartama, Y. Sun, L. Mercier and M. Romantschuk, Hydrotreated Vegetable Oil Migrates through Soil and Degrades Faster than Fossil Diesel and Hydrotreated Vegetable Oil-Fossil Diesel Blend, *Environ. Sci. Pollut. Res. Int.*, 2024, **31**(40), 53177–53192.
122. A. K. Sharma, P. K. Sharma, V. Chintala, N. Khatri and A. Patel, Environment-Friendly Biodiesel/Diesel Blends for Improving the Exhaust Emission and Engine Performance to Reduce the Pollutants Emitted from Transportation Fleets, *Int. J. Environ. Res. Public Health*, 2020, **17**(11), 1–18.
123. C. M. Granadeiro, S. O. Ribeiro, M. Karmaoui, R. Valença, J. C. Ribeiro, B. De Castro, L. Cunha-Silva and S. S. Balula, Production of Ultra-Deep Sulfur-Free Diesels Using a Sustainable Catalytic System Based on UiO-66(Zr), *Chem. Commun.*, 2015, **51**(72), 13818–13821.
124. K. Alles and Y. Demirel, Measuring Risk of Renewable Diesel Production Processes Using a Multi-Criteria Decision Strategy, *Chemosphere*, 2024, **354**(24), 1–15.
125. G. Monteverde, F. Bianco, P. Papetti, D. Komínková, D. Spasiano, G. Paolella, M. Muscetta, S. Varjani, N. Han, G. Esposito and M. Race, Reuse of Polymeric Waste for the Treatment of Marine Water Polluted by Diesel, *J. Environ. Manage.*, 2024, **356**(24), 120529.

126. E. Puspawiningtiyas, O. Muraza, H. Devianto, M. Pratiwi, S. Subagjo, T. Prakoso, V. E. Krisnawan, U. Zaki, L. Elizabeth, T. H. Soerawidjaja, Y. A. Situmorang and A. Indarto, Sustainable Diesel from Pyrolysis of Unsaturated Fatty Acid Basic Soaps: The Effect of Temperature on Yield and Product Composition, *Molecules*, 2022, **27**(3), 1–13.
127. Y. Sulbaran-Bracho, M. Orellana-Saez, J. Castro-Severyn, C. Galbán-Malagón, E. Castro-Nallar and I. Poblete-Castro, Continuous Bioreactors Enable High-Level Bioremediation of Diesel-Contaminated Seawater at Low and Mesophilic Temperatures Using Antarctic Bacterial Consortia: Pollutant Analysis and Microbial Community Composition, *Environ. Pollut.*, 2023, **321**(23), 1–15.
128. A. Anderson, V. Kumar M, B. Gavurová, H. Li, C. Xia and X. Zhang, Optimizing Engine Performance and Reducing Emissions of Greenhouse Gases through Spirulina Microalgae and Nano-Additive Blends, *Environ. Res.*, 2023, **231**(1), 1–18.
129. U. Ulagarjun, V. V. Varma, A. K. Menon, N. Gobinath, A. R. Palanivelrajan, T. M. Yunus Khan, R. U. Baig, N. Almakayeel and F. M, Investigation on Effect of Cerium Oxide Additive in Waste Plastic Oil Fueled CI Engine, *Heliyon*, 2024, **10**(4), 1–16.
130. M. Dhanarasu, K. A. RameshKumar and P. Maadeswaran, Effect of Acetone as an Oxygenated Additive with Used Sunflower Oil Biodiesel on Performance, Combustion and Emission in Diesel Engine, *Environ. Technol.*, 2022, **43**(24), 3682–3692.
131. K. R. Landwehr, R. Mead-Hunter, R. A. O'Leary, A. Kicic, B. J. Mullins and A. N. Larcombe, Respiratory Health Effects of In Vivo Sub-Chronic Diesel and Biodiesel Exhaust Exposure, *Int. J. Mol. Sci.*, 2023, **24**(6), 1–23.
132. U. Rajak, P. K. Chaurasiya, T. N. Verma, A. Dasore, U. Ağbulut, K. Meshram, C. Saleel, S. Saboor, E. Cuce and Z. Mian, Optimizing Soybean Biofuel Blends for Sustainable Urban Medium-Duty Commercial Vehicles in India: An AI-Driven Approach, *Environ. Sci. Pollut. Res. Int.*, 2024, **31**(22), 32449–32463.
133. H. H. Fayyaz, M. A. Mujtaba, S. Jahangir, S. Imran, M. A. Ijaz Malik, H. Fayaz, C. A. Saleel, U. Hassan, S. Quershi and H. farooq, A Comprehensive Study on the Performance and Emission Analysis in Diesel Engine via Optimization of Novel Ternary Fuel Blends: Diesel, Manganese, and Diethyl Ether, *Heliyon*, 2023, **9**(10), 1–16.
134. S. Eswaran, S. Subramaniam, S. Geleynse, K. Brandt, M. Wolcott and X. Zhang, Dataset for Techno-Economic Analysis of Catalytic Hydrothermolysis Pathway for Jet Fuel Production, *Data Brief*, 2021, **39**(21), 1–11.
135. Q. Liu, G. Xu, X. Wang, X. Liu and X. Mu, Highly Selective Upgrading of Biomass-Derived Alcohol Mixtures for Jet/Diesel-Fuel Components, *ChemSusChem*, 2016, **9**(24), 3465–3472.
136. S. Kesharvani, G. Dwivedi and T. N. Verma, Optimization of Performance Characteristics in Diesel Engine Utilizing Chlorella Vulgaris Fuel-

a Green Approach towards Sustainable Development, *Environ. Sci. Pollut. Res. Int.*, 2023, **32**(25), 1–27.

137. M. M. Khan, S. M. M. Hasnain, A. K. Kadian, R. P. Sharma, M. Abbas and S. Pandey, Reduction of Fossil Fuel Usage and Emissions of Diesel Engines by the Application of Quaternary Blends and Varied Fuel Injection Pressures, *ACS Omega*, 2024, **9**(24), 26388–26399.
138. A. Giuliano, E. Catizzone and C. Freda, Process Simulation and Environmental Aspects of Dimethyl Ether Production from Digestate-Derived Syngas, *Int. J. Environ. Res. Public Health*, 2021, **18**(2), 1–21.
139. B. Das, P. V. Bhave, S. P. Puppala, S. Adhikari, S. Sainju, E. Mool and R. M. Byanju, Emission Factors and Emission Inventory of Diesel Vehicles in Nepal, *Sci. Total Environ.*, 2022, **812**(22), 1–12.
140. J. Garg, A. A. Mohite, P. Sharma, D. Barik, B. J. Medhi, H. Deka and B. J. Bora, Energizing Tomorrow: Unleashing Spirulina's Potential in Engine Performance Optimization and Emission Reduction, *Environ. Sci. Pollut. Res. Int.*, 2024, **32**(23), 1–19.
141. J. Feng, Y. Li, T. J. Strathmann and J. S. Guest, Characterizing the Opportunity Space for Sustainable Hydrothermal Valorization of Wet Organic Wastes, *Environ. Sci. Technol.*, 2024, **58**(5), 2528–2541.
142. K. B. Bhumula and G. N. Kumar, Using CRITIC-TOPSIS and Python to Examine the Effect of 1-Hepatnol on the Performance and Emission Characteristics of CRDI CI Engine with Split Injection, *Heliyon*, 2024, **10**(11), 1–20.

CHAPTER 3

Major Resources for Renewable-diesel Production: Challenges and Opportunities

KHALIL UL REHMAN, IFTIKHAR HUSSAIN, ISHFAQ MAJEED MIR, RAQEEBA AZIZ AND MOHAMMAD ASLAM*

Biofuels Laboratory, Department of Chemistry, National Institute of Technology Srinagar (J&K), India
*Email: maslam@nitsri.ac.in

3.1 Introduction

The world continues to rely largely on fossil fuels for energy. Due to increased industrialization and urbanization, the requirement for energy is quickly increasing.[1] As reported in the Statistical Review of World Energy (2023),[2] fossil fuels remained the dominant source in global energy consumption (Figure 3.1), comprising approximately 82% of the world's primary energy use in 2023. Fossil fuels are non-renewable and finite, with depletion expected in the near future. Sadly, our significant dependence on these fuels for energy production results in CO_2 emissions and other harmful gases, contributing to environmental issues like global warming, air pollution, climate change, and health concerns.[3] As reported by the International Energy Agency (IEA), in 2023 global CO_2 emissions (Figure 3.2) increased by 1.1%, amounting to a total of 37.4 billion tonnes (Gt).[4] This is a new all-time high, with emissions rising mostly because of an increase in coal usage, which accounts for more than 65% of the entire growth. The transportation industry relies significantly

Green Chemistry Series No. 89
Biorefining of Renewable Diesel: From Production to Market
Edited by Mohammad Aslam, Sanjeev Mishra and Jorge Aburto

Published by the Royal Society of Chemistry, www.rsc.org

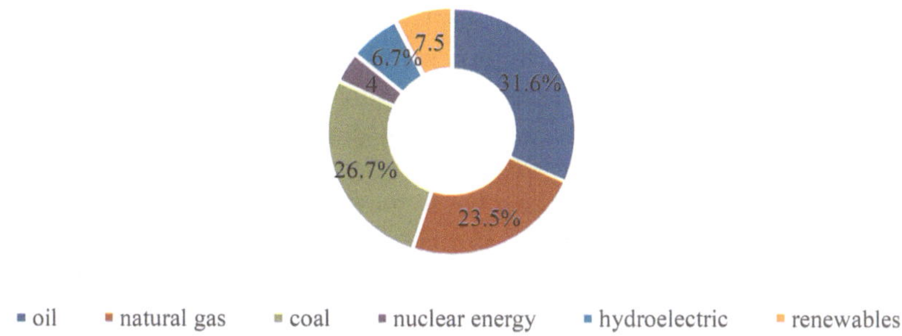

Figure 3.1 Primary global energy consumption. Data from ref. 2.

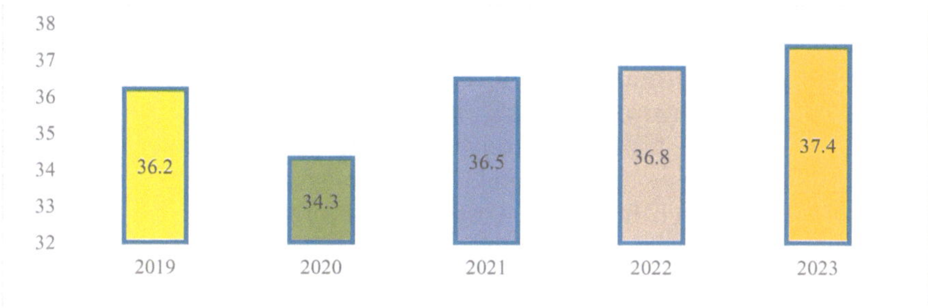

Figure 3.2 Global carbon dioxide emissions (Gt). Data from ref. 4.

on petroleum, and its consumption is increasing. It is expected that by 2070–2080, the world will confront a petroleum crisis.[5] Crude-oil prices have risen due to increased demand, causing economic instability and inflation. Currently, the growing cost of fossil-fuel energy, the depletion of crude-oil supplies, and the detrimental environmental repercussions of their usage are the key issues driving the quest for sustainable energy sources, which must be economical, ecologically benign, and renewable.[6]

In this context, biofuels produced from biomass offer a compelling alternative to fossil fuels due to their renewable, cost-effective, and environmentally friendly characteristics.[7] Biofuels are derived from environmentally friendly and biodegradable sources such as plants, animals, and microbes (biomass), and they are non-toxic, free of sulfur, and biodegradable. Depending on the feedstock utilized, biofuels are categorized into four types: first-, second-, third-, and fourth-generation biofuels.[8] First-generation biofuels are generally derived from food crops such as sugarcane, maize, and soybean. Second-generation biofuels are derived from non-food sources such as lignocellulosic biomass. The complicated structure of lignin leads to low yields and significant pre-treatment expenses, limiting its economic viability. Third-generation biofuels are obtained from algal biomass, and fourth-generation biofuels use genetically engineered algae.[9,10] Algal biomass is considered a suitable feedstock for third-generation biofuel production due to its quick growth, high productivity, and significant lipid content.[11] Green diesel is a third-generation (3-G) advanced liquid biofuel, also known as super cetane diesel. It is regarded one of the most efficient fuel alternatives to petrodiesel for compression ignition (CI) engines because its cetane number (cetane no. ~90) is more than two times that of biodiesel and petrodiesel. Green diesel has been marketed by Neste Oil (Finland), Diamond Green Diesel (USA), and Honeywell (USA).

Renewable diesel is produced using a variety of feedstocks, mainly vegetable oils, animal fats, and waste oils, each with its own set of advantages and challenges. Vegetable oils, such as soybean oil and palm oil, are plentiful but might compete with food supplies, causing sustainability problems.[12] Despite their sustainability benefits, animal fats and waste oils present technological challenges due to compositional variability and impurities, necessitating advanced processing to ensure consistent quality, protect catalysts, and maximize conversion efficiency. The production process often includes hydroprocessing, in which catalysts play an important role in turning these feedstocks into high-quality diesel. However, the dependence on expensive, noble metal-based catalysts presents economic issues, spurring research into more cost-effective alternatives.[13]

3.2 Green Diesel

Green diesel is a third-generation (3-G) advanced liquid biofuel also known as super cetane diesel, and is considered a highly efficient alternative fuel for CI engines. Green diesel, also known as renewable diesel, has a higher cetane number (cetane no. ~90) than biodiesel and petrodiesel, making it a very efficient alternative fuel for CI engines.[14] Renewable diesel is a non-oxygenated fuel primarily composed of carbon and hydrogen. It has a similar composition to petroleum diesel and has a high cetane number due to a high percentage of straight-chain hydrocarbons equivalent to cetane ($C_{16}H_{34}$). Green diesel has no aromatics, which is a positive attribute; however, the acid value is higher and requires special attention.[15] Green diesel has a lower pour point than petroleum diesel, making it suitable for use in internal combustion engines in cold environments. Green diesel is characterized similarly to petroleum diesel, using ASTM, IP, or EN criteria.[16] However, green diesel significantly differs from biodiesel with regard to its structural chemistry, composition and production routes. Table 3.1 shows the differences in physicochemical properties between green diesel, petroleum diesel, and biodiesel.

Green diesel is less dense than biodiesel and petrodiesel, resulting in a lower volumetric heating value and higher volumetric fuel usage. Nonetheless, green diesel's greater lower heating value (LHV) per mass than biodiesel and petrodiesel compensates for the disadvantage of its lower density. Green diesel is more stable than biodiesel in oxidation and storage, and when treated with a hydroisomerization technique, it demonstrates remarkable cold weather properties (*e.g.* cloud point). Green diesel is

Table 3.1 Physicochemical properties of green diesel, petroleum diesel and biodiesel.[14,17,18]

Property	Petrodiesel	Biodiesel	Green diesel
Carbon content (wt%)	86.8	76.2	84.9
Oxygen content (wt%)	0	11.2	0
Hydrogen content (wt%)	13.2	12.6	15.1
Density at 15 °C ($kg\,m^{-3}$)	841	880	790
Cetane number	44–67	45–65	70–90
Specific gravity	0.84	0.88	0.78
Flash point (°C)	60–170	100–180	59–138
Lower heating value ($kJ\,kg^{-1}$)	43	37.2–38	43.7–44.5
Cloud point (°C)	−5	20	−10
Viscosity at 40 °C ($mm^2\,s^{-1}$)	1.9–4.1	2.9–11	2–4
Sulfur content (ppm)	<10	<5	<10

chemically comparable to petrodiesel (both hydrocarbons) and is expected to someday replace petrodiesel. It represents a potential option for sustainable energy, providing a renewable alternative to fossil fuels. The hydroprocessing of vegetable oils primarily produces green diesel with a high cetane number (cetane no. 90) with hydrocarbons in the C_{15}–C_{18} range.[19]

3.3 Resources for Green Diesel

Green diesel is sometimes referred to as renewable diesel owing to its renewable sources. Sources are divided into three generations based on their source and availability. Renewable sources improve sustainability and process utilization in business settings. The majority of the first-generation feedstocks are edible oils or fat sources, whereas the second-generation comprises non-edible oils, plant biomass, and so on. The third-generation category focuses on non-competitive feeds such as microalgae, waste oil, and animal fat. An overview of biofuel generations in terms of feedstock, method, merits and demerits is presented in Table 3.2. All types of feedstocks are explained in depth in the following sections.

3.3.1 First-generation Feedstocks

3.3.1.1 Vegetable Oils

One of the primary sources of energy for the production of renewable diesel is vegetable oil, which offers an effective and high-yield raw material that can be transformed into hydrocarbons that resemble traditional diesel. Soybean oil, canola oil, palm oil, sunflower oil, and corn oil are the most widely used vegetable oils.[20] Because it is widely available and derives advantages from well-established farming methods, soybean oil is especially important in the United States. Canola oil is preferred due to its low saturated fat level and favorable fatty acid composition, making it an ideal feedstock for high-quality renewable diesel. Palm oil, noted for its excellent output per acre, is

Table 3.2 Overview of biofuel generations in terms of feedstock, method, merits and demerits.

Generation	Feedstock	Fuel production techniques	Merits	Demerits
First	Edible oils, starch, sugar	Transesterification, fermentation, hydrotreating	Decreased CO_2 emissions, reduced fossil-fuel consumption	Adverse effects on the availability of food and agricultural land, and on food prices
Second	Non-edible oils, non-food biomass (*e.g.* forestry residue, agricultural residue), aquatic plants, waste cooking oil	Hydrothermal liquefaction, pyrolysis, gasification, Fischer–Tropsch, hydroprocessing, enzymatic hydrolysis	Low-cost feedstock, no threat to hunger, uses waste or non-food sources, potentially higher yields	Additional processing stages and the need for land
Third	Microalgae, macroalgae, and cyanobacteria	Transesterification, hydrothermal liquefaction	Less space and simple requirements for growth, ecofriendly, high oil content, high productivity	Expensive and unstable biofuel
Fourth	Microalgae, genetically modified crops, and synthetic biology-derived organisms	Genetic modification of algae, algal bioengineering, photobiological water splitting	High growth rate, high oil content, low structural complexity	Still in the early stages of development

mostly supplied from Southeast Asia, but it offers sustainability issues because of its relationship with deforestation and biodiversity loss.[21] Sunflower oil, which has significant quantities of unsaturated fats, is another potential choice, particularly in Europe. Maize oil, a byproduct of maize ethanol synthesis, expands the range of feedstocks accessible for renewable-diesel generation. The use of vegetable oils as feedstocks has several advantages, including well-established supply networks, relatively simple processing needs, and the possibility for large reductions in greenhouse gas emissions compared to fossil fuels. However, the use of edible oils raises questions regarding food security and land usage, needing careful management and sustainable agricultural techniques in order to balance energy production with food consumption.[22]

3.3.1.2 Soybean Oil

Soybean oil is a prominent fuel for production of renewable diesel, owing to its high lipid content and broad availability. This vegetable oil is refined to eliminate contaminants, allowing it to be further processed.[23] Hydrotreating is commonly used to produce green diesel from soybean oil, in which the oil combines with hydrogen in the presence of a catalyst to remove oxygen and other impurities, yielding long-chain hydrocarbons. To increase fuel efficiency, especially in colder areas, the hydrocarbon molecules are hydroisomerized, resulting in branched chains with improved cold flow characteristics. The end result is a high-quality renewable diesel that may be used interchangeably with traditional diesel in current engines.[24] Soybean is the feedstock with the most potential for diesel generation in the US and Brazil, with 18–20% of oil content. The growing demand for green diesel has led to an increase in soybean output in the US, with approximately 120 million acres of land and 11 billion pounds used. To meet this need, the total edible oil cover must be doubled within the next ten years.[25,26] According to a survey, China, the United States, Brazil, and India would use the most soybean oil in 2024, accounting for 17 100 thousand metric tonnes (TMT), 12 247 TMT, 8525 TMT, and 4935 TMT, respectively.[27] As additional plants come online, there will be a higher need for soybean oil as feedstock. Soybean oil has five unique fatty acids: palmitic (12%), oleic (26%), stearic (4%), linoleic (52%), and linolenic acid (6%).[28] Hydrogenation is a method for producing green diesel that improves oil stability and processing. The hydrogenation of soybean oil with niobium catalyst was investigated to yield jet-fuel and green-diesel range hydrocarbons. Soybean feedstock has shown promise in creating drop-in bio-hydrocarbons with petro-fuel characteristics.[29] Green diesel may be produced using renewable resources like soybean oil. Because of the high cost of large-scale manufacturing, soybean oil is a preferred feedstock for renewable diesel synthesis in Brazil.[30]

3.3.1.3 Palm Oil

Palm oil is the world's most commonly used vegetable oil and is extracted from both the fruit and the seed of the palm tree. However, its high saturated fatty acid content has raised controversies over consumption of the oil. The plant is a good source of edible oil and each palm fruit contains 30–35% oil. Palm oil is extracted from the outer mesocarp of palm kernel oil from the endosperm.[31] Palm oil is an edible vegetable oil obtained from the mesocarp (reddish pulp) of oil palm fruits. The oil is utilized in food processing, cosmetics, and biofuel. Indonesia and Malaysia are major producers, accounting for 60% and 30% of global palm oil output, respectively.[32] Palm oil contains 50% saturated fatty acids (mostly palmitic acid), 40% unsaturated fatty acids (oleic acid), and roughly 10% polyunsaturated fatty acids (linoleic and linolenic acids).[33] Palm oil's consistent mix of saturated and unsaturated fatty acids makes it suitable for producing high-quality diesel.

A study found that crude palm oil may provide around 68% yield of green diesel by catalytic hydrogenation.[34]

3.3.1.4 Corn Oil

Corn oil, also known as maize oil, is a kind of oil derived from corn germ. Corn oil is an edible oil. Large-scale manufacturing uses the inedible component to produce green fuel. In the United States, Diamond Green Diesel and Valero built a system for green-diesel manufacturing based on corn oil. Corn oil has more polyunsaturated fatty acids, particularly linoleic acid, at around 60%.[35] In 2024, the United States is predicted to be the world's top maize producer, with a production output of around 389.7 million metric tonnes. China and Brazil are among the biggest corn producers.[36] The fatty acid composition of corn oil is: stearic acid (2%), palmitic acid (11%), linoleic acid (62%), and oleic acid (28%).[37]

3.3.1.5 Coconut Oil

Coconut oil is derived from the kernel of mature coconuts harvested from the coconut palm. The high saturated-fat content makes it slow to oxidize and hence resistant to rancidification, lasting up to six months at 24 °C (75 °F) without spoilage. Coconut oil, like other vegetable oils, is made up of triglyceride fatty acids, with about 90% saturated (C_8–C_{12}), particularly lauric and myristic acids. Coconut oil offers benefits over other vegetable oils due to its high content of medium-chain fatty acids, which can be converted into fuels like kerosene and jet fuel. Coconut oil contains a variety of fatty acids, including caprylic acid (8%), capric acid (7%), lauric acid (49%), myristic acid (8%), palmitic acid (8%), stearic acid (2%), oleic acid (6%), and linoleic acid (C-18:2) (2%).[38] The Philippines produces (1.73 MMT) of edible oil, followed by Indonesia (1.03 MMT), India (0.57 MMT), and the EU (0.44 MMT).[39]

3.3.1.6 Rapeseed Oil

Rapeseed oil is widely used as a household cooking oil. It has a significant competitive disadvantage if not produced in large quantities to balance food inequalities. Rapeseed oil, commonly known as canola oil in some countries like European Union, has a high production of almost 20.05 million metric tons (MMT). Aside from European Union, Canada, China, and India all produce this as a staple oil supply, with production capacities of 18.8 MMT, 15.4 MMT, and 12.02 MMT, respectively.[40] Rapeseed oil contains a variety of fatty acids, including, palmitic acid (4%), stearic acid (2%), oleic acid (64.4%), linoleic acid (22.3%) and linolenic acid (8.2%).[41] Rapeseed oil has been widely explored for green-diesel generation using several hydrotreating procedures, including direct hydrotreating with Pt/zeolite and subsequent hydrogenation and decarboxylation.

3.3.1.7 Sunflower Oil

Sunflower oil, commonly used for cooking, may be converted into biodiesel by an alkali-catalyzed transesterification process. Sunflower is a major oil seed crop grown globally. As fossil fuels deplete, biodiesel is becoming a popular alternative fuel.[42] Sunflower oil is a non-volatile oil squeezed from the sunflower seeds. Sunflower oil consists largely of linoleic acid (64.35%), a polyunsaturated fatty acid, and oleic acid (19.81%), a monounsaturated fatty acid. Oils with varying fatty acid compositions are created through selective breeding and production techniques.[43] Russia leads sunflower oil production with 17.1 MMT, followed by Ukraine (14.5 MMT) and the European Union (10.2 MMT).[41] Raney nickel catalyst was used for hydroconversion of sunflower oil, reducing side reactions and increasing green-diesel output.

3.3.1.8 Olive Oil

Olive oil contains 83% oleic acid and 20% palmitic acid among its total fatty acids. Olive oil is mostly used in medicine, soap production, and lighting.[44] With 1080 TMT, the European Union is the world's largest producer of olive oil, followed by the US (350 TMT), Turkey (150 TMT), and China (38 TMT).[45] Studies show that green diesel from olive oil is not as efficiently processed as biodiesel. This feedstock is undervalued in green-diesel production due to its low production and consumption across the world.

3.3.1.9 Animal Fats

Fats derived from animals make up almost one-third of all oils and fats generated in the US. This includes chicken fat, hog lard, and cattle tallow. For biodiesel, animal fats are an appealing feedstock due to their significantly cheaper cost compared to vegetable oil. This is partially because, compared to the market for vegetable oil, the animal fat industry is far more restricted in the United States because the majority of the fat generated there is not deemed edible by humans.[46] Animal fats are highly saturated, which means they solidify at relatively high temperatures. Therefore, biodiesel generated from animal fat has a high cloud point. For example, biodiesel manufactured from beef tallow and hog fat has a cloud point between 55 °F

Table 3.3 The percentage composition, or fatty acid profile, of various animal fats.

Fatty acid	Beef tallow	Hog lard	Chicken fat
Myristic (14:0)	1.4–6.3	0.5–2.5	1
Palmitic (16:0)	20–37	20–32	25
Palmitoleic (16:1)	0.7–8.8	1.7–5	8
Stearic (18:0)	6–40	5–24	6
Oleic (18:1)	26–50	35–62	41
Linoleic (18:2)	0.5–5	3–16	18

and 60 °F. B100 (pure biodiesel) derived from animal fat should only be utilized in hot climates. However, animal fat biodiesel may be combined with petroleum diesel. At lesser blends, such as B5 (5% biodiesel and 95% petrodiesel), the high cloud point of the animal fat biodiesel has no effect on the blend's cloud point.

Beef tallow and hog lard are generally around 40% saturated (a combination of myristic, palmitic and stearic acids). Chicken fat has around 30–33% lower saturated fatty acid content (Table 3.3). For comparison, soybean oil is around 14% saturated and canola oil is just 6%. Thus, tallow and lard are often solid at ambient temperature, but chicken fat, while normally liquid, is extremely viscous and virtually solid.[47] When animal fat is converted into biodiesel, the problem regarding solidification at lower temperatures persists. The methyl esters of the saturated fatty acids, namely methyl stearate and methyl palmitate, have relatively high melting points.[48]

3.3.2 Second-generation Feedstocks

3.3.2.1 Castor Oil

Castor oil is a vegetable oil derived from the *Ricinus communis* plant, which is common in the Eastern areas of the world. Castor beans are used to make castor oil by pressing. It is mostly used for lubricant and therapeutic applications. The fatty acid composition of castor oil includes ricinoleic acid, which makes up around 90% of its total fatty acid content, with the remaining 10% consisting primarily linoleic acid (4%), oleic acid (3%) and stearic acid (1%). There is tremendous commercial potential for using the entire castor bean plant as animal feed, fertilizer, biofuel, and for phytoremediation.[49] India is the world's largest producer and exporter of castor oil, with 1618.76 TMT produced there last year, the most in the Asia-Pacific region.[50]

3.3.2.2 Jatropha Oil

Jatropha curcas, a perennial plant from the Euphorbiaceae family, yields seed with 31–37% oil, which may be combusted as fuel without refinement.[51] With catalytic hydroprocessing, jatropha oil has the potential to produce green diesel with a conversion rate of over 95%, and using this hydrotreated oil as fuel does not require any engine modifications.[52] Jatropha oil contains traces of oleic acid along with palmitic acid (35%) and linoleic acid (50%). Despite its great potential for producing diesel, jatropha has difficulties with high output, more acreage for cultivation, untapped expertise, and unstable weather patterns. The oil's fatty acid composition includes oleic acid (47%), linolenic acid (31%), and palmitic acid (14%), with saturated fatty acids making up the largest proportion.[53]

3.3.2.3 Neem Oil

Neem oil is a natural vegetable oil with a yellow-brown hue, a bitter flavor, and a strong aroma that is derived from the seeds of the neem tree (*Azadirachta indica*), a native of the Indian subcontinent. The fatty acid composition of neem oil includes the following: stearic acid (11.4%), palmitic acid (23.4%), linoleic acid (15.94%), oleic acid (34.09%) and arachidic acid (0.73%). When squeezed in the machine, neem kernels release over half of their oil content.[54]

3.3.2.4 Mahua Oil

Mahua is a tropical tree that grows in the northern plains and central woodlands of the Indian subcontinent. It is also called a mohul or mahua (*Madhuca indica*) and grows to a height of 55–75 feet. It belongs to the family of Sapotaceae. *Madhuca indica* seed oil is produced in India with an annual production capacity of 181 thousand metric tonnes. Oleic acid (41.1–51%), palmitic acid (16–28%), linoleic acid (8.9–13.7%) and arachiare acid (0.0–3.3%) are present in mahua oil, which is an alternative feedstock. The possibility of generating hydrocarbon fuel was demonstrated by the catalytic cracking of mahua oil using coal fly ash catalyst. Because waste oil and animal fat are less expensive than fresh vegetable oil, they are also used as feedstocks; nevertheless, obtaining them remains difficult.[55]

3.3.2.5 Karanja Oil

Karanja oil, also known as pongamia oil, is derived from the seeds of the *Pongamia pinnata* tree, a species native to India and other parts of Southeast Asia. This tree is often referred to as the Indian beech or karanja tree. It belongs to the family of Fabaceae. The fatty acid composition of karanja oil includes the following: stearic acid (2.4–8.9%), palmitic acid (3.7–7.9%), linoleic acid (10.8–18.3%), oleic acid (44.05–71.3%) and arachidic acid (2.2–4.7%).[56]

3.3.3 Third-generation Feedstocks

Third-generation biofuels are produced from non-food feedstocks, with algae (also called oilgae) being the primary source. Owing to their high lipid content, rapid growth rate, high rate of CO_2 absorption, and low land requirement, algae are regarded as a viable alternative energy source. Advanced resources that are not used in crop-type cultivation are known as third-generation feedstocks. Microalgae are the most popular and widely utilized third-generation feedstock; through photosynthesis, they aid in the absorption of CO_2 from the atmosphere. These are better than feedstocks from the first and second generations since they produce fewer greenhouse gases, can be grown on insignificant land, grow and produce more, do not compete with other crops for food, *etc.*[57] Microalgae are particularly produced in regulated *in situ* or *ex situ* conditions as fatty sources since they are a rich source of lipids. It is possible to extract the lipids and prepare diesel from them.[58] Microalgae are microscopic algae that thrive in ponds and other bodies of water. Microalgae have more lipids and grow faster in nature than macroalgae. When compared to other energy crops, algae may produce 30 times greater yields per acre.

3.3.4 Fourth-generation Feedstocks

The fourth generation of biofuels involves the metabolic engineering of algae for biofuel production. The fourth generation of biofuels offers various benefits, including a high yield from lipid-containing algae, improved CO_2-collection capabilities, and a rapid production rate, among others. The initial investment, on the other hand, is exceedingly high, which is a major drawback.[59] Table 3.2 outlines several biofuel generations, including feedstock, production technique, and advantages and downsides.

3.4 Catalysts and Support Materials for Green-diesel Production

In heterogeneously catalyzed systems, solid catalysts generally comprise an active phase supported by an inert substance (Figure 3.3). In supported-metal catalysts, metals serve as active sites that facilitate processes like hydrogen splitting, insertion, and C–C cleavage. The metal active sites control the deoxygenation pathways (hydrodeoxygenation (HDO), decarbonylation (DCO), and decarboxylation (DCO_2)). Metals used in deoxygenation processes typically form the active phase of solid catalysts and are categorized as noble or non-noble. Noble metals in deoxygenation catalysts include Pt, Pd, Rh, Os, Ru, Ir, and Re, while non-noble metals

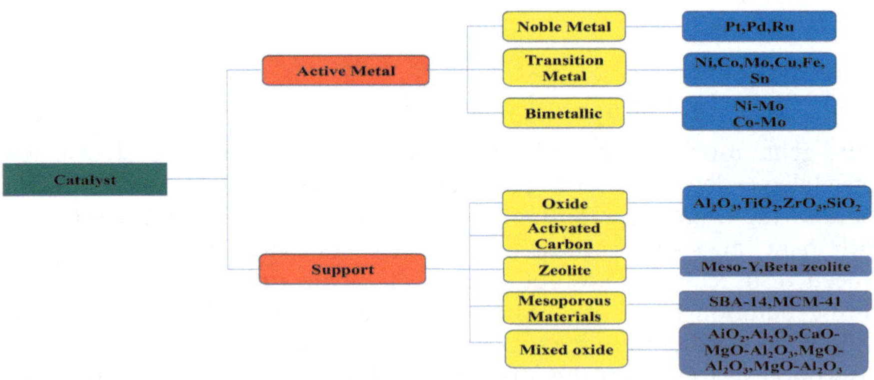

Figure 3.3 Catalysts employed for the generation of green diesel, including active metals and supports.

comprise Ni, Mo, Cu, Co, Zn, V, and W. Noble-metal catalysts, particularly palladium (Pd) and platinum (Pt), are favored for their high activity and product selectivity. Noble metal-based catalysts offer greater selectivity for deoxygenation pathways like DCO and DCO_2, requiring less H_2 compared to non-noble metal catalysts.[60] The HDO method also makes use of bimetallic MoS_2 catalysts such as $CoMoS_2$ and $NiMoS_2$. These catalysts exhibit considerable DO conversion activity for triglycerides (TG) and model compounds, but they are prohibitively expensive and pose issues for equipment, safety, and industrial scalability. It is noted that Pd- and Ni-based catalyst are widely used for green-diesel production.[61]

The support materials used in the deoxygenation process have a substantial influence on catalyst activity. Several variables should be considered when choosing an appropriate support material, such as interaction with the active phase, high surface area, correct pore size, pore structure, active metal dispersion, and high stability.[62] The HDO method uses a variety of support materials, including mesoporous materials, activated carbon, zeolites, and metal oxides such as γ-Al_2O_3, SiO_2, Nb_2O_5, TiO_2, ZrO_2, and CeO_2. Zeolites were initially considered promising catalysts for green-diesel production due to their high breaking capacity, unique porous structure, enhanced acidity, and inherent stability. When using Ni/zeolite Y as a catalyst for deoxygenating microalgal biodiesel, the pore size of zeolite Y is crucial for reaction efficiency. The larger pores of zeolite Y (3.9 nm) facilitate effective mass transfer, as they exceed the size of methyl palmitate molecules (2.5 nm).[63]

Mesoporous materials are chosen because to their high thermal stability, ease of surface modification, moderate pore size, vast surface area, and inert nature. Mesoporous silica-based materials like Santa Barbara Amorphous-15 (SBA-15), SBA-16, and Mobil Composition of Matter No. 41 (MCM-41) are popular because of their moderate acidity, high specific surface area, and large pore diameter, making them ideal for processing large fatty-acid molecules in reactants.[64,65] Mixed metal oxides such as TiO_2, Al_2O_3, SiO_2, ZrO_2, and CeO_2 are interesting supports for green-diesel production *via* HDO. It was found that a mixed metal oxide comprising Al_2O_3 and TiO_2 improves catalyst stability, activation, surface area, reducibility, metal dispersion, and acidic centers compared to individual oxides. SiO_2 is utilized as a support because it has lower acidity than Al_2O_3.[66]

3.5 Conversion Technologies for Production of Green Diesel

Bio-crude is a liquid product derived from biomass through high-temperature, high-pressure liquid-phase thermal processing. It is an intricate blend of various oxygenated organic compounds of various molecular weights. Although bio-crude shares some properties with petroleum, it has distinct differences. It is not easily miscible with petroleum and contains 5–10% moisture. Additionally, bio-crude has an oxygen content of 5–15%, which is

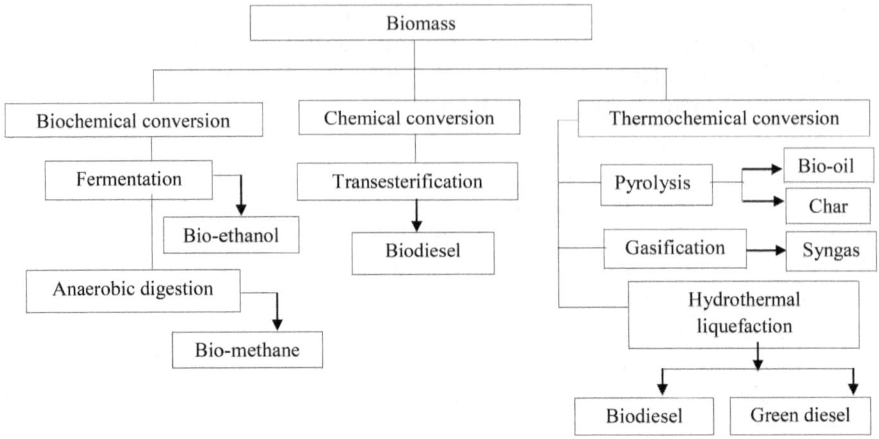

Figure 3.4 Classification of conversion techniques for biomass into green fuels.

higher than petroleum. Its density is also slightly greater than that of petroleum (~10%). Moreover, bio-crude usually has higher nitrogen content (3–8 wt% for bio-crudes from organic wet wastes) and sulfur levels, depending on the biomass used.[67] Biomass can be converted into bio-crude using several methods (Figure 3.4). These methods fall into three main categories: thermochemical, biochemical, and chemical. Each method produces a different main product, such as syngas, bio-oil, or char. Each of these technologies has its distinct mechanism, advantages and challenges. However, in this chapter, we will focus exclusively on thermochemical processes because they offer higher efficiency in converting biomass to bio-crude, provide faster reaction times, and are more scalable for industrial applications. Additionally, thermochemical methods offer greater diversification compared to biochemical processes. While biochemical methods are primarily used for bio-ethanol production, thermochemical processes produce bio-crude oil, which can be further refined into biofuel and various petrochemical value-added products.

Moreover, thermochemical methods are better suited for handling the heterogeneous nature of lignocellulosic biomass and can produce a broader range of valuable by-products. Another benefit of thermochemical conversion methods is that they are not limited to one type of feedstock only; rather, they are versatile in processing various feedstocks, including combinations of different types.[68]

3.5.1 Thermochemical Conversion

Thermochemical conversion is the use of heat and chemical processes to transform biomass or other organic resources into energy-dense fuels, chemicals, or other useful products. This conversion process is a key mechanism for converting biomass into biofuels and bioenergy. It usually requires high temperatures and, in certain situations, pressures to break down complex organic molecules into simpler ones.[69]

3.5.2 Pyrolysis

Pyrolysis is a thermochemical process where biomass is heated at 300–800 °C in an inert atmosphere to produce bio-crude oil, syngas, and char without combustion (Figure 3.5). Different biomasses, like agricultural residues, forest residues, and aquatic biomass, require heat in the range of 400–800 kJ kg^{-1} for their pyrolysis. Rapid heating of biomass in an oxygen-free atmosphere produces organic vapours. These vapours are made up of fragments of cellulose, hemicellulose, and lignin polymers in biomass. These vapours condense to form bio-oil, an organic free-flowing liquid. The process emits non-condensable gases that can be utilized for heating the pyrolysis by directing them outside the reaction system. The leftover carbon-rich substance is called bio-char. Bio-oil, the primary product of pyrolysis, can be refined into drop-in fuels such as diesel and gasoline.[70]

3.5.3 Hydrothermal Liquefaction

Hydrothermal liquefaction (HTL) is an advanced thermal depolymerization method used to convert macromolecules and wet biomass into bio-crude. The processing takes place in a heated and pressurized water environment, where complex macromolecules are decomposed into products with lower molecular weight, which then polymerize to generate bio-crude oil.

During the liquefaction process, the biomass undergoes a series of depolymerization activities, including hydrolysis, dehydration, and decarboxylation, which result in the production of insoluble compounds like bio-crude. HTL is comparable to other procedures as it does not require pre-drying of the biomass. HTL crude has a greater energy density than crude from pyrolysis or gasification. Several sources in the literature estimate liquefaction temperatures ranging from 200 °C to 350 °C, pressures between

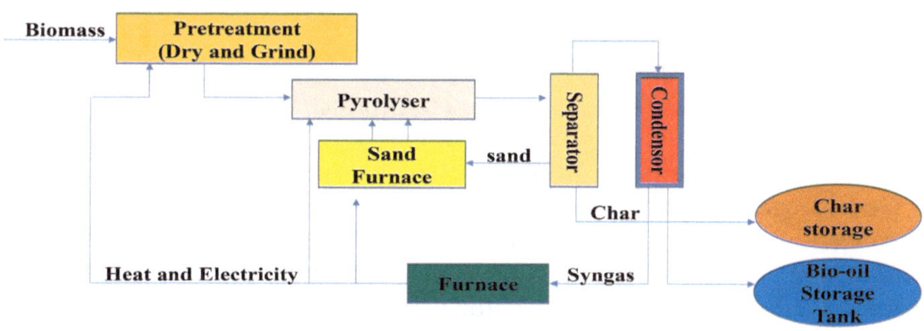

Figure 3.5 Schematic representation of the pyrolysis process.

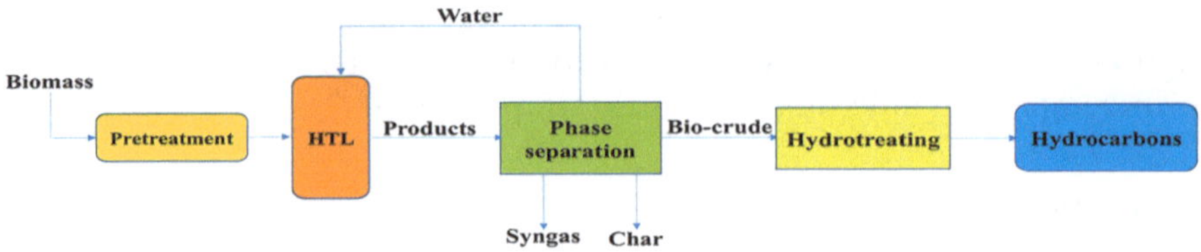

Figure 3.6 Schematic representation of the hydrothermal liquefaction process.

5 MPa and 10 MPa, and residence periods of 5–60 minutes.[71] Figure 3.6 shows a schematic representation of working of hydrothermal liquefaction.

3.5.4 Hydroprocessing Process

Hydroprocessing is a technique for converting liquid biomass into biofuels, alongside transesterification. It is a crucial method in petroleum refineries and petrochemical complexes for breaking down larger molecules and removing heteroatoms (S, N, O, and metals) from feedstocks like heavy gas oil or vacuum gas oil. This process transforms non-edible seed oils, vegetable oils, used cooking oil, and fats into valuable liquid fuels (straight-chain alkanes) by leveraging existing refinery infrastructure and equipment. Today, hydroprocessing produces high-quality bio-based diesel fuels, often referred to as renewable diesel or green diesel rather than biodiesel.[72]

During the hydroprocessing of vegetable oil, three main chemical reactions take place:[73] (i) hydrogenation, which involves the addition of hydrogen to double bonds in unsaturated fatty acids; (ii) HDO, which removes oxygen atoms from the carboxylate group *via* water; and (iii) hydrodecarboxylation (HDC), which results in the removal of carboxylate groups as carbon dioxide. In this process, hydrogenation occurs prior to HDO and HDC (Scheme 3.1). The reaction rates for HDO and HDC of saturated triglycerides can vary based on the reaction conditions and the catalyst used. The primary products of vegetable oil hydroprocessing are typically *n*-alkanes that match the number of carbon atoms in the original fatty acids (produced by HDO) and *n*-alkanes that have one fewer carbon atom (produced by HDC).[74] The hydroprocessing of triglycerides involves hydrogenating the double bonds in the side chains and removing oxygen at the metal sites of the catalysts. Most vegetable oils (such as rapeseed, sunflower, soybean, and palm oil) are hydroprocessed to produce hydrocarbon mixtures in the C_{15}–C_{18} range.

This liquid mixture is commonly known as green diesel, renewable diesel, or bio-hydrogenated diesel due to its boiling point range being similar to that of conventional diesel. It is chemically similar to petroleum-based diesel. Furthermore, the main byproduct of vegetable oil hydroprocessing is straight-chain paraffins (*n*-alkanes), which may be hydroisomerized to make high-quality diesel or hydrocracked to produce kerosene. Propane, a gaseous byproduct, is a valuable feedstock for the petrochemical sector and is also

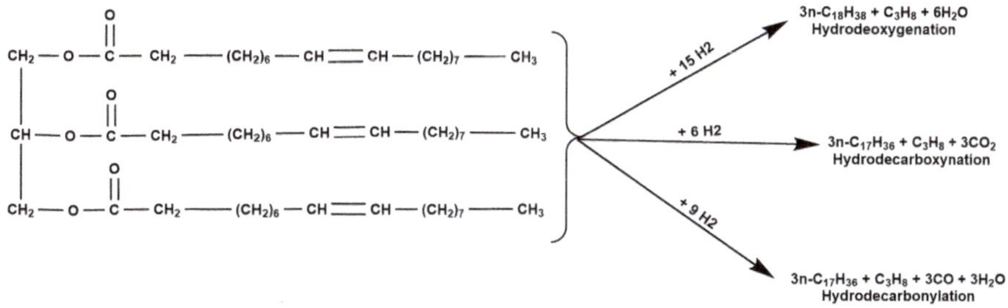

Scheme 3.1 Reaction pathways for oxygen removal by H_2 from trioleic triglycerides.

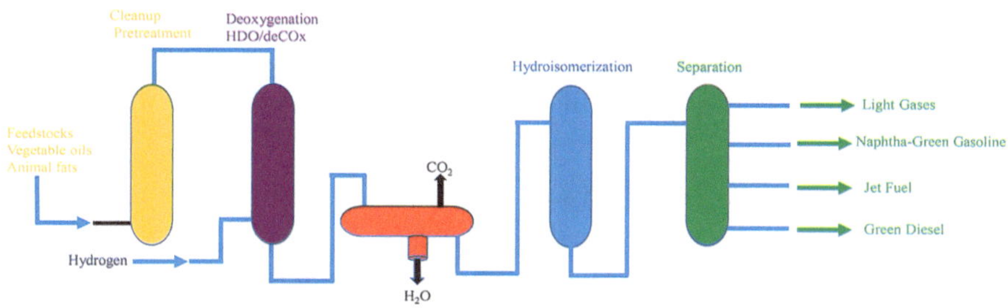

Figure 3.7 Schematic representation of a hydroprocessing plant.

Table 3.4 List of the main green-diesel producers currently in operation. Data from ref. 76–81.

Company name	Plant location	Plant production capacity (million tons per year)
Neste Oil Corporation	Netherlands	1.4
	Singapore	2.6
	Finland	12.0
Diamond Green Diesel	USA	3.9
Renewable Energy Group	USA	1.3
UOP/Eni	Italy	~1.0

utilized as an LPG substitute.[75] Hydrotreating is widely used in refinery units to remove sulfur, nitrogen, and metals from petroleum-based feedstock. The facilities used for the production of petroleum-based fuel, including catalysts, types of reactors and distillation facilities, can be used for the production of green diesel as well (Figure 3.7).

A list of the main green-diesel producers is given in Table 3.4. Most of the producers have developed proprietary technologies including a pretreatment process for raw materials, the hydrotreatment process, a hydroisomerization reactor, and finally separation columns to obtain green diesel.[82] The choice of feedstock, catalyst selection, reaction time, reaction temperature, and pressure are considered to be critical for the production of green diesel.[83]

3.6 Challenges and Opportunities

Renewable diesel has emerged as a possible alternative to conventional diesel, providing a sustainable solution to the growing demand for cleaner fuels. Renewable diesel is completely compatible with existing diesel engines and infrastructure, since it has no chemical differences compared to petroleum diesel, in contrast to biodiesel, which is an ester-based fuel. However, producing renewable fuel has a number of obstacles. The route to large-scale manufacturing entails traversing a complicated environment that includes feedstock availability, technological challenges, and economic feasibility. At the same time, numerous opportunities exist to optimize production processes, expand feedstock options, and leverage policy incentives, positioning renewable diesel as a key player in the transition to a low-carbon future.

3.6.1 Feedstock Challenges

One of the major issues in renewable-diesel production is the availability and sustainability of feedstocks. Traditional edible feedstocks such as vegetable oils (*e.g.* soybean, canola, corn, coconut, rapeseed, sunflower, olive, palm) and animal fats and non-edible feedstocks (*e.g.* castor, jatropha, neem, mahua, karanja) are in high demand for both food and fuel production, leading to competition that can drive up prices and strain supply chains. The conflict can lead to ethical considerations, particularly with the exploitation of food crops for fuel production during a time when global food security is at risk. These alternative feedstocks also come with their own set of challenges, including inconsistent supply, variable quality, and the need for extensive collection and processing infrastructure. Oils from algae, with their high lipid content and quick growth rate, are a potential future feedstock, but they are still in the experimental stage, with scalability and economic feasibility yet to be completely realized.

3.6.2 Technological and Process Challenges

Renewable-diesel generation requires advanced hydroprocessing technology as well as considerable financial investment and operating experience. Hydroprocessing of vegetable oils and animal fats to make renewable diesel includes the employment of catalysts to enhance hydrogenation, deoxygenation, and hydrocracking processes. The choice of catalysts is crucial for improving product output and quality. Noble metal catalysts such as platinum (Pt) and palladium (Pd) are extremely effective but expensive, necessitating the development of cheaper replacements based on non-noble metals such as nickel (Ni) and cobalt (Co). Additionally, renewable-diesel production requires significant energy input, particularly in the form of hydrogen, which is used to remove oxygen from the feedstock during hydrodeoxygenation and hydrodecarboxylation processes. Hydrogen generation is energy-intensive and often reliant on fossil fuels, which might counteract the environmental benefits of renewable diesel. As a result, creating more efficient catalysts, optimizing hydrogen consumption, and incorporating renewable energy sources into the manufacturing process are critical areas for technical advancement.

3.6.3 Economic and Market Challenges

Economic viability is another important barrier for renewable-diesel generation. The high costs associated with feedstock procurement, technological infrastructure, and catalyst development can make renewable diesel more expensive than conventional diesel, particularly in the absence of strong policy incentives. Market fluctuation in feedstock pricing exacerbates the problem, making it harder for companies to maintain constant profitability. However, there are significant opportunities to improve the economic outlook for renewable diesel. Government policies and regulations, such as carbon pricing, renewable-fuel standards, and clean-energy subsidies, can give vital financial assistance to the business. Moreover, the growing demand for low-carbon fuels from both consumers and industries presents a market opportunity for renewable-diesel producers. As companies seek to reduce their carbon footprint and comply with sustainability goals, renewable diesel could see increased adoption, particularly in sectors like transportation and aviation, where decarbonization options are limited.

3.6.4 Environmental and Sustainability Opportunities

Despite the hurdles, renewable diesel has considerable environmental advantages that correspond with global efforts to mitigate climate change. Renewable-diesel production from waste-based feedstocks like used cooking-oil greases and non-edible oils not only reduces reliance on fossil fuels but also helps to manage waste streams that would otherwise contribute to environmental pollution. Renewable diesel emits much less greenhouse gas (GHG) across its life cycle than conventional diesel, especially when sustainable feedstocks and renewable energy inputs are employed. Furthermore, renewable diesel is chemically similar to petroleum-derived diesel, making it a drop-in fuel that can be smoothly incorporated into existing fuel-distribution systems and utilized in current diesel engines with no modifications. This compatibility eliminates the need for additional infrastructure expenditures and speeds up the shift to greener fuels.

3.6.5 Research and Innovation Opportunities

Ongoing research and development initiatives provide several possibilities to address the issues connected with renewable-diesel generation. Advances in catalyst design, such as the creation of bimetallic catalysts and the use of sophisticated materials like activated carbon and silica supports, are increasing process efficiency and lowering costs. Furthermore, improvements in feedstock pretreatment and upgrading technologies are helping to increase the spectrum of feasible feedstocks, allowing the use of lower-quality or more-diversified raw materials. The investigation of novel feedstocks, such as algal oils and industrial byproducts like tall oil, is also creating new opportunities for renewable-diesel generation.

3.7 Conclusion

Renewable-diesel production is a key component of the global shift to sustainable energy. Renewable diesel is commonly produced from feedstocks such as animal fats, used cooking oils, and vegetable oils like soybean and canola. These feedstocks are transformed into renewable diesel *via* processes like hydroprocessing, which eliminates oxygen and impurities, producing a fuel chemically comparable to petroleum diesel but with a much lower carbon footprint. However, producing renewable diesel presents several challenges. Key challenges include the availability and cost of feedstocks, which become more pressing as production expands. Additionally, the competition between using these resources for food or fuel creates ethical and economic dilemmas, particularly in regions facing food insecurity. Technological challenges, including the development of cost-effective and efficient catalysts, especially those using non-noble metals, are crucial for lowering production costs and enhancing process efficiency. To maintain the overall sustainability of renewable diesel, it is crucial to carefully manage both the environmental impact of feedstock cultivation and the energy consumption of production processes. Despite these challenges, the renewable-diesel industry offers significant opportunities. Continued research in catalyst technology and progress in feedstock diversification could reduce production costs and further improve the environmental benefits of renewable diesel. As global efforts to reduce greenhouse-gas emissions intensify, renewable diesel offers a promising alternative to conventional fossil fuels, paving the way for a more sustainable and resilient energy future.

Declarations

Mohammad Aslam is an editorial board member for this book titled *Biorefining of Renewable Diesel: From Production to Market* and was not involved in the editorial review of, or the decision to publish, this chapter. All authors declare that there are no competing interests.

Acknowledgements

The authors would like to express their gratitude to Anusandhan National Research Foundation (formerly known as Science and Engineering Research

Board, Department of Science and Technology), Government of India, for support of this research through CRG/2022/003994.

References

1. H. Burhan, K. Arikan, M. H. Alma, M. S. Nas, H. Karimi-Maleh, F. Şen, F. Karimi and Y. Vasseghian, Highly efficient carbon hybrid supported catalysts using nano-architecture as anode catalysts for direct methanol fuel cells, *Int. J. Hydrogen Energy*, 2023, **48**(17), 6657–6665.
2. https://www.energyinst.org/statistical-review.
3. S. Sri Shalini, K. Palanivelu, A. Ramachandran and V. Raghavan, Biochar from biomass waste as a renewable carbon material for climate change mitigation in reducing greenhouse gas emissions—a review, *Biomass Convers. Biorefin.*, 2021, **11**(5), 2247–2267.
4. https://www.iea.org/reports/co2-emissions-in-2023.
5. S. Sharma, A. Kundu, S. Basu, N. P. Shetti and T. M. Aminabhavi, Sustainable environmental management and related biofuel technologies, *J. Environ. Manage.*, 2020, **273**, 111096.
6. S. Manikandan, R. Subbaiya, M. Biruntha, R. Y. Krishnan, G. Muthusamy and N. Karmegam, Recent development patterns, utilization and prospective of biofuel production: Emerging nanotechnological intervention for environmental sustainability–A review, *Fuel*, 2022, **314**, 122757.
7. M. A. Sanhoob, U. Khalil, E. N. Shafei, K. H. Choi, T. Yokoi and O. Muraza, Steam cracking of green diesel (C12) to BTX and olefins over silane-treated hierarchical BEA, *Fuel*, 2020, **263**, 116624.
8. N. Banerjee, P. Sukichandran, P. Chaudhari, A. K. Thakur and R. Kumar, Energy analysis and feasibility studies for algal biomass and biofuels, *Mater. Today: Proc.*, 2022, **57**, 1448–1454.
9. Z. Rahimi, A. Anand and S. Gautam, An overview on thermochemical conversion and potential evaluation of biofuels derived from agricultural wastes, *Energy Nexus*, 2022, 7, 100125.
10. C. Debnath, T. K. Bandyopadhyay, B. Bhunia, U. Mishra, S. Narayanasamy and M. Muthuraj, Microalgae: Sustainable resource of carbohydrates in third-generation biofuel production, *Renewable Sustainable Energy Rev.*, 2021, **150**, 111464.
11. X. B. Tan, M. K. Lam, Y. Uemura, J. W. Lim, C. Y. Wong and K. T. Lee, Cultivation of microalgae for biodiesel production: a review on upstream and downstream processing, *Chin. J. Chem. Eng.*, 2018, **26**(1), 17–30.
12. S. Kumar, M. K. Singhal and M. P. Sharma, Utilization of mixed oils for biodiesel preparation: a review, *Energy Sources, Part A*, 2021, 1–34.
13. A. E. F. Abomohra, M. Elsayed, S. Esakkimuthu, M. El-Sheekh and D. Hanelt, Potential of fat, oil and grease (FOG) for biodiesel production: A critical review on the recent progress and future perspectives, *Prog. Energy Combust. Sci.*, 2020, **81**, 100868.

14. M. Aslam, N. C. Kothiyal and A. K. Sarma, True boiling point distillation and product quality assessment of biocrude obtained from Mesua ferrea L. seed oil *via* hydroprocessing, *Clean Technol. Environ. Policy*, 2015, **17**, 175–185.
15. S. L. Douvartzides, N. D. Charisiou, K. N. Papageridis and M. A. Goula, Green diesel: Biomass feedstocks, production technologies, catalytic research, fuel properties and performance in compression ignition internal combustion engines, *Energies*, 2019, **12**(5), 809.
16. S. A. Farooqui, R. Kumar, A. K. Sinha and A. Ray, Green Diesel Production by Hydroprocessing Technology, in *Green Diesel: An Alternative to Biodiesel and Petrodiesel*, Springer Nature Singapore, Singapore, 2022, pp. 109–148.
17. H. I. Mahdi, A. Bazargan, G. McKay, N. I. W. Azelee and L. Meili, *Chem. Eng. Res. Des.*, 2021, **174**, 158–187.
18. L. Cao, I. K. M. Yu, X. Xiong, D. C. W. Tsang, S. Zhang, J. H. Clark, C. Hu, Y. H. Ng, J. Shang and Y. S. Ok, Biorenewable hydrogen production through biomass gasification: a review and future prospects, *Environ. Res.*, 2020, **186**, 109547.
19. R. Zarchin, M. Rabaev, R. Vidruk-Nehemya, M. V. Landau and M. Herskowitz, Hydroprocessing of soybean oil on nickel-phosphide supported catalysts, *Fuel*, 2015, **139**, 684–691.
20. U. S. Umana, M. S. Ebong and E. O. Godwin, Biomass production from oil palm and it value chain, *J. Hum., Earth, Future*, 2020, **1**(1), 30–38.
21. M. Mahdavi, E. Abedini and A. H. Darabi, Biodiesel synthesis from oleic acid by nano-catalyst (ZrO_2/Al_2O_3) under high voltage conditions, *RSC Adv.*, 2015, **5**(68), 55027–55032.
22. D. S. Kim, M. Hanifzadeh and A. Kumar, Trend of biodiesel feedstock and its impact on biodiesel emission characteristics, *Environ. Prog. Sustainable Energy*, 2018, **37**(1), 7–19.
23. P. Maheshwari, M. B. Haider, M. Yusuf, J. J. Klemeš, A. Bokhari, M. Beg, A. Al-Othman, R. Kumar and A. K. Jaiswal, A review on latest trends in cleaner biodiesel production: Role of feedstock, production methods, and catalysts, *J. Cleaner Prod.*, 2022, **355**, 131588.
24. J. McFarlane, *Processing of Soybean Oil into Fuels*, INTECH Open Access Publisher, 2011.
25. Fuel the Crush: Renewable Diesel Pumps Up Soybean Demand|AgWeb. https://www.agweb.com/news/crops/soybeans/fuel-crush-renewable-diesel-pumps soybean-demand. Accessed 22 May 2024.
26. Renewable diesel boom is wild card for U.S. soybeans|Successful Farming. https://www.agriculture.com/news/business/renewable-diesel-boom-is-wild-card-for-us-soybeans. Accessed 22 May 2024.
27. Top countries of soybean oil Production worldwide, 2024, Statista. https://www.statista.com/statistics/1199485/domestic-production-of-soybean-oil worldwide-by-country/.
28. L. G. Woyann, D. Meira, A. D. Zdziarski, G. Matei, A. S. Milioli, A. C. Rosa, L. A. Madella and G. Benin, Multiple-trait selection of soybean for biodiesel production in Brazil, *Ind. Crops Prod.*, 2019, **140**, 111721.

29. C. A. Scaldaferri and V. M. D. Pasa, Production of jet fuel and green diesel range biohydrocarbons by hydroprocessing of soybean oil over niobium phosphate catalyst, *Fuel*, 2019, **245**, 458–466.
30. T. J. Hilbers, L. M. Sprakel, L. B. Van Den Enk, B. Zaalberg, H. Van Den Berg and L. G. Van Der Ham, Green diesel from hydrotreated vegetable oil process design study, *Chem. Eng. Technol.*, 2015, **38**(4), 651–657.
31. M. M. Urugo, T. A. Teka, P. G. Teshome and T. T. Tringo, Palm oil processing and controversies over its health effect: overview of positive and negative consequences, *J. Oleo Sci.*, 2021, **70**(12), 1693–1706.
32. Which Countries Produce The Most PalmOil?|Statista. https://www.statista.com/chart/23097/amount-of-palm-oil-produced-in-selected-countries/. Accessed 22 May 2024.
33. F. R. Mohamed, Oils From Fruit Nuts, Coconut (Cocos nucifera) Oil, in *Fruit Oils: Chemistry and Functionality*, ed. A. Osman, 2019, pp. 209–221.
34. A. Zikri and M. Aznury, Green diesel production from Crude Palm Oil (CPO) using catalytic hydrogenation method, *IOP Conf. Ser.: Mater. Sci. Eng.*, 2020, **823**, 012026.
35. V. B. Veljković, M. O. Biberdžić, I. B. Banković-Ilić, I. G. Djalović, M. B. Tasić, Z. B. Nježić and O. S. Stamenković, Biodiesel production from corn oil: A review, *Renewable Sustainable Energy Rev.*, 2018, **91**, 531–548.
36. Corn production by country 2023/24|Statista. https://www.statista.com/statistics/254292/global-corn-production-by-country/. Accessed 22 May 2024.
37. L. Boateng, R. Ansong, W. Owusu and M. Steiner-Asiedu, Coconut oil and palm oil's role in nutrition, health and national development: A review, *Ghana Med. J.*, 2016, **50**(3), 189–196.
38. Coconut production by country 2023/24|Statista. https://www.statista.com/statistics/1199479/domestic-production-of-coconut-oil-worldwide-by-country/. Accessed 22 May 2024.
39. Production of rapeseed by main producing countries, 2023/2024|Statista. https://www.statista.com/statistics/263930/worldwide-production-of rapeseed-by-country/. Accessed 22 May 2024.
40. M. A. Rahim, H. Ayub, A. Sehrish, S. Ambreen, F. A. Khan, N. Itrat, A. Nazir, A. Shoukat, A. Shoukat, A. Ejaz and F. Özogul, Essential components from plant source oils: A review on extraction, detection, identification, and quantification, *Molecules*, 2023, **28**(19), 6881.
41. M. Thirumarimurugan, V. M. Sivakumar, A. M. Xavier, D. Prabhakaran and T. Kannadasan, Preparation of biodiesel from sunflower oil by transesterification, *Int. J. Biosci., Biochem. Bioinf.*, 2012, **2**(6), 441.
42. A. Petraru, F. Ursachi and S. Amariei, Nutritional characteristics assessment of sunflower seeds, oil and cake. Perspective of using sunflower oilcakes as a functional ingredient, *Plants*, 2021, **10**(11), 2487.
43. Major producer countries of sunflower seed 2023/2024|Statista. https://www.statista.com/statistics/263930/worldwide-production-of-rapeseed-by-country/. Accessed 22 May 2024.
44. Olive oil—Wikipedia. https://en.wikipedia.org/wiki/Olive_oil. Accessed 26 July 2024.

45. Leading consumers of olive oil worldwide 2023/24|Statista. https://www.statista.com/statistics/940532/olive-oil-consumption-worldwide-by-leading-country/. Accessed 22 May 2024.
46. P. Andreo-Martinez, V. M. Ortiz-Martinez, M. J. Salar-Garcia, J. M. Veiga-del-Bano, A. Chica and J. Quesada-Medina, Waste animal fats as feedstock for biodiesel production using non-catalytic supercritical alcohol transesterification, *Energy Sustainable Dev.*, 2022, **69**, 150–163.
47. https://farm-energy.extension.org/animal-fats-for-biodiesel-production/.
48. R. Jambulingam, V. Shankar, S. Palani and G. R. Srinivasan, Effect of dominant fatty acid esters on emission characteristics of waste animal fat biodiesel in CI engine, *Front. Energy Res.*, 2019, 7, 63.
49. M. Landoni, G. Bertagnon, M. Ghidoli, E. Cassani, F. Adani and R. Pilu, Opportunities and challenges of castor bean (Ricinus communis L.) genetic improvement, *Agronomy*, 2023, **13**(8), 2076.
50. APAC: castor oil seed production by country|Statista. https://www.statista.com/statistics/658666/asia-pacific-castor-oil-seed-production-by-country/. Accessed 22 may 2024.
51. W. Abobatta, Jatropha curcas: an overview, *J. Adv. Agric.*, 2019, **10**, 1650–1656.
52. N. Asikin-Mijan, N. A. Rosman, G. AbdulKareem-Alsultan, M. S. Mastuli, H. V. Lee, N. Nabihah-Fauzi, I. M. Lokman, F. A. Alharthi, A. A. Alghamdi, A. A. Aisyahi and Y. H. Taufiq-Yap, Production of renewable diesel from Jatropha curcas oil *via* pyrolytic-deoxygenation over various multi-wall carbon nanotube-based catalysts, *Process Saf. Environ. Prot.*, 2020, **142**, 336–349.
53. M. Moniruzzaman, Z. Yaakob, M. Shahinuzzaman, R. Khatun and A. K. M. Aminul Islam, Jatropha biofuel industry: The challenges, *Front. Bioenergy Biofuels*, 2017, **1**(12), 23–256.
54. O. P. Adigwe, J. A. Ibrahim, A. Adamu, S. E. Okhale, A. Elias and H. O. Egharevba, Chemometric analysis of volatile compounds of NIPRINEEM oil and other brands of neem seed oils sold in Nigeria, *J. Biophys. Chem.*, 2022, **13**(01), 1–12.
55. N. Mahato, Cracked Madhuca Indica (Mahua) Oil as a Diesel Engine Application: An Overview, *Advances in Science & Technology*, 2020, p. 30.
56. M. Bala, T. N. Nag, S. Kumar, M. Vyas, A. Kumar and N. S. Bhogal, Proximate composition and fatty acid profile of Pongamia pinnata, a potential biodiesel crop, *J. Am. Oil Chem. Soc.*, 2011, **88**(4), 559–562.
57. D. Singh, D. Sharma, S. L. Soni, S. Sharma, P. K. Sharma and A. Jhalani, A review on feedstocks, production processes, and yield for different generations of biodiesel, *Fuel*, 2020, **262**, 116553.
58. A. Veeramuthu and C. Ngamcharussrivichai, Potential of microalgal biodiesel: challenges and applications, in *Renewable Energy-Technologies and Applications*, 2020.
59. A. Mishra, M. Kumar, K. Medhi and I. S. Thakur, Biomass energy with carbon capture and storage (BECCS), in *Current developments in biotechnology and bioengineering*, Elsevier, 2020, pp. 399–427.

60. L. G. Woyann, D. Meira, A. D. Zdziarski, G. Matei, A. S. Milioli, A. C. Rosa, L. A. Madella and G. Benin, Multiple-trait selection of soybean for biodiesel production in Brazil, *Ind. Crops Prod.*, 2019, **140**, 111721.
61. C. A. Scaldaferri and V. M. D. Pasa, Production of jet fuel and green diesel range biohydrocarbons by hydroprocessing of soybean oil over niobium phosphate catalyst, *Fuel*, 2019, **245**, 458–466.
62. T. J. Hilbers, L. M. Sprakel, L. B. Van Den Enk, B. Zaalberg, H. Van Den Berg and L. G. Van Der Ham, Green diesel from hydrotreated vegetable oil process design study, *Chem. Eng. Technol.*, 2015, **38**(4), 651–657.
63. M. M. Urugo, T. A. Teka, P. G. Teshome and T. T. Tringo, Palm oil processing and controversies over its health effect: overview of positive and negative consequences, *J. Oleo Sci.*, 2021, **70**(12), 1693–1706.
64. Which Countries Produce The Most PalmOil?|Statista. https://www.statista.com/chart/23097/amount-of-palm-oil-produced-in-selected-countries/. Accessed 22 May 2024.
65. F. R. Mohamed, Oils From Fruit Nuts, Coconut (Cocos nucifera) Oil, in *Fruit oils: Chemistry and Functionality*, ed. A. Osman, 2019, pp. 209–221.
66. A. Zikri and M. Aznury, Green diesel production from Crude Palm Oil (CPO) using catalytic hydrogenation method, *IOP Conf. Ser.: Mater. Sci. Eng.*, 2020, **823**, 012026.
67. Q. Zhang, J. Chang, T. Wang and Y. Xu, Review of biomass pyrolysis oil properties and upgrading research, *Energy Convers. Manage.*, 2007, **48**(1), 87–92.
68. A. V. Bridgwater, Review of fast pyrolysis of biomass and product upgrading, *Biomass Bioenergy*, 2012, **38**, 68–94.
69. S. M. Farouk, A. M. Tayeb, S. M. S. Abdel-Hamid and R. M. Osman, Recent advances in transesterification for sustainable biodiesel production, challenges, and prospects: a comprehensive review, *Environ. Sci. Pollut. Res.*, 2024, **31**, 12722–12747.
70. V. Dhyani and T. Bhaskar, A comprehensive review on the pyrolysis of lignocellulosic biomass, *Renewable Energy*, 2018, **129**, 695–716.
71. L. Grande, I. Pedroarena, S. A. Korili and A. Gil, Hydrothermal liquefaction of biomass as one of the most promising alternatives for the synthesis of advanced liquid biofuels: a review, *Materials*, 2021, **14**(18), 5286.
72. A. Dahiya, Cutting-edge biofuel conversion technologies to integrate into petroleum-based infrastructure and integrated biorefineries, in *Bioenergy*, Academic Press, 2020, pp. 649–670.
73. R. Sotelo-Boyás, F. Trejo-Zárraga and F. D. J. Hernández-Loyo, Hydroconversion of triglycerides into green liquid fuels. *Hydrogenation*, 2012, p. 338.
74. S. Chen, Green oil production by hydroprocessing, *Int. J. Clean Coal Energy*, 2012, **1**(4), 43–55.
75. F. Pinto, F. T. Varela, M. Gonçalves, R. N. André, P. Costa and B. Mendes, Production of bio-hydrocarbons by hydrotreating of pomace oil, *Fuel*, 2014, **116**, 84–93.

76. https://www.neste.com/about-neste/how-we-operate/production/rotterdam-refinery.
77. https://www.neste.com/en-sg/about-neste/how-we-operate/production/singapore-refinery.
78. https://www.neste.com/about-neste/how-we-operate/production/porvoo-refinery.
79. https://www.diamondgreendiesel.com/about-us.
80. https://www.regi.com/about/our-company.
81. https://www.biobased-diesel.com/post/italy-s-eni-enters-growing-us-biofuels-market-through-renewable-diesel-joint-venture-with-pbf-energy?
82. Y. Liu, D. Zheng, H. Yu, X. Liu, S. Yu, X. Wang, L. Li, J. Pang, X. Liu and Z. Yan, Rapid and green synthesis of SAPO-11 for deoxygenation of stearic acid to produce bio-diesel fractions, *Microporous Mesoporous Mater.*, 2020, **303**, 110280.
83. M. C. Vásquez, E. E. Silva and E. F. Castillo, Hydrotreatment of vegetable oils: A review of the technologies and its developments for jet biofuel production, *Biomass Bioenergy*, 2017, **105**, 197–206.

CHAPTER 4

Catalytic Systems for Renewable-diesel Production

D. VALENCIA* AND J. ABURTO*

Dirección de Investigación, Instituto Mexicano del Petróleo, Eje Central Lázaro Cárdenas Norte 152, Col. San Bartolo Atepehuacan, C.P. 07730, Mexico City, Mexico
*Emails: dvalencia@imp.mx; jaburto@imp.mx

4.1 Heterogeneous Catalysis in Hydrotreating

Hydrotreating (HDT) processes comprise a series of chemical transformations using hydrogen in the presence of a heterogeneous catalyst at high temperature and pressure. This process is quite developed in the petroleum-refining industry as it has been applied in the elimination of targeted atoms within specific molecules. The main aim of this process is the production of specific mixtures without undesired atoms such as metals, sulfur, nitrogen or oxygen. After HDT, the mixture of products can be processed or used as fuels or commodities.

Biomass transformation to produce green fuels or sustainable chemicals has been explored to provide commodities for society from renewable feedstocks. The composition of biomass is complex and it is related to its origin and processing. However, the presence of oxygen-containing compounds in biomass feedstocks is a challenge for its further utilization. The change of feedstock from petroleum to biomass requires the selection or development of proper operating conditions and catalytic materials to obtain the desired products.[1] Since HDT processes are used in several types of refinery infrastructure, they have been adapted to biomass transformation for the elimination of oxygen atoms, namely hydrodeoxygenation (HDO).[2] During earlier research steps, HDO conditions and catalysts were adapted

Green Chemistry Series No. 89
Biorefining of Renewable Diesel: From Production to Market
Edited by Mohammad Aslam, Sanjeev Mishra and Jorge Aburto

Published by the Royal Society of Chemistry, www.rsc.org

complete to biomass feedstocks, so that the feedstocks were treated under the same chemical conditions with the hope for the same results. As a result, this adaptation failed and produced by-products that required treatment of biomass feedstocks similar to that for petroleum-based ones, highlighting the need to shift towards a different approach to deal with the specific nature of biomass feedstocks.

Green diesel, consisting of straight-chain hydrocarbons or paraffins, is produced through the HDO of bio-based feedstocks such as vegetable oils and bio-oils. Its commercial utilization can be an economically viable solution to address the current energy demands. Traditionally, transition metal sulfides have been employed for hydroprocessing reactions. However, the use of transition metals or their oxides as catalysts has recently gained prominence in biomass transformation.[3,4]

4.2 Mo-based Catalysts

Catalysts developed for hydrodesulfurization (HDS) and hydrodenitrogenation (HDN) have been used in refineries for decades. They are based on a very mature technology for the elimination of sulfur and nitrogen atoms from organic molecules within petroleum. Sulfur and nitrogen in petroleum have negative consequences when their products are in the distillate or in fuels, such as poisoning catalysts in subsequent refinery processes or, due to combustion, the release of SO_x and NO_x to the atmosphere, which are responsible for acid rain and environmental pollution.[5,6]

HDS and HDN processes take place at high temperature and pressure under a hydrogen atmosphere. A heterogeneous catalyst is essential for the elimination of sulfur and nitrogen from the organic molecules. The catalyst in this process contains three main components, namely (i) active phase, (ii) support, and (iii) promoter. The most-used industrially available catalyst is composed of MoS_2 as the active phase, Al_2O_3 as the support, and Ni or Co as promoters, depending on the fraction of the petroleum-derived distillate.[7–10] In these catalysts, these promoters donate electrons to the molybdenum atoms. This weakens the bond between molybdenum and sulfur, thereby creating a sulfur vacancy site. Both HDS and HDO involve the breaking of C–heteroatom bonds (C–S in HDS and C–O in HDO) followed by hydrogenation. The catalysts that effectively break these bonds in HDS often demonstrate good activity in HDO as well. Despite the correlation, some key differences exist between HDS and HDO reactions. Optimizing catalysts for HDO may require further research into factors like oxygen-containing functional groups in bio-oil and their interaction with the catalyst surface. However, the existing knowledge base from HDS research has served as a valuable starting point for developing next-generation HDO catalysts.[11]

In the HDO of methyl oleate, all essayed catalysts converted the feedstock to C_{17} and C_{18} olefins and paraffins. While $NiMo/Al_2O_3$ and NiMo/ASA (ASA = amorphous silica–alumina) initially displayed high HDO activity, they deactivated over time, reaching similar activity to $NiMo/SiO_2$ by the end of the test. Conversely, NiMo/C and NiMo/SiO_2 maintained stable activity. The initial deactivation of NiMo/Al_2O_3 and NiMo/ASA is attributed to the Lewis acidity of surface Al species promoting methyl oleate hydrolysis, followed by deactivation from heavy product deposition. These findings highlight the importance of catalyst selection for green-diesel production *via* HDO. While some catalysts may exhibit high initial activity, their long-term stability is essential for industrial feasibility. Catalysts like NiMo/C, demonstrating both high initial activity and stable performance, hold promise for the efficient production of green diesel from renewable sources.[12]

A study explores the influence of sulfidation temperature on the performance of Mo-based catalysts for palmitic acid HDO. A series of catalysts were synthesized with different Mo oxidation states and surface species controlled by sulfidation temperature (100–400 °C). Spectroscopic techniques revealed the presence of mixed MoO_3 and MoS_2 phases, with the optimal ratio found to be crucial for high activity. Sulfidation significantly enhanced the performance compared to the unsulfided catalysts. The most-active catalyst was obtained at 200 °C, exhibiting optimal balance between MoO_3 and MoS_2 species. Higher sulfidation temperatures (400 °C) led to a decrease in activity due to excessive sulfide formation.[13]

These findings highlight the importance of both Mo–S and Mo–O active species for efficient HDO. The transition with increasing sulfidation temperature from MoS_2 to MoO_3 as the catalytically active species suggests a synergy between these species for oxygen removal from biomass feedstock. This mixture of oxide and sulfide species[13] leads to higher yields of oxygen-free biofuels due to the combination of different active chemical species, as shown in Figure 4.1.

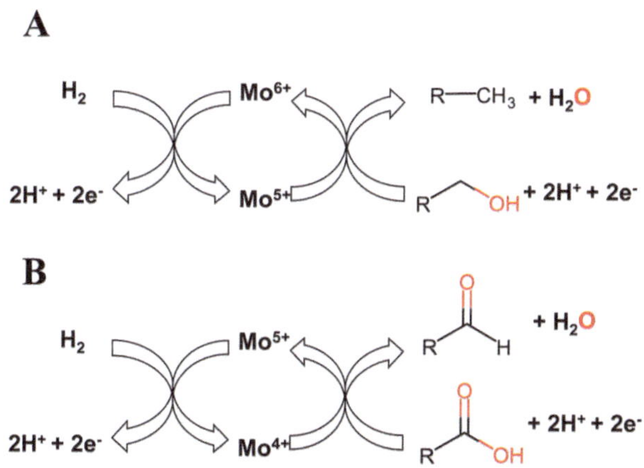

Figure 4.1 Catalytic cycles for the active Mo species: (A) activation of H_2 and hydrodeoxygenation of fatty alcohol by Mo^{5+}/Mo^{6+}, and (B) activation of H_2 and reduction of fatty acid by the Mo^{4+}/Mo^{5+} active species. Reproduced from ref. 13 with permission from the Royal Society of Chemistry.

MoO_3–MoS_2 catalysts supported on TiO_2 or ZrO_2 were synthesized and evaluated for HDO of palmitic acid. The impact of sulfidation temperature (200 °C or 400 °C) on Mo speciation was also explored. Their characterization revealed significant changes in the chemical composition of Mo, S, and O depending on the support and sulfidation temperature. Structure–activity relationships were established: the presence of a mixture of Mo^{6+}, Mo^{5+}, and Mo^{4+} species enhanced activity across all materials. While MoO_3/TiO_2 exhibited higher activity initially, ZrO_2-supported catalysts yielded superior performance upon sulfidation. Notably, sulfidation on ZrO_2 led to the formation of oxygen vacancies that are active in the elimination of O atoms in palmitic acid. DFT calculations provided insights into the interaction between palmitic acid and the catalyst surface. The results highlight the importance of a combination of Mo^{n+} species with S-containing compounds and oxygen vacancies to develop highly active HDO catalysts for fatty acids.[14]

MoO_3 supported on TiO_2 and ZrO_2 without any extent of sulfurization has been studied as catalyst for biomass transformation. MoO_3 as the active phase is promising since the Mo oxidation states and oxygen vacancies must play important roles in the catalytic transformation of biomass feedstocks. A study suggests a prominent HDO pathway facilitated by the presence of oxygen vacancies within the MoO_3 lattice.[15] These vacancies can lead to the partial reduction of Mo ions, creating Mo species in lower oxidation states (Mo^{5+} and Mo^{3+}).[16] MoO_3 catalytic materials exhibited a crucial role for the support in stabilizing partially reduced and coordinatively unsaturated sites (CUS) on Mo within surface oligomeric structures. An important correlation was also observed between higher activity and a larger presence of Mo in intermediate oxidation states (Mo^{5+} and Mo^{3+}). Conversely, catalysts with lower activity display different Mo oxidation states: bulk MoO_3 shows primarily Mo^{4+} and metallic Mo, while MoO_3/CeO_2 has a high proportion of Mo^{6+}.[17]

Another study showed HDO of palmitic acid over MoO_3 catalysts supported on SiO_2. This support is interesting for biomass transformation since it is resistant to water environments and capable of adsorbing polar products, while converting them to the desired hydrocarbon molecules. Two organic acids, citric acid (CA) and malonic acid (MA), were employed during the impregnation process to manipulate Mo oxidation states. The catalyst prepared with CA exhibited superior performance in HDO of palmitic acid. Characterization revealed that the oxidation state of Mo plays a critical role in this reaction. The outstanding performance of the CA-based catalyst is attributed to a well-balanced distribution of Mo oxidation states and the presence of oxygen vacancies. The impregnation method is also important in controlling Mo oxidation states within MoO_3/SiO_2 catalysts. A proper balance of oxidation states, facilitated by the selection of appropriate organic acids during impregnation, promotes the reduction of fatty acids and their derivatives while enhancing hydrogen activation during the HDO process.[18]

Bio-inspired Ca-modified SiO_2 supports for MoO_3-based catalysts in biomass conversion have been developed. $CaCO_3$ and $CaSO_4$ were employed to introduce basic and acidic sites, respectively, onto high-surface-area SiO_2 using L-alanine or L-glycine. These amino acids mimicked natural calcium-binding processes.[19] The introduction of Ca resulted in the formation of crystalline phases on the support surface. $CaCO_3$ reacted with MoO_3 during catalyst preparation to form $CaMoO_4$, while $CaSO_4$ remained intact. Surface characterization revealed minimal influence on basicity across all materials, but significant differences in acidity were observed. The $CaSO_4$-based supports displayed a combination of Lewis and Brønsted acid sites, while $CaCO_3$-modified supports possessed only Lewis acid sites. The choice of amino acid further influenced acid characteristics. The resulting MoO_3 catalysts exhibited superior HDO activity for palmitic acid when supported on $CaSO_4$-based materials compared to $CaCO_3$ counterparts. L-Alanine modification proved beneficial for both Ca sources, whereas L-glycine enhanced activity only for $CaCO_3$-supported catalysts.

Hydroxylation of surface atoms in the MoO_3 active phase is important due to H-bonding interactions between fatty acids and the catalytic surface. A theoretical study revealed that H-bonding serves as the primary driving force for interactions.[20] The H atom within the carboxylic acid moiety interacts with surface oxygen atoms, establishing a crucial anchor point. As the degree of hydroxylation (σ_H) increases, the number of available H-bonding sites on the MoO_3 surface grows, leading to structural modifications within the adsorbed organic molecule. The analysis of electron density (ρ) and Laplacian of electron density ($\nabla^2\rho$) further underscores the critical role of H-bonding in adsorption strength. The strongest H-bond interaction, characterized by the highest ρ (0.302) and $\nabla^2\rho$ (5.504) values, occurs at $\sigma_H = 11\%$, coinciding with the peak adsorption energy. These findings offer valuable insights into the complex interplay between surface chemistry and the adsorption behaviour of biomass-derived molecules on oxide materials, paving the way for further exploration of tailored catalyst design for biomass-conversion processes.

4.3 Noble-metal Catalysts

Noble metals, such as Rh, Pt and Pd, possess an inherent ability to accelerate chemical reactions.[21] Their exceptional catalytic activity stems from their unique electronic structure, which allows them to weaken bonds within the triglyceride molecules and facilitate their rearrangement into diesel-like hydrocarbons. While supported Pd and Pt catalysts demonstrate exceptional activity in converting fatty acids and their derivatives into fuel-like hydrocarbons through decarboxylation, their high cost due to their precious metal nature presents a significant economic hurdle.[22]

Fresh Pt/ZIF-67/zeolite 5A bead catalysts exhibited high heptadecane yields (~81%) and minimal activity loss after two reaction cycles. Compared to existing catalysts, these Pt/ZIF-67/zeolite 5A bead catalysts demonstrate superior performance, particularly under mild reaction conditions. Additionally, this type of catalyst offers significant advantages in terms of recyclability and scalability. Unlike powder catalysts, which typically experience material loss during recycling, the bead format allows for complete catalyst recovery, minimizing waste.[23]

Coke formation, a major catalyst deactivation pathway in HDO of microalgae oil, was most significant for NiMo, followed by Pt and Rh.[24] Interestingly, coke accumulation on NiMo increased with reaction time, suggesting progressive catalyst deactivation. In contrast, Pt and Rh exhibited no such time dependence, indicating better stability. Rhodium displayed high selectivity towards methane (CH_4) production through hydrocracking, particularly in its freshly reduced state. The highest hydrocarbon yield (76.5%) was obtained using 1% Pt at a lower reaction temperature (310 °C) compared to NiMo (360 °C). This 50 °C reduction presents a potential for energy savings during green-diesel production. Therefore, Pt demonstrated superior performance in terms of hydrocarbon yield and potentially lower reaction temperature for the green-diesel production.

Pt/SAPO-11-Al_2O_3 catalyst was studied in a continuous process for production of green diesel from soybean oil.[25] The Pt/SAPO-11-Al_2O_3 catalyst promotes various deoxygenation pathways, including hydrodeoxygenation, decarboxylation and decarbonylation. Notably, the catalyst exhibited minimal deactivation, reaching steady-state after 150 hours on stream. This suggests good long-term stability, crucial for industrial applications. Pd/γ-Al_2O_3 displayed slightly higher activity than Pt, with a greater preference for decarboxylation/decarbonylation over hydrogenation. In contrast, Ni/γ-Al_2O_3 showed significantly lower activity and favoured hydrogenation pathways. This highlights the unique selectivity of Pt for these oxygen-removal reactions.[26]

Studies using various Pd/C catalysts with different acidities revealed that the most alkaline catalyst produced the highest yield of the desired *n*-heptadecane during the decarboxylation of ethyl stearate, but also generated unwanted aromatic byproducts.[27] Pd/C catalysts were also synthesized with varying degrees of metal dispersion using a precipitation deposition method. The deoxygenation reaction was conducted in a semi-batch reactor under various temperatures (260–300 °C) and a hydrogen atmosphere. The results revealed an optimal metal dispersion for achieving the highest reaction rate. These catalysts offer a sufficient concentration of active Pd atoms with the appropriate metal structure, maximizing the deoxygenation activity. It is important to tailor metal dispersion in Pd catalysts for efficient green-diesel production. By optimizing this parameter, the catalysts can achieve high conversion rates for renewable feedstocks to clean fuels.[28]

Pd catalysts supported on SAPO-11 showed the influence of the Si/Al ratio and Pt loading on the performance of Pt/SAPO-11 catalysts for jatropha oil hydroconversion. The results revealed a correlation between the Si/Al ratio and the presence of medium-strength acid sites within the SAPO-11 framework. As the Si/Al ratio increased, the number of these acid sites also increased. The deoxygenation, isomerization, and cracking activities of the Pt/SAPO-11 catalysts were all significantly influenced by the Pt loading. The catalyst with 3 wt% Pt loading exhibited the best overall performance, generating a high yield (83%) of iso-C_{15}–C_{18} hydrocarbons. This highlights the importance of Pt as an active metal component for promoting various hydroconversion pathways.[25]

Pd/AC (palladium on activated carbon) catalysts for green-diesel production through HDO were also studied, using an Al_2O_3 coating on the composite material. This study demonstrated the effectiveness of Al_2O_3 coatings in enhancing the stability of Pd/AC catalysts for green-diesel production. The critical difference lies in the deactivation kinetics. The coated catalyst exhibited a significantly higher activation energy (76.6 kJ mol^{-1}) for deactivation compared to the uncoated one (60.6 kJ mol^{-1}). By hindering deactivation, the coating allows the catalyst to maintain its high deoxygenation activity for a longer period, contributing to a more sustainable and efficient biofuel production process.[29]

4.4 Ni-based Catalysts

Noble metal and sulfided NiMo, CoMo, and NiW catalysts supported on high-surface-area carriers have shown promise in HDO processes. However, their high cost and potential sulfur contamination in the final product necessitate the development of cost-effective alternatives. Beyond their lower cost compared to noble metal catalysts, Ni-based non-sulfided catalysts offer several advantages that make them attractive for green-diesel production. It is a relatively abundant and readily available element compared to noble metals like Pt or Pd, making it a more sustainable and cost-effective choice for large-scale production. The catalytic activity and selectivity of Ni-based catalysts can be tailored through various strategies. For instance, using appropriate support materials (*e.g.* oxides, carbons) can influence nickel dispersion, accessibility, and interaction with reactants, ultimately impacting activity and product distribution.[30]

The influence of various factors, including catalyst pretreatment, reaction atmosphere, and hydrogen pressure in HDO of stearic acid over Ni/γ-Al_2O_3 catalyst was investigated. High conversion (99%) and selectivity (97%) towards heptadecane were achieved at 300 °C under 30 bar total pressure with hydrogen. Increasing hydrogen pressure led to a rise in the yield of stearyl alcohol, an intermediate hydrogenation product.[31] Support composition is also important in the activity of Ni-based catalysts for green-diesel production. Another study, showed the impact of support morphology and composition on the intrinsic activity of Ni catalysts for the HDO of various feedstocks, including methyl stearate, tristearate, and algal oil (a triglyceride mixture). MFI-type zeolite serves as the support material, with its morphology manipulated to explore the interplay between surface characteristics and catalytic performance. Notably, Ni catalysts supported on self-pillared nanosheet zeolites exhibit superior activity compared to those on conventional zeolites. This enhanced activity is attributed to a higher degree of Ni dispersion within the nano-structured support. Interestingly, the acidity of the support plays a significant role in boosting the intrinsic activity of Ni.

The catalyst demonstrating the highest activity combines Ni dispersion with the presence of abundant acid sites. These acid sites likely facilitate Ni-catalyzed C–O hydrogenolysis, promoting efficient deoxygenation. Self-pillared nanosheet zeolites offer a promising approach for catalysts processing bulky reactants. Their hierarchical structure, characterized by small particle sizes and a well-developed mesopore network, provides significant advantages.[32] In this line, C–O bond cleavage in biomass-derived products is crucial when using Ni catalysts since hydrogenation is only one step in the HDO process, then HZSM-5 nanostructured support can produce linear oxygen-free hydrocarbons.[33]

A study on the influence of calcination temperature on the performance of Ni catalysts supported on MgO–Al_2O_3 (70 wt% MgO) for oleic acid decarboxylation showed that the catalyst with a calcination temperature of 300 °C demonstrated the highest conversion of oleic acid at the reaction temperature of 300 °C. This superior performance is attributed to the presence of NiO species, which exhibit a weaker interaction with the support. Weaker interaction likely facilitates easier reduction and generation of more active Ni sites for decarboxylation.[34]

New supports have also been studied for Ni catalysts. For instance, the case of mesoporous silica KIT-5 is remarkable. The characteristics and topology of KIT-5 promote effective dispersion of Ni particles, preventing agglomeration and maximizing the available active sites for the reaction. The catalytic activity HDO of palmitic acid increased with Ni loading, reaching a maximum at 1.8 wt% NiO. This suggests an optimal balance between the amount of active Ni sites and the reduction of potential particle agglomeration. Ultra-low Ni loadings on KIT-5 achieve similar activity compared to higher loadings on carbon supports. This suggests that KIT-5 facilitates better utilization of the available Ni, potentially due to its chemical characteristics and ordered mesoporous structure.[35]

Dispersion of Ni particles within KIT-5 was further studied by a novel approach for developing highly active and selective catalysts for HDO of biomass feedstocks. The key innovation lies in the utilization of Ni–PhN complexes (where PhN represents 1,10-phenanthroline).[36] PhN, being a bidentate ligand, forms strong bonds with Ni^{2+} ions in aqueous solution. This creates stable Ni–PhN complexes in the impregnation solution. The study observed exceptional catalytic performance of the Ni-based catalysts prepared with PhN on KIT-5. The formation of oxygen-free molecules in higher yields was also remarkable, highlighting the attractive selectivity of this material in producing green diesel. Compared to traditional metal oxide supports, KIT-5 facilitates superior activity and selectivity, and potentially offers additional benefits through metal–support interactions modified during the impregnation step (Figure 4.2). This study paves the way for further exploration of the potential of KIT-5 in designing efficient and selective catalysts for various biomass conversion processes.

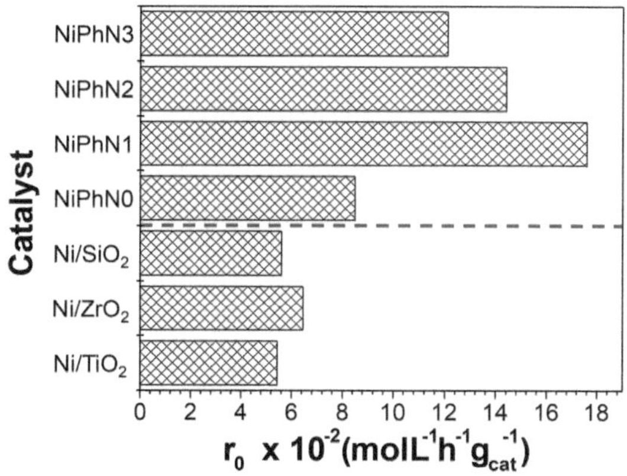

Figure 4.2 Comparison of initial rate of reaction (r_0) in the HDO of palmitic acid over new and reference supported Ni catalysts. Reproduced from ref. 36 with permission from the Royal Society of Chemistry.

4.5 Future Perspectives in Developing Catalytic Materials

Biomass is a renewable energy carrier that is sustainable since it is a product of CO_2 fixation and utilization. Within the biospheric carbon cycle, bioenergy can be carbon neutral because the carbon that is released during combustion has previously been utilized from the atmosphere and will be fixed again as plants grow. The use of biomass feedstocks to produce green fuels or chemicals is also supported by legislation for the reduction of greenhouse gas (GHG) emissions to the atmosphere. This is also aligned worldwide with the 2030 Agenda for Sustainable Development, adopted by all members of the United Nations, and goals 7, 12 and 13 of the Sustainable Development Goals (SDGs).

Recent research presents a groundbreaking approach using sulfonated naphthalene carbonized at 200–250 °C as a metal-free solid acid catalyst for green-diesel production.[37] The carbonized sulfonated naphthalene exhibits superior stability compared to its non-carbonized counterpart. This research is a step towards the development of environmentally friendly, metal-free solid acid catalysts for green-diesel production. Additionally, these metal-free catalysts could find applications in a wide range of other acid-catalysed reactions, further promoting sustainable chemical processes. This is in line with a novel approach for new catalytic materials without metallic species that are active and selective towards desired products.[38]

Another important challenge in HDO is the high pressure of H_2 needed for the hydrogenation and hydrogenolysis reactions to produce green diesel. Some efforts have been conducted to produce hydrogen *in situ* during the chemical transformation.[39] Sacrificial compounds or organic molecules that can be oxidized within the process can release hydrogen that will be used in the HDO. For instance, limonene as a biomass by-product in some processes related to the citrus industry can be converted into *p*-cymene by dehydrogenation reactions.[40] Water can also act as a readily available and *in situ* source of hydrogen, eliminating the need for external high-pressure gas. New multifunctional catalysts serve as the key that unlocks this alternative hydrogen source within water molecules.[41] Those approaches are intended to close the gap towards having green diesel with green chemistry incorporated within the process.

Acknowledgements

The authors wish to thank Projects Y.61023, D.62020 and D.61126 (Instituto Mexicano del Petróleo) for financial support.

References

1. G. W. Huber and A. Corma, *Angew. Chem., Int. Ed.*, 2007, **46**, 7184–7201.
2. T. Ennaert, J. Van Aelst, J. Dijkmans, R. De Clercq, W. Schutyser, M. Dusselier, D. Verboekend and B. F. Sels, *Chem. Soc. Rev.*, 2016, **45**, 584–611.
3. N. Arun, R. V. Sharma and A. K. Dalai, *Renewable Sustainable Energy Rev.*, 2015, **48**, 240–255.
4. A. Srifa, K. Faungnawakij, V. Itthibenchapong and S. Assabumrungrat, *Chem. Eng. J.*, 2015, **278**, 249–258.
5. S. S. Bello, C. Wang, M. Zhang, H. Gao, Z. Han, L. Shi, F. Su and G. Xu, *Energy Fuels*, 2021, **35**, 10998–11016.
6. S. Brunet, D. Mey, G. Pérot, C. Bouchy and F. Diehl, *Appl. Catal., A*, 2005, **278**, 143–172.
7. F. P. Daly, H. Ando, J. L. Schmitt and E. A. Sturm, *J. Catal.*, 1987, **108**, 401–408.
8. D. Valencia, I. García-Cruz, V. H. Uc, L. F. Ramírez-Verduzco and J. Aburto, *ChemistrySelect*, 2018, **3**, 8849.
9. A. S. Rocha, A. C. Faro, L. Oliviero, J. Van Gestel and F. Maugé, *J. Catal.*, 2007, **252**, 321–334.
10. A. Tanimu and K. Alhooshani, *Energy Fuels*, 2019, **33**, 2810–2838.
11. P. M. Mortensen, J.-D. Grunwaldt, P. A. Jensen, K. G. Knudsen and A. D. Jensen, *Appl. Catal., A*, 2011, **407**, 1–19.
12. A. E. Coumans and E. J. M. Hensen, *Catal. Today*, 2017, **298**, 181–189.
13. D. Valencia, L. Díaz-García, L. F. Ramírez-Verduzco, A. Qamar, A. Moewes and J. Aburto, *RSC Adv.*, 2019, **9**, 18292–18301.
14. L. Díaz-García, D. Valencia, J. Navarrete-Bolaños, E. Terrés, L. F. Ramírez-Verduzco and J. Aburto, *Mater. Today Commun.*, 2019, **20**, 100543.
15. K. Murugappan, C. Mukarakate, S. Budhi, M. Shetty, M. R. Nimlos and Y. Román-Leshkov, *Green Chem.*, 2016, **18**, 5548–5557.

16. C. Ranga, R. Lødeng, V. I. Alexiadis, T. Rajkhowa, H. Bjørkan, S. Chytil, I. H. Svenum, J. Walmsley, C. Detavernier, H. Poelman, P. Van Der Voort and J. W. Thybaut, *Chem. Eng. J.*, 2018, **335**, 120–132.
17. M. Shetty, K. Murugappan, T. Prasomsri, W. H. Green and Y. Román-Leshkov, *J. Catal.*, 2015, **331**, 86–97.
18. D. Valencia, L. Díaz, L. F. Ramírez-Verduzco, M. A. Amezcua-Allieri and J. Aburto, *Mater. Lett.*, 2019, **251**, 226–229.
19. D. Valencia, R. I. Conde, B. García, L. F. Ramírez-Verduzco and J. Aburto, *Renewable Energy*, 2020, **148**, 1034–1040.
20. D. Valencia, I. García-Cruz, L. F. Ramírez-Verduzco and J. Aburto, *ACS Omega*, 2018, **3**, 14165–14172.
21. N. Chen, S. Gong, H. Shirai, T. Watanabe and E. W. Qian, *Appl. Catal., A*, 2013, **466**, 105–115.
22. E. Santillan-Jimenez and M. Crocker, *J. Chem. Technol. Biotechnol.*, 2012, **87**, 1041–1050.
23. L. Yang, K. L. Tate, J. B. Jasinski and M. A. Carreon, *ACS Catal.*, 2015, **5**, 6497–6502.
24. L. Zhou and A. Lawal, *Catal. Sci. Technol.*, 2016, **6**, 1442–1454.
25. M. Herskowitz, M. V. Landau, Y. Reizner and D. Berger, *Fuel*, 2013, **111**, 157–164.
26. A. T. Madsen, E. H. Ahmed, C. H. Christensen, R. Fehrmann and A. Riisager, *Fuel*, 2011, **90**, 3433–3438.
27. P. Mäki-Arvela, I. Kubickova, M. Snåre, K. Eränen and D. Y. Murzin, *Energy Fuels*, 2007, **21**, 30–41.
28. I. Simakova, O. Simakova, P. Mäki-Arvela, A. Simakov, M. Estrada and D. Y. Murzin, *Appl. Catal., A*, 2009, **355**, 100–108.
29. S. T. Mohammed, K. I. Hamad, S. A. Gheni, D. Y. Aqar, S. M. R. Ahmed, M. A. Mahmood, S. Ceylan and G. H. Abdullah, *Chem. Eng. Sci.*, 2022, **251**, 117489.
30. C. Kordulis, K. Bourikas, M. Gousi, E. Kordouli and A. Lycourghiotis, *Appl. Catal., B*, 2016, **181**, 156–196.
31. K. Jeništová, I. Hachemi, P. Mäki-Arvela, N. Kumar, M. Peurla, L. Čapek, J. Wärnå and D. Y. Murzin, *Chem. Eng. J.*, 2017, **316**, 401–409.
32. M. W. Schreiber, D. Rodriguez-Niño, O. Y. Gutiérrez and J. A. Lercher, *Catal. Sci. Technol.*, 2016, **6**, 7976–7984.
33. W. Song, Y. Liu, E. Baráth, C. Zhao and J. A. Lercher, *Green Chem.*, 2015, **17**, 1204–1218.
34. H.-S. Roh, I.-H. Eum, D.-W. Jeong, B. E. Yi, J.-G. Na and C. H. Ko, *Catal. Today*, 2011, **164**, 457–460.
35. D. Valencia, C. Zenteno, P. Morales-Gil, L. Díaz-García, D. Gómora-Herrera, E. Palacios-González and J. Aburto, *New J. Chem.*, 2020, **44**, 2435–2441.
36. D. Valencia, C. Zenteno, D. Gómora-Herrera, M. T. Gómez-Pérez and J. Aburto, *Green Chem.*, 2020, **22**, 3470–3480.
37. M. Toda, A. Takagaki, M. Okamura, J. N. Kondo, S. Hayashi, K. Domen and M. Hara, *Nature*, 2005, **438**, 178.

38. B. Lou, N. Shakoor, M. Adeel, P. Zhang, L. Huang, Y. Zhao, W. Zhao, Y. Jiang and Y. Rui, *J. Cleaner Prod.*, 2022, **363**, 132523.
39. N. Krobkrong, V. Itthibenchapong, P. Khongpracha and K. Faungnawakij, *Energy Convers. Manage.*, 2018, **167**, 1–8.
40. J. Zhang and C. Zhao, *Chem. Commun.*, 2015, **51**, 17249–17252.
41. W. Jin, L. Pastor-Pérez, J. J. Villora-Picó, A. Sepúlveda-Escribano, S. Gu and T. R. Reina, *ACS Sustainable Chem. Eng.*, 2019, 7, 16041–16049.

CHAPTER 5

Processing Technologies for Renewable-diesel Production

JORGE ABURTO,[a] ELÍAS MARTÍNEZ-HERNÁNDEZ[b] AND ARICK CASTILLO-LANDERO*[b]

[a] Independent Researcher, Alc. Coyoacán, Mexico City 04360, Mexico;
[b] Energy Efficiency and Biomass Research Division, Instituto Mexicano del Petróleo, Mexico City 07730, Mexico
*Email: acastillo@imp.mx

5.1 Introduction

The implementation of renewable resources has become more important with the goal of using non-fossil fuel sources for the decarbonization of the energy sector, in order to mitigate the effects that the use of fossil fuels has on the environment. One example where a non-fossil alternative has been explored is the transport sector, as it represents a global problem due to the increasing amount of fuel used to meet society's needs.[1]

This increase in fuel use has negative environmental responses, which today are manifested globally due to climate change. This concern has led to the definition of objectives at international level such as those set out in Agenda 2023, which promotes the implementation of strategies that have a sustainable basis recognizing the link between the economic, environmental and social spheres,[2] that have had a great impact in driving the development of new policies such as the use of biofuels like biodiesel and renewable diesel obtained from fats and oils.[3] The production of these fuels has characteristics that need to be considered when choosing a biofuel that can begin to replace fossil fuels. For example, biodiesel can be produced by transesterification reactions between triacylglycerides and an alcohol,

Green Chemistry Series No. 89
Biorefining of Renewable Diesel: From Production to Market
Edited by Mohammad Aslam, Sanjeev Mishra and Jorge Aburto

Published by the Royal Society of Chemistry, www.rsc.org

resulting in a mixture of fatty acid alkyl esters, while renewable diesel is an oxygen-free hydrocarbon mixture obtained by hydrotreating.[4]

Currently, certain countries have defined a roadmap in order to achieve the implementation of these biofuels to mitigate the negative effects on the global environment. One of the countries is the United States (US), where a 32% increase in demand for liquid fuels between 2020 and 2050 is predicted to be a problem for the environment.[5] Other countries include Brazil, which is working on developing strategies to achieve the country's new environmental policies based on renewable-diesel production,[6] and Malaysia, where renewable diesel is considered a good option to implement due to high cetane number, good cold-weather performance, and lower impact of greenhouse-gas emissions, and represents 6% of renewable fuels in the transport sector.[7] On the European continent, the German federal government started to implement new political strategies for climate protection; one of these strategies is the implementation of renewable fuels by 2050, which is part of the target to increase energy consumption from renewable sources by 32%.[8,9]

In order to consider biodiesel and renewable diesel as good substitute fuels, one must take into account the advantages and disadvantages in their properties at the time of obtaining the fuel. Among the reported advantages of biodiesel over automotive diesel are that it is renewable, non-toxic, has higher lubricity, lower ignition delays due to higher cetane number, higher thermal braking efficiency, and produces lower CO and CO_2 emissions. On the other hand, there are some disadvantages such as higher NO_x emissions, higher production cost, restriction in the use of raw materials, lower energy density, and higher viscosity, stability, and freezing point.

As a solution to the aforementioned disadvantages, renewable diesel was proposed for its improved properties over biodiesel, such as better cold-flow properties, higher energy content due to the absence of double bonds and oxygen, lower viscosity, absence of sulfur, and ease of customization.[10,11] Renewable diesel gives reduced emissions compared to fossil diesel of approximately 28–46% for particulate emissions, 7–14% for NO_x, and 5–78% for total hydrocarbons.[12] Some other examples of renewable diesel have been reported, for example green diesel, which is obtained from catalytic hydroprocessing of vegetable oils, white diesel, which is a paraffinic fuel obtained *via* catalytic hydrotreatment of 100% waste cooking oil, and Fischer–Tropsch diesel (FT diesel), which is produced in a slurry reactor or a fixed-bed reactor using iron or cobalt catalysts.[13]

5.2 Biomass Used to Produce Renewable Diesel

The selection of a technology to produce renewable diesel must depend on the type of biomass used, due to the required treatment and reactions for the components of that biomass. The biomass used to produce renewable diesel is divided into lignocellulose, sugar/starch, and lipids/fats.

- Lignocellulose biomass is the most abundant biorenewable material on the earth. It is either plants or plant-based materials, and mainly includes agricultural residues, energy crops, forestry residues, and yard trimmings. It started to be used to produce second-generation biofuels using thermochemical (combustion, gasification or pyrolysis) or biochemical (extraction or hydrolysis) routes.[14,15] The major components of this biomass are cellulose (35–50%), hemicellulose (20–25%), and lignin (10–25%).[16]
- Sugar and starch crops are biomasses used to produce first-generation biofuels. The sugar crop is considered a good candidate because of the high concentrations of sucrose in the stalk. Some examples of sugar/starch crops are sugarcane, sugar beet, sugar sorghum, cereals and corn.[17] It is reported that sugarcane is composed of cellulose (14.5–31.3%), hemicellulose (9.9–25.5%), lignin (7.9–32.2%), and the rest of ash.[18]
- Animal fats as raw materials are a more abundant waste than frying oils. The use of animal fats for biodiesel production is a good option for recycling this waste. Different animal fats, such as lard, beef tallow, chicken fat, yellow fat, and brown fat, have been used for biodiesel production.[19]

5.3 Processing Technologies to Produce Renewable Diesel

Nowadays, the production of renewable diesel is a task that has increasing importance. This is due to the need for routes to improve the efficiency of the process and to have, in the future, a renewable alternative that can compete with fossil sources. A number of technologies have been described that aim to obtain renewable diesel from biomass, such as hydrotreating, hydrocracking, pyrolysis, the Fischer–Tropsch process and liquid transesterification. Table 5.1 summarises each technology used to obtain renewable diesel.

The combination of at least two of the technologies mentioned in Table 5.1 are required to get renewable diesel with properties that meet the requirements of diesel fuel, such as density, sulfur, cetane number, flash point, carbon residue, viscosity, copper strip corrosion, polyaromatic hydrocarbons, oxidation stability, heating value, cold filter plugging point (CFPP), cloud point, and pour point. A good strategy to combine the technologies into synthetic pathways will have consequences in economic results.

On the other hand, their application at industrial level is limited due to the required characteristics of the fuel to be produced, which are standardised by ASTM D 6751 in order to use the biofuel commercially, such as density, cetane number, flash point, carbon residue, viscosity, to name a

Table 5.1 Definitions of the processing technologies used to obtain renewable diesel.

Process technology	Definition
Hydrotreating	This technology is a primary pretreatment process whose main role is to hydrogenate unsaturated hydrocarbons and remove heteroatoms from the feedstock. Hydrotreating removes oxygen, sulfur and nitrogen.[20]
Hydrocracking	Hydrocracking is a route that typically operates at high pressure and temperature to prevent the aromatisation reaction from deactivating the catalyst. The main objective of this process is to convert heavier hydrocarbons into lighter hydrocarbons.[21]
Pyrolysis	Pyrolysis is the thermochemical conversion of organic matter to produce a bio-oil as a precursor for renewable fuels. The most commonly used biomass sources are agricultural residues, wood chips or algae to obtain liquid bio-oil, gases and solid charcoal.[22]
Fischer–Tropsch	In this technology, syngas from a biomass transformation is converted into liquid hydrocarbons such as diesel fuel. This process is well accepted because the FT products are predominantly linear, so the quality of the diesel is very high, with cetane numbers of up to 75.[23,24]
Liquid transesterification	This process uses the catalytic reaction of fats and oils with an alcohol to produce renewable fuels. The main objective of this process is to transform triglycerides into fatty acid methyl esters. The transesterification reaction is usually catalysed by acids, bases and enzymes, which makes the process economically limited.[25]

few.[13] To meet the standards and to have a biofuel that is competitive with fossil fuels, strategies have been created by combining these technologies to produce renewable diesel from biomass. The possible routes and the combination of technologies create synthetic pathways and thus strategies that guarantee a product that meets the stipulations of the international standard and that are economically competitive and provide improvements in environmental impact compared to the process for obtaining diesel from fossil sources.

5.4 Synthetic Pathways to Produce Renewable Diesel

Renewable diesel can be obtained through different synthetic pathways including but not limited to: (1) gas to liquids; (2) hydrogenation of oils and fats; (3) sugars to hydrocarbons; and (4) pyrolysis oil upgrading. A mixture of renewable fuel and petrodiesel can also be produced by co-processing biomass with petroleum fractions. These pathways are presented in Figure 5.1 and described below.

- *Gas-to-liquids.* In this route (see Figure 5.2), syngas ($CO + H_2$) is generated by gasification of lignocellulosic biomass. The gasification occurs

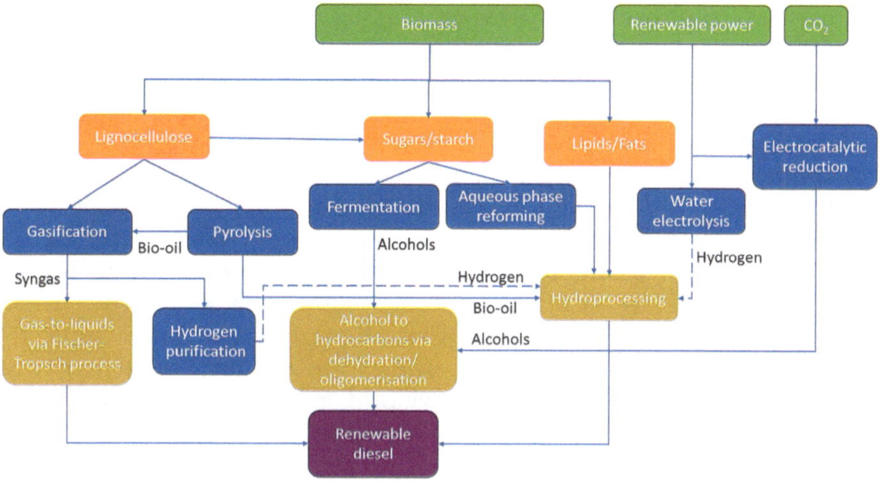

Figure 5.1 Pathways for renewable-diesel production from biomass, CO_2 and hydrogen.

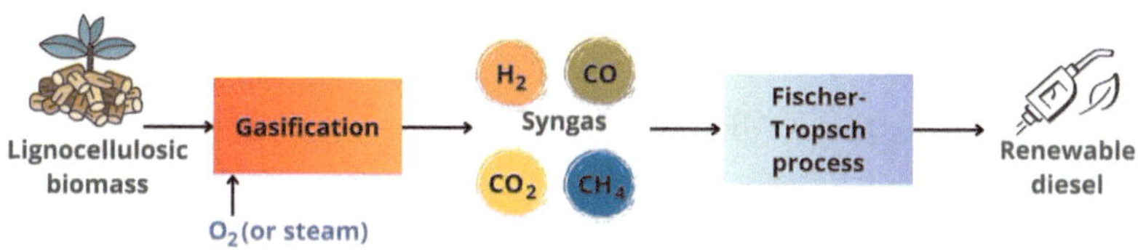

Figure 5.2 Gas-to-liquids route to produce renewable diesel.

by partial oxidation in the presence of a limited amount of oxygen or steam to produce syngas as well as CO_2, CH_4 and water, as shown in eqn (5.1).

$$\text{biomass} + O_2 \text{ (or steam)} \rightarrow CO, CO_2, H_2, CH_4 \tag{5.1}$$

Then, the syngas is transformed by Fischer–Tropsch reactions into saturated or unsaturated hydrocarbons (C_1–C_{40}) with catalysts such as Fe and Co. The resulting product is generally named Fischer–Tropsch diesel.[26] General Fischer–Tropsch reactions are described in eqn (5.2)–(5.5).[27,28]

$$(2n+1)H_2 + nCO \rightarrow C_nH_{2n+2} + nH_2O \tag{5.2}$$

$$2nH_2 + nCO \rightarrow C_nH_{2n} + nH_2O \tag{5.3}$$

$$2nH_2 + nCO \rightarrow C_nH_{2n} + 2O + (n-1)H_2O \tag{5.4}$$

$$CO + H_2O \rightarrow CO_2 + H_2 \tag{5.5}$$

FT diesel has a density range of 0.72–0.82 g mL^{-1}, matching international standards for diesel. In comparison to fossil diesel, it presents cleaner combustion and lower NO_x emissions due to a very low aromatic content. Its cetane number is in the range 55–99, compared to the fossil diesel cetane number of around 50, resulting a better auto-ignition quality. The flash point is low (55–78 °C) and the heating value is 43–45 MJ kg^{-1}, higher than the biodiesel standard (37.1–40.4 MJ kg^{-1}), so FT diesel can be considered one of the most commercially advanced fuels available to replace fossil diesel.[13,29]

- *Hydrogenation of oils and fats.* The hydrogenation of triacylglycerides present in vegetable oils or animal fats by the hydrodeoxygenation reaction allows production of propane from the glycerol moiety and linear paraffins from the fatty acid part, with water as by-product. Decarbonylation and/or decarboxylation reactions can take place, generating linear paraffins with, respectively, CO and/or CO_2.[30] A general reaction scheme is presented in eqn (5.6)–(5.9).[31]

  Depropanation reaction:

$$C_3H_5(C_nH_{2n+1-2m}COO)_3+(m+3)H_2 \rightarrow C_3H_8+3C_nH_{2n+1}COOH \quad (5.6)$$

  Hydrodeoxygenation reaction:

$$C_nH_{2n+1-2m}COOH+(m+3)H_2 \rightarrow C_{n+1}H_{2n+4}+2H_2O \quad (5.7)$$

  Decarbonylation reaction:

$$C_nH_{(2n+1-2m)}COOH+(m+1)H_2 \rightarrow C_nH_{2n+2}+CO+H_2O \quad (5.8)$$

  Decarboxylation reaction:

$$C_nH_{2n+1-2m}COOH+mH_2 \rightarrow C_nH_{2n+2}+CO_2 \quad (5.9)$$

  In these equations, n is the number of carbon atoms in the aliphatic chain and m is the number of carbon–carbon double bonds in the fatty acid molecule. Hence, hydrogen saturation is implicit in the equations.[31] Figure 5.3 shows a schematic representation of the hydroprocessing pathway.

  The diesel obtained is known as green diesel and consists mainly on paraffins, which means that is free of aromatics, oxygen and sulfur. The cetane number of green diesel is between 80 and 99, higher than fossil diesel. The density is 0.77–0.83 g mL^{-1}, the net heating value is 42–44 mJ kg^{-1}, and its combustion is cleaner due its low aromatic content. Additionally, its flash point is 68–120 °C, which allows for safer handling and storage.[13]

- *Aqueous-phase reforming.* Starches and sugars from biomass are especially suited for conversion through Virent's BioForming® process that involves several steps (see Figure 5.4). The biomass is treated by hydrolysis, hydrogenation, dehydration, and isomerization into

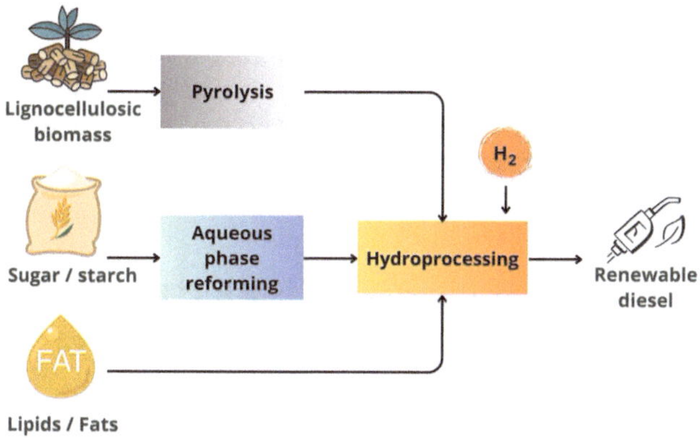

Figure 5.3 Hydroprocessing pathway to produce renewable diesel.

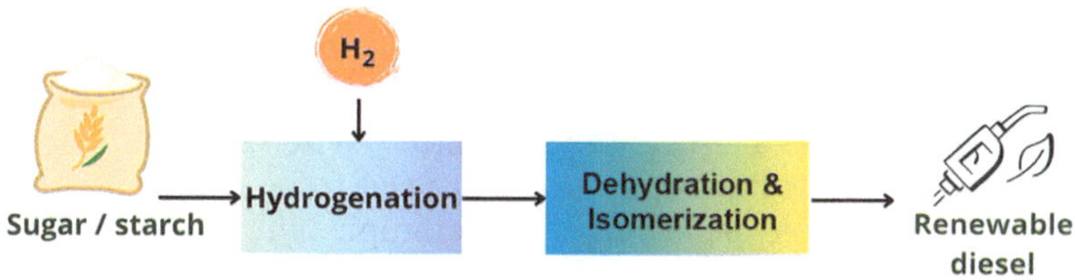

Figure 5.4 Virent's BioForming® process to produce renewable diesel.

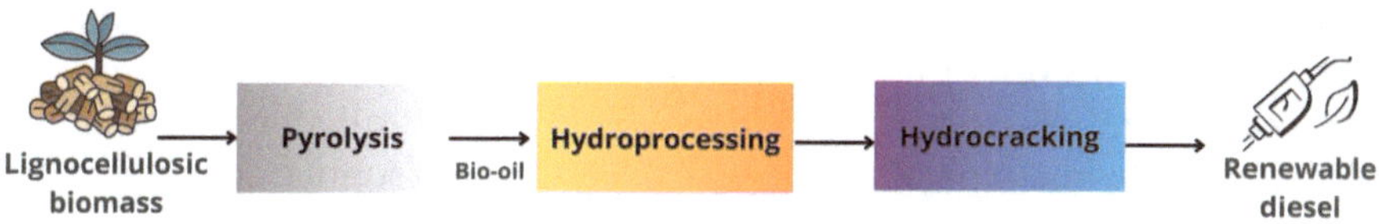

Figure 5.5 Upgrading of pyrolysis oil to produce renewable diesel.

hydrocarbons.[32] Nevertheless, the technology readiness levels (TRL) of this approach was set at 3 which means that the proof of concept through analytical and experimental function was reached.

- *Upgrading of pyrolysis oil.* Another thermochemical approach is *via* the pyrolysis of biomass to produce bio-oil and followed by hydrotreating and hydrocracking to produce renewable diesel and gasoline (see Figure 5.5).[33]

An alternative approach is the conversion of captured CO_2 into alcohols such as ethanol, and then using alcohol dehydration and oligomerization processes to generate hydrocarbons.[34,35] Figure 5.1 shows that hydrogen plays a key role in several conversion routes. This hydrogen is usually externally supplied and is generally produced from a non-renewable source such as natural-gas or naphtha reforming. Figure 5.1 shows that hydrogen from water electrolysis using renewable power should be used. Alternatively, hydrogen can also be produced from gasification of by-product streams from biomass processing such as propane, bio-naphtha, bio-oil, lignin, and unconverted biomass. In this way, the diesel produced maintains a high degree of its renewable and low-carbon nature.

5.5 Fuel Specifications and Properties of Renewable Diesel

Another criterion for technology selection is the properties of the biofuels obtained. Table 5.2 shows a comparison between the properties of the different types of renewable diesel and the standard range of diesel considered for marketing biofuel. Biofuels obtained by hydroprocessing have improved properties compared to those obtained by transesterification, such as FAME biodiesel. This is because of the hydrocarbon nature of the hydroprocessed products, due to oxygen removal through hydrodeoxygenation (HDO) reactions. White diesel, obtained from the hydroprocessing of waste cooking oil, also has properties that meet the standard.[13] When some properties are not met, the renewable-diesel product may need to be blended with fossil fuel up to an allowable percentage. For example, the green diesel from hydrotreated vegetable oil (HVO) may need to be treated by isomerization reactions to meet cold-flow properties, which makes the process more complex.

Another important aspect of diesel fuel quality is the cetane number, which is a measure of the ignition quality of a fuel in a compression engine. A higher cetane number implies a shorter delay in the ignition after injection to the combustion chamber, and thus better quality. While the ignition delay is determined by the type of engine and overall maintenance conditions, the cetane number is highly affected by the chemical composition of a fuel. In general, renewable diesel produced by the different pathways meet the minimum standard specification, as seen in Table 5.2. However, the search for cetane-enhancing additives has been a topic of research in recent years, which is combined with the use of renewable feedstocks such as other biomass feedstocks or even derivatives such as the glycerol coproduced in a biodiesel process.

5.6 Process Simulation and Economic Evaluation of a Green-diesel Process from *Jatropha curcas* Oil

The selection of the type of biomass and the technology pathway to be implemented require detailed evaluation of different aspects to facilitate a decision on investment and implementation. The economic feasibility and environmental impact for the development of an industrial-scale plant must be evaluated, as each technology has different characteristics and

Table 5.2 Properties of different types of renewable diesel.[13] Reproduced from ref. 13 with permission from Elsevier, Copyright 2013.

Analysis	Units	White diesel	FT diesel	FAME biodiesel	Green diesel hydrode-oxygenation of vegetable oils (HDO-VO)	Hybrid diesel vaccum gas oil + vegetable oil (VGP + VO)	Fossil diesel	Diesel standard Min	Diesel standard Max
Density	$g\,mL^{-1}$	0.79	0.72–0.82	0.855–0.9	0.77–0.083	0.781–0.85	0.85	0.8	0.845
Sulfur	$mg\,kg^{-1}$ (ppm wt)	1.54	<10	0–0.012	<10	3–13	12		10
Cetane index	—	77.23	70	58.3	50–105	51–64	54.57	46	
Cetane number	—		55–99	45–72.7	80–99	50–101	50	51	
Flash point	°C	116	55–78	96–188	68–120	74–105	52–136	60	170
Water	$mg\,kg^{-1}$	13	19	28.5–500	42–95	10–50	0.5		200
Micro carbon residue test (MCRT)	wt%	0.0066	0.02–4.5	0.02–0.3		85.8			0.3
Viscosity (40 °C)	cSt	3.5	2.1–3.5	3.89–7.9	2.5–4.15	2.7–5.5	2.71	2	4.5
Copper strip corrosion (3 h, 50 °C)	—	1N		1			<3	Class 1	
Colour (ASTM)	—	0			~2				
Polyaromatic hydrocarbons (PAH) based on a HPLC analysis	wt%		0		<0.1	0.1–1.2		<11	
Induction time (oxidation stability time) (100 °C)	h	>22	>22	0.9–10.9	>22	>22		6	
Distillation temperature (99 vol%)	°C	302.6	295–335		298–342	300–332	341	85–360	
Net heating value	$MJ\,kg^{-1}$	49	43–45	37.1–40.4	42–44	43.3–47	34.97	35	
Cold filter plugging point (CFPP)	°C	20	(−22) to 0	(−13) to 15	>20	(−24) to 22	−6	−5	5
Cloud point	°C		(−25) to 0	(−3) to 17	(−25) to 30	(−23) to 20	−5	−5	12
Pour point	°C	23		(−15) to 16	(−3) to 29	(−26) to 20	−21	−13	10

h = hours.

development maturity. Here, the case of green-diesel production from *Jatropha curcas* oil is discussed.

5.6.1 Process Description, Integration and Simulation

The process flowsheet presented in Figure 5.6 was simulated using SuperPro Designer® for a plant processing 1680 metric tonnes per day (11 890 bbl per day) and the mass balance results are presented in Table 5.3. The simulation of the green-diesel production section is based on previous works.[31] In this process, refined oil is fed into the HDO reactor for hydrodeoxygenation together with hydrogen at a ratio of 0.046 kg H_2 per kg oil and at reactor conditions of 310 °C and 40 bar. In this reactor, the reactions in eqn (5.6)–(5.9) occur yielding a mixture of hydrocarbons, CO, and CO_2. This mixture passes through a heat exchanger (HX-102) to preheat the reactor feed, and is cooled down for gas, hydrocarbon and water separation using a three-phase separator at 1 bar. This results in a liquid hydrocarbon mixture which is cooled down further and then passed through a distillation column to separate a light short-chain hydrocarbon mixture containing mainly propane at the top and the green-diesel product at the bottom. This mixture is also known as hydrodeoxygenation of vegetable oils (HVO or HDO-VO) and contains a range of C_{15}–C_{18} hydrocarbons. This range is different for each vegetable oil because of its initial composition range of fatty acids. The overall green-diesel yield is 0.84 kg diesel per kg oil.

The propane gas generated can be used as fuel to generate the heat required for heating the reactor feed or for the separation units. Proper heat and utility integration will yield a highly efficient process. Alternatively, this propane can be used in a steam-reforming process to generate fresh hydrogen (0.03 kg H_2 per kg oil) to recover the hydrogen consumed in the HDO reactions, as shown in Figure 5.7. This alternative was explored here and simulated using a steam-reformer reactor followed by a water–gas shift reactor (WGSR) to increase hydrogen generation, which is separated using a

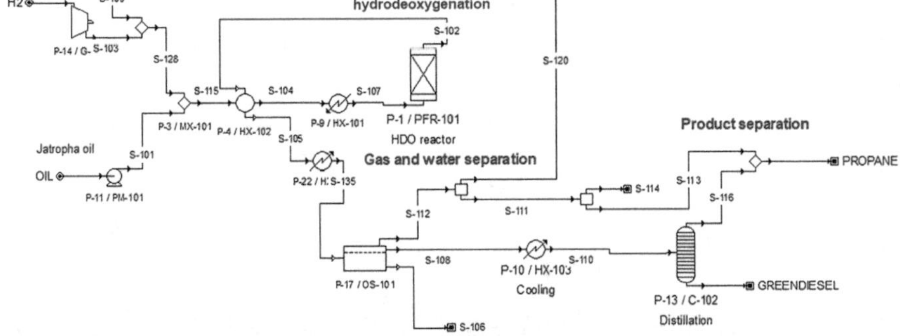

Figure 5.6 Process simulation of green-diesel production from *Jatropha curcas* oil.

Table 5.3 Mass balance of green-diesel production from *Jatropha curcas* oil.

Variable	Units	Green diesel	H2	OIL	S-128	Propane	S-106	S-114
Temperature	°C	343.5	25	25	149.6	89.6	100	100
Pressure	bar	3	5	1.013	40	3	40	35
Volume flow	L h^{-1}	102 582.87	0	78 766.2084	0	181.994	8.0995	0
Ratio		0.84145726	0.03	1	0.04624773	0.05369597	0.07595174	0.05900663
Mass flow	kg h^{-1}	58 902.0085	2100	70 000	3237.3414	3758.7178	5316.6221	4130.4644
Carbon dioxide	kg h^{-1}	0	0	0	0	0	0	3821.2697
Carbon monoxide	kg h^{-1}	0	0	0	0	0	0	256.0035
Heptadecane	kg h^{-1}	20 230.0544	0	0	0	0	2.0232	0
Hexadecane	kg h^{-1}	4035.2664	0	0	0	0	0.4036	0
Hydrogen	kg h^{-1}	0	2100	0	3237.009	126.3343	0	0
Octadecane	kg h^{-1}	32 249.2999	0	0	0	0	3.2253	0
Oleic acid	kg h^{-1}	0	0	721	0	0	0	0
Pentadecane	kg h^{-1}	2387.387	0	0	0	125.6519	0.2513	0
Propane	kg h^{-1}	0	0	0	0.3323	3497.7879	0	0
Trilinolein	kg h^{-1}	0	0	28 210	0	0	0	0
Triolein	kg h^{-1}	0	0	31 500	0	0	0	0
Tripalmitin	kg h^{-1}	0	0	7980	0	0	0	0
Tristearin	kg h^{-1}	0	0	1589	0	0	0	0
Water	kg h^{-1}	0.0008	0	0	0	8.9436	5310.7187	53.1913

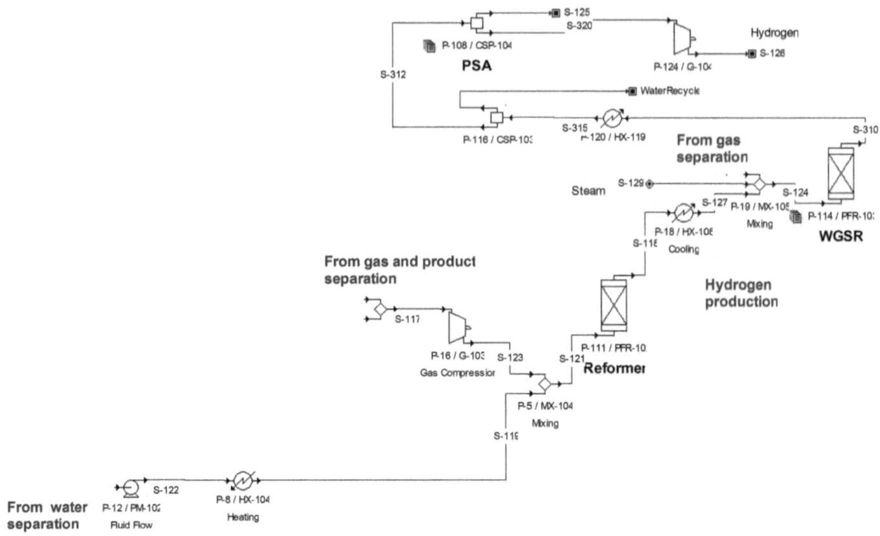

Figure 5.7 Process simulation of the on-site hydrogen generation from propane and off-gases to supply the green-diesel process.

Table 5.4 Energy and utility demands for the integrated green-diesel process.

Utilities	Total demand	Units	Per kg oil	Units	Cost
Cooling water	1255.6	ton per h	0.018	ton per kg oil	US$0.05 per ton
Direct heat	46 889.9	$MJ\,h^{-1}$	0.67	MJ per kg oil	US$6 per GJ
Steam for WGSR	3000	$kg\,h^{-1}$	0.043	kg per kg oil	US$22 per ton
Electricity	28 457 311	kW h per year	0.051	kW h per kg oil	US$0.072 per kW h

pressure swing adsorption (PSA) unit. This alternative applies material recovery principles by using the water separated from the reactor effluent in the reforming reactions and using the off-gases containing CO, CO_2 and water in the water–gas shift reactor. With this configuration, the process generates about 1523 $kg\,h^{-1}$ of hydrogen, which means it can cover 72.5% of the fresh hydrogen demand (2100 $kg\,h^{-1}$), thus reducing costs and emissions from imported hydrogen production. However, the process requires 3000 $kg\,h^{-1}$ of steam as a raw material for the WGSR. The energy requirements of the integrated process with on-site hydrogen generation are presented in Table 5.4.

5.6.2 Economic Assessment

A preliminary economic assessment of the simulated process includes estimating the cost of production of the green-diesel product and the

Table 5.5 Economic-assessment parameters.

Parameter	Value
Oil price	US$0.6 per kg
Hydrogen price	US$2 per kg
Green-diesel price	US$1.111 per kg
Operating time	7920 h per year
Unlisted purchase cost (PC)	20% of estimated purchase cost
Direct plant cost factor (DC)	4 times the estimated purchase cost
Indirect plant cost factor (IC)	60% of estimated direct cost
Other costs factor	15% of the sum of DC + IC
Working capital	30 days of expenses for operating costs
Startup and validation costs	5% of fixed capital cost
Maintenance	6% of fixed capital cost
Insurance, local taxes and miscellaneous	8% of fixed capital cost
Average labour costs	US$15 per h
Quality-control costs	15% of total labour costs
Project life time	15 years
Inflation rate	4%
Interest rate	10%
Construction period	3 years
Depreciation	10 years straight line
Salvage value	5% of fixed capital costs
Income tax	30%

investment cost, as well as feasibility indicators. The assumptions for the parameters used in the assessment are summarised in Table 5.5. With these, the estimated net cost of production is US$0.875 per kg of green diesel (US$0.7 per L), which is a cost competitive with the current diesel price in Mexico (approx. US$1.4 per L).

The purchased equipment cost is US$39.272 million, which yields a direct fixed capital cost of US$244.23 million. Considering all the initial costs (working capital, startup and validation), the total investment amounts to US$290.441 million for an oil-processing capacity of 1680 metric tonnes per day (11 890 bbl per day). An internal rate of return of 21.9% is estimated with a payback period of 3 years and net present value (NPV) of US$214.9 million.

The cost of production and feasibility indicator results are subject to a relatively low price of the vegetable oil at US$0.6 per kg. A 50% increase in price increases the cost of production to US$1.232 per kg and makes the plant unfeasible. The oil price should be lower than US$0.7 per kg to obtain a minimum IRR of 10%. Alternatively, the product price requires subsidisation to become attractive and competitive with fossil-fuel prices.

With new requirements regarding the development of sustainable processes that seek to select a process from an environmental, economic and social approach in a balanced way, all the factors mentioned above are taken into account. At the research level, different sources have reported quantitative and qualitative economic and environmental costs of production and costs of raw materials. In environmental terms, it has been shown that CO_2 emissions could be reduced by up to 110% compared to the conventional diesel-production process, but there are still considerable improvements that could be made for mitigation of the environmental impact by implementing biomass-transformation technology, by integrating services from renewable sources, or by implementing advanced process-improvement tools such as intensification or energy integration.[36–39]

5.7 Conclusions

Selection of the route to produce renewable diesel is a task that must be done carefully because there are many factors to take into account, starting with the type of biomass, such as lignocellulose, sugar and starch, or lipids, which will depend on the availability of biomass in the area of investment and the type of technology available, as each type of biomass has requires initial pre-treatment and subsequent route selection based on the needs of the investment.

The routes mentioned in this chapter show the relationships they can have with each other and the importance of not only focusing on the use of biomass, but also of taking into account aspects such as energy services or steam from renewable sources, which would support moving the industry towards decarbonisation and impact the energy sector in economic, environmental and social terms.

The need has arisen for future work to take into account the considerations mentioned in this chapter and implement them in bioprocesses, including the technologies described. It is necessary to consider the economic and resultant aspects to have a global vision of the process from the agro-industrial production of the biomass to the sustainability benefits of the product and its integration in national and international markets.

Acknowledgements

This work was supported by a Newton Fund Impact Scheme (NFIS) grant, ID 540821111. The grant is funded by the UK Department for Business, Energy and Industrial Strategy and delivered by the British Council. For further information, please visit www.newtonfund.ac.uk. The authors would also like to acknowledge Project Y.62001 (IMP) for financial support. Dr Castillo acknowledges financial support for his postdoctoral stay at IMP from the NFIS grant.

References

1. H. Xu, U. Lee and M. Wang, Life-cycle energy use and greenhouse gas emissions of palm fatty acid distillate derived renewable diesel, *Renewable Sustainable Energy Rev.*, 2020, **134**, 110144.
2. G. Assembly, Resolution adopted by the General Assembly, *Agenda*, 2002, **21**(7), 11.
3. G. Knothe, Biodiesel and renewable diesel: a comparison, *Prog. Energy Combust. Sci.*, 2010, **36**(3), 364–373.

4. A. Arguelles-Arguelles, M. A. Amezcua-Allieri and L. F. Ramírez-Verduzco, Life cycle assessment of green diesel production by hydrodeoxygenation of palm oil, *Front. Energy Res.*, 2021, **9**, 690725.
5. H. Xu, L. Ou, Y. Li, T. R. Hawkins and M. Wang, Life cycle greenhouse gas emissions of biodiesel and renewable diesel production in the United States, *Environ. Sci. Technol.*, 2022, **56**(12), 7512–7521.
6. A. A. V. Julio, T. S. Milessi, E. A. O. Batlle, E. E. S. Lora, D. M. Y. Maya and J. C. E. Palacio, Techno-economic and environmental potential of Renewable Diesel as complementation for diesel and biodiesel in Brazil: A comprehensive review and perspectives, *J. Cleaner Prod.*, 2022, 133431.
7. S. R. Chia, S. Nomanbhay, M. Y. Ong, K. W. Chew and P. L. Show, Renewable diesel as fossil fuel substitution in Malaysia: A review, *Fuels*, 2022, **314**, 123137.
8. T. Pregger, J. Nitsch and T. Naegler, Long-term scenarios and strategies for the deployment of renewable energies in Germany, *Energy Policy*, 2013, **59**, 350–360.
9. D. E. Aliabadi, K. Chan, N. Wulff, K. Meisel, M. Jordan and I. Österle, *et al.*, Future renewable energy targets in the EU: Impacts on the German transport, *Transp. Res. D: Transp. Environ.*, 2023, **124**, 103963.
10. D. Ogunkoya, W. L. Roberts, T. Fang and N. Thapaliya, Investigation of the effects of renewable diesel fuels on engine performance, combustion, and emissions, *Fuels*, 2015, **140**, 541–554.
11. D. Kumar, B. Singh and J. Korstad, Utilization of lignocellulosic biomass by oleaginous yeast and bacteria for production of biodiesel and renewable diesel, *Renewable Sustainable Energy Rev.*, 2017, **73**, 654–671.
12. K.-H. Hsu, W.-C. Wang and Y.-C. Liu, Experimental studies and techno-economic analysis of hydro-processed renewable diesel production in Taiwan, *Energy*, 2018, **164**, 99–111.
13. S. Bezergianni and A. Dimitriadis, Comparison between different types of renewable diesel, *Renewable Sustainable Energy Rev.*, 2013, **21**, 110–116.
14. F. Xu and Y. Li, Biomass Digestion, in *Encyclopedia of Sustainable Technologies*, ed. M. A. Abraham, Elsevier, Oxford, 2017, pp. 197–204. Available from: https://www.sciencedirect.com/science/article/pii/B9780124095489101083.
15. A. Yousuf, D. Pirozzi and F. Sannino, Fundamentals of lignocellulosic biomass, in *Lignocellulosic Biomass to Liquid Biofuels*, ed. A. Yousuf, D. Pirozzi and F. Sannino, Academic Press, 2020, ch. 1, pp. 1–15. Available from: https://www.sciencedirect.com/science/article/pii/B9780128159361000010.
16. H. Wei, Y. Yingting, G. Jingjing, Y. Wenshi and T. Junhong, Lignocellulosic Biomass Valorization: Production of Ethanol, in *Encyclopedia of Sustainable Technologies*, ed. M. A. Abraham, Elsevier, Oxford, 2017, pp. 601–604. Available from: https://www.sciencedirect.com/science/article/pii/B9780124095489102398.
17. E. Ottum, S. E. Baker and E. L. Bredeweg, Production of Biofuels From Biomass by Fungi, in *Encyclopedia of Mycology*, ed. Ó. Zaragoza and

A. Casadevall, Elsevier, Oxford, 2021, pp. 555–576. Available from: https://www.sciencedirect.com/science/article/pii/B9780128199909000627.
18. P. J. Mason, A. Furtado, A. Marquardt, K. Hodgson-Kratky, N. V. Hoang and F. C. Botha, *et al.*, Variation in sugarcane biomass composition and enzymatic saccharification of leaves, internodes and roots, *Biotechnol. Biofuels*, 2020, **13**(1), 1–19.
19. J. M. Encinar, N. Sánchez, G. Martínez and L. García, Study of biodiesel production from animal fats with high free fatty acid content, *Bioresour. Technol.*, 2011, **102**(23), 10907–10914. Available from: https://www.sciencedirect.com/science/article/pii/S0960852411013563.
20. M. Patel and A. Kumar, Production of renewable diesel through the hydroprocessing of lignocellulosic biomass-derived bio-oil: A review, *Renewable Sustainable Energy Rev.*, 2016, **58**, 1293–1307.
21. W. Attaphaiboon, A. Neramittagapong, S. Neramittagapong and S. Theerakulpisut, Potential of vegetable oils for producing green diesel *via* hydrocracking process, *Thai Environ. Eng. J.*, 2021, **35**(2), 1–11.
22. S. Spatari, V. Larnaudie, I. Mannoh, M. C. Wheeler, N. A. Macken and C. A. Mullen, *et al.*, Environmental, exergetic and economic tradeoffs of catalytic-and fast pyrolysis-to-renewable diesel, *Renewable Energy*, 2020, **162**, 371–380.
23. D. Leckel, Diesel production from Fischer—Tropsch: the past, the present, and new concepts, *Energy Fuels*, 2009, **23**(5), 2342–2358.
24. M. E. Dry, High quality diesel *via* the Fischer–Tropsch process–a review, *J. Chem. Technol. Biotechnol.*, 2002, **77**(1), 43–50.
25. J. Calero, D. Luna, E. D. Sancho, C. Luna, F. M. Bautista and A. A. Romero, *et al.*, An overview on glycerol-free processes for the production of renewable liquid biofuels, applicable in diesel engines, *Renewable Sustainable Energy Rev.*, 2015, **42**, 1437–1452.
26. S. L. Douvartzides, N. D. Charisiou, K. N. Papageridis, M. A. Goula, Green Diesel: Biomass Feedstocks, Production Technologies, Catalytic Research, Fuel Properties and Performance in Compression Ignition Internal Combustion Engines, *Energies*, 2019, **12**, 809. Available from: https://www.mdpi.com/1996-1073/12/5/809/htm.
27. J. Hu, F. Yu and Y. Lu, Application of Fischer–Tropsch synthesis in biomass to liquid conversion, *Catalysts*, 2012, **2**(2), 303–326.
28. S. S. Gill, A. Tsolakis, K. D. Dearn and J. Rodríguez-Fernández, Combustion characteristics and emissions of Fischer–Tropsch diesel fuels in IC engines, *Prog. Energy Combust. Sci.*, 2011, **37**(4), 503–523.
29. M. Cadrazco, A. Santamaría and J. R. Agudelo, Chemical and nanostructural characteristics of the particulate matter produced by renewable diesel fuel in an automotive diesel engine, *Combust. Flame*, 2019, **203**, 130–142.
30. S. Lycourghiotis, E. Kordouli, K. Bourikas, C. Kordulis and A. Lycourghiotis, The role of promoters in metallic nickel catalysts used for green diesel production: A critical review, *Fuel Process. Technol.*, 2023, **244**, 107690.

31. E. Martinez-Hernandez, L. F. Ramírez-Verduzco, M. A. Amezcua-Allieri and J. Aburto, Process simulation and techno-economic analysis of bio-jet fuel and green diesel production—Minimum selling prices, *Chem. Eng. Res. Des.*, 2019, **146**, 60–70.
32. Virent, Fractional Multistage Hydrothermal Liquefaction of Biomass and Catalytic Conversion into Hydrocarbons Virent Final Report for, 2017. Available from: https://www.osti.gov/servlets/purl/1392550.
33. S. Jones, D. Elliott, C. Kinchin, C. Valkenburg, J. Holladay and S. Czernik, *et al.*, Production of Gasoline and Diesel from Biomass via Fast Pyrolysis, Hydrotreating, and Hydrocracking, [cited 2024 Mar 19]. Available from: https://www.energy.gov/eere/bioenergy/articles/design-case-summary-production-gasoline-and-diesel-biomass-fast-pyrolysis.
34. D. R. Sánchez, K. Khalilpour and A. F. A. Hoadley, How sustainable is CO_2 conversion to ethanol?–A life cycle assessment of a new electro-catalytic carbon utilisation process, *Sustainable Energy Fuels*, 2021, **5**(22), 5866–5880.
35. M. D. Garba, M. Usman, S. Khan, F. Shehzad, A. Galadima and M. F. Ehsan, *et al.*, CO_2 towards fuels: A review of catalytic conversion of carbon dioxide to hydrocarbons, *J. Environ. Chem. Eng.*, 2021, **9**(2), 104756.
36. V. B. Borugadda, G. Kamath and A. K. Dalai, Techno-economic and life-cycle assessment of integrated Fischer–Tropsch process in ethanol industry for bio-diesel and bio-gasoline production, *Energy*, 2020, **195**, 116985.
37. Y. Zhang, M. A. Dubé, D. D. McLean and M. Kates, Biodiesel production from waste cooking oil: 2. Economic assessment and sensitivity analysis, *Bioresour. Technol.*, 2003, **90**(3), 229–240.
38. A. Arguelles-Arguelles, M. A. Amezcua-Allieri and L. F. Ramírez-Verduzco, Life cycle assessment of green diesel production by hydrodeoxygenation of palm oil, *Front. Energy Res.*, 2021, **9**, 690725.
39. A. Castillo-Landero, D. Dominguillo-Ramirez, J. Aburto, J. Sadhukhan and E. Martinez-Hernandez, Improving the Economic, Environmental, and Safety Performance of Bio-Jet Fuel Production through Process Intensification and Integration Using a Modularity Approach, *ACS Sustainable Chem. Eng.*, 2023, **11**(2), 660–669.

CHAPTER 6

Hydrodeoxygenation of Triglycerides into Renewable Diesel

AJAIKUMAR SAMIKANNU,[a] MARIAPPAN MANI,[b] LAKHYA JYOTI KONWAR,*[c,d] PÄIVI MÄKI-ARVELA,*[e] PASI VIRTANEN[e] AND JYRI-PEKKA MIKKOLA*[a,e]

[a] Technical Chemistry, Department of Chemistry, Chemical-Biological Centre, Umeå University, SE-901 87, Umeå, Sweden; [b] Department of Chemical Engineering, Indian Institute of Technology Madras, Chennai 600036, India; [c] Inorganic Materials and Catalysis Division, CSIR-Central Salt and Marine Chemicals Research Institute, GB Marg, Bhavnagar-364 002, India; [d] Academy of Scientific and Innovative Research (AcSIR), Ghaziabad 201002, India; [e] Laboratory of Industrial Chemistry and Reaction Engineering, Åbo Akademi University, 20100 Turku/Åbo, Finland
*Emails: lakhyak@csmcri.res.in; pmakiarv@abo.fi; jyri-pekka.mikkola@umu.se; jpmikkol@abo.fi

6.1 Introduction

The global energy landscape is undergoing a transformative shift towards sustainable and renewable sources to address the dual challenges of energy security and environmental sustainability. Among the various renewable energy options, biofuels have emerged as a promising alternative to conventional fossil fuels. Particularly, within the realm of biofuels, triglycerides and fatty acids derived from natural sources (*i.e.* plants, animals or algae)

Green Chemistry Series No. 89
Biorefining of Renewable Diesel: From Production to Market
Edited by Mohammad Aslam, Sanjeev Mishra and Jorge Aburto

Published by the Royal Society of Chemistry, www.rsc.org

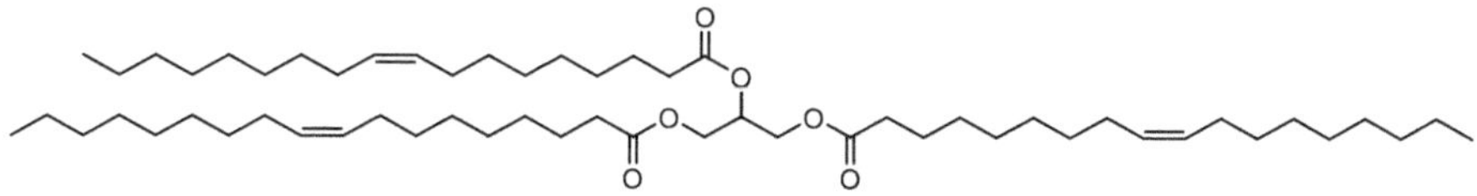

Figure 6.1 Simplified structure of the model triglyceride triolein containing the mono-unsaturated C_{18} fatty acid oleic acid.

have received considerable attention as important renewable feedstocks for the production of biofuels.[1–5] Chemically, triglycerides are composed of long-chain fatty acids esterified to a glycerol backbone, which is responsible for its relatively high energy content in comparison to other renewable feedstocks (Figure 6.1). Although global vegetable oil production has increased from 140 million metric tonnes in 2009–2010 to 200 million metric tonnes in 2020–2021, it is still significantly smaller than lignocellulosic biomass estimated at 181.5 billion tons.[6,7] Despite the limited availability, a sizable fraction (*ca.* 20%) of natural triglycerides and fats is utilized industrially for various non-food applications such as production of oleochemicals and biofuels (such as biodiesel and hydrogenated vegetable oil), mainly due to its compatibility with the existing technological platforms.[7] In fact, the use of triglycerides or vegetable oils as fuel is not new and can be dated back to the early days of the diesel engine. Despite this, the direct use of triglycerides as a fuel is mainly restricted by inherently higher viscosity, which reduces fuel atomization that in turn contributes to the formation of engine deposits, piston ring sticking, injector coking, and thickening of oil. Therefore several methods have been developed for upgrading these triglycerides into fuels.[7–13]

6.2 Overview of Triglycerides: Structure, Physicochemical Properties and Fuel Properties

Triglycerides (or triacylglycerols) are a class of lipids that play a fundamental role in energy storage within living organisms, as well as other important functions including insulation, waterproofing and protection. Structurally they are made up of three long-chain fatty-acid molecules esterified to a glycerol molecule, with saturated C_{16} and unsaturated C_{18} fatty acids being the most common/dominant fatty-acid chain lengths.[6,7] Saturated fatty acids have no double bonds between carbon atoms and possess a straight structure, while unsaturated fatty acids contain one or more double bonds, introducing kinks in the hydrocarbon chain. Figure 6.1 shows the simplified model triglyceride triolein comprising three units of the mono-unsaturated C_{18} fatty acid oleic acid linked to the glycerol backbone. However, unlike triolein most natural triglycerides are structurally unsymmetrical being derived from mixtures of fatty acids, and their resulting physicochemical properties (molecular weight, solubility, melting

Table 6.1 Fatty acid composition and physicochemical and fuel properties of triglycerides (compiled from ref. 14–21).

Property	Rapeseed	Coconut	Palm	Jatropha	Soybean
Major fatty acid components of natural triglycerides (wt%)					
C8: 0 caprylic	0	44–45	0	0	0
C10: 0 capric	0	5.93	0	0	0
C12: 0 lauric	0	44–51	0	0	0
C14: myristic	1.5	13–18.5	0.6–2.4	0	0
C16: 0 palmitic	3.2–5.0	7.5–10.5	32–46.3	11.5	2.3–11
C18: 0 stearic	1.0–2.5	1–3	4–6.3	7.8	2.4–6
C18: 1 oleic	52.6–64	5–8.2	37–53	44	22–30.8
C18: 2 linoleic	20–28.1	1.0–2.6	6–12	31.9	49–53
C18: 3 linolenic	9–15.5	0	0	0.3	2–10.5
C22: 1 erucic	0.0–1.7	0	0	0	0
Important physicochemical and fuel properties					
Calorific value/MJ kg^{-1}	36.89	35.80	39.15	39.23	39.12
Density at 15 °C kg^{-1} m^{-3}	0.914	0.910	0.9180	0.86	0.925
Viscosity at 40 °C/cSt	34.8	28.80	39.6	46.60	31
Oxidation stability/h	11	44	n/a	n/a	n/a
Pour point/°C	−20 to −32	−12.2	21–24	4	−12
Cetane number	41.6	50.3	42.0	41.8	n/a

point, boiling point, vapour pressure, *etc.*) largely depend on the incorporated fatty-acid side chains (Table 6.1). For example, triglycerides comprised mainly of saturated fatty acids (*e.g.* coconut oil, palm oil) present higher melting points compared to those comprised of unsaturated fatty acids (*e.g.* olive oil, sunflower oil, rapeseed oil). Similarly, triglycerides comprised of hydroxy acids, such as castor oil (which is made up of *ca.* 90% ricinoleic acid), present high solubilities in polar solvents like ethanol and methanol, while triglycerides lacking hydroxy acids are generally immiscible with alcohols (*e.g.* sunflower oil, rapeseed oil, olive oil, coconut oil).[8,9,14,15]

Similar to the physicochemical properties, the important fuel properties of triglycerides, such as viscosity, calorific value, cetane number, and cold-flow properties, are also dependent directly on their fatty acid composition (Table 6.1). As outlined in Table 6.1, triglycerides containing large amounts of saturated fatty acids exhibit superior cetane numbers and oxidative stabilities, but poor cold-flow properties which can be attributed to the straight-chain structure of saturated acids. In terms of energy density, those containing a larger amount of longer-chain fatty acids presented higher heating values when compared to triglycerides composed mainly of short-chain fatty acids.[9,14–22] Nevertheless, despite the inherent drawbacks, such as limited availability, high cost and food *vs.* fuel competition, triglycerides have remained important feedstocks for manufacturing of biofuels and chemicals, mainly due to their structural simplicity and compatibility with the existing technological platforms.[7]

6.3 Outline of Upgrading Technologies for Triglycerides

Due to the inherent difficulties associated with the direct utilization of triglycerides as a fuel, several methods have been developed to reduce viscosity and improve the fuel properties of natural triglycerides or vegetable oils to enable their use as alternative fuels. These methods include both physical methods of upgrading such as dilution/blending and micro-emulsification and chemical methods such as pyrolysis (thermal as well as catalytic cracking), co-processing (with crude oil), transesterification (alcoholysis), hydrodeoxygenation (HDO), decarboxylative/decarbonylative deoxygenation (deCOx) or their combinations.[4–14,23–25] Figure 6.2 gives a schematic overview of the different upgrading technologies available for the conversion of triglycerides into alternative fuels. Nevertheless, a common feature of all upgrading methods is the reduction in viscosity of the end-product, with some approaches even producing materials with fuel specifications on par with or even superior to petroleum-based counterparts (Table 6.3). However, it should be emphasized here that the presence of oxygen in triglycerides (or upgraded products) in the form of ester groups in addition to the high viscosity also leads to poor fuel properties, particularly low energy density, undesirable cold-flow properties and poor oxidative stability, thus requiring modifications in existing infrastructure or addition additives for effective use.

6.3.1 Physical Methods

Physical methods of upgrading involve either (a) direct mixing of triglycerides with petro-diesel or gasoline (referred to as direct blending) or (b) mixing of triglycerides with short-chain alcohols such as ethanol and butanol in the presence of surface-active agents (referred to as micro-

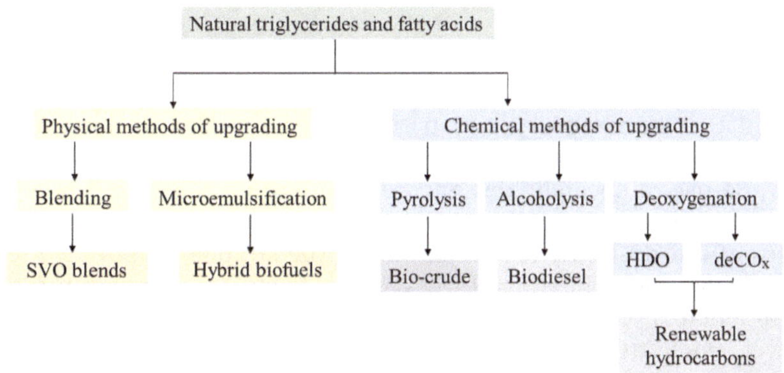

Figure 6.2 Overview of the different technologies or methods available for upgrading triglycerides and natural fatty acids to alternative fuels.

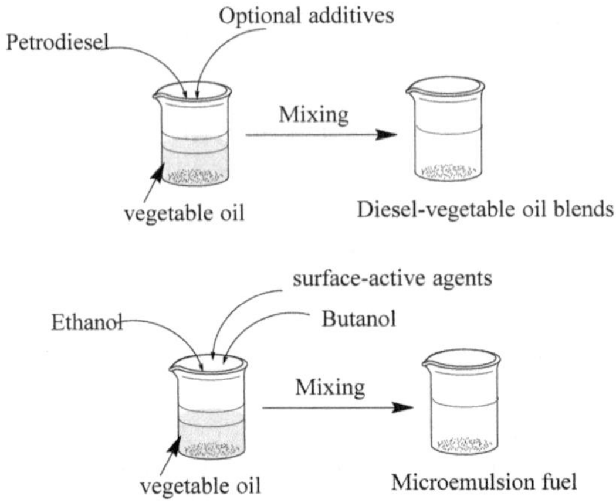

Figure 6.3 Simplified schematic representation of blending and microemulsification methods used in upgrading triglycerides to alternative fuels. Adapted from ref. 9 with permission from Elsevier, Copyright 2014.

emulsification) under ambient conditions producing a synthetic formulation with reduced viscosity and the majority of fuel parameters comparable to petroleum counterparts (Figure 6.3).[9–11,26,27] However, one major drawback of the former approach is the relatively low percentage of triglycerides (*ca.* 10–20%) that can be blended without compromising fuel specifications.[27] Although with the latter approach it is possible to increase triglyceride content up to 70%, such an approach requires the use of expensive surfactants/emulsifiers and cetane boosters to produce stable fuel-grade formulations.[28] Some alternative approaches have been developed which utilize less-expensive compounds such as fatty esters and acids as surfactants in the formulations, the performance of such formulations were found to be typically inferior to the corresponding petroleum counterparts (petroleum blends and petroleum diesel).[9,29]

6.3.2 Chemical Methods

Unlike physical methods, chemical upgrading methods involve breaking of C–O and C–C bonds present in triglyceride molecules to form products with low molecular weight and reduced viscosity. This approach often requires the use of specially designed equipment such as reactors, specific catalysts and specific conditions (such as elevated temperature and pressure) to afford desired yields. Pyrolysis is the simplest chemical method, which involves rapid heating of triglycerides to elevated temperatures in an oxygen-deficient environment in the presence or absence of catalysts. However, such treatment affords non-selective C–O and C–C cleavage, producing a complicated cocktail of products usually referred to as biocrude that consists of hydrocarbons (aromatics and aliphatics), acids and partially deoxygenated molecules such as aldehydes and ketones, requiring additional energy-intensive processing steps such as distillation prior to use.[23–25] The composition of the product depends on several factors including the presence and nature of the catalyst used, the reaction temperature, and the fatty-acid composition of the triglyceride feedstock used. On the other hand, alcoholysis (or transesterification) and deoxygenation (HDO, deCOx), by reaction (usually in the presence of a catalyst) of triglycerides with, respectively, either monohydric alcohols (usually methanol) or pressurized H_2, are more selective and respectively produce either fatty-acid alkyl esters, usually referred to as biodiesel (alcoholysis), or hydrocarbons from deoxygenation through C–O and C–O/C–C cleavage. The former process produces *ca.* 10% glycerol as by-product, while the later process produces equimolar amount of water as by-product.[4–8,12–15,30,31] While it is possible to conduct both pyrolysis and alcoholysis in the presence as well as absence of catalysts, the catalyst-free approach typically results in an inferior product due to incomplete conversion of triglycerides.[23,24,32] Figure 6.4, presents a summary of the important reactions involved in the chemical upgrading of triglycerides.

Among chemical methods, alcoholysis or transesterification produces fatty-acid esters of monohydric alcohols (biodiesel) at relatively low cost. However, biodiesel cannot be regarded as an ideal drop-in replacement due to its inferior energy density, poor oxidative stability and cold-flow properties. On the other hand, deoxygenative processing (HDO, deCOx) results in production of a hydrocarbon mixture (mainly paraffinic and olefinic hydrocarbons) which may be regarded as a potential drop-in replacement for direct use or further upgrading through hydrocracking and hydroisomerization.[4,5,7,12,13,30,31] Nevertheless, all upgrading methods have their inherent advantages and disadvantages which are summarized in Table 6.2. From the information in Table 6.2, it can be seen that, while physical methods of upgrading offer considerable advantages in terms of cost due to the requirement of minimal infrastructure, they lead to a product which may not be deemed suitable as a drop-in replacement, while chemical methods typically produce products with improved fuel characteristics (Tables 6.2 and 6.3).

Overall, among all upgrading (or conversion) methods discussed above (Figure 6.2 and Table 6.2), only HDO and deCOx are able to completely remove oxygen from triglycerides and mitigate the shortcomings of all other methods producing synthetic hydrocarbon products with equivalent or improved fuel properties compared to their petroleum counterparts (drop-in replacements).[4,5,7,12,13,30,31,33,34] The deoxygenation of triglycerides into drop-in biofuels involves various complex chemical processes, with HDO and deCOx steps playing a pivotal role in enhancing the fuel quality (energy density, viscosity and associated fuel properties) of the resulting hydrocarbon products.[4,5,7,12,13,30,31,33–35]

Accordingly, among the various upgrading (or conversion) technologies outlined in Figure 6.2, deoxygenative transformation of triglycerides (HDO, deCOx) has gained significant attention for industrial use, as it enables the

Table 6.2 Brief comments for each upgrading approach with respect to cost, hazards, and quality of the product fuel.

Upgrading method	Conditions	Hazard	Cost	Product composition	Product quality
Blending	Ambient	None	+	Triglycerides, hydrocarbons (aliphatic and aromatic)	++
Micro-emulsification	Ambient	None	++	Triglycerides, alcohols, surfactants (ionic or non-ionic)	+++
Pyrolysis	Harsh (400 °C and above)	Minimum	+++	Aliphatic and aromatic hydrocarbons, acids, aldehydes, ketones	+
Transesterification (alcoholysis)	Mild (60–200 °C)	Moderate (acid, bases)	++	Fatty-acid alkyl esters	+++
Deoxygenation	Harsh (300–450 °C, high pressure)	High (high pressure H_2, flammable solvents and products)	++++	Aliphatic hydrocarbons	++++ (highest drop-in replacement)

Table 6.3 Overview of fuel properties of triglyceride-derived fuels obtained by the different methods of upgrading (compiled from ref. 14–35).

Property	Vegetable oil–diesel blends	Microemulsion fuel	Bio-crude[a]	Biodiesel	Hydrotreated vegetable oil	Petro-diesel
Calorific value/MJ kg^{-1}	41.27	38.81–39.18	45.265–45.754	36.50–40.49	43.91–44.45	45.82
Density at 15 °C kg^{-1} m^{-3}	0.84	0.86–0.882	0.83–0.84	0.865–0.889	0.772–0.775	0.839
Viscosity at 40 °C/cSt	3.85	4.72–9.94	2.34–2.8	3.6–4.9	2.3–2.5	2.91
Cetane number	49.55	32–35	—	49–63	58–74.1	46
Flash point/°C	—	37–43	74–78	—	65–101	75
Cloud point/°C	—	−3 to −7	—	6.1	−25	2
Pour point/°C	—	−7 to −14	—	−1.2	—	−16

[a] Data shown for diesel-range distillate (180–370 °C).

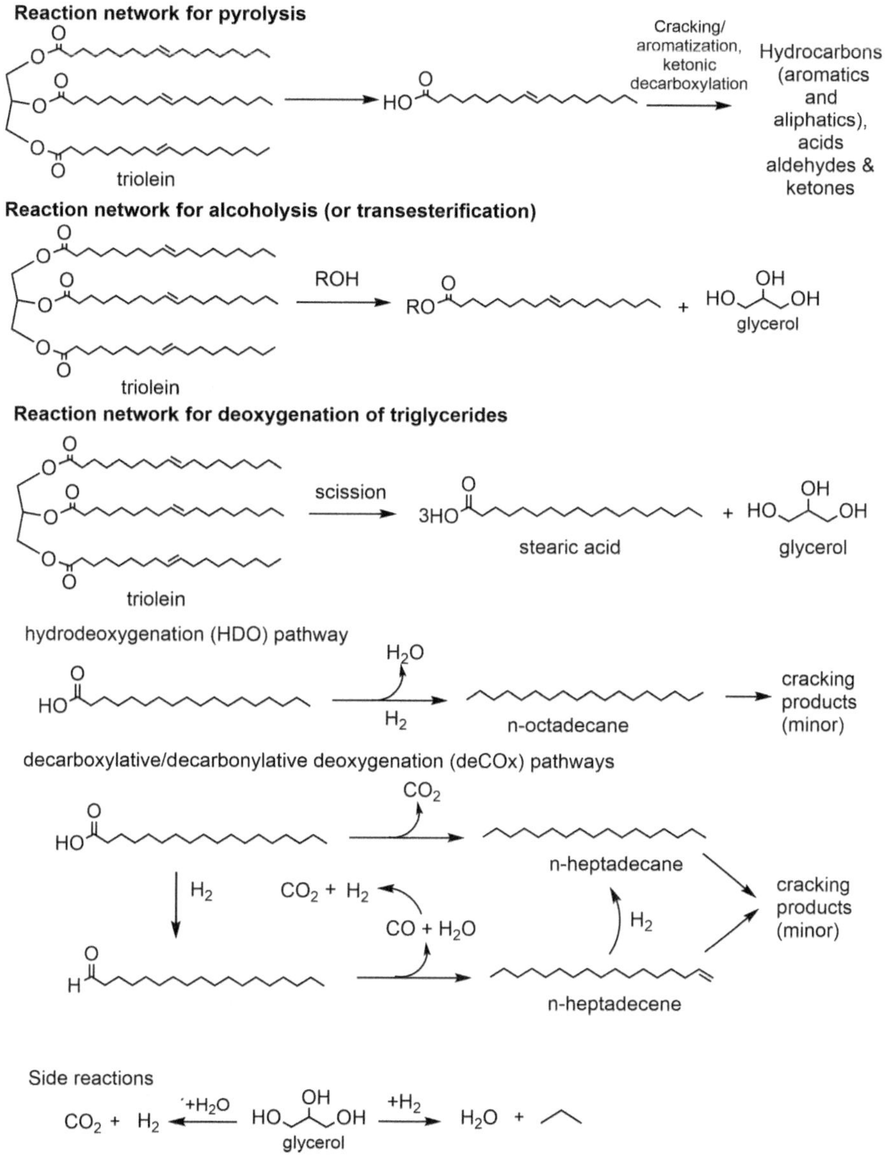

Figure 6.4 Simplified schematic representation of reactions involved in pyrolysis, alcoholysis (transesterification), and deoxygenative upgrading of triglycerides. Adapted from ref. 5 and 23 with permission from Elsevier, Copyright 2022 and 2015.

direct production of high-quality drop-in hydrocarbons suitable for use both in diesel and jet engines (usually with an intermediate upgrading step such as hydroisomerization). In fact, catalytic deoxygenation of fatty acids and triglycerides has been successfully commercialized using HDO technology (Neste NEXBTL™ and UPM BioVerno).[5,30,31,33]

6.4 Deoxygenation and Hydrodeoxygenation of Triglycerides to Fuel-range Hydrocarbons

In a simplistic sense, HDO is a catalytic process that involves the removal of oxygen from triglycerides with the aid of molecular H_2, resulting in the production of hydrocarbons suitable for use as transportation fuels. The process also produces an equimolar amount of H_2O as co-product (Figure 6.4). While the analogous deCOx involves removal of oxygen from fatty acids (formed *in situ* upon hydrolysis of triglycerides) in the form of CO_2 or/and CO, producing hydrocarbons with 1 fewer carbon atoms as the main products. Both processes typically require the use of a suitable catalyst, with heterogeneous Pt-group transition metals, metal sulfides, phosphides, and nitrides being the most commonly employed catalysts. It should also be emphasized, however, that in practice both HDO and deCOx reactions proceed in parallel, where the selectivity of the HDO and deCOx products is dependent on several factors including H_2 pressure, temperature and choice of catalyst (Figure 6.4). In fact, various multifunctional hydro-conversion catalysts have recently been designed, which enable tandem upgrading of triglycerides and fatty acids to either aviation-grade or diesel-grade hydrocarbons in a single-step process.[5,31,35] Overall, deoxygenative upgrading of triglycerides offers considerable advantages for production of biofuels, including high-quality (drop-in) products, tuneable product composition, feedstock versatility and, most importantly, compatibility with the existing technological platforms and catalysts used in petroleum refinerining.[4,5,12,13,30,31,33–35]

6.4.1 Catalytic Deoxygenation and Hydrodeoxygenation

Deoxygenation of triglycerides was first investigated nearly three decades ago[36,37] and restarted in 2006[38,39] when Pd/C was used as a catalyst for HDO of tristearin at 300 °C in a semi-batch reactor, giving mainly heptadecane as the product. HDO is a potential technology to produce long-chain hydrocarbons, which are also found in conventional diesel. Thereafter, research has been very much focused on hydroprocessing or HDO of triglycerides.[5,40–60] Several characteristics of triglyceride conversion have been investigated, including the reaction mechanism,[36,37,40,45] influence of feed type,[36] catalyst selection[45,55] and reaction conditions.[54] In addition, thermodynamic analysis have been performed[60] and, finally, some challenges for HDO in industrial use have been identified.[52] These include, for example, the reactor size, which should be small enough to achieve high yields of the desired components. It is also well-known that coke formation is much larger in comparison to petroleum-based feedstocks and the catalyst stability is of utmost importance. Also, the catalyst regeneration process should be well-planned and simple enough. Water is one of the products of hydrodeoxygenation and, thus, water tolerance of the catalyst is also crucial. In addition to hydrocarbons, water, CO, CO_2 and propane can be formed, which can cause some challenges in large-scale operations (see below).[59] Catalyst selection is discussed in more detail in Section 6.5 and process flowsheets and scale-up are discussed in Section 6.7.

Several reactions occur upon HDO of triglycerides including hydrogenation, which leads to the saturation of unsaturated double bonds[55,59] and degradation of triglycerides[37] followed by hydrogenation of intermediates.[55,59] For example, when the formed stearic acid reacts further, it can be directly decarboxylated to *n*-heptadecane or, alternatively, it can form stearyl aldehyde, which in turn can be decarbonylated to *n*-heptadecane or hydrogenated to stearyl alcohol (Figure 6.4). In the final step *n*-octadecane is formed *via* dehydration of stearyl alcohol in the presence of hydrogen.[55,59] In the formation of *n*-octadecane from stearic acid, the intermediates are stearyl aldehyde and stearyl alcohol.

The mechanisms for triglyceride deoxygenation were discussed in ref. 40, and three different types of transformations, namely β-scission, β-elimination and γ-*H*-migration, prevail (Figure 6.5). In the γ-*H*-migration route, C_{n-2} products are formed, while in the case of β-scission a selective C–C bond scission occurs between the β- and γ-carbon atoms, forming C_{n-3} products. Furthermore, it was proposed that triglycerides are thermally decomposed *via* β-elimination to form fatty acids. Direct deoxygenation forming C_{18} hydrocarbons occurs typically over metal sulfide catalysts.[40]

The effect of feed type, reaction conditions and catalyst selection are the most important parameters defining product distribution in HDO of triglycerides. The feed type, *i.e.* the degree of unsaturated fatty acids in the feed, can affect the product distribution in HDO. The highest yields from HDO of different feedstocks are reported in Table 6.4, which also gives the reaction conditions and catalyst. Typically, HDO of triglycerides has been performed in the temperature range of 260–425 °C under 1–5 MPa hydrogen. The highest green-diesel yields, *ca.* 97 wt% of C_{15}–C_{18} hydrocarbons, have been obtained over highly dispersed NiMo particles on mildly acidic Al_2O_3[45] or NiMoC/Al-SBA-15, using sunflower oil or soybean oil, respectively, as the feedstock.[45,50] In the former case, high metal dispersion was obtained *via* the coprecipitation method, while for the latter one the mesoporous SBA-15 support exhibited large surface area and pore size and thus also giving high metal dispersion and easy diffusion of molecules through the pores. These two feedstocks contain high amounts of polyunsaturated fatty acids in triglycerides; however, they are still efficient for green-diesel production. In another study,[54] the optimal conditions for HDO of sunflower oil were

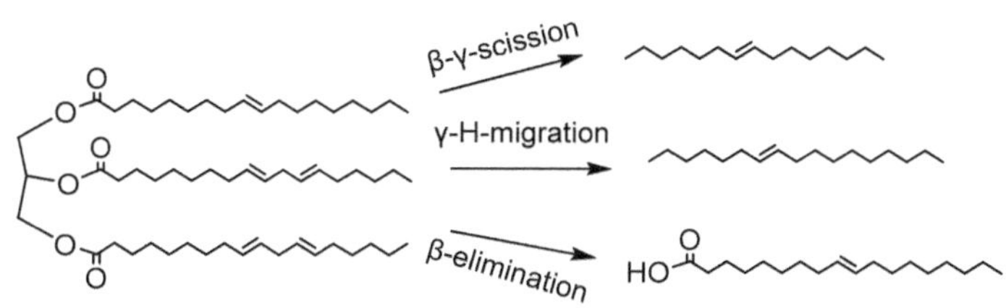

Figure 6.5 Generalised pathway for decomposition of triglycerides.

Table 6.4 Results from hydroprocessing (HDO) of different triglycerides.

Feedstock	Conditions	Catalyst	Conversion (%)	Green-diesel yield	Ref.
Palm oil	425 °C, 5 MPa H_2, H_2/oil = 1000 v/v, WHSV = 0.9 h^{-1}	Ni_2P–$Ni1_2P_5$/NaMOR	100	83.5 mol% n-C_{15}–C_{18}	47
	300 °C, 5 MPa H_2, H_2/oil = 1000 N cm^3 cm^{-3}, LHSV = 2 h^{-1}	$NiAl_2O_4$	100	94 mol% n-C_{15}–C_{18}	46
Sunflower oil	310 °C, 4 MPa H_2, H_2 flow = 100 ml min^{-1}, oil/catalyst = 100 ml g^{-1}	$NiMo/Al_2O_3$	99	97 wt% n-C_{15}–C_{18}	44
	380 °C, 4–6 MPa, H_2/oil = 500–600 N m^3 m^{-3}, GHSV 1 h^{-1}	$CoMo/Al_2O_3$	100	73.7–73.9 wt%	54
Jatropha oil	370 °C, 3.5 MPa H_2, H_2/oil = 1000 v/v, LHSV = 0.9 h^{-1}	NiMoLa(X)/Al_2O_3	83	78 wt% n-C_{15}–C_{18}	43
	370 °C, 3.5 MPa H_2, H_2/oil = 1000 v/v, LHSV = 0.9 h^{-1}	NiMoCe(X)/Al_2O_3	89	80 wt% n-C_{15}–C_{18}	58
Soybean oil	400 °C, 4.5 MPa H_2, H_2 flow = 50 ml min^{-1}, LHSV = 1 h^{-1}	NiMoC/Al-SBA-15	100	97 wt% n-C_{15}–C_{18}	50
	360 °C, 5 MPa H_2, H_2/oil = 1800 ml ml^{-1}, WHSV = 7 h^{-1}	$NiMo/Al_2O_3$	100	81.5 wt% n-C_{15}–C_{18}	45
	350 °C, 1 MPa H_2, catalyst/oil 25 wt%, 5 h	NiMo/USY-Al_2O_3	100	81.9 wt% n-C_{15}–C_{18}	45
Rapeseed oil	350 °C, 3.1 MPa H_2, H_2/oil = 600 ml ml^{-1}, LHSV = 1 h^{-1}	Sulfided NiMo/γ-Al_2O_3	100	69 wt% n-C_{15}–C_{18}	48
Microalgae oil	260 °C, 4 MPa H_2, catalyst/oil = 0.2, dodecane	Ni/H-Beta	100	72 mol% n-C_{15}–C_{18}	49

reported to be 380 °C, 4–6 MPa, using 500–600 N m^3 m^{-3} ratio of H_2 to sunflower oil, and GHSV (gas hourly space velocity) of 1 h^{-1} over a $CoMo/Al_2O_3$ catalyst. Under these conditions, complete triglyceride conversion occurred and the yield of paraffins was in the range of *ca.* 73.7–73.9 wt%, which was lower than the theoretical yield of 81.4–86.5 wt%.[54] When theoretical yield was not achieved, it was necessary to separate the paraffins and recirculate the heavy components. It is also important to note that the degree of unsaturated fatty acids in the feed affects the product distribution. In addition, due to high reaction temperatures used in HDO of sunflower oil,[54] lower liquid yield was obtained in comparison with ref. 44, in which the reaction temperature was only 310 °C. Large amounts of aromatic products can also be formed from the unsaturated fatty acid fraction of triglycerides,

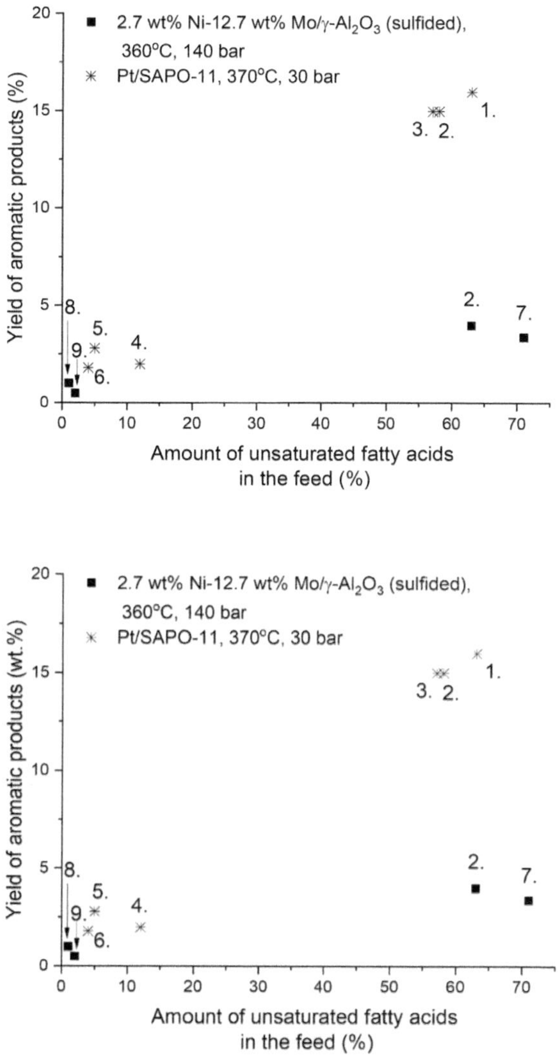

Figure 6.6 Yield of aromatic products produced during hydroprocessing of different oils as a function of the weight fraction of unsaturated fatty acids in triglycerides. Oils: (1) sunflower, (2) soybean, (3) camelina, (4) palm, (5) castor, (6) tallow, (7) maracuja, (8) babassu, and (9) buriti. Adapted from the data in ref. 36.

which was the case especially for soybean oil when the reaction was performed at 370 °C (Figure 6.6).[36] It was observed that the yield of aromatic components obtained *via* the Diels–Alder reaction upon the HDO of triglycerides increased with increasing amount of polyunsaturated fatty acids in the feed.[36]

Palm oil was also successfully applied as a feedstock in HDO over nickel aluminate catalyst at 300 °C.[46] It can also be seen from Table 6.4 that it is

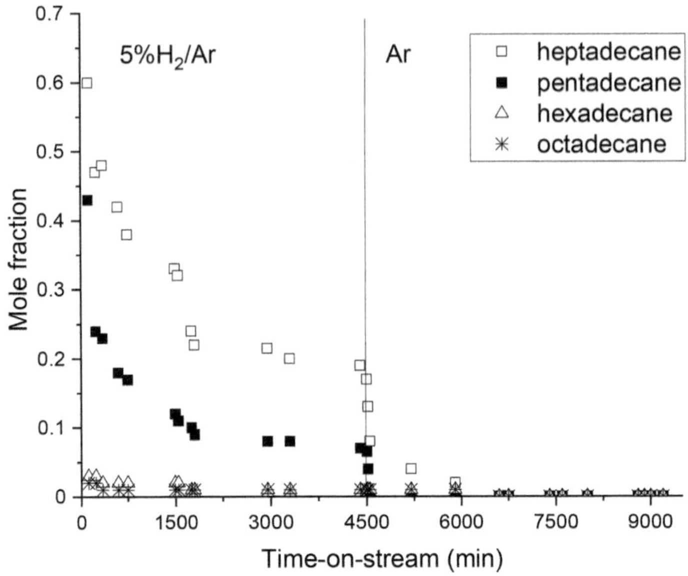

Figure 6.7 Continuous deoxygenation of technical tristearin (containing 65 wt% stearic and 35 wt% palmitic acid) over mesoporous Pd/Sibunit at 0.075 ml min^{-1} liquid flow rate, 300 °C and 2 MPa. Reproduced from ref. 56 with permission from Springer Nature, Copyright 2013.

possible to produce more than 80 wt% yield of green diesel from jatropha oil.[58] The highest reported green-diesel yield from microalgae oil was only 72 mol%. This result might be explained by the relatively low temperature and the use of an acidic Ni/H-Beta catalyst.[49]

The effect of the gas atmosphere has also been investigated in hydroprocessing (HDO) of triglycerides.[56] Technical tristearin was also successfully deoxygenated in a continuous reactor to heptadecane, pentadecane and only traces of hexadecane and octadecane under 5 vol% hydrogen in argon. When the gas was changed to argon, rapid deactivation occurred (Figure 6.7).

The influence of pressure is important in tuning the activity and product distribution; for example, the content of C_{18} increased while C_{17} decreased with increasing pressure (Figure 6.8).[54] This can be partially explained by thermodynamic analysis using Joback's method being in line with the experimental data (Figure 6.8).[61] As can be seen, a larger amount of C_{18} is formed by hydrodeoxygenation in comparison to hydrodecarboxylation, thus showing that the latter reaction is promoted by higher pressure. Analogously, the C_{18}/C_{17} ratio was investigated as a function of temperature in hydrotreating of sunflower oil over $CoMo/Al_2O_3$ catalyst under 8 MPa using LHSV of 1 h^{-1} and H_2/sunflower oil ratio of 600 N m^3 m^{-3}.[54] The results showed that C_{18}/C_{17} ratio decreased with increasing temperature and an analogous trend was obtained *via* thermodynamic analysis (Figure 6.9).[60]

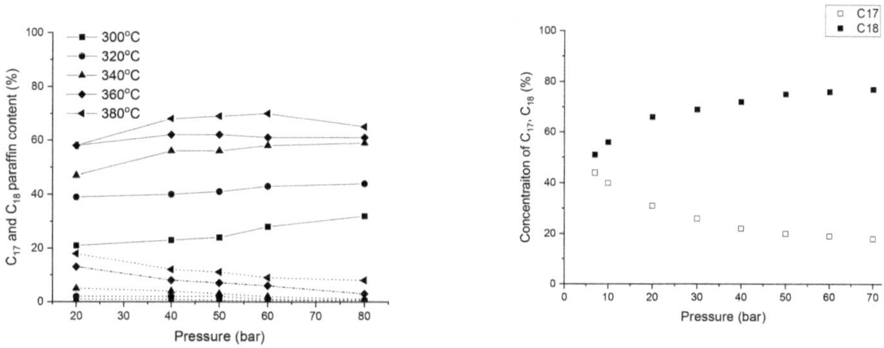

Figure 6.8 (left) Concentration of C_{17} (dashed line) and C_{18} (solid line) paraffins in total organic liquid content following hydrotreatment of sunflower oil over $CoMo/Al_2O_3$ catalyst. Conditions: 1.0 h^{-1} and H_2/feed volume ratio 600 $N\,m^3\,m^{-3}$. (right) Calculated content of C_{17} and C_{18} hydrocarbons as a function of pressure at 310 °C in the hydroprocessing of tristearate using 7 mol hydrogen per 1 mole tristrearate. Adapted from ref. 54 and 60 with permission from Elsevier, Copyright 2010 and 2009.

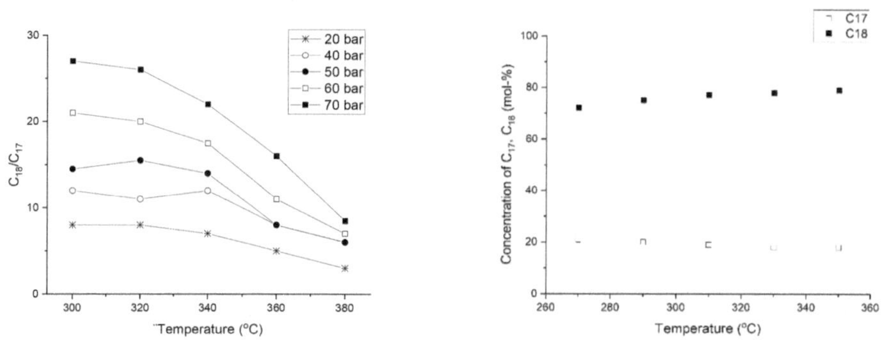

Figure 6.9 (left) Experimental C_{18}/C_{17} ratio as a function of temperature in hydrotreating sunflower oil over $CoMo/Al_2O_3$ catalyst under 8 MPa using LHSV of 1 h^{-1} and H_2/sunflower oil ratio of 600 $N\,m^3\,m^{-3}$.[54] (right) Calculated concentration of C_{17} and C_{18} hydrocarbons as a function of temperature at 7 MPa in hydroprocessing tristearate using 7 mol hydrogen per mol tristrearate and feed dilution of 1 : 100 by hydrogen. Adapted from ref. 54 and 60 with permission from Elsevier, Copyright 2010 and 2009.

Hydrodeoxygenation of triglycerides has, however, some challenges, such as high hydrogen consumption[41] and the product containing, in addition to hydrocarbons, large amounts of propane, water, CO and CO_2,[41] which can affect catalytic performance during the process. The hydrodecarboxylation reaction of tristearin is written as:[60]

$$(C_3H_5)\left(OOC(CH_2)_{16}CH_3\right)_3 + 3H_2 \rightarrow 3C_{17}H_{36} + C_3H_8 + 3CO_2 \tag{6.1}$$

If the additional CO_2 shift reaction is taken into account:

$$CO_2 + H_2 \rightarrow CO + H_2O \tag{6.2}$$

and, thereafter, methanation of CO:

$$CO + 3H_2 \rightarrow CH_4 + H_2O \tag{6.3}$$

the hydrogen consumption is 15 mol per mol of triglyceride.

On the other hand, the hydrodeoxygenation the reaction is:

$$(C_3H_5)\left(OOC(CH_2)_{16}CH_3\right)_3 + 12H_2 \rightarrow 3C_{18}H_{36} + C_3H_8 + 6H_2O \tag{6.4}$$

showing that more product molecules are formed per molecule of triglyceride from hydrodeoxygenation (10) than from decarboxylation (7), which explains the result of thermodynamic analysis that C_{18}/C_{17} increases with increasing hydrogen pressure. The drawback of having the by-product gases in the reactor is that they lower hydrogen partial pressure and result in a decrease in catalytic activity.[41]

Catalyst selection is crucial for product distribution, especially C_{18}/C_{17} product ratio.[55,59] Some examples are given below, while more discussion of catalyst selection is given in Sections 6.4.2, 6.4.3 and 6.5, respectively. For example, large differences in yields were found over Ni/H-Y-80 and Pd/C catalysts during transformation of animal fat to hydrocarbons.[55] The tested animal fat contained 93.6 wt% triglycerides, 5.2 wt% diglycerides and 1.4 wt% monoglycerides, its fatty acid fraction was composed of C_{16}–C_{18} acids, and it contained both saturated and monounsaturated fatty acids. When comparing the performance of these two catalysts, it was observed that Ni–H-Y-80, with a higher acidity in comparison to Pd/C, efficiently promoted the formation of hydrocarbons with the selectivity at complete conversion being close to 100 mol%. On the other hand, over Pd/C primarily fatty acids were formed, which very rapidly transformed at very high conversion levels (92%) in the consecutive step to hydrocarbons with 48 mol% yield (Figure 6.10).

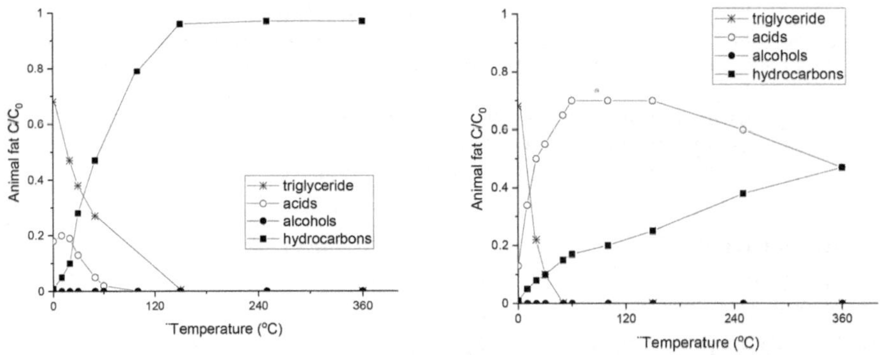

Figure 6.10 Results from hydroprocessing of animal fat under 3 MPa hydrogen at 300 °C over 5 wt% Ni–H-Y-80 (left) and 5 wt% Pd/C (right). Adapted from ref. 55 with permissions from Elsevier, Copyright 2017.

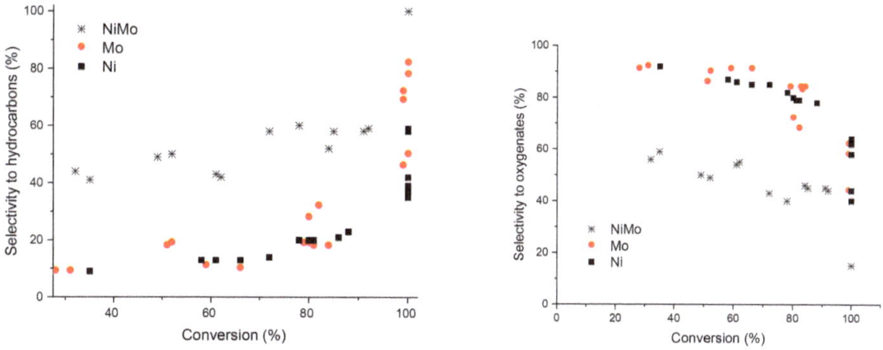

Figure 6.11 Selectivity to hydrocarbons (left) and oxygenates (right) as a function of conversion in hydrotreatment of rapeseed oil over NiMo (✶), Mo (■) and Ni (●) catalysts in the temperature range of 260–280 °C under 3.5 MPa using V/F ratio of 0.25–4 h^{-1}. Adapted from ref. 59 with permission from Elsevier, Copyright 2010.

Another example is deoxygenation of rapeseed oil, for which a comparative study was performed in the temperature range of 260–280 °C under 3.5 MPa using either nickel–molybdenum, nickel or molybdenum catalysts supported on Al_2O_3.[59] It can be clearly seen from these results that only the bimetallic NiMo catalyst is efficient in the production of hydrocarbons (Figure 6.11). Under these conditions, the products are mainly saturated hydrocarbons and the extent of skeletal isomerization is low. In conclusion, it can be stated that cold-flow properties of the product are not sufficient and a further isomerization step might be required.

6.4.2 Decarboxylative/Decarbonylative (deCOx) Deoxygenation

Decarboxylative/decarbonylative deoxygenation is an alternative deoxygenation route to HDO in which oxygen is removed from the fatty acid chains as CO_2 or as CO and H_2O, and as such requires no net H_2 input to convert fully saturated fatty acids (*e.g.* lauric, palmitic, stearic and the like) (Figure 6.4). Typically, supported metal catalysts based on noble metals (Pt, Pd, Rh, Ru, and Ir) are reported to be the most active catalysts for decarboxylation, with activity and selectivity influenced by both the nature of the metal and the support (carrier).[5,11,12,62–64] Similar to HDO, deCOx also requires the presence of low-boiling organic solvents (dodecane, hexane, mesitylene, *etc.*) for improved fluidity and mass transfer and to inhibit catalyst deactivation. Interestingly, unlike HDO where H_2O is regarded as a well-known poison, it is possible to perform deCOx in aqueous media under both sub- and supercritical conditions.[62–67] In fact, recent studies have shown deCOx to proceed under much milder conditions than HDO (temperature 160–260 °C) when using supported Ru/C catalysts, and accordingly there is growing interest in developing alternative deCOx-based processes for green-diesel

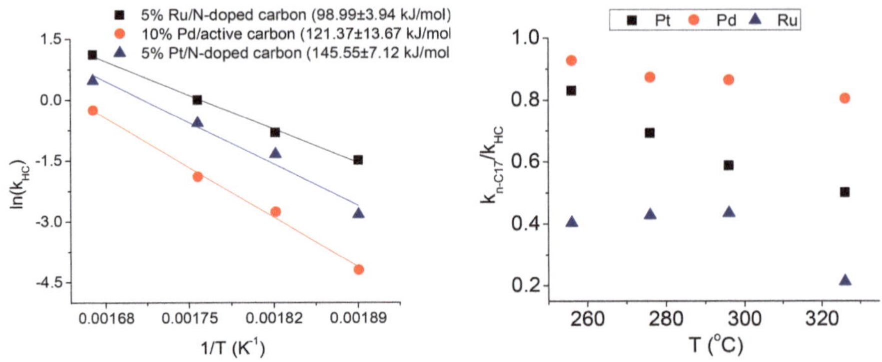

Figure 6.12 (left) Arrhenius plots (ln k *vs.* $1/T$) with apparent activation energies and (right) $k_{n\text{-}C_{17}}/k_{HC}$ (the ratio of rate of formation of deCOx product to rate of formation of all deoxygenation products) as a function of temperature for tristearin deCOx over Pt-, Pd- and Ru-based catalysts. Adapted from ref. 5 with permission from Elsevier, Copyright 2022.

production from fats and oils.[5,63] An additional advantage of deCOx-based processes include the ability to work with a wide range of H_2-donating molecules (such as polyols, monohydric alcohols and formic acid) as well as metal-free catalysts such as active carbon and nitrogen-doped carbon.[65–67]

Similar to HDO, the distribution of hydrocarbon products is largely dependent on the nature of the feedstock, the catalyst, and the H_2 pressure and/or concentration of H_2 donor.[65–68] For example, when using tristearin as feedstock, kinetic analysis revealed the apparent activation energy for hydrothermal deCOx to be, in decreasing order: Pd/active carbon (145.55 ± 7.12 kJ mol^{-1}) > Pt/N-doped carbon (121.37 ± 13.67 kJ mol^{-1}) > Ru/N-doped carbon (98.99 ± 3.94 kJ mol^{-1}) (Figure 6.12). Moreover, the authors also observed considerable influence of reaction temperature on $k_{n\text{-}C_{17}}/k_{HC}$ (the ratio of rate of formation of deCOx product n-heptadecane to rate of formation of all deoxygenation products), where decreasing $k_{n\text{-}C_{17}}/k_{HC}$ with increasing temperature indicated suppression of the deCOx pathway at elevated temperatures as also observed for HDO transformation of triglycerides.

6.4.3 Tandem Process for Direct Upgrading of Triglycerides

A tandem process is a process in which various deoxygenation processes HDO and deCOx or their combination are accompanied by hydrocracking, cracking and/or hydroisomerization.[5,31,67,68] Unlike HDO and deCOx that produce paraffinic long-chain n-alkanes as the main product, tandem upgrading leads to a complex mixture of short-chain, long-chain and branched hydrocarbons which are suitable as diesel, kerosene and jet fuel.[5,31,67,68] For example, using NiMo-based HDO catalysts supported on acidic zeolites such as ZSM-5 produced up to 77 wt% jet fuel-range hydrocarbons with tuneable isomerization selectivity (iso/n = 2.5–13.5) from jatropha oil,[31] while a NiMo catalyst supported on SAPO-11 produced 84 wt% liquid hydrocarbons in the range of diesel (40%), aviation kerosene (40%) and gasoline (20%).[68] Recently, deCOx-based tandem processes have also been discovered which enable direct production of C_5–C_{20} paraffinic hydrocarbons from various triglycerides and fatty acids, including tristearin, sunflower oil, coconut oil and tall oil fatty acids under hydrothermal conditions.[5,67] The results showed that the product distribution was dependent on the metal selection, *i.e.* supported Ru catalysts produced *ca.* 29–44 wt% *n*-C_{17} hydrocarbons, while supported Pt and Pd catalysts were more selective towards C_{17} products. The high activity of, especially, supported Ru was interpreted to be due to *in situ* hydrogen formation from glycerol, when it reacts with water forming CO_2 and hydrogen *via* aqueous-phase reforming or *via* gasification of carbon followed by water–gas shift reaction. The overall amount of diesel-like *n*-C_5–C_{20} produced by Ru supported on carbon was 92–93%, with *ca.* 7% iso-C_5–C_{20} paraffins also obtained. The main benefits with this approach was elimination of the use of external hydrogen and also the use of less-harsh reaction conditions, *i.e.* 256 °C and 5% Ru on NAC (carbon containing basic N-groups) could be successfully recirculated under these hydrothermal conditions.[5] Using N-doped carbon as a metal-free catalyst also directly produced similar diesel-range *n*-C_5–C_{20} hydrocarbons as the product from various unsaturated triglyceride feedstocks at 15–30 bar in the presence of aqueous HCOOH as the H_2 donor,[68] as opposed to an active carbon catalyst producing only deCOx products.[65,66]

6.5 Catalysts for Triglyceride Deoxygenation

As already outlined in Section 6.4, in triglyceride deoxygenation (HDO and deCOx), catalysts play the most important role next to the feedstcok.[59] In general, in the absence of a catalyst triglyceride gets converted into fatty acids, propane, and glycerol.[69] Therefore, the extent of conversion and product composition/selectivity could be tailored by the proper choice of metal and support combination. In a similar fashion, poor selection of experimental conditions could be detrimental for the deoxygenation performance of the catalyst towards the triglyceride.[70] Overall, the basic requirements of the catalyst include the ability to perform hydrogenation (*via* activating the H_2 molecule) and to activate oxygen in the triglyceride molecule (*via* deCOx, HDO pathways; Figure 6.4). Further, thermal stability and reusability of the catalyst is expected from economical perspective.

The choice of feedstock is important in developing a versatile hydrotreating catalyst for biodiesel production (Table 6.1). For example, palm oil contains two different carbon lengths (C_{16} and C_{18}). On the contrary, the average number of unsaturated C=C bonds per fatty acid is higher in soybean oil. The presence of unsaturated carbon bonds can lead to formation of heavy products which can result in coke on the catalyst.[71] The catalyst must be stable against the changes observed in the feedstock due to different geographic location.[71] Observing the interaction between the catalyst and the feedstock such as soybean oil or palm oil could aid in understanding the effect of unsaturated carbon–carbon bonds.

6.5.1 Support

The purpose of the support is to combine its catalytic advantage with efficient dispersal of the active metal species to enhance the deoxygenation performance. Depending on the experimental conditions, for example, γ-Al_2O_3 is reported to be active in the conversion of palm oil into free fatty acids and propane.[75] However, deoxygenated diesel-range products are not produced without the aid of active metal species.[72] Instead, heavy products and coke formation are reported due to the presence of abundant Lewis-acid sites.[71] The heavy products are the result of oligomerization of unsaturated fatty acids *via* a Diels–Alder reaction.[71] On the other hand, the importance of the pore-size distribution of the alumina support is crucial to maximize the production of deoxygenated products. It has been reported that mass-transfer limitations exist up to alumina pore sizes of approx. 12 nm. In a similar fashion, support pore size and acidity can tweak the selectivity between short-chain (gasoline) and long-chain (diesel and jet fuel) products.[50] In a recent study, support effects were reported for deoxygenation of palm oil on NiMoC, where mesoporous supports such as Al-SBA-15 and γ-Al_2O_3 produced hydrocarbons mainly in the diesel and jet fuel ranges.[73] Meanwhile, the strongly acidic small-pore zeolites USY, ZSM5, and zeolite B produced higher short-chain gasoline-range products due to their microporous nature.[50] The constrained channels lead to longer diffusion distances and provide more possibilities for cracking of reactant and product species.[50] Furthermore, dispersion of the metal nanoparticles is also important in catalytic experiments. The abundance of penta-coordinated Al sites in γ- and θ-Al_2O_3 led to Pt dispersion of approx. 1–1.5 nm. On the other hand, irregular large Pt chunks were observed on α-Al_2O_3, which contained no penta-coordinated Al sites.[71] Interestingly, NiMo supported on an acidic support such as Al_2O_3 show higher deoxygenation performance compared to the inert support SBA-15. However, the HDO selectivity is higher when supported on SBA-15 compared to Al_2O_3. This could be due to improved mass transfer due to high specific surface and organized structure compared to amorphous Al_2O_3.[76]

Basic supports are used due to their non-acidic nature, in order to reduce C–C cracking. Unfortunately, the deoxygenation performance of NiMo was reported to be 2–3 times lower compared to NiMo/Al_2O_3.[76] Another disadvantage of basic supports such as dolomite in deoxygenation of triglycerides is the possibility of creating difficulties in downstream processing due to the formation of soap.[77] The interaction between the $CaCO_3$ with released carboxylic acids from triglycerides leads to formation of fatty acid soaps ($Ca(RCOO)_2$). The basic support is active in conversion (36%); however, coking was very high on the basic support. Unfortunately, the soap formation still persisted despite promotion with transition metals such as

Ni, Cu, Zn, Co and Fe. It should be noted that an increase in conversion and decrease in coke formation was observed after doping with transition metals.[77,78]

6.5.2 Precious Metal Catalysts

Generally, precious metals such as Pt, Pd and Ru have been the choice of catalyst for deoxygenation (HDO and deCOx) of triglyceride feedstocks (Table 6.5) where the reaction pathway is dependent on two main factors, *viz.* degree of unsaturation in the feedstock and H_2 pressure, as illustrated in Figures 6.8 and 6.9. Ru (5 wt%) supported on commercial alumina demonstrated higher performance than Pt and Pd for deoxygenation of fatty-acid mixtures derived from rapeseed oil at 340 °C and 10 MPa,[79] and an identical trend has also been obtained for carbon-supported precious metal catalysts.[71,72,79] Furthermore, the deoxygenation varies directly as a function of metal loading, with higher loadings leading to higher yields of the desired hydrocarbons.[71,72] In a recent study, lower metal loading of noble metals

Table 6.5 Deoxygenation performances of supported precious metal catalysts with different feedstocks.

S. no	Catalyst	Feedstock	Conditions	Conversion (%)	Selectivity	Ref.
1	$5Pd/Al_2O_3$	Palm oil	330 °C, 4 MPa H_2, flow reactor	100	$C_{17} = 79.1\%$, $C_{18} = 4.7\%$	72
2	$5Pt/Al_2O_3$	Palm oil	330 °C, 4 MPa H_2, flow reactor	100	$C_{17} = 71.9\%$, $C_{18} = 7.6\%$	72
3	$2Pd/Al_2O_3$	Palm oil	330 °C, 4 MPa H_2, flow reactor	99	$C_{17} = 36.6\%$, $C_{18} = 2.7\%$	72
4	$2Pt/Al_2O_3$	Palm oil	330 °C, 4 MPa H_2, flow reactor	95	$C_{17} = 27.5\%$, $C_{18} = 3.7\%$	72
5	$1Pt/\theta\text{-}Al_2O_3$	Soybean oil	360 °C, 2 MPa, fixed-bed reactor	—	$C_{13,15,17} \approx 70$, $C_{14,16,18} \approx 5$	71
6	$5Pd/Al_2O_3$	Fatty-acid mixture derived from rapeseed oil	340 °C, 10 MPa, approx. 2 h, batch reactor	—	$C_{18}/C_{17} \approx 0.2$	79
7	$5Pt/Al_2O_3$	Fatty-acid mixture derived from rapeseed oil	340 °C, 10 MPa, approx. 2 h, batch reactor	—	$C_{18}/C_{17} \approx 0.58$	79
8	$5Ru/Al_2O_3$	Fatty-acid mixture derived from rapeseed oil	340 °C, 10 MPa, approx. 2 h, batch reactor	—	$C_{18}/C_{17} \approx 0.06$	79

such as Pt and Pd (2 wt%) on alumina shows moderate conversion of palm oil.[72] However, the desired product yield is lower, and the liquid phase is solidified. On the other hand, increasing the metal promotion to 5 wt% Pt or Pd on alumina improves the deCOx performance by 2–3 times, which results in high product yield in the final liquid product.[72] Similarly, bimetallic combinations of noble metals such as $PtPd/Al_2O_3$ are also reported to afford improved catalytic activity for deoxygenation of jatropha oil at 350 °C and 3 MPa.[80] Therefore, promotion of noble metals with transition metals to form bimetallic catalysts is a viable strategy to tune the deoxygenation performance as well as the selectivity for HDO or deCOx products. For example, the promotion of Sn to Pt on alumina shifted the deoxygenation pathway from deCOx to HDO for rapeseed oil at 400 °C and 5 MPa H_2, with HDO (*i.e.* C_{18}) selectivity reaching a maximum of 83% when the molar ratio of Sn to Pt was 5 : 1.[81]

6.5.3 Non-precious Metal Catalysts

Cheap and abundantly available transition metals (Ni, Mo, Co, Fe, Cu, Zn, W) are an alternative to the high-cost precious metals, and therefore have also been extensively studied in the deoxygenation of triglycerides. However, major drawbacks of these transition-metal catalysts are the requirements for higher metal loading and harsher conditions to afford a high conversion of triglycerides (approx. 90–100%) during catalytic deoxygenation of vegetable oils. Another challenge lies in tailoring the product distribution towards diesel-range products.[69] For instance, Ameen *et al.* observed a linear increase in hydrocarbon yield upon increasing Ni or Mo loading on Al_2O_3 from 3 wt% to 12 wt% to 15 wt%[69] (Table 6.6).

In terms of activity, the performance of Ni-based catalysts appears to be superior to other transition metals for deoxygenation of triglycerides. For example, Hafriz *et al.* reported transition metal-doped dolomite for deoxygenation of waste cooking oil at 390 °C.[77] Superior performance for monometallic promoted Ni on a basic support was observed compared to Fe, Co, Cu and Zn.[77] From the reported molecular size distribution, very low amounts C_{18} and C_{16} products were observed compared to C_{17}, C_{14} and C_{11} selectivity in the liquid product.[77] Especially, Ni/dolomite was highly selective to C_{17} compared to rest, which was attributed to the high deCOx activity of Ni.[77] For example, Ni/Al_2O_3 prefers the deCOx pathway, whereas over Co/Al_2O_3 a combination of HDO and deCOx pathways are followed and over Mo/Al_2O_3 deoxygenation proceeds *via* HDO.[69] In another study, a similar decarboxylation pathway was dominant on Fe/AC catalyst.[82] However, over Fe/AC cracking reactions were prevalent, which gave a maximum of 52 wt% of $C_{15} + C_{17}$ hydrocarbons. In fact, up to 50 wt% of gaseous products were produced when the Fe loading on AC was increased from 2.5 wt% to 15 wt%.[82]

Although the cheap Ni-based catalyst is observed to give superior performance compared to other transition metals, unfortunately the undesired

Table 6.6 Deoxygenation performance of supported non-precious metal catalysts with different feedstocks.

S. no	Catalyst	Feedstock	Conditions	Conversion (%)	Selectivity	Ref.
1	$5Co/Al_2O_3$	Palm oil	330 °C, 4 MPa H_2, flow reactor	100	$C_{17} = 33.8\%$, $C_{18} = 50.1\%$	72
2	$10Co/Al_2O_3$	Palm oil	330 °C, 4 MPa H_2, flow reactor	100	$C_{17} = 35.9\%$, $C_{18} = 42.9\%$	72
3	$5Ni/Al_2O_3$	Palm oil	330 °C, 4 MPa H_2, flow reactor	100	$C_{17} = 65.3\%$, $C_{18} = 2.5\%$	72
4	$10Ni/Al_2O_3$	Palm oil	330 °C, 4 MPa H_2, flow reactor	100	$C_{17} = 80.1\%$, $C_{18} = 1.8\%$	72
5	$3Ni/Al_2O_3$	Rubber seed oil	350 °C, 3.5 MPa, fixed-bed reactor	>95	$C_{17} \approx 12\%$, $C_{18} \approx 4\%$	69
6	$12Ni/Al_2O_3$	Rubber seed oil	350 °C, 3.5 MPa, fixed-bed reactor	>95	$C_{17} \approx 29\%$, $C_{18} \approx 11\%$	69
7	$15Ni/Al_2O_3$	Rubber seed oil	350 °C, 3.5 MPa, fixed-bed reactor	>95	$C_{17} = 41\%$, $C_{18} = 13\%$	69
8	$3Mo/Al_2O_3$	Rubber seed oil	350 °C, 3.5 MPa, fixed-bed reactor	>95	$C_{17} \approx 8\%$, $C_{18} \approx 10\%$	69
9	$12Mo/Al_2O_3$	Rubber seed oil	350 °C, 3.5 MPa, fixed-bed reactor	>95	$C_{17} \approx 7\%$, $C_{18} \approx 35\%$	69
10	$15Mo/Al_2O_3$	Rubber seed oil	350 °C, 3.5 MPa, fixed-bed reactor	>95	$C_{17} = 22\%$, $C_{18} = 45\%$	69

C–C bond breakage reduces the selectivity for diesel- and jet fuel-range hydrocarbons compared to noble metal catalysts.[83] Therefore, promoting the Ni-based catalyst to modify its electronic and geometric properties and tailor its activity and selectivity is a hot topic.[83] According to mechanistic studies, Ni and NiO ensembles are required for the C–C bond breaking and by forming NiFe, NiCu, or NiPt alloys the continuity in adjacent Ni atoms can be prevented, thereby decreasing undesirable C–C bond cleavage.[83] In this regard, the promotional effect of a hydrogenating metal such as Cu or Pt is to reduce NiO back to Ni, which actively catalyses the deCOx reaction.[83]

For W/Al_2O_3, its promotion with Co does not improve the deoxygenation compared to promotion with Ni as observed for rapeseed oil deoxygenation (270 °C, 5 MPa)[76] (Table 6.7). In a similar fashion, promoting Mo with Ni on Mo/Al_2O_3 was superior relative to promotion with Co.[76] Overall, the order of performance for various promoted catalyst was: $NiMo/Al_2O_3 > NiW/Al_2O_3 > CoMo/Al_2O_3 > CoW/Al_2O_3$.[76]

6.5.4 Sulfide, Phosphide, Carbide and Nitride Catalysts

Conventional hydrotreating catalysts such as sulfided NiMo and CoMo supported on Al_2O_3 have also been investigated for deoxygenation of vegetable oils. In order to understand hydrotreating over such catalysts, a critical comparison of the product distributions from sulfided Ni, Mo and NiMo has been evaluated. When rapeseed oil conversion was compared at similar

Table 6.7 Effect of bimetallic catalysts on deoxygenation of triglycerides.

S. no	Catalyst	Feedstock	Conditions	Conversion (%)	Selectivity	Ref.
1	3.9Ni–16.3Mo/ Al_2O_3	Rapeseed oil	270 °C, 5 MPa, batch reactor, W/F = 1	100	$C_{18}/C_{17} = 30$	76
2	3.9Co–13.2Mo/ Al_2O_3	Rapeseed oil	270 °C, 5 MPa, batch reactor, W/F = 1	100	$C_{18}/C_{17} < 5$	76
3	3.7Ni–15.4W/ Al_2O_3	Rapeseed oil	270 °C, 5 MPa, batch reactor, W/F = 1	100	$C_{18}/C_{17} < 5$	76
4	3.7Co–15.9W/ Al_2O_3	Rapeseed oil	270 °C, 5 MPa, batch reactor, W/F = 1	100	$C_{18}/C_{17} < 5$	76
5	3Ni–15.5Mo/ TiO_2	Rapeseed oil	270 °C, 5 MPa, batch reactor, W/F = 1	—	$C_{18}/C_{17} \approx 1$	76
6	4Ni–15.4Mo/ hydrotalcite	Rapeseed oil	270 °C, 5 MPa, batch reactor, W/F = 1	—	$C_{18}/C_{17} \approx 4$	76
7	3.3Ni–11.9Mo/ SiO_2	Rapeseed oil	270 °C, 5 MPa, batch reactor, W/F = 1	100	$C_{18}/C_{17} \approx 1$	76
8	3.7Ni–9.7Mo/ SBA-15	Rapeseed oil	270 °C 5 MPa, batch reactor, W/F = 1	100	$C_{18}/C_{17} \approx 9$	76
9	3.7Ni–3.7Co–13.8Mo/Al_2O_3	Rapeseed oil	250 °C, 5 MPa, batch reactor, W/F = 1	—	$C_{18}/C_{17} \approx 6$	76
10	3.6Co–3.6Ni–15.5W/Al_2O_3	Rapeseed oil	250 °C, 5 MPa, batch reactor, W/F = 1	—	$C_{18}/C_{17} = 3$	76
11	20Ni–5Cu/ Al_2O_3	Waste cooking oil	375 °C, 4 MPa, 76 h, fixed-bed reactor	100	$C_{18}/C_{17} = 0.8$	83
12	20Ni–5Fe/Al_2O_3	Waste cooking oil	375 °C, 4 MPa, 76 h, fixed-bed reactor	100	$C_{18}/C_{17} = 0.4$	83
13	20Ni–0.5Pt/ Al_2O_3	Waste cooking oil	375 °C, 4 MPa, 76 h, fixed-bed reactor	100	$C_{18}/C_{17} = 0.4$	83

concentrations of active sites for Ni, Mo and NiMo supported on Al_2O_3 the conversion and deoxygenation yield decreased in the order of $NiMo/Al_2O_3 > Mo/Al_2O_3 > Ni/Al_2O_3$ at all studied temperatures (260 °C, 270 °C and 280 °C) and 3.5 MPa.[59] The conversion of rapeseed oil at Ni/Ni + Mo ratios of 0.2, 0.3, and 0.4 did not reveal any stark difference between these 3 catalysts.[59] Most importantly, the sulfided $NiMo/Al_2O_3$ was observed to be approx. 19 times more active than the corresponding sulfided Ni/Al_2O_3, which indicates the importance of synergy between NiS and MoS_2 to increase the deoxygenation (especially HDO) activity.[59]

Furthermore, there was a selectivity difference between the sulfided Mo, NiMo and Ni supported on Al_2O_3 (Table 6.8). Unlike with Mo and NiMo, the formation of alcohols was completely absent over NiS and, therefore. fatty esters (formed by the chemical interaction between fatty acids and alcohols) were not observed.[59] A possible explanation could be the preference for the deCOx route in the latter, while in the former deoxygenation mainly followed the HDO route.[59] This difference was attributed to the electronic properties of Ni and Mo which controlled the rate of hydrogenation, HDO and deCOx.[59] Although, the formation of fatty esters and alcohols were observed on $NiMoS/Al_2O_3$ irrespective of Ni/Ni + Mo, their concentration was insignificant due to the high HDO activity. The difference in electronic properties was further reflected in the higher selectivity of C_{17} hydrocarbon (deCOx route) over Ni-rich catalyst compared to C_{18} (HDO) over NiMo catalyst.[59] The study points out the importance of Ni as a promoter for Mo/Al_2O_3 in enhancing selectivity to hydrocarbons *via* both HDO and deCOx pathways.[59] Also, the $NiMoS/Al_2O_3$ with lower content retains higher C_{18} selectivity, even at high temperature. It was also observed that increasing the reaction temperature increases the decarboxylation route irrespective of the Mo content in the commercial sulfided $NiMo/Al_2O_3$ catalyst.[74]

Table 6.8 Deoxygenation performance of supported metal sulfide catalysts with different feedstocks.

S. no	Catalyst	Feedstock	Conditions	Conversion (%)	Selectivity	Ref.
1	2.6Ni–15.7MoS/Al_2O_3	Rapeseed oil	Approx. 325 °C, MPa, flow reactor	100	$C_{16}/C_{15} \approx 4$	74
2	2.6-Ni8.8–MoS/Al_2O_3	Rapeseed oil	Approx. 325 °C, MPa	100	$C_{16}/C_{15} \approx 4$	74
3	$NiMoS_2/Al_2O_3$	Palm oil	300 °C, 5 MPa, LHSV = 1	≥90	$C_{18}/C_{17} = 4.34$	75
4	0.2Ni–MoS_2	Waste palm oil	300 °C, 40 MPa, 2 h	—	$C_{18}/C_{17} = 1.9$	93
5	0.2Ni–MoS_2	Waste SBO	300 °C, 4 MPa, 3 h	—	$C_{18}/C_{17} = 1.9$	93
6	0.2Ni–MoS_2/Al_2O_3	Waste palm oil	300 °C, 4 MPa, 2 h	—	$C_{18}/C_{17} = 1.6$	93
7	0.2Ni–MoS_2/Al_2O_3	Waste SBO	300 C, 4 MPa, 3 h	—	$C_{18}/C_{17} = 1.7$	93
8	$NiMoC/Al_2O_3$	Soybean oil	400 °C, 4.5 MPa, flow reactor	100	$C_{18}/C_{17} \approx 0.07$	91
9	10Ni–2P/SiO_2	Palm oil	400 °C, 4 MPa, 36 h, trickle-bed reactor	—	$C_{18}/C_{17} \approx 0.08$	88

In another study, the HDO route was dominant over deCOx over $NiMoS_2/Al_2O_3$.[75] The influence of temperature and pressure was evaluated for deoxygenation of palm oil over $NiMoS_2/Al_2O_3$. Unlike deCOx, an increase in temperature from 270 °C to 330 °C boosted the HDO selectivity from 21.9% to 74.9%. On the contrary, the deCOx selectivity increased only marginally from 4.6% to 16.8%.[75] As expected, the cracking tendency of the catalyst was observed to be proportional to the hydrotreating temperature. For example, an increase in temperature from 330 °C to 420 °C decreased the liquid yield (*ca.* 90% to 40%) and enhanced cracking, isomerization and cyclization reactions. Unlike temperature, the HDO performance (67.8% to 78.1%) was little affected by pressure increase from 1.5 MPa to 8 MPa.[75]

Beside the catalyst deactivation from coking and sintering, sulfided catalysts also deactivate because of sulfur leaching. To compensate for such leaching, usually an additional step of sulfidation is required to activate the catalyst. Sulfur leaching is rigorous when water as a byproduct of HDO is produced. Additionally, sulfur leaching into the final product is considered undesirable.[72] In a recent study, continuous feeding of sulfur-containing feed was beneficial compared to reactivation of the $CoMoS/Al_2O_3$ by sending a pulse of sulfiding agent.[84] Additionally, natural impurities in the feedstock such as Ca, Mg and Na have also been shown to have a negative impact on the activity and stability of such catalysts.[84] Similarly, the phospholipids in the oil feedstock decompose to phosphoric acid which could catalyse the unwanted oligomerization reactions, thereby resulting in coking on the catalyst surface.[84] Especially, the balance between the cationic (Ca, Mg, Na) and charge-compensating anionic (P) impurities in the oil feedstock could significantly deteriorate the deoxygenation performance of the catalyst during the course of reaction.[84] For example, waste rapeseed oil contained an imbalance of cationic ($Ca > Na > Mg$) to anionic P impurities compared to neat rapeseed oil, which resulted in an accelerated drop in the deoxygenation activity for the former at 310 °C and 3.5 MPa.[84]

Similarly, cracking and isomerization activity was observed to be enhanced by optimizing the acidity of the support for active NiMoS. A recent study reported a major increase in the isomerization yield from *ca.* 2% to 23% when the $Si/Si + Al$ ratio was increased from 0.002 to 0.99. In the case of cracked products (C_5–C_{14}), a slight increase to 2% was observed on $NiMoS/SiO_2$–Al_2O_3 upon deoxygenation of jatropha oil at 350 °C and 3 MPa H_2.[85] On the other hand, the cracking behaviour significantly increased when NiMoS was supported on an acidic Al_2O_3-ZSM-5 support. A gradual addition of ZSM-5 from 0% to 30% increased the amount of C_5–C_{14} up to 40%.[85] Alternatively, the acidity of the catalyst could also be enhanced by the addition of boron oxide onto the $NiMoS/Al_2O_3$, whereupon the isomerization activity was increased up to 78% (10 wt% isoalkane) when boron oxide loading was *ca.* 30 wt% compared to an unmodified $NiMoS/Al_2O_3$.[86]

Unlike sulfided catalysts, literature available on phosphide-, carbide- and nitride-based catalysts for vegetable oil deoxygenation is scarce. In a recent study, Al_2O_3-supported NiP was shown to exhibit relatively higher HDO performance when compared to NiS and NiC.[87] Especially, deoxygenation by the NiP catalyst followed the HDO pathway as opposed to the deCOx and methanation pathways observed for NiS and NiC catalysts.[87] Similar studies of the hydrotreatment of soybean oil reported deCOx as the major deoxygenation pathway for a Ni_2P/SiO_2 at 370 °C and 3 MPa H_2.[88] On the contrary, methanation was observed to be dominant (22–30 wt%) for MoC_x, MoN_x and MoN_x catalysts at 350 °C and 5.5 MPa upon hydrotreatment of rapeseed oil.[89] The deCOx route was also observed to be dominant for $NiMoC/Al_2O_3$ and Mo_2C/AC during the deoxygenation of soybean oil at approx. 350 °C.[90,91] In a recent study, deoxygenation of soybean oil in the absence of hydrogen with $NbOPO_4$ yielded up to 91 wt% liquid hydrocarbons compared to FCC (88 wt%), H-Beta (50 wt%) and ZSM-5 (70 wt%) at 350 °C and 1 MPa.[92] Interestingly, the aromatic and isoalkane percentage in the liquid phase was significantly higher for $NbOPO_4$ (25 wt%, 19 wt%) compared to FCC (22 wt%, 1 wt%). On the other hand, the highest aromatic production was observed over ZSM-5 with up to 83 wt%.[92] It must be noted that addition of phosphorous increases the overall acidity of the catalyst, thereby increasing the cracking and isomerization products as also observed on $PtPdP/Al_2O_3$.[80]

6.5.5 Other Catalysts, Including Metal-free Catalysts

In addition to the various noble and transition metal-based catalysts discussed above, numerous examples are available in the literature of deoxygenation of triglycerides and triglyceride-derived fatty acids over metal-free catalysts such as activated carbon or nitrogen-doped carbon.[5,36–41] Over such catalysts, deCOx has been identified as the dominant pathway for deoxygenation, with negligible to no formation of HDO products, and is often referred to as hydrothermal deoxygenation. Such reactions typically require the use of sub- or supercritical water as the solvent and have emerged as an alternative, organosolvent-free, green method for producing diesel-range synthetic hydrocarbons. It has been postulated that under near-critical/supercritical conditions water becomes a highly reactive reaction media due to its reduced dielectric constant and the increase in interphase mass-transfer effects, thus leading to improved solubility of non-polar reactants like fats and oils.[5,39–41] Although hydrothermal deoxygenation was originally demonstrated for deoxygenation of saturated fatty acids like palmitic acid, under supercritical conditions (370–400 °C), with large catalyst-to-substrate (w/w) ratios (≥ 1),[39] hydrothermal deoxygenation has been successfully demonstrated with both unsaturated fatty acids as well as triglycerides using carbon-based catalysts under milder sub-critical conditions (350–370 °C, 3 MPa). In fact, introduction of heteroatoms and the use of H_2 donors such as HCOOH, MeOH and glycerol are shown to be beneficial for improving catalyst stability, reducing aromatic- and methane-forming side reactions. Using HCOOH as the H_2 donor and high surface-area N-doped activated carbon as catalyst, natural triglycerides have been deoxygenated into diesel-range hydrocarbons in a batch reactor as well as a fixed-bed reactor, and the N-doped carbon catalysts showed two-fold longer lifetime than activated carbon-based catalysts.[41]

6.6 Other Technical Aspects: Solvent and Reactor Considerations

As outlined in the sections above, deoxygenation (HDO and deCOx) of triglycerides and triglyceride-derived fatty acids has been extensively studied in the literature. In addition to feedstock and choice of catalyst, other technical parameters such as choice of solvent or reactor type are also of considerable significance as they influence catalyst lifetime and economics. At the commercial scale, sulfided metals (NiMo/Al_2O_3, NiMo/zeolite) have been the choice of catalyst due to their relative lower cost and proven stability/durability under industrial settings. Despite this, almost all deoxygenation catalysts require the use of low-boiling organic solvent like dodecane, mesitylene or hexane for improved fluidity and mass transfer, to inhibit of catalyst deactivation due to heavy organic deposition and coking, and also for higher solubility of H_2 in the liquid phase.

Although triglyceride deoxygenation has been demonstrated under solvent-free conditions, it usually results in a drastic loss of deoxygenation (HDO and deCOx) activity.[94] Solvents also influence H_2-donating ability, as illustrated in a comparison between dodecane and tetralin, with the latter affording improved HDO performance at lower H_2 pressure, which was attributed to the improved hydrogen-donating ability of the latter.[95] In another study, the influence of solvent (hexane, dodecane, isopropanol, water, and solvent-free) on deCOx of the model triglyceride-derived compound stearic acid was investigated for a Ru/C catalyst, which revealed hexane as the most suitable solvent followed by dodecane and solvent-free conditions, respectively. The improved performance was attributed to supercritical hexane which decreased the mass-transfer resistances and favoured the rapid desorption of hydrocarbon products from the catalyst surface.[36] Similar results were also obtained for deoxygenation of sunflower oil over Pd-based catalysts.[96] It is also worth mentioning that the use of non-polar hydrocarbons as solvents also supressed HDO, methanation and cracking reactions leading to improved mass balance.[36] Recently, deCOx of triglyceride-derived C_{18} fatty acids such as oleic acid has also been demonstrated using supercritical CO_2 as the solvent with commercial $MoO_3/\gamma\text{-}Al_2O_3$; however, the catalyst was found to show poor stability in supercritical CO_2.[97] In another work, triglyceride deoxygenation was also investigated using waste polypropylene-derived hydrogenated liquid as a solvent over Ni–Cu/ZrO_2, which afforded alkane yields up to 77.4 wt%, confirming the suitability of waste-derived products as low-cost solvents.[98] H_2O (sub- as well as supercritical) is another emerging solvent for deoxygenation of triglycerides and triglyceride-derived fatty acids and it can be particularly useful for upgrading waste and low-grade oils, which contain considerable amounts of H_2O. However, the biggest challenge with the use of H_2O is the poor stability of most known deoxygenation catalysts under hydrothermal conditions: although using carbon-based catalysts such as Ru/C, Ru/N-doped carbon and metal-free catalysts (*e.g.* activated carbon, N-doped carbon) improved stability in batch as well as fixed-bed reactors, it was impossible to entirely eliminate Ru leaching and/or degradation of carbon material under the harsh hydrothermal conditions.[5,39–41] In fact, deCOx of the model compound butyric acid over Pt/C in hot compressed water at 350 °C over the course of 24 h in a fixed-bed reactor confirmed the contribution of poisoning, coking, and structural degradation to deactivation of the carbon-based catalyst during hydrothermal decarboxylation.[99] While for deCOx reactions performed in hydrocarbon solvents, poisoning and organic deposition were most likely the main contributing factors towards the deactivation of carbon-supported Pd catalysts, the use of an H_2 environment considerably delayed the onset of catalyst deactivation.[56] In addition to deactivation, other drawbacks of carbon catalysts include challenges in scale-up such as mechanical stability, large pressure drop and difficulties in regeneration and fabrication of shaped carbon catalysts.

Unlike the noble metal-based catalysts predominately used in batch deoxygenation of triglyceride feedstocks, sulfided metal catalysts, and in particular NiMo-based catalysts, have been extensively employed for deoxygenation of triglycerides in fixed/flow reactors.[31,33,42] Usually such catalysts present exceptional stability when operated in a fixed-bed reactor and when co-feeding a sulfiding agent such as dimethyl disulfide (DMDS) in low concentrations with the triglyceride feed.[42] Co-feeding of sulfiding agents (either continuously or in pulsed doses) is necessary to compensate for sulfide leaching from the catalyst, and without such treatments these catalysts often undergo deactivation, requiring intermittent regeneration or reactivation. However, it is to be emphasized that when operated under a kinetic regime and/or with low H_2 partial pressure even sulfided catalysts undergo rapid deactivation similar to noble metal catalysts, due to deposition of heavy organic molecules and coke formation.[100,101]

6.7 Large-Scale Process Examples for Renewable-diesel Production

All the known commercial large-scale processes that are presented here are, in general, very similar to each other. All the processes can utilize a wide variety of feedstocks, including waste fats, oils and greases, vegetable oils, animal fats, used cooking oil and tall oil. The basic hydrotreating process is based on old patents.[102,103] In principle the processes consist of five main steps with small variations (Figure 6.13):[104–109]

Pretreatment: The feedstock is pretreated to remove impurities such as water, metals, and, if applicable, free fatty acids that could interfere with the subsequent reactions.

Hydrotreatment: hydrotreatment can consist of one or several steps during which the feedstock is treated with hydrogen at elevated temperature

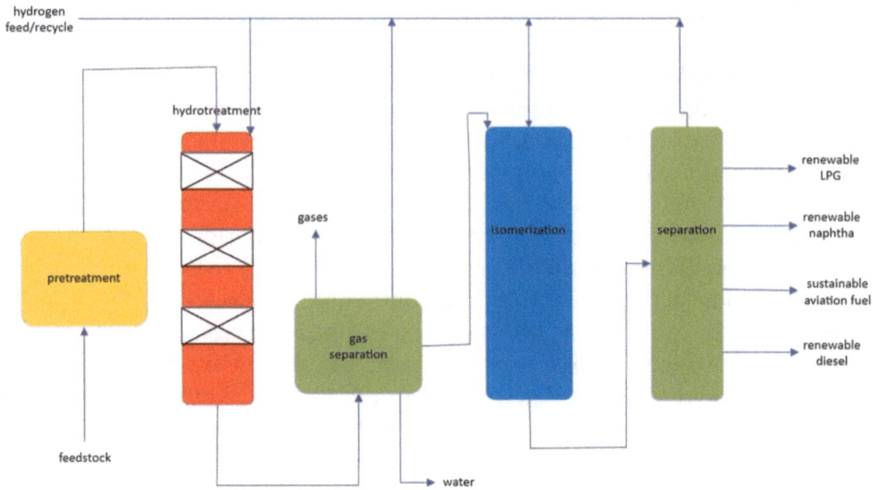

Figure 6.13 Simplified process scheme for production of renewable fuels from bio-based feedstocks.

and pressure over a proprietary catalyst to remove oxygen and possibly sulfur and nitrogen to achieve pure hydrocarbons by the methods described here and in more detail in Sections 6.4 and 6.5.

Gas separation: formed gaseous products and wastes such as propane, carbon dioxide, carbon monoxide, water, H_2S, and NH_3 are separated from hydrocarbons.

Isomerization: the intermediate hydrocarbons produced in the hydrotreatment step are fed to an isomerization reactor where they are converted into branched paraffins.

Separation: the final product is separated into different fractions depending on the process, such as renewable diesel, sustainable aviation fuel, renewable naphtha, and renewable LPG, through distillation.

After distillation the products can be used as such or blended with additives and/or fossil fuels to meet specific fuel quality specifications.

Even though this seems to be a straightforward process, there are many challenges to make it work in real life. Variations and impurities in feedstock can cause the formation of unwanted side products as well as coking and deactivation of the catalyst. There is also lot to improve in the economics of the processes, such as optimising co-processing of bio-based feedstock with fossil feedstock and combining the isomerization step and hydrotreatment process in same reactor with a series of different catalyst beds.[52]

6.7.1 Bio-Synfining® Process by Chevron Renewable Energy Group Inc.

Bio-Synfining is a proprietary thermochemical conversion technology originally developed by Syntroleum Co. through a joint venture with Tyson

Foods to produce renewable diesel fuel from a variety of bio-feedstocks. The process is designed to be flexible and can be used to produce a variety of renewable diesel fuels with different properties, such as renewable hydrocarbon diesel (RHD), renewable naphtha, and renewable liquefied petroleum gas (LPG).[104,110–113] The hydrotreating process consists of two-steps, where pretreated feedstock is hydrotreated in series over two catalyst beds. In the first step, a molybdenum catalyst with low hydrogenation activity is utilized. In the second step, typically a bimetallic catalyst, such as NiMo, CoMo or NiW, can be used for hydrotreatment. The process is carried out at a temperature of 290–379 °C and under hydrogen pressure of 8–11 MPa.[113] Hydrocarbons achieved from the hydrotreating process are isomerized to branched hydrocarbons over various catalysts, including Pt, Pd or Ni supported on many possible supports at 305–399 °C and 7–14 MPa with LHSV 0.2–4 h^{-1}. In the final step, the produced hydrocarbon mixture is distilled to three main products: renewable diesel, renewable naphtha and renewable LPG.[104,111,112]

Renewable Energy Group (REG) acquired Syntroleum and all its facilities, including the bio-refinery at Geismar, Louisiana, USA, that has produced renewable fuels since 2014.[110] Today the unit has a production capacity of approx. 300 000 tons per year. In 2022, Chevron acquired REG and its product portfolio and has plans to increase the production of renewable fuels.[114]

6.7.2 BP's Renewable-diesel Process

British Petroleum (BP) has developed its own renewable-diesel process since the beginning of 2000. The process is based on several patents describing hydroprocessing of carboxylic acids, such as fatty acids, in two steps to diesel fuels.[105,115] In a simplified description, in the first step of the process, a stream of hydrogen and carboxylic acid is reacted over CoMo or NiMo supported on γ-Al_2O_3, ZrO_2 or TiO_2 at temperatures between 250–430 °C with 2–4 MPa pressure. In the next step, this partially deoxygenated hydrocarbon stream is treated over sulfided CoMo/γ-Al_2O_3 at 300–400 °C under pressure of 3–7 MPa. Depending on the raw material, the stream is then hydrocracked and separated to desired products; formed H_2S is also separated.[105,115] BP has also developed a process to treat fossil feedstock and bio-feedstock at the same time.[116] The benefit of this process is that the same catalyst can be applied in hydrotreatment of bio-feedstock and desulfurization of fossil feedstock. BP started renewable-diesel production at the Bulver Island refinery in Brisbane, Australia, in 2007, but that was terminated in 2015.[117,118] Today, BP is producing renewable diesel in Cherry Point refinery in Blaine, Washington, USA, with a capacity of approx. 350 000 tons per year. Even though BP has developed its own technology, in 2023 they announced that in the future they will use Honeywell's Ecofining™ technology to increase sustainable production of aviation fuel and renewable diesel worldwide.[119] Ecofining™ technology will be installed at the following BP sites: Cherry

Point refinery in Blaine, Washington, USA; Rotterdam II refinery in Rotterdam, Netherlands; Lingen refinery in Lower Saxony, Germany; Castellón de la Plana refinery in Castellón, Spain; and Kwinana Oil refinery in Kwinana, Australia.

6.7.3 Cetan Energy

Cetan Energy LLC in Carlsbad, New Mexico, USA, has developed a hydrodeoxygenation process which can produce renewable diesel from various bio-feedstocks. They have a pilot plant in Carlsbad where they can produce renewable diesel. In the process, pretreated and filtered bio-feedstock is reacted with hydrogen over sulfided Mo/Al_2O_3 catalyst at 270 °C under 4.5 MPa. The active catalyst contains sulfur bonded to molybdenum, which is consumed during the reaction. In order to maintain the catalyst activity, di-*tert*-butyl polysulfide is continuously fed to the reaction mixture. This also causes formation of H_2S, which can be removed from the mixture in the separation phase. The separation proceeds at 50 °C under 4 MPa pressure. During the separation phase, other formed gases, such as methane, ethane and propane, as well as water, are separated from the hydrocarbon mixture. The hydrocarbon mixture is then distilled to renewable diesel and naphtha fractions. The fractions can be used as such or blended with fossil products or processed further.[106,120]

6.7.4 Ecofining™ by ENI/Honeywell-UOP

The Ecofining™ process to produce green diesel from a range of renewable feedstocks derived from oil-seed crops (vegetable oils) or waste animal fats and greases, as well as second generation feedstocks such as algal oils, was developed as a collaboration between ENI SpA of Italy and Honeywell-UOP.[121] In general, the Ecofining™ process is a two-stage process. In the first stage, after possible cleaning and pre-treatment of the feedstock, the feedstock is hydroprocessed to C_{16}–C_{18} linear paraffins. Hydrodeoxygenation or decarboxylation is usually performed in a multistage adiabatic reactor at approx. 310 °C, over sufided NiMo or CoMo supported on, for example, Al_2O_3 under 3–5 MPa pressure. Since the sulfur is consumed during the reaction, a small amount of sulfidizing agent dimethyldisulfide is also fed to the reactor. Optionally a co-feed of fossil-based gas oil containing sulfur can be utilized. The reactor is preferably operated with LSHV of 0.5–1 h^{-1}. Before the second stage, formed gaseous products propane, CO_2 and water, as well as unreacted hydrogen, are separated from the paraffins. During the second stage, linear paraffins are isomerized with hydrogen to branched hydrocarbons which have better properties to be utilized as liquid fuels. Isomerization is conducted over a catalyst of either Pd, Pt, or Ni by themselves or of Ni or Co with Mo or W, supported on acidic silico–alumina. Hydroisomerization is preferably performed at 280–380 °C under 3–5 MPa pressure with LHSV of 0.5–2 h^{-1}. Product distribution of different hydrocarbons can be affected by choosing the desired temperature, pressure and catalyst combination. Finally, the product mixture is separated by distillation to renewable diesel, naphtha and sustainable aviation fuel.[107,122]

The first production unit to use the Ecofining™ process was started in 2013 in Des Plaines, Illinois, USA, by Valero Energy Corp. and Darling Ingredients Inc. Today the unit has a capacity of approx. 900 000 tons per year.[123] Eni started renewable-diesel production in 2014 in Venice, Italy, with a second unit in 2019 in Gela, Italy. The combined production capacity of these units is 1.1 Mtons per year with plans to increase the capacity to 2 Mtons per year by 2025.[124] Ecofining™ technology has also been taken into use by other companies producing renewable diesel and sustainable aviation fuel. One of the companies is AltAir Fuels in Paramount, California, USA, having production capacity of 150 000 tons per year.[125] Another company that has used this technology since 2015 is Petrixo Oil & Gas in Fujairah, United Arab Emirates. This unit has a production capacity of 1 Mtons per year.[126] In the second half of the 2020s, Total in France,[127] ECB Group in Paraguay,[128] and BP in USA, Netherlands, Germany, Spain and Australia[119] will also use the Ecofining™ process in their refineries to produce sustainable fuels.

6.7.5 Hydroflex™ Hydrotreating Process by Topsoe

The HydroFlex™ process, developed by Danish company Topsoe,[129] was originally designed to treat a mixture containing 1–35% oil from biological origin such as vegetable oil, animal fats and tall oil and their derivatives such as fatty acid methyl esters. The rest of the flow comes from fossil origin. The process happens in two steps. During the first step, the mixture flow is streamed with hydrogen over a catalyst containing 0.1–20 wt% Mo supported on either alumina, silica, titania or combinations of them at 50–350 °C under pressure of 0.1–20 MPa with LHSV of 0.1–10 h^{-1}. The conversion of bio-based oil to hydrocarbons during this step is 50–70%. In next step, the flow is led over a catalyst bed containing a conventional hydrotreating catalyst such as CoMo, NiMo or NiW supported on, for example, alumina. Since the mixture also contains fossil-based oil, this step will remove sulfur and nitrogen as well as remining oxygen from the compounds. This step is conducted at 200–500 °C under up to 20 MPa. The product from this phase is pure straight-chain hydrocarbons. These two steps can be carried out in the same reactor so that the first step takes place in the top 20 vol% of the catalyst bed and second step in the bottom 80 vol% of the catalyst bed. The third part of the whole process is a hydroisomerization of the straight paraffins to branched hydrocarbons over NiW supported on, for example, alumina at 200–500 °C under up to 20 MPa. Hydroisomerization can be carried out in the same reactor as the first two steps or in a separate unit. Finally, the liquid products are separated from gaseous products.[130]

UPM in Finland has licensed the HydroFlex™ process and developed it further to produce renewable diesel from crude tall oil.[131] UPM opened its biorefinery in Lappeenranta in 2015, which produces approx. 100 000 tons per year of renewable diesel.[132] Preem in Gothenburg, Sweden, has decided to start a unit in 2024 which utilises the HydroFlex™ technology to produce 800 000 tons per year of renewable diesel and sustainable aviation fuel from tall oil, tallow and other renewable feedstocks.[133] Tesoro oil refinery in Dickinson, North Dakota, USA, has planned to design a co-process unit to produce a renewable-diesel blend by the HydroFlex™ technology. The plan was to co-process up to 20 000 tons per year of renewable feedstock.[134] SkyNRG has planned to start the production of renewable diesel and sustainable aviation fuel (SAF) in Delfzijl, Netherlands, based on the HydroFlex™ technology. As a strategic partner, KLM Royal Dutch Airlines has committed to buy 75 000 tons of SAF per year for a ten-year period.[135]

6.7.6 NEXBTL™ Technology by Neste

NEXBTL™ stands for Next Generation Biomass to Liquid and is a proprietary hydrotreating technology developed by Neste, a Finnish oil refining and marketing company. It is a versatile process that can produce various high-quality renewable fuels, including renewable diesel, renewable naphtha, sustainable aviation fuel and renewable liquefied petroleum gas (LPG), from a wide range of feedstocks, including vegetable oils, animal fats, waste fats, oils, and greases, as well as tall oil and fatty acid alkyl esters.[136,137]

After pretreatment of the feedstock, pre-hydrogenation under mild conditions can be performed to hydrogenate carbon–carbon double bonds and avoid side reactions. It is performed over Pd, Pt, Ni, NiMo or CoMo supported on alumina or silica at 150–250 °C under 1–10 MPa of hydrogen pressure. During the next step, hydrotreatment to remove oxygen, sulfur and nitrogen is conducted. In optimal process, the feedstock is diluted 12–25 : 1 with paraffinic hydrocarbons either from fossil sources or recycled from the product stream. This dilution is done to prevent unwanted side reactions and catalyst coking. Hydrotreating is typically conducted over NiMo or CoMo supported on alumina. However, Pd, Pt, Ni, NiMo or CoMo supported on alumina and/or silica can be utilized. The temperature can be set to 200–400 °C and the pressure to 2–15 MPa. In order to boost the decarbonisation reaction before hydrodeoxygenation, a small amount of sulfur, either inorganic (H_2S or elemental sulfur) or organic (dimethyl disulfide or carbon disulfide), is added to the feedstock stream. The hydrotreatment is performed in a series of catalyst beds, which can be located in the same reactor or different reactors. Dilution agent is introduced on top of the first catalyst bed and fresh feed, recycled liquid and hydrogen on top of each catalyst bed. This is done in order to control the temperature, minimize side reactions and increase catalyst lifetime. After the hydrotreatment, light hydrocarbons and formed gases such as propane, water, CO_2, CO, H_2S and NH_3 are removed from the product. Straight-chain paraffins are then isomerized to branched hydrocarbons. The isomerization is conducted over a catalyst containing Pt, Pd or Ni supported on a mixture of SAPO-11, SAPO-41, ZSM-22, ZSM-23 or ferrierite with Al_2O_3

or SiO_2. The temperature during isomerization is set to 200–500 °C and pressure to 2–15 MPa. A small amount of hydrogen is fed to the reactor during isomerization to reduce catalyst deactivation. The main product, renewable diesel (or jet fuel), is separated from small amounts gasoline and gaseous hydrocarbons.[39,108,138–141]

Neste is the largest producer of renewable diesel in the world, applying its NEXBTL™ technology in three countries. Production started in Porvoo, Finland in 2007. Today, the production capacity is approx. 1 Mton per year. In 2010, Neste started the production of renewable fuels in Singapore. In 2023 the production capacity was doubled to approx. 2.6 Mton per year. A biorefinery in Rotterdam, Netherlands, started in 2011 and has a production capacity of 1.4 Mton per year with plans to increase the capacity to 2.7 Mton per year during 2026.[142]

6.7.7 Vegan® Technology by Axens and IFP

Vegan® Technology is a proprietary HVO solution developed by Axens, a French engineering and technology company, as a collaboration with IFP. With this technology it is possible to produce high-quality renewable diesel, sustainable aviation fuel, renewable naphtha and renewable propane from various bio-feedstocks such as vegetable oils, animal fats and waste-based oils. The processes can be in a stand-alone unit or fitted to existing refineries.[143,144]

The Vegan® process starts with pretreatment of the feedstock to remove all impurities. The next step is optional mild pre-hydrogenation. It is preferably carried out at a temperature of 150–200 °C over a catalyst containing Pd, Pt, Ni, NiMo or CoMo supported on alumina and/or silica. The next step is hydrodeoxygenation, which is conducted preferably at a temperature of 220–310 °C under pressure of 1–4 MPa with LHSV of 0.1–10 h^{-1}. For this step, the feedstock is mixed with hydrogen so that hydrogen/feedstock ratio is 70–1000 N m^3 of hydrogen per m^3 of feedstock. The reaction takes place over a catalyst containing Pd, Pt or Ni with NiMo, CoMo or NiW supported on, for example, silica, alumina or a mixture of them. The catalyst may contain only one metal or a mixture. The catalyst may also contain Si with P, B, and/or F. It is also possible to use a mixture of these catalysts in a single bed or series of catalyst beds. One option is to add dimethyl sulfide to the hydrogen flow, which converts to H_2S and keeps the catalyst in the sulfided form. After dehydrogenation, the formed gaseous products, such as CO_2, CO, H_2S and water, are removed from the product stream. The next step in the process is hydroisomerization where straight-chain paraffins are converted to branched hydrocarbons with better fuel properties. Hydroisomerization is performed over a catalyst containing Pd, Pt or Ni supported on amorphous silica–alumina at a temperature preferably in the range 320–420 °C, under 1–9 MPa pressure with LHSV 0.5–5 h^{-1} and hydrogen content of 150–1500 N m^3 per N m^3 of feedstock. Throughout the process, the unspent hydrogen is cleaned and recycled.[109]

So far, the Vegan® process has been applied in one production unit in the La Mède biorefinery in France by TotalEnergies. The unit started in 2019 and it has a production capacity for renewable diesel on the scale of 500 000 ton per year.[145] Sandong Haike Chemical Co. Ltd in China has announced that it will start a unit based on Vegan® technology to produce sustainable aviation fuel in 2024.[146] In 2021, Aemetis started to plan a production unit for renewable diesel and sustainable aviation fuel based on Vegan® technology in Riverbank, California, USA.[147]

6.8 Conclusions, Prospective and Outlook

Among the various upgrading technologies discussed, deoxygenative transformation of triglycerides is the most effective process for direct production of drop-in hydrocarbon fuels (green diesel and green jet fuel). The deoxygenation of triglycerides into such drop-in biofuels involves various complex chemical processes, with HDO and deCOx steps playing a pivotal role in removing oxygen from the feedstocks and improving the fuel quality (energy density, viscosity and associated fuel properties) of hydrocarbon products. In deoxygenation processes, next to feedstock, the choice of catalyst plays the most important role as it determines the overall reaction pathway (HDO, deCOx, cracking, isomerization, methanation), which has a direct impact on the product yield, composition and fuel properties as well as process economics. In catalysts, the choice of support is most crucial as it is responsible for determining the dispersion, distribution and stability of the catalyst active phase. Similarly, the flexibility of using different triglyceride feedstocks is also important, as the degree of unsaturated bonds have a marked impact on both catalyst performance and product quality. Nevertheless, from the perspective of H_2 economics and carbon footprint, the deCOx route is more preferred than HDO. Methanation is the major side reaction that is detrimental from the viewpoint of carbon as well as H_2 loss. In practice, however, the actual process is a combination of HDO and deCOx. Among the various catalysts reported, *i.e.* noble metals, transition metals, sulfides, phosphides, carbides, nitrides and metal-free catalysts, sulfided catalysts are preferred industrially due to their high activity and long-term stability (usually in the order of years). As discussed above, for industrial sulfided NiMo catalysts, HDO pathways is usually the more dominant deoxygenation pathway, which is subsequently followed by upgrading, cracking and hydroisomerization steps, either *in situ* or in a separate unit, to produce fuel-grade synthetic hydrocarbons. For sulfided NiMo catalysts, the ratio of Ni/Ni + Mo and support acidity are crucial since they directly influence the selectivity towards deoxygenated (HDO *vs.* deCOx) as well as cracking/isomerized products.

Thus, significant scope still exists for further development of improved deoxygenation technologies, and catalytic processes aimed at minimizing H_2 consumption (supressed HDO and methanation), minimizing/eliminating use of organic solvents, and/or working without the use of expensive noble metal catalysts are of particular importance. Within this context, the deoxygenation of triglycerides and triglyceride-derived products under hydrothermal conditions holds considerable promise, as such processes have been successfully demonstrated at the laboratory scale with inexpensive metal-free catalysts in a H_2-free environment under conditions which may be deemed industrially feasible. Even so, further studies are needed to gain insights into the nature of the active sites present and to improve the long-term stability of such catalysts, as well as validating their performance in pilot-scale studies.

Acknowledgements

This work is part of the activities of Technical Chemistry, Department of Chemistry, Chemical-Biological Centre, Umeå University, Sweden. The Swedish Bio4Energy program and KEMPE Foundations are gratefully acknowledged. This work is also a part of the activities of the Johan Gadolin Process Chemistry Centre at Åbo Akademi University, Finland. The authors LJK and MM also gratefully acknowledge financial support from SERB (SRG/2020/000325) and MLP0028 (CSIR, New Delhi). The article is assigned PRIS registration number 201/2024.

References

1. M. J. B. Kabeyi and O. Olanrewaju, *Front. Energy Res.*, 2022, **9**(1032), 2296.
2. A. K. Akella, R. Saini and M. Sharma, *Renewable Energy*, 2009, **34**(2), 390.
3. O. G. Austvik, *Energy Policy*, 2016, **96**, 372.
4. S. Zhao and K. Wang, *Energy Fuels*, 2023, **37**(23), 18899.
5. L. J. Konwar and J.-P. Mikkola, *Appl. Catal., A*, 2022, **638**, 118611.
6. M. Mujtaba, L. F. Fraceto, M. Fazeli, S. Mukherjee, S. M. Savassa, G. A. Medeiros, A. E. S. Pereira, S. D. Mancini, J. Lippon and F. Vilaplana, *J. Cleaner Prod.*, 2023, **402**, 136815.
7. L. J. Konwar, J.-P. Mikkola, N. Bordoloi, R. Saikia, R. S. Chutia and R. Kataki, Sidestreams From Bioenergy and Biorefinery Complexes as a Resource for Circular Bioeconomy, *Waste Biorefinery*, 2018, ch. 3, p. 85.
8. L. J. Konwar, J. Boro and D. Deka, *Renewable Sustainable Energy Rev.*, 2014, **29**, 546.
9. P. Bora, L. J. Konwar, J. Boro, M. M. Phukan, D. Deka and B. K. Konwar, *Appl. Energy*, 2014, **135**(15), 450.
10. C. Attaphong and D. A. Sabatini, *Energy Fuels*, 2013, **2711**, 6773.
11. C. Attaphong, L. Do and D. A. Sabatini, *Fuel*, 2012, **94**, 606.
12. P. Mäki-Arvela, I. Kubičkova, M. Snåre, K. Eränen and D. Y. Murzin, *Energy Fuels*, 2007, **21**(1), 30.
13. M. Snåre, I. Kubičkova, P. Mäki-Arvela, K. Eränen and D. Y. Murzin, *Ind. Eng. Chem. Res.*, 2006, **45**, 5708.

14. B. K. Barnwal and M. P. Sharma, *Renewable Sustainable Energy Rev.*, 2005, **9**(4), 363.
15. G. Knothe, R. O. Dunn and M. O. Bagby, Biodiesel: The Use of Vegetable Oils and Their Derivatives as Alternative Diesel Fuels, in *Fuels and Chemicals from Biomass*, ACS Symposium Series, 1997, ch. 10, vol. 66, p. 172
16. C.-H. Tan, H. M. Ghazali, A. Kuntom, C.-P. Tan and A. A. Ariffin, *Food Chem.*, 2009, **113**(2), 645.
17. J. J. Lafont, A. A. Espitia and J. R. Sodré, *Mater. Renewable Sustainable Energy*, 2015, **4**(1), 1–7.
18. P. Mazumdar, V. B. Borugadda, V. V. Goud and L. Sahoo, *Biomass Bioenergy*, 2012, **46**, 546.
19. T. Laza and Á. Bereczky, *Fuel*, 2011, **90**(2), 803.
20. J. Čedík, M. Pexa, M. Holúbek, Z. Aleš, R. Pražan and P. Kuchar, *Energies*, 2020, **13**, 3796.
21. C.-Y. Lin and X.-E. Wu, *Processes*, 2022, **10**, 1502.
22. D. Alviso, C. Zárate and T. Duriez, *Fuel*, 2021, **284**, 119026.
23. M. Aslam, L. J. Konwar, A. K. Sarma and N. C. Kothiyal, *J. Anal. Appl. Pyrolysis*, 2015, **115**, 401.
24. D. Konwer, S. E. Taylor, B. E. Gordon, J. W. Otvos and M. Calvin, *J. Am. Oil Chem. Soc.*, 1989, **66**, 223.
25. A. K. Sarma and D. Konwer, *Energy Fuels*, 2005, **19**(4), 1755.
26. Z. Franco and Q. D. Nguyen, *Fuel*, 2011, **90**(2), 838.
27. S. C. Mat, M. Y. Idroas, M. F. Hamid and Z. A. Zainal, *Renewable Sustainable Energy Rev.*, 2018, **82**(1), 808.
28. H. Kumar, A. K. Sarma and P. Kumar, *Renewable Sustainable Energy Rev.*, 2020, **117**, 109498.
29. H. Kumar, L. J. Konwar, M. Aslam and A. K. Sarma, *RSC Adv.*, 2016, **6**, 37646.
30. R. Sotelo-Boyas, F. Trejo-Zarraga and F. de Jesus Hernandez-Loyo, *Hydrogenation*, 2021, **338**, 31.
31. D. Verma, R. Kumar, B. S. Rana and A. K. Sinha, *Energy Environ. Sci.*, 2011, **4**, 1667.
32. S. Saka and D. Kusdiana, *Fuel*, 2001, **80**, 225.
33. P. Kumar, D. Verma, M. G. Sibi, P. Butolia and S. K. Maity, Hydrocarobn biorefinery, Elsevier, Chapter 4 – Hydrodeoxygenation of triglycerides for the production of green diesel: Role of heterogeneous catalysis, in *Hydrocarbon Biorefinery Sustainable Processing of Biomass for Hydrocarbon Biofuels*, 2022, pp. 97–126.
34. F. Long, W. Liu, X. Jiang, Q. Zhai, X. Cao, J. Jiang and J. Xu, *Renewable Sustainable Energy Rev.*, 2021, **148**, 111269.
35. M. W. Schreiber, D. Rodriguez-Niño, O. Y. Gutiérrez and J. A. Lercher, *Catal. Sci. Technol.*, 2016, **6**, 7976.
36. G. N. da Rocha Filho, D. Brodzki and G. Djéga-Mariadassou, *Fuel*, 1993, **72**(4), 543.
37. E. Vonghia, D. G. Boocock, S. K. Konar and A. Leung, *Energy Fuels*, 1995, **9**(6), 1090.

38. I. Kubičková, M. Snåre, K. Eränen, P. Mäki-Arvela and D. Y. Murzin, *Catal. Today*, 2005, **106**(1–4), 197.
39. D. Y. Murzin, I. Kubičkova, M. Snåre, P. Mäki-Arvela and J. Myllyoja, *US Pat.*, US7491858B2, 2009.
40. S. A. Scaldaferri and V. M. D. Pasa, *Fuel*, 2019, **245**, 458.
41. B. Donnis, R. G. Egeberg, P. Blom and K. G. Knudsen, *Top. Catal.*, 2009, **52**, 229.
42. D. Kubička and V. Tukač, *Adv. Chem. Eng.*, 2013, **42**, 141.
43. J. Liu, C. Liu, G. Zhou, S. Shen and L. Rong, *Green Chem.*, 2012, **14**(9), 2499.
44. E. Kordouli, B. Pawelec, K. Bourikas, C. Kordulis, J. L. G. Fierro and A. Lycourghiotis, *Appl. Catal., B*, 2018, **229**, 139.
45. A. Ishihara, N. Fukui, H. Nasu and T. Hashimoto, *Fuel*, 2014, **134**, 611.
46. A. Srifa, R. Kaewmeesri, C. Fang, V. Itthibenchapong and K. Faungnawakij, *Chem. Eng. J.*, 2018, **345**, 107.
47. S. Rakmae, N. Osakoo, M. Pimsuta, K. Deekamwong, C. Keawkumay, T. Butburee, K. Faungnawakij, C. Geantet, S. Prayoonpokarach, J. Wittayakun and P. Khemthong, *Fuel Process. Technol.*, 2020, **198**, 106236.
48. A. A. Taromi and S. Kaliaguine, *Fuel Process. Technol.*, 2018, **171**, 200.
49. B. Peng, Y. Yuan, Z. Chen and J. A. Lercher, *Angew. Chem., Int. Ed.*, 2012, **51**(9), 2072.
50. H. Wang, S. Yan, S. O. Salley and K. S. Ng, *Fuel*, 2013, **111**, 81.
51. C. Wang, Z. Tian, L. Wang, R. Xu, Q. Liu, W. Qu, H. Ma and B. Wang, *ChemSusChem*, 2012, **5**(10), 1974.
52. T. V. Choudhary and C. B. Phillips, *Appl. Catal., A*, 2011, **397**(1–2), 1.
53. B. Veriansyah, J. Y. Han, S. K. Kim, S. A. Hong, Y. J. Kim, J. S. Lim, Y. W. Shu, S. G. Oh and J. Kim, *Fuel*, 2012, **94**, 578.
54. M. Krár, S. Kovács, D. Kalló and J. Hancsók, *Bioresour. Technol.*, 2010, **101**(23), 9287.
55. I. Hachemi, N. Kumar, P. Mäki-Arvela, J. Roine, M. Peurla, J. Hemming, J. Salonen and D. Y. Murzin, *J. Catal.*, 2017, **347**, 205.
56. A. T. Madsen, B. Rozmysłowicz, P. Mäki-Arvela, I. L. Simakova, K. Eränen, D. Y. Murzin and R. Fehrmann, *Top. Catal.*, 2013, **56**, 714.
57. M. Rabaev, M. V. Landau, R. Vidruk-Nehemya, V. Koukouliev, R. Zarchin and M. Herskowitz, *Fuel*, 2015, **161**, 287.
58. J. Liu, K. Fan, W. Tian, C. Liu and L. Rong, *Int. J. Hydrogen Energy*, 2012, **37**(23), 17731.
59. D. Kubička and L. Kaluža, *Appl. Catal., A*, 2010, **372**(2), 199.
60. Q. Smejkal, L. Smejkalová and D. Kubička, *Chem. Eng. J.*, 2009, **146**(1), 155.
61. K. G. Joback, Unified Approach to Physical Property Estimation Using Multivariate Statistical Techniques, MS thesis, MIT, Cambridge, MA, 1984.
62. Z. Zhang, F. Okejiri, Y. Li, J. Li and J. Fu, *New J. Chem.*, 2020, **44**, 7642.
63. J. Fu, X. Lu and P. E. Savage, *Energy Environ. Sci.*, 2010, **3**(3), 311.

64. J. Fu, X. Lu and P. E. Savage, *ChemSusChem*, 2011, **4**(4), 481.
65. J. Fu, F. Shi, L. T. Thompson, X. Lu and P. E. Savage, *ACS Catal.*, 2011, **1**(3), 227.
66. S. Popov and S. Kumar, *Energy Fuels*, 2015, **29**(5), 3377.
67. J.-P. Mikkola, K. Vanklint, W. Siljebo, L. J. Konwar and A. Samikannu, *EP Pat.*, EP3841186A1, 2020.
68. D. Verma, B. S. Rana, R. Kumar, M. G. Sibi and A. K. Sinha, *Appl. Catal., A*, 2015, **490**, 108.
69. M. Ameen, M. T. Azizan, A. Ramli, S. Yusup and B. Abdullah, *Catal. Today*, 2020, **355**, 51.
70. Z. Rahmawati, L. Santoso, A. McCue, N. L. Azua Jamari, S. Y. Ninglasari, T. Gunawan and H. Fansuri, *RSC Adv.*, 2023, **13**, 13698.
71. M. Oh, M. Jin, K. Lee, J. C. Kim, R. Ryoo and M. Choi, *Chem. Eng. J.*, 2022, **439**, 135530.
72. A. Srifa, K. Faungnawakij, V. Itthibenchapong and S. Assabumrungrat, *Chem. Eng. J.*, 2015, **278**, 249.
73. M. Makcharoen, A. Kaewchada, N. Akkarawatkhoosith and A. Jaree, *Energy Convers. Manage: X*, 2021, **2**, 100125.
74. P. Šimáček, D. Kubička, G. Šebor and M. Pospíšil, *Fuel*, 2009, **88**, 456.
75. A. Srifa, K. Faungnawakij, V. Itthibenchapong, N. Viriya-empikul, T. Charinpanitkul and S. Assabumrungrat, *Bioresour. Technol.*, 2014, **158**, 81.
76. J. Horáček, Z. Tišler, V. Rubáš and D. Kubička, *Fuel*, 2014, **121**, 57.
77. R. S. R. M. Hafriz, I. Nor Shafizah, A. Salmiaton, N. A. Arifin, R. Yunus, Y. H. Taufiq Yap and S. Abd Halim, *Arabian J. Chem.*, 2020, **13**, 8146.
78. G. Petropoulos, J. Zafeiropoulos, E. Kordouli, A. Lycourghiotis, C. Kordulis and K. Bourikas, *Energies*, 2023, **16**, 4333.
79. K. Malins, *Fuel*, 2021, **285**, 119129.
80. S. Gong, A. Shinozaki, M. Shi and E. W. Qian, *Energy Fuels*, 2012, **26**, 2394.
81. A. V. Chistyakov, V. V. Kriventsov, A. V. Naumkin, A. Y. Pereyaslavtsev, P. A. Zharova and M. V. Tsodikov, *Pet. Chem.*, 2016, **56**, 607.
82. T. Thangadurai and C. T. Tye, *Period. Polytech., Chem. Eng.*, 2021, **65**, 350.
83. G. C. R. Silva, D. Qian, R. Pace, O. Heintz, G. Caboche, E. Santillan-Jimenez and M. Crocker, *Catalysts*, 2020, **10**, 91.
84. D. Kubička and J. Horáček, *Appl. Catal., A*, 2011, **394**, 9.
85. S. Gong, A. Shinozaki and E. W. Qian, *Ind. Eng. Chem. Res.*, 2012, **51**, 13953.
86. A. A. Nepomnyashchii, E. A. Buluchevskii and A. V. Lavrenov, *Pet. Chem.*, 2019, **59**, 1049.
87. S. Phimsen, W. Kiatkittipong, H. Yamada, T. Tagawa, K. Kiatkittipong, N. Laosiripojana and S. Assabumrungrat, *Energy Convers. Manage.*, 2017, **151**, 324.
88. R. Zarchin, M. Rabaev, R. Vidruk-Nehemya, M. V. Landau and M. Herskowitz, *Fuel*, 2015, **139**, 684.

89. J. Horáček, U. Akhmetzyanova, L. Skuhrovcová, Z. Tišler and H. de Paz Carmona, *Appl. Catal., B*, 2020, **263**, 118328.
90. J. L. C. W. Pimenta, R. D. T. Barreto, O. A. A. dos Santos and L. M. de Matos Jorge, *Environ. Prog. Sustainable Energy*, 2020, **39**, 1.
91. F. Wang, J. Xu, J. Jiang, P. Liu, F. Li, J. Ye and M. Zhou, *Fuel*, 2018, **216**, 738.
92. C. A. Scaldaferri and V. M. D. Pasa, *Chem. Eng. J.*, 2019, **370**, 98.
93. D. Prangklang, D. Tumnantong, B. Yoosuk, C. Ngamcharussrivichai and P. Prasassarakich, *ACS Omega*, 2023, **8**, 40921.
94. S. Fangkoch, S. Boonkum, S. Ratchahat, W. Koo-amornpattana, A. Eiad-Ua, W. Kiatkittipong, W. Klysubun, A. Srifa, K. Faungnawakij and S. Assabumrungrat, *ACS Omega*, 2020, **5**(12), 6956.
95. Z. Li, Z. Huang, S. Ding, F. Li, Z. Wang, H. Lin and C. Chen, *Energy*, 2008, **157**, 270.
96. J. Han, H. Sun, Y. Ding, H. Lou and X. Zheng, *Green Chem.*, 2010, **12**, 463.
97. A. S. Morgan, Z. Hossain, M. B. I. Chowdhury and P. Charpentier, *J. Supercrit. Fluids*, 2024, **205**, 106120.
98. D. Supramono, J. Tjioe and A. Kiswanto, *S. Afr. J. Chem. Eng.*, 2024, **47**, 220.
99. T. Yeh, S. Linic and P. E. Savage, *ACS Sustainable Chem. Eng.*, 2014, **2**(10), 2399.
100. D. Kubička, P. Šimáček and N. Žilková, *Top. Catal.*, 2009, **52**, 161.
101. R. K. Sharma, M. Anand, B. S. Rana, R. Kumar, S. A. Farooqui, M. G. Sibi and A. K. Sinha, *Catal. Today*, 2012, **198**(1), 314.
102. W. K. Craig and D. W. Soveran, *US Pat.*, US 4992605A, 1989.
103. J. Monnier, G. Tourigny, D. W. Soveran, A. Wong, E. N. Hogan and M. Stumborg, *US Pat.*, US5705722A, 1995.
104. S. P. Pyl, C. M. Schietekat, M.-F. Reyniers, R. Abhari, G. B. Marin and K. M. Van Geem, *Chem. Eng. J.*, 2011, **176–177**, 178.
105. N. J. Gudde and J. A. Townsend, *WO Pat.*, WO2007/138254A1, 2007
106. R. Aves and J. Smith, *US Pat.*, US8563792B2, 2013.
107. C. Perego, L. M. F. Sabatino, F. Baldiraghi and G. Faraci, *US Pat.*, US8608812B2, 2013.
108. J. Myllyoja, P. Aalto, P. Savolainen, V.-M. Purola, V. Alopaeus and J. Grönqvist, *WO Pat.*, WO2007/003709A1, 2007.
109. T. Chapus and N. Dupassieux, *US Pat.*, US7880043B2, 2011.
110. http://mlppubsonline.com/publication/?m=52840&i=446449&p=1&ver=html5, 30.11.2023.
111. R. Abhari, L. Tomlinson, P. Havlik and N. Jannasch, *US Pat.*, US2008/0244962A1, 2008.
112. R. Abhari, L. Tomlinson, G. Roth and M. Smith, *WO Pat.*, WO2009/148909A2, 2009.
113. R. Abhari and P. Havlik, *US Pat.*, US8026401B2, 2011.
114. https://www.regi.com/about/our-company, 1.12.2023.
115. BP Oil International Ltd., *EP Pat.*, EP1911734A1, 2008.

116. J. W. Shabaker, *US Pat.*, US8884086B2, 2014.
117. https://www.fuelsandlubes.com/knowledge-base/bp-australia-to-produce-renewable-diesel-fuel-from-tallow-hydrogen/, 5.12.2023.
118. https://www.bp.com/content/dam/bp-country/en_au/media/media-releases/facts-about-closure-bulwer-island-refinery.pdf, 5.12.2023.
119. https://www.honeywell.com/us/en/press/2023/05/bp-to-use-honeywell-s-ecofining-technology-to-increase-sustainable-aviation-fuel-production-worldwide, 5.12.2023.
120. http://cetaneenergy.com/home.html, 5.12.2023.
121. https://uop.honeywell.com/en/industry-solutions/renewable-fuels/honeywell-renewable-diesel, 7.12.2023.
122. T. N. Kalnes, K. P. Koers, T. Marker and D. R. Shonnard, *Environ. Prog. Sustainable Energy*, 2009, **28**, 111.
123. https://uop.honeywell.com/en/news-events/2017/03/diamond-green-diesel-to-expand-renewable-fuel, 7.12.2023.
124. https://oilproducts.eni.com/uploads/media/default/0001/04/fuels-automotive.html, 7.12.2023.
125. https://biorrefineria.blogspot.com/2016/03/AltAir-Fuels-Honeywell-UOP-biorefinery-California-HVO.html, 8.12.2005.
126. https://www.hydrocarbononline.com/doc/honeywell-uop-green-fuels-technology-petrixo-jet-fuel-diesel-0001, 8.12.2023.
127. https://uop.honeywell.com/en/news-events/2020/11/total-to-use-honeywell-uop-technology, 8.12.2023.
128. https://uop.honeywell.com/en/news-events/2021/april/honeywell-renewable-fuels-technology-chosen-for-first-advanced-biofuels-plant-in-paraguay-for-ecb-group, 8.12.2023.
129. https://renewables.topsoe.com/hydroflex, 11.12.2023.
130. R. G. Egeberg, K. G. Knudsen, N. J. Blom and J. A. Hansen, *WO Pat.*, WO2010/028717A2, 2010.
131. P. Knuuttila, P. Kukkonen and U. Hotanen, *WO Pat.*, WO2010/097519A2, 2010.
132. https://www.upm.com/about-us/for-media/releases/2015/01/upm-lappeenranta-biorefinery-is-in-commercial-production/, 11.12.2023.
133. https://www.topsoe.com/blog/preem-selects-hydroflex-for-renewable-fuels-plant-with-potential-to-save-2.5-million-tons-co2, 11.12.2023.
134. https://biodieselmagazine.com/articles/north-dakota-oil-refinery-plans-to-co-process-renewable-diesel-2516083, 11.12.2023.
135. https://www.topsoe.com/blog/skynrg-chooses-topsoes-hydroflex-technology-for-europes-first-sustainable-aviation-fuel-plant, 11.12.2023.
136. https://www.neste.fi/konserni/tietoa-meista/tuotanto/nexbtl-teknologia, 12.12.2023.
137. https://www.neste.com/sites/default/files/attachments/neste_renewable_diesel_handbook.pdf, 12.12.2023.
138. P. Aalto, O. Piirainen and U. Kiiski, *FI Pat.*, FI100248B, 1997.
139. J. Jakkula, V. Niemi, J. Nikkonen, V.-M. Purola, J. Myllyoja, P. Aalto, J. Lehtonen and V. Alopaeus, *EP Pat.*, EP1396531B1, 2007.

140. E. Koivusalmi and J. Jakkula, *US Pat.*, US7459597B2, 2008.
141. J. Myllyoja, P. Aalto and E. Harlin, *US Pat.*, US8278492B2, 2012.
142. https://www.neste.com/about-neste/who-we-are/production, 13.12.2023.
143. https://www.axens.net/markets/renewable-fuels-bio-based-chemicals/renewable-diesel-and-jet#2652, 14.12.2023.
144. https://www.ifpenergiesnouvelles.com/innovation-and-industry/our-expertise/renewable-energies/biofuels/our-solutions, 14.12.2023.
145. https://totalenergies.com/energy-expertise/projects/bioenergies/la-mede-a-forward-looking-facility, 14.12.2023.
146. https://www.axens.net/resources-events/news/pr-axens-will-provide-veganr-technology-haike-chemicals-its-sustainable, 14.12.2023.
147. https://www.aemetis.com/aemetis-selects-axens-vegan-technology-for-its-carbon-zero-project-in-riverbank-california-for-the-production-of-renewable-diesel-and-sustainable-aviation-fuel/, 14.12.2023.

CHAPTER 7

Turning Lignocellulose Biomass into Etheric Biofuel

QI YANG, XINYUE ZHAO, JILONG WANG, HAO XU, YEJUN GUAN* AND PENG WU*

State Key Laboratory of Petroleum Molecular & Process Engineering, Shanghai Key Laboratory of Green Chemistry and Chemical Processes, School of Chemistry and Molecular Engineering, East China Normal University, Shanghai 200062, China
*Emails: yjguan@chem.ecnu.edu.cn; pwu@chem.ecnu.edu.cn

7.1 Introduction

Current economic development is undoubtedly based on carbon energy and this will continue for many years.[1] Presently, carbon energy is dominated by fossil resources including coal, crude oil and natural gas.[2] The unprecedented world population growth has led to the fast depletion of fossil fuels as well as a huge amount of CO_2 emissions, which put great pressure on the sustainable development of the world. The key to solving these problems lies in tackling the trilemma (see Figure 7.1) between sustainable development, carbon energy, and CO_2 emissions.[3]

The concept of "green carbon science" has well illustrated how to optimize the transformation of carbon-containing compounds, utilize carbon resources efficiently, and minimize net CO_2 emissions.[4,5] Biomass conversion is one of the important platforms which can overcome such challenges and is thought to be carbon neutral.[6] Non-edible lignocellulosic biomass as a sustainable source of organic carbon is available in enormous quantities from agricultural resources across the globe, such as forestry and agricultural residues, aquatic plants, and industrial waste streams (*e.g.* paper

Green Chemistry Series No. 89
Biorefining of Renewable Diesel: From Production to Market
Edited by Mohammad Aslam, Sanjeev Mishra and Jorge Aburto

Published by the Royal Society of Chemistry, www.rsc.org

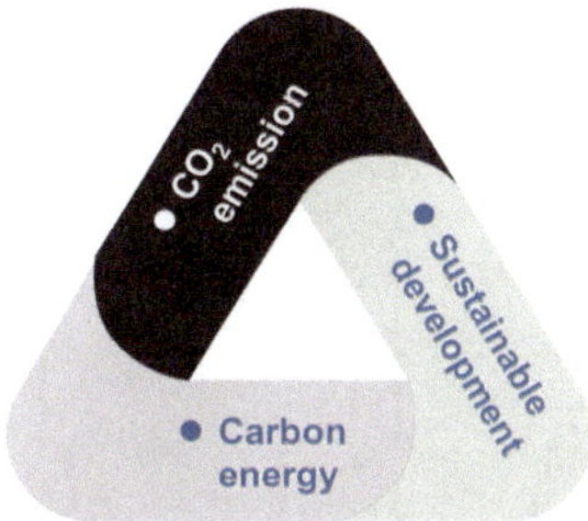

Figure 7.1 The trilemma between sustainable development, carbon energy, and CO_2 emissions.

manufacture).[7] The chemical components of biomass contain a wide variety of functional groups which makes it a promising resource to produce not only bio-fuels but also fine chemicals.[8] Unlike petroleum feeds, which are a mixture of hydrocarbons with a low oxygen content, biomass generally includes complex oxygen-rich polymeric components (such as lignin, cellulose, and hemicellulose) which put additional demands on the transformation processes. The polysaccharides of cellulose and hemicellulose can be converted into monosaccharide units such as glucose, fructose and soluble intermediates *via* fermentation or hydrolysis.[9] These compounds can be further converted into several so-called platform compounds such as 5-hydroxymethylfurfural (HMF), furfural, and levulinic acid/levulinate esters (LA/LE).[10–13] Based on these platform compounds, many value-added organic chemicals with various functional groups (alkyoxymethylfurfural, bis(alkyoxymethyl)furan, dimethylfuran, methylfuran, alkyoxymethyfuran, valerate esters, and etheric valerate esters) are produced as fuels and fine chemicals, as shown in Figure 7.2.[14–20]

7.2 The Merits of Etheric Compounds as Fuel

Gasoline and diesel fuel are crude oil-derived hydrocarbon mixtures containing approximately 86 wt% of carbon and 14 wt% of hydrogen. Gasoline contains mainly alkanes (paraffins), alkenes (olefins), and aromatics with typically 4–12 carbon atoms in the boiling point range of 30–210 °C. Diesel contains mainly paraffins, aromatics and naphthenes with approximately 12–20 carbon atoms in the boiling point range of 170–360 °C. Gasoline exhaust emissions from spark-ignition engines include CO, hydrocarbon (HC), and NO_x mainly produced at cold start or heavy acceleration.[21–24] Diesel engines always show higher combustion efficiency but increased NO_x emissions and higher particulate matter (PM) emissions.[25] The retention of oxygen atoms in biomass-derived etheric/ester compounds could solve these problems to some extent, for example, by reduction in CO/HC emissions as well as NO_x emissions.[26] So far, commonly used oxygenates are alcohols (methanol, ethanol, isopropyl alcohol (IPA), *tert*-butyl alcohol (TBA),

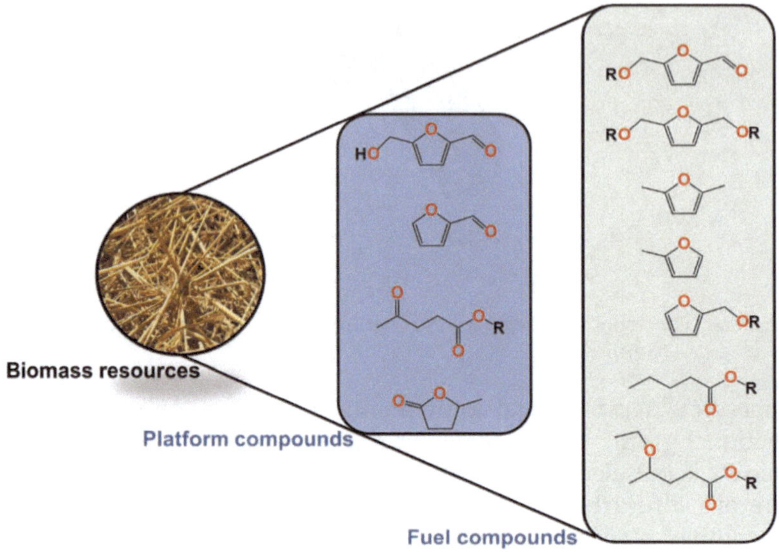

Figure 7.2 Promising biomass-derived fuels and fine chemicals.

isobutyl alcohol (IBA)) and ethers thereof. Compared with alcohols, ethers possess lower boiling points and melting points. Linear ethers, such as dimethyl ether, diethyl ether, and dibutyl ether have relatively high cetane numbers. Branched ethers, such as methyl *tert*-butyl ether (MTBE), are widely used as octane boosters. The oxygenates should meet the requirements of fuel quality, *e.g.* density, viscosity, volatility, flash point, cloud point, energy density, oxidative degradation, *etc.*[27,28] In this regard, short-chain ethers have some obvious drawbacks. For example, dimethyl ether (DME) has a rather low energy density (28.6 MJ kg^{-1}), while diethyl ether (DEE) is light and air-sensitive, tending to form explosive peroxides. Moreover, DEE is a known anesthetic which has negative effects on the human nervous system. Therefore, research on the synthesis of oxygenates with long carbon chains has attracted great attention.[27] In 2009, Shell disclosed a patent which provided a process for the preparation of alkyl furfuryl ethers, wherein dehydration of furfuryl alcohol (FA) and an alkyl alcohol in the range of 1–4 carbon atoms occurred in liquid phase over an acidic zeolite catalyst such as ZSM-5 or Beta at a temperature of 100–150 °C.[29] The boiling points of several typical furanic ethers range from 134 °C to 189 °C, which meets the requirement of gasoline fuel. This study has inspired many studies on the efficient synthesis of furanic ethers by different methods and various catalysts.[30] Another advantage of using furanic ethers as fuel additives is that they possess a low heat of vaporisation and high volumetric heat of combustion. Therefore, they hold great promise as components in high-octane gasoline. Here, we will summarize the major progress that has been achieved recently starting with furanic platform compounds (FAL, FA, and HMF). This chapter comprises three main sections: (1) direct etherification based on an acid catalyst starting with FA and HMF, (2) hydrogenative etherification based on bifunctional catalysts starting with FAL and HMF, and (3) transfer hydrogenative etherification based on bifunctional catalysts also starting with FAL and HMF.

7.3 Direct Etherification of Bio-derived Alcohols, FA and HMF

Direct etherification involves the dehydration of two alcohols, namely FA/HMF and a low-carbon alcohol. The low-carbon alcohol serves as both reactant and solvent. Acid is generally required in this process. Several types of homogeneous and heterogeneous acids (both Brønsted and Lewis acids) have been explored (Figure 7.3). The high selectivity of furan ethers appears to be linked to the presence of acid sites with moderate strength. In this chapter, we mainly focus on three types of catalysts: inorganic acid catalysts, heteropolyacids (HPAs; also called polyoxometalates, POMs) and oxides, and zeolite catalysts. Each type of catalyst has its advantages and disadvantages. Particularly, heterogeneous catalysts are commonly utilized due to their adjustable acidity and recoverability without impacting the reaction pathway, especially HPAs, oxides and zeolites. They are typically preferred for their non-corrosive, environmentally friendly, and renewable nature as well as their stability and longevity. The choice of catalyst depends on the properties of catalysts and the nature of the reactants and products.

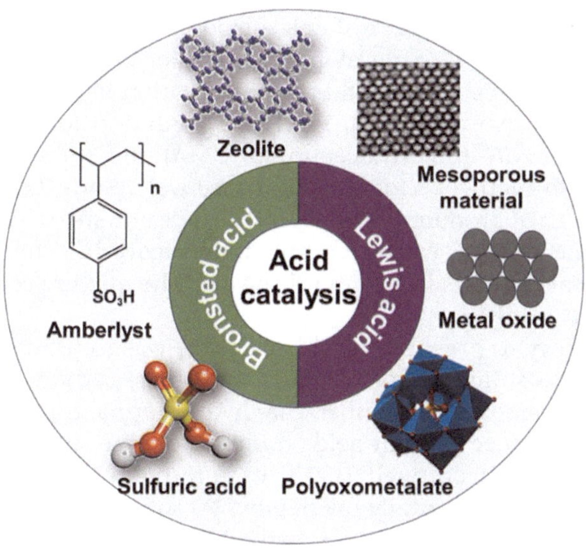

Figure 7.3 Summary of acid catalysts used in direct etherification of FA/HMF and alcohol.

7.3.1 Inorganic Acid Catalysts

The intermolecular dehydration of two alcohol molecules is a commonly employed method to prepare ethers that has attracted widespread attention in biofuel production.[31,32] This strategy is widely used for primary alcohols. In this regard, furfuryl ethers with different alkyl chains can be prepared by etherification of FA/HMF with different alcohols, *i.e.* methanol, ethanol, and IPA. Table 7.1 lists the catalytic performances in direct etherification of FA or 5-HMF of some mineral acid and supported sulfonic acid catalysts. Sulfuric acid and *p*-toluenesulfonic acid (*p*-TSA) are the most used homogeneous acids. Balakrishnan *et al.*[33] tested the direct etherification of 5-HMF with ethanol and butanol in the presence of sulfuric acid or *p*-TSA (5 mol%) at 75–120 °C. The reaction products included 5-(alkoxymethyl)furfural, alkyl levulinate and trace amounts of dialkylacetal (see Figure 7.4). The product distribution depended on the reaction conditions employed. At 75 °C, alkyl ethers were the major products with a selectivity of *ca.* 80%. At 120 °C, alkyl levulinate was formed exclusively with the selectivity of *ca.* 60%, which resulted from the alcoholysis of 5-HMF. It should be noted that the acid-catalysed reaction at higher temperatures generally led to low product yield because of polymerization.

Similar to sulfuric acid, diluted HCl produced by hydrolysis of metal chlorides (MCl_x) was also able to catalyse the etherification. Nguyen *et al.*[34] demonstrated that the HCl formed by hydrolysis of $CrCl_3$ and YCl_3 was active in etherification of FA and IPA with FA conversion of 80%, although with rather low ether yield (35%).

The inherent drawbacks of homogeneous acids are obvious, such as non-reusability, high reaction temperature, corrosion in the reactor, non-regeneration of the catalyst, difficult disposal of waste, *etc.* Therefore, researchers are devoted to developing reusable acid catalysts. Brønsted acidic ionic liquids (ILs) have found their application in etherification. Alam *et al.*[35] prepared two ILs, $[DMA]^+[CH_3SO_3]^-$ and $[NMP]^+[CH_3SO_3]^-$ by reaction between *N*,*N*-dimethylacetamide (DMA) and *N*-methyl-2-pyrrolidinone (NMP) with methyl sulfonic acid, respectively. Both ILs were effective for high-purity EMF production from HMF. The catalysts can be recycled by separating HMF by DEE extraction and then recovering the bottom layer containing spent catalysts in DMA-LiCl solvent. The mixture could be reused for at least four catalytic cycles.

Another strategy to prepare reusable catalysts was to immobilize the sulfonic group onto solid supports such as polymer, carbon, silica or their mixtures. Traditionally, high-temperature hydrothermal treatment (>150 °C) is required to introduce sulfonic acid groups onto these supports while using large amounts of concentrated H_2SO_4 or fuming H_2SO_4. This process is hazardous and environmentally unfriendly. To solve these problems, Russo and coworkers[36] developed a simple method by chemical activation of *p*-TSA with H_2SO_4 at 250 °C. This procedure allows tuning of the surface composition, texture and acid properties of the materials, leading

Table 7.1 The acid catalytic performances of inorganic acid catalysts in direct etherification of FA or HMF.

Entry	Substrate	Catalyst	Solvent	Temp./°C	Time/h	Product	Conv. (%)	Yield (%)	Ref.
1	HMF	H_2SO_4 (5 mol%)	Ethanol	75	24	EMF	100	81	33
2	HMF	*p*-TSA (5 mol%)	Ethanol	75	24	EMF	100	75	33
3	FA	$CrCl_3$	2-Propanol	90	8	iPFE	80.0	35	34
4	HMF	$[NMP]^+[CH_3SO_3]^-$	Ethanol	120	2	EMF	100	76.5	35
5	HMF	$[DMA]^+[CH_3SO_3]^-$	Ethanol	120	2	EMF	100	84.7	35
6	HMF	GC-SO_3H/SiO_2	Ethanol	110	6	EMF	99.0	84	36
7	HMF	PDVTA-SO_3H	Ethanol	110	0.5	EMF	99.8	87	37
8	HMF	HNTs-SO_3H	Ethanol	130	0.58	EMF	97.3	88	38
9	FA	Amberlyst-15	Methanol	40	24	MFE	100	25	39
10	HMF	Amberlyst-15	Ethanol	75	24	EMF	90.0	55	33
11	FA	Amberlyst-70	2-Propanol	130	1	iPFE	100	22	40

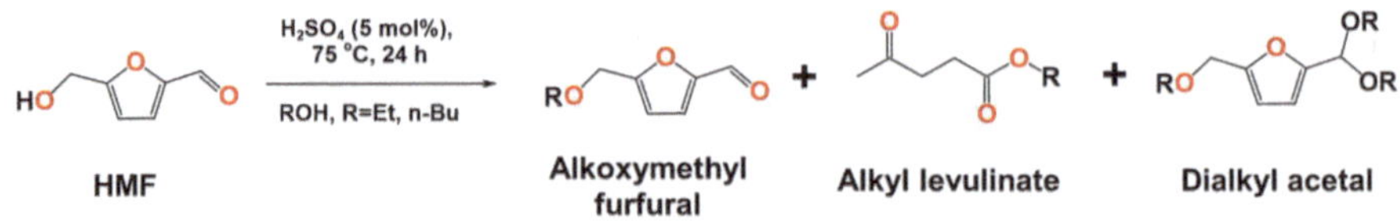

Figure 7.4 Etherification of 5-(hydroxymethyl)furfural (HMF) in alcohols. Reaction conditions: HMF (1 mmol), H_2SO_4 (5 mol%), ROH (2 g), 75 °C, 24 h. Adapted from ref. 33 with permission from the Royal Society of Chemistry.

simultaneously to solids (C-SO_3H/SiO_2) that are stable towards leaching phenomena. By increasing the ratio between the activating agent and the organic precursor, the amount and strength of the acid sites increased, and this was accompanied by the creation of mesoporosity, which enhanced the accessibility of active sites. Over the optimized catalyst, an HMF conversion of 99% and an ether yield of 84% were achieved. This catalyst also worked well for the etherification of FA. Xiang *et al.*[37] investigated the catalytic potential of a functionalized porous organic polymer bearing sulfonic acid groups (PDVTA-SO_3H) for the etherification of HMF to EMF under solvent-free conditions. The PDVTA-SO_3H catalyst exhibited efficient catalytic activity with 99.8% 5-HMF conversion and 87.5% 5-EMF yield within 30 min at 110 °C. This type of catalyst was superior to homogeneous *p*-TSA. Sulfonic acid-grafted halloysite nanotubes (HNTs-SO_3H) were also synthesized by functionalizing the halloysite nanotubes with 1,3-propanesultone (see Figure 7.5).[38] An excellent EMF yield of 86.4% and HMF conversion of 99.3% could be achieved with these organic–inorganic mesoporous materials in specific reaction conditions. Other commercially used solid acid catalysts are highly acidic cation-exchange resins, such as Amberlyst-15,[33,39] Amberlyst-70,[40] *etc.* These catalysts have been used as the benchmark for comparison of catalytic performance. It is generally accepted that the Amberlysts possess large amounts of acid sites with strong strength, favouring the formation of levulinates under the reaction conditions compared with carbon-based acidic catalysts. In this regard, sulfonic acid-containing catalysts are unsuitable for etherification because of their strong Brønsted acidity.

7.3.2 Polyoxometalates and Oxides

Polyoxometalates are a special type of acidic catalysts that combine metals like tungsten, vanadium and molybdenum with p-block non-metals like phosphorous, silicon and arsenic having acidic hydrogen and oxygen.[41] POM-based composites containing tuneable Brønsted and Lewis acidity, have found wide applications in biomass conversion, such as hydrolysis, dehydration, etherification, alcoholysis, *etc.*[42] Because POMs are normally soluble in both water and polar organic solvents, which hinders their practical application, one commonly used method is introducing inorganic cations (*e.g.* Cs^+, Ag^+, *etc.*) as counter-cations to enhance their insolubility or

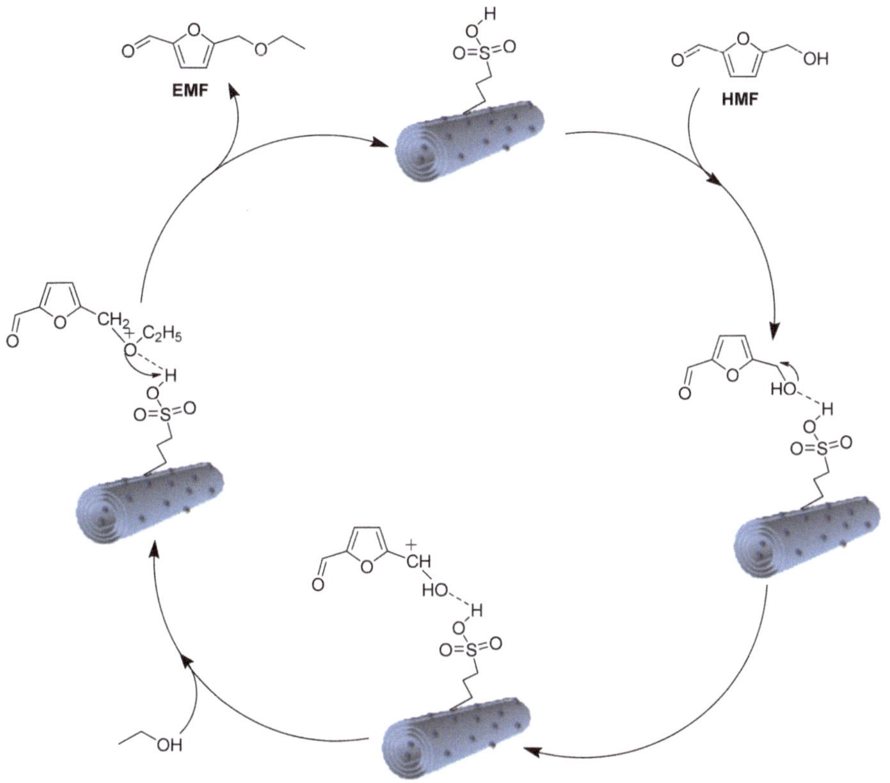

Figure 7.5 The proposed reaction mechanism for the conversion of HMF to EMF in ethanol catalysed by HNTs-SO_3H. Adapted from ref. 38 with permission from the Royal Society of Chemistry.

immobilization onto solid supports.[43] In this section, we will give some examples of etherification catalysed by HPAs and POMs and most of the results are shown in Table 7.2.

Mulik *et al.*[44] prepared Cs-exchanged heteropoly acid ($H_{3-x}Cs_xPW_{12}O_{40}$, $x = 0.5$, 1, 2, 2.5) by ion exchange, which significantly decreased the solubility of HPW. They then tested the synthesis of ethyl furfuryl ether (EFE) by etherification of FA with ethanol over such heterogenized HPAs. The results suggested that $H_1Cs_2PW_{12}O_{40}$ catalyst having moderate acidity showed 65% EFE yield and 98% FA conversion and the catalyst was reusable for four cycles. Liu *et al.*[45] found that partially silver-exchanged HPW (Ag_1H_2PW) showed the best catalytic activity among silver-exchanged HPWs. The highest EMF yield of 88.7% was obtained from HMF at 100 °C after 10 h.

Other strategies have also been proposed to increase the reusability of POMs. Yuan and coworkers[46] reported the synthesis of niobium molybdate catalyst with precisely controlled Nb/Mo compositions for upgrading HMF to EMF (see Figure 7.6). Increased Nb content is beneficial for producing more accessible acid sites and enlarged interlayer spaces in the catalyst. As a

Table 7.2 The acid catalytic performances of polyoxometalates and Zr-based oxides in direct etherification of FA or HMF.

Entry	Substrate	Catalyst	Solvent	Temp./°C	Time/h	Product	Conv. (%)	Yield (%)	Ref.
1	FA	$H_1Cs_2PW_{12}O_{40}$	Ethanol	120	2.5	EFE	98.0	65.0	44
2	HMF	Ag_1H_2PW	Ethanol	100	10	EMF	94.4	88.7	45
3	HMF	$H_{1.15}Nb_{1.15}Mo_{0.85}O_6$	Ethanol	100	1	EMF	100	>99.0	46
4	HMF	$[MIMPS]_3PW_{12}O_{40}$	Ethanol	70	24	EMF	98.1	90.7	47
5	HMF	$H_4SiW_{12}O_{40}$/MCM-41	Ethanol	90	2	EMF	92.0	77.4	48
6	HMF	HPW/MCM-41	Ethanol	100	12	EMF	96.1	83.4	49
7	HMF	PWMo(12.5%)@SiO_2	*n*-Butanol	100	1	BMF	89.0	69.4	50
8	HMF	Fe_3O_4@SiO_2-HPW	Ethanol	100	11	EMF	97.9	83.6	51
9	FA	ZrAl-mp	Ethanol	140	0.5	EFE	100	71.0	53
10	FA	ZrAlW-mp	Ethanol	140	0.5	EFE	100	61.0	53
11	FA	Zr-SBA-15(0.09)	Ethanol	78	5	EFE	95.0	87.4	54
12	HMF	Z-SBA-15	Ethanol	140	5	EMF	100	76.0	55
13	HMF	SZ-SBA-15	Ethanol	140	5	EMF	100	62.0	55
14	FA	Zr5-ST-C	Ethanol	100	1	EFE	95.5	90.2	56
15	HMF	SO_4^{2-}/Al_1-Zr_5/KIT-6	Ethanol	120	2.5	EMF	99.0	89.8	57

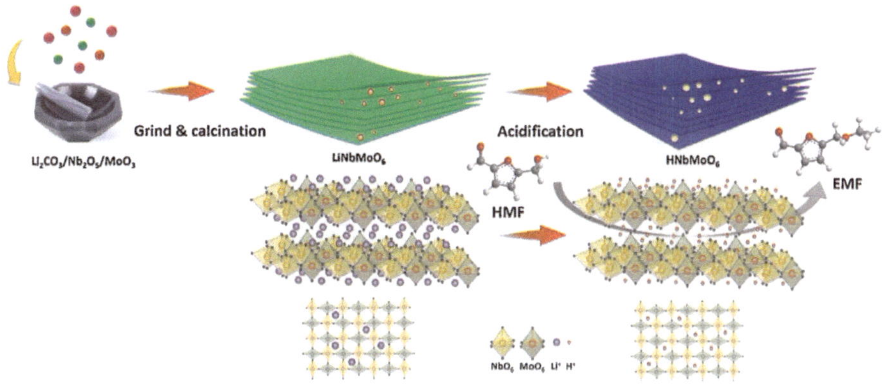

Figure 7.6 Synthetic description for multilayer polyoxometalate niobium molybdate solid acid catalyst. Adapted from ref. 46 with permission from Elsevier, Copyright 2019.

result, the full conversion of HMF into the EMF with selectivity of >99% in ethanol was realized within only 60 min under solvent-free conditions. Zhang *et al.*[47] reported the etherification of HMF by ethanol over an organic–inorganic hybrid solid catalyst $[MIMBS]_3PW_{12}O_{40}$. The recycling test demonstrated that the catalyst could be reused several times without losing catalytic activity, with an average EMF yield of approximately 90%. In addition to the usage of ILs, there are a variety of examples of etherification with HPAs immobilized on mesoporous materials, SiO_2, and zeolites. MCM-41 has been widely investigated as a support of POMs. Che *et al.*[48] prepared $H_4SiW_{12}O_{40}$/MCM-41 nanospheres by the wet impregnation method. This catalyst exhibited 84.1% selectivity to EMF when HMF conversion reached 92.0%. Liu *et al.*[49] also reported the application of 12-tungstophosphoric acid (HPW/MCM-41) for etherification. A high EMF yield of 83.4% was obtained. Besides mesoporous MCM-41, SiO_2 can be also act as a support. Portilla-Zuñiga *et al.*[50] have included Preyssler-type HPAs into the SiO_2 matrix, which can decrease the acidity strength. As a result, higher selectivity toward ether was achieved due to the inclusion effect in the etherification reaction of HMF and *n*-BuOH. Under optimal reaction conditions, the PWMo(12.5%)@SiO_2 catalyst exhibited the highest catalytic activity with 89% 5-HMF conversion and 75% BMF selectivity. HPAs and supported HPA catalysts have significant advantages, such as suitably regulated acidity, sufficient acid sites, recyclability, environmental friendliness, *etc.* In addition, some oxide materials can catalyse the direct etherification of HMF. When a magnetic oxide is included, a magnetically recoverable catalyst can be obtained. Wang *et al.*[51] prepared a magnetically-recoverable catalyst by supporting HPW onto silica-coated Fe_3O_4 nanoparticles (Fe_3O_4@SiO_2-HPW). Under optimal reaction conditions, EMF was obtained in a high yield of 83.6% by the etherification of HMF.

Zirconia-based oxides have been widely applied either as catalysts themselves or as supports in acid catalysis.[52] ZrO_2 and mixed oxides containing

ZrO_2 have received great attention in etherification. Neves *et al.*[53] reported a class of mesoscopic zirconium-based mixed oxides (MZM) that catalysed the etherification of FA with ethanol. They compared the performance of the MZM catalyst with that of zirconium–(tungsten and/or aluminium) mixed oxides (without templates) synthesized by conventional co-precipitation. Under the same reaction conditions (140 °C, 30 min reaction), the MZM catalysts (ZrAl-mp and ZrAlW-mp) showed higher activities, yields of EFE and total yields of bio-additives than the catalysts prepared by co-precipitation. Another type of zirconium-based catalyst, mesoporous Zr-incorporated SBA-15 catalyst was reported by Patil *et al.*[54] The co-existence of Lewis and Brønsted acid sites in Zr-SBA-15 was responsible for catalysing etherification of FA with high selectivity. These Zr-SBA-15 catalysts were prepared by varying the Zr/Si ratio from 0.09 to 0.19 by a simple sol–gel hydrothermal method involving self-generated acidic conditions. Various characterizations confirmed the incorporation of Zr into the framework of SBA-15. Catalyst-screening studies showed that the highest FA conversion of 95% with 92% selectivity towards EFE was obtained for lowest the Zr loading of 0.09. Similarly, zirconia or sulfated zirconia supported on SBA-15 (Z-SBA-15 and SZ-SBA-15) catalysed the direct etherification of 5-HMF was reported by Lanzafame *et al.*[55] The selectivity to EMF was closely related to the presence of Lewis and/or Brønsted acidity on the catalyst. Xiong *et al.*[56] prepared a porous carbon-based material with adjustable acid strength, acid content and acid species ratio to solve the problem of high reaction temperature and poor product selectivity during FA etherification. They employed zirconium as a Lewis acid and adjusting its load is the key approach to driving the etherification reaction. The Zr5-ST-C catalyst showed excellent catalytic performance of 95.5% FA conversion and 94.5% EFE selectivity under mild reaction conditions. The abundant and exposed acid sites induce lower apparent activation energy (30.25 kJ mol^{-1}). Zr5-ST-C also exhibited reliable heterogeneity and recyclability with broad potential for industrial applications. Hafizi *et al.*[57] have synthesized a series of mesoporous composites of SO_4^{2-}/Al1-Zr5/KIT-6, with varied ratios of ZrO_2 to Al_2O_3 and degrees of sulfation, as an efficient, clean, facile and environmental-friendly nano-catalyst for the selective etherification of biomass-derived HMF to produce EMF. The resulting catalyst has been further tuned through tailored reaction conditions to produce an EMF yield of 89.8% and an HMF conversion of 99%. The above-mentioned studies emphasized the importance of the Lewis acidity of Zr-containing amorphous/mesoporous siliceous materials in catalysing direct etherification.

7.3.3 Zeolites

Zeolites are aluminosilicates that consist of interconnected tetrahedrons of aluminium, silicon, and oxygen atoms. Zeolites are microporous materials with high surface areas and a crystalline ordered network of channels due to which they are widely used as solid acid catalysts for etherification.[58,59]

For the direct etherification of FA and HMF with alcohols, the total amount/strength of acidity, type of acid sites, framework topology, and the size of reactants and products are important parameters that should be taken into account. The catalytic results of zeolites are listed in Table 7.3. As early as 2009, Haan *et al.*[29] attempted to catalyse the etherification of FA with ethanol using ZSM-5 ($SiO_2/Al_2O_3 = 30$). Although the reaction conditions are somewhat harsh, the conversion of FA can reach 67% and the yield of EFE is 27%, pointing to a low selectivity of the target ether compound. Cao *et al.*[39] tried the acidic zeolite HZSM-5 (Si/Al = 25) in 2015. After lowering the reaction temperature, it achieved 90.6% conversion of FA and 44.8% selectivity of EFE from etherification of FA with ethanol. Both studies showed that strong Brønsted acid sites are not selective in direct etherification. Other zeolites like H-Beta and H-USY have also been tested for the direct etherification of FA. Their strong acid sites led to unsatisfactory catalytic performance, mainly due to the degradation of furans.[40] To solve this challenge, Chaffey *et al.*[60] reported a low-temperature strategy for the generation of furan ethers from FA using orthoesters, such as trimethyl orthoformate (TMOF) or triethyl orthoformate (TEOF) as sacrificial reagents. With the aid of sacrificial reagents, the etherification on ZSM-5 could occur at temperatures as low as 40 °C (see Figure 7.7). Under the optimized reaction conditions, the FA conversion was 93% and the EFE yield was 73%. However, the use of expensive orthoesters limits the practical application of this strategy. Therefore, major efforts have been devoted to preparing zeolites with suitable acidity. Hu *et al.*[61] prepared the hierarchical catalysts by modifying commercial microporous zeolites with sequential alkali and acid post-synthesis treatments. The thus-obtained catalysts showed highly selective conversion of furanic molecules to biofuels. The yield of EMF and EFE could reach 90.6% and 58.2%, respectively. Barbera *et al.*[62] compared the HMF etherification activity of NH_4^+-exchanged zeolites (BEA, MFI and Silicalite-1) and their proton counterparts. The ammonium-exchanged form of zeolites

Table 7.3 The catalytic performances of various zeolites in direct etherification of FA or 5-HMF with alcohols.

Entry	Substrate	Catalyst	Solvent	Temp./ °C	Time/ h	Product	Conv. (%)	Yield (%)	Ref.
1	FA	ZSM-5	Ethanol	125	17	EFE	67.0	27.0	29
2	FA	ZSM-5	Ethanol	55	18	EFE	90.6	40.6	39
3	FA	ZSM-5	Ethanol (with TEOF)	40	2	EFE	92.0	73.0	60
4	FA	ZSM-5	Ethanol	80	3	EFE	90.8	58.2	61
5	HMF	ZSM-5	Ethanol	140	6	EMF	97.0	90.6	61
6	FA	H-Beta	Ethanol	110	1.5	EFE	100	14.4	40
7	FA	H-USY	2-Propanol	130	1	iPFE	100	40.0	40
8	HMF	NH_4-BEA	Ethanol	140	5	EMF	100	70.4	62
9	HMF	H-BEA	Ethanol	140	5	EMF	100	62.2	62
10	FA	Al-TUD-1	Ethanol	140	0.5	EFE	97.0	52.0	63
11	HMF	Al-TUD-1	Ethanol	140	24	EMF	88.0	81.0	63
12	FA	Kaolinite	2-Propanol	180	2	iPFE	95.0	91.2	64

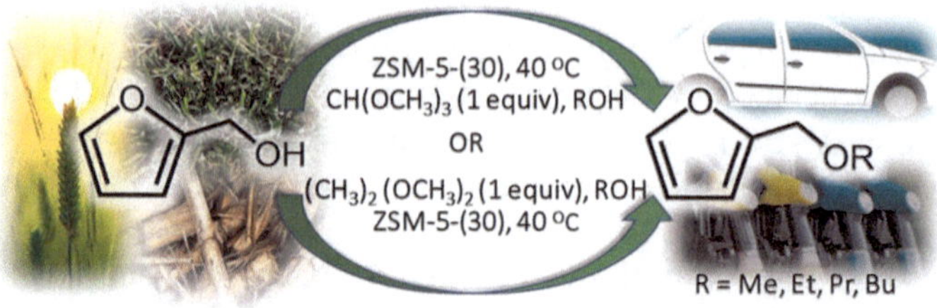

Figure 7.7 ZSM-5-catalysed orthoester-exchange reaction of TMOF or TEOF to produce the corresponding furfuryl ethers at 40 °C. Adapted from ref. 60 with permission from American Chemical Society, Copyright 2018.

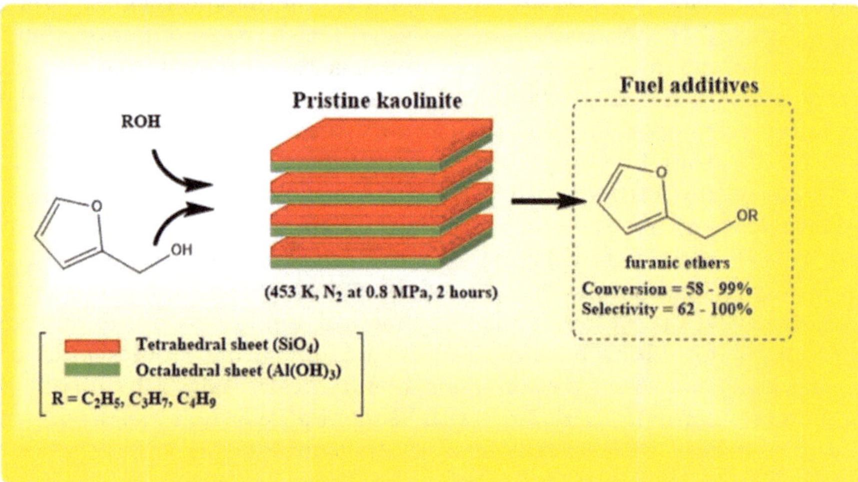

Figure 7.8 Synthesis of commercially available kaolinite-catalysed furanic ethers from furfuryl alcohol (FFA). Adapted from ref. 64 with permission from American Chemical Society, Copyright 2018.

presented a higher selectivity in comparison with the corresponding protonic form. The ammonium-exchanged form of the zeolite was active in the acid-catalysed etherification reaction of HMF in anhydrous ethanol, leading to improved selectivity and activity, in particular for the BEA zeolite form. Neves *et al.*[63] reported that mesoporous aluminosilicates of the type Al-TUD-1, prepared *via* "green", low-cost, non-surfactant templating routes, were effective and versatile heterogeneous acid catalysts. The acid-treated Al-TUD-1(4)-at catalyst led to an EFE yield of 52% at 97% conversion with a reaction temperature of 140 °C and reaction time of 30 min. The yield of furanic ethers decreased and the yield of LE increased gradually after the reaction time was prolonged. At 140 °C for 24 h, Al-TUD-1(4)-at could catalyse the etherification of HMF with an 88% conversion and an 81% EMF yield. Kaolinite also served as a good catalyst for the etherification of FA by IPA, according to Natsir *et al.*,[64] with 95% FA conversion and 96% ether selectivity (see Figure 7.8). The reason behind the uniqueness of kaolinite is likely associated with the high number of Lewis acid sites and the low number of Brønsted acid sites. Very recently, Zhang and co-workers[65] found that by manipulating the concentration of silanol nests on ZSM-5, it is possible to tune the selectivity of FA etherification with alcohol.

In summary, the types and strengths of acidity of zeolites should be precisely tuned to achieve satisfactory selectivity. Brønsted acid sites with weak-to-moderate strength and Lewis acid sites are preferred for this process.

7.4 Reductive Etherification of 5-Hydroxymethylfurfural and Furfural in H_2

As we have mentioned above, acid-catalysed etherification can efficiently convert biomass-derived alcohols into ethers. However, the direct etherification catalysed by acids is restricted to unhindered primary alcohols. The synthesis of furanic ethers with branched alkoxyl groups is challenging. Therefore, it will be interesting to directly start with furfural or HMF, which contain a carbonyl group. In this case, the transformation of the carbonyl group needs hydrogenation sites along with acid sites. In this section, we will focus on the reductive etherification by bifunctional catalysts or a mixture of metal catalysts and acids.

There are generally two reaction pathways to convert FAL to its corresponding ether. As shown in Figure 7.9, one way involves the hydrogenation of the carbonyl group to hydroxyl groups as the first step, then two alcohols undergo inter-dehydration in the presence of acid. Another method involves the formation of acetal as an intermediate, and then the hydrogenolysis of acetal will lead to the formation of ether. This strategy was first reported in 1929 by Sigmund and Uchann, who found that nickel–kieselguhr catalyst

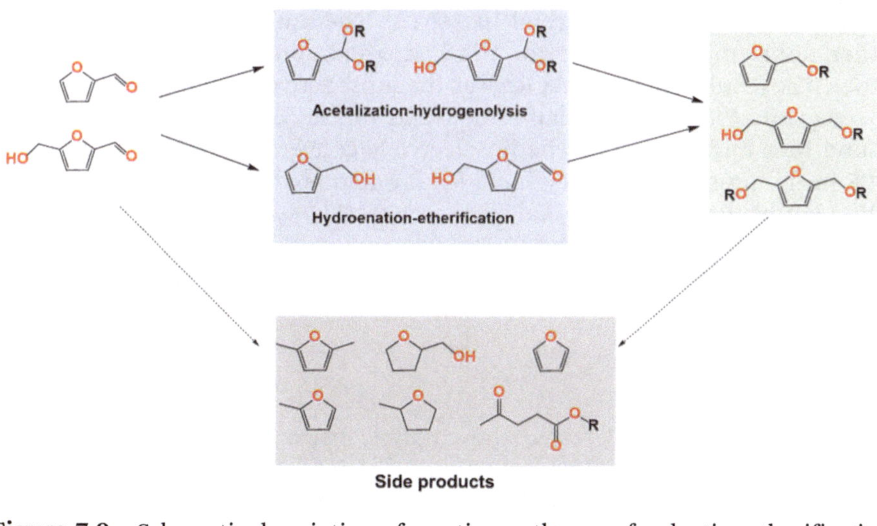

Figure 7.9 Schematic description of reaction pathways of reductive etherification of furans and the prominent side products.

could convert ketals into ethers when heated at 180 °C under an H_2 atmosphere.[66,67] This reaction was not well studied until 1961. Howard and Brown in 1961 found that in the presence of both acid and Rh/Al_2O_3 catalysts, the hydrogenolysis of ketals to ether could be realized at temperatures as low as 25 °C.[68] They proposed that the acidification of ketals results in a reaction equilibrium whose components are ketal, unsaturated ether and alcohol. Hydrogenation then removes the unsaturated ether and favours the further conversion of the ketal. Since the 2010s, this process has attracted much attention because of the search for biofuel additives from furfural-based compounds.[30] Different to acetal, the multifunctionality of furfural and HMF has led to many competitive reactions under the reaction process, for instance, the complete hydrogenation of the furan ring, the decomposition of the furan ring, the hydrogenolysis of hydroxyl group, and decarbonylation (see Figure 7.9). All of these undesired reactions lead to the detriment of ether yield. Hence, developing a catalyst with balanced hydrogenation activity as well as acidity to improve the yield of furanic ethers is of paramount importance.[30]

7.4.1 Noble-metal Catalysts

Noble metals possess unique electronic structures and surface properties that enable efficient hydrogen adsorption and dissociation and always result in higher catalytic activity and selectivity in the hydrogenation process. In this context, Rh,[68,69] Pt,[33,70,71] Ir[71,72] and Pd[73–80] have been explored.

7.4.1.1 *Rh*

Supported Rh nanoparticles were first recognized to be a selective catalyst for ether formation from acetal in 1961.[68] The authors also tested several other noble metal catalysts for the hydrogenation of isopropenyl methyl ether. The results showed that Rh was the most active. Pt and Ru were almost inactive and Pd was about half as active as Rh. Recently, Shaikh *et al.*[69] also found that furfural can be chemoselectively converted in a single step into EFE in the presence of a recyclable Rh-impregnated red mud (RM) catalyst. The 1% Rh@RM composite, calcined at 400 °C (1% Rh@RM-400), produced 75% EFE selectivity with the >99% conversion of furfural. The prepared catalyst retains its stability for the multiple cycles of the furfural hydrogenation reactions.

7.4.1.2 *Pt*

Verzele *et al.*[70] in 1963 tested the catalytic performance of supported Pt catalyst for ether formation starting with a variety of ketones and aldehydes. The alkane and alcohol related to the carbonyl compound are also formed, with varying yields depending on the reaction conditions. Higher HCl concentration favours the formation of alkanes. Possible intermediates are the hemi-ketal, $R^1R^2C(OH)(OR^3)$, the ketal, $R^1R^2C(OR^3)_2$, and the enol ether $R^2CH{=}CR^1OR^3$. The authors considered that the hemi-ketal seemed to be the most likely intermediate. Based on this research, Balakrishnan *et al.*[33] prepared a series of Pt/Al_2O_3 and $PtSn/Al_2O_3$ catalysts and applied them to one-pot reductive etherification of HMF. To better complete the etherification step, the Pt/Al_2O_3 or $PtSn/Al_2O_3$ and Amberlyst-15 were added to the reaction system in a certain proportion (5 mol% Amberlyst-15, 0.25 mol% Pt/Al_2O_3 or $PtSn/Al_2O_3$). The selectivity for BBMF of $PtSn/Al_2O_3$ is generally higher than Pt/Al_2O_3. When the Pt/Sn ratio is 1, the catalyst (0.25 mol% $Pt1Sn1/Al_2O_3$, 5 mol% Amberlyst-15) exhibited the highest yield of BBMF (47%) at 60 °C, 200 psi H_2, reaction time of 18 h, with *n*-butanol as solvent. Pizzi *et al.*[71] also tested the catalytic performance of supported Pt (on Al_2O_3, TiO_2, and carbon) catalysts. They found that in the absence of acid (Amberlyst-15), Pt showed high conversions but not with significant selectivity to the desired ether.

7.4.1.3 *Ir*

According to Pizzi and coworkers,[71] the catalytic performance of Ir is quite similar to Pt under identical reaction conditions, although with high selectivity toward FA. Recently, Sun *et al.*[72] presented an Ir/Na-ZSM-5 bifunctional catalyst for efficient synthesis of 2,5-bis(methoxymethyl)furan (BMMF) ascribing to the cooperative interaction between selective hydrogenation enabled by ultrasmall Ir nanoparticles and appropriate surface acidity resulting from alkali exchange. The authors identified H-ZSM-5 with a Si/Al molar ratio of 11.5 as a potential catalyst capable of delivering notable etherification activity, while the serious carbon loss limited its practical application. When exchanged by alkaline, a good carbon balance of about 92% was achieved yet with satisfactory etherification activity for 0.8Na-ZSM-5. The Ir nanoparticles with an average diameter of around 1.2 nm were uniformly dispersed on the external surface of 0.8Na-ZSM-5. These Ir nanoparticles exhibited higher efficiency in C=O hydrogenation and lower in –C=C– hydrogenation and –OH hydrogenolysis activity. The authors also compared the catalytic activity of Ir/Na-ZSM-5 with Pt, Pd and Ru/Na-ZSM-5. Under identical reaction conditions, Pt/0.8Na-ZSM-5 exhibited much lower selectivity, likely due to the strong acidity formed *in situ via* rapid H_2 dissociation over Pt particles. Both Pd/0.8Na-ZSM-5 and Ru/0.8Na-ZSM-5 favoured the formation of over-hydrogenated products. The 0.75Ir/0.8Na-ZSM-5 catalyst achieved an impressive BMMF yield (91%) in methanol at 50 °C, 1 MPa H_2 and reaction time of 12 h.

7.4.1.4 *Pd*

Despite good catalytic performance of Pt, Rh and Ir in hydrogenation, their very high cost reduces their practical application in industry. Therefore, there has been a continuous effort to investigate Pd-based catalysts with various supports such as carbon, Al_2O_3, SiO_2, zeolite, and metal oxides. Table 7.4 lists recent studies on supported Pd catalysts for reductive etherification. The introduction of Pd into reductive etherification was first made by Howard and Brown,[68] who found that Pd showed half the activity of Pt catalysts. Pizzi *et al.*[71] have recently screened several commonly used metals (Ni, Cu, Co, Ir, Pd, and Pt) on various supports in a high-throughput reactor. Among all samples tested, palladium on charcoal catalyst showed outstanding activity, with the highest selectivity (77%) to FME (furfuryl methyl ether) under the same conditions. This study unambiguously emphasized the superior performance of supported Pd catalysts in this application. Inspired by this study, Wang *et al.*[74] successfully synthesized EFE from furfural and ethanol *via* reductive etherification. By altering the catalyst support, Pd loading, and reaction conditions, an EFE yield of up to 81% was realized over 0.7Pd/C and under mild reaction conditions (0.3 MPa H_2, 60 °C, 2 h). They proposed that the palladium hydride (Pd–H) formed in aqueous solution can express a protonic nature to some extent, which allows Pd–H to catalyse several reactions that are usually catalysed with protonic catalysts. Through the analysis of experimental results, Wang and co-workers suggested that the Pd–H not only can realize the hydrogenolysis of the intermediate (2-(diethoxymethyl)furan; DEOF) but also is beneficial to the acetalization of furfural. Carbon has been proven to be a unique support in reductive etherification.[71,74] However, the reusability of Pd/C catalysts is problematic due to coke deposition. It is difficult to remove the coke deposited by calcination. Therefore, searching for oxide supports other than carbon is of great interest. However, the challenges lie in that the metal–support

Table 7.4 The catalytic performances of supported Pd catalysts in reductive etherification of FAL in various alcohols.

Substrate	Catalyst	Reaction conditions	Solvent	Product	Conv. (%)	Yield (%)	Ref.
FAL	Pd/C	50 bar H_2, 100 °C, 1 h	Methanol	MFE	95.0	77.0	71
FAL	0.7Pd/C	0.3 MPa H_2, 60 °C, 2 h	Ethanol	EFE	98.0	81.0	74
FAL	Pd–Na/SiO_2	0.3 MPa H_2, 120 °C, 2 h	Ethanol	EFE	100	0	75
FAL	I–Pd/Al_2O_3	20 bar H_2, 60 °C, 1 h	Isopropyl alcohol	iPFE	100	85.0	76
FAL	Pd/ECNU-35	0.5 MPa H_2, 120 °C, 1.5 h	Ethanol	EFE	100	74.0	77
FAL	Pd/Nb_2O_5	1 MPa H_2, 180 °C, 6 h	Ethanol	EFE	98.3	87.7	78
FAL	Pd(TPE)/HZSM-5	0.3 MPa H_2, 120 °C, 1 h	Ethanol	EFE	88.0	84.0	79
FAL	Pd/SiO_2–Al_2O_3	0.2 MPa H_2, 100 °C, 16 h	Isopropyl alcohol	iPFE	100	86.2	80

interaction between Pd and the oxide generally favours side reactions such as total hydrogenation as well as hydrogenolysis.

The reaction pathway on Pd nanoparticles is controlled by the electronic properties of Pd, the metal–support interaction between Pd and the support, and the acidity of the support. Long and co-workers[75] found that when SiO_2 was used as support, the Pd nanoparticles should be loaded by an impregnation method. This method led to Pd nanoparticles around 10 nm, which showed moderate selectivity toward ether formation. In contrast, the conventional deposition–precipitation using NaOH as precipitate resulted in a Pd catalyst which selectively catalyses the total hydrogenation of furan ring. This study showed that a tiny amount of alkaline on the surface will alter the reaction pathway. Another study by Wu and co-workers[76] showed that a halogen anion, especially I^-, on the Pd surface also influenced the catalytic performance of Pd nanoparticles (see Figure 7.10). They have constructed a Pd–I/Al_2O_3 bi-functional catalyst for selective reductive etherification of furfural. It has been reported that the introduction of iodine led to the formation of new active species (PdI_2). The presence of PdI_2 enhanced the selectivity to 2-(isopropoxymethyl)furan of the catalyst. These results suggested that the electronic properties of Pd are affected by the metal–support interaction and may tune the reaction pathway. Therefore, by adjusting the morphology or nature of the oxide support, it can achieve a satisfactory ether yield. Wang *et al.*[77] introduced a delaminated derivative (DECNU-35, see Figure 7.11) comprising ferrierite (FER) nanosheets with 4 nm thickness as a novel support for the Pd catalyst. The delaminated DECNU-35 had hierarchical porosity and moderate acidity, which showed a 74.0% yield of EFE with 0.5 MPa H_2, 120 °C and reaction time of 90 min.

Zhang and coworkers[78] prepared a Pd/Nb_2O_5 catalyst and applied it to reductive etherification of furfural into FEE. The reaction results indicated that Pd/Nb_2O_5 was extremely effective for the reductive etherification of furfural into FEE, the yield of FEE was 87.7% under 180 °C, 1 MPa H_2 and reaction time of 6 h. The authors suggested such excellent catalytic activity of Pd/Nb_2O_5 was due to the suitable synergistic effect between the reductive Pd^0

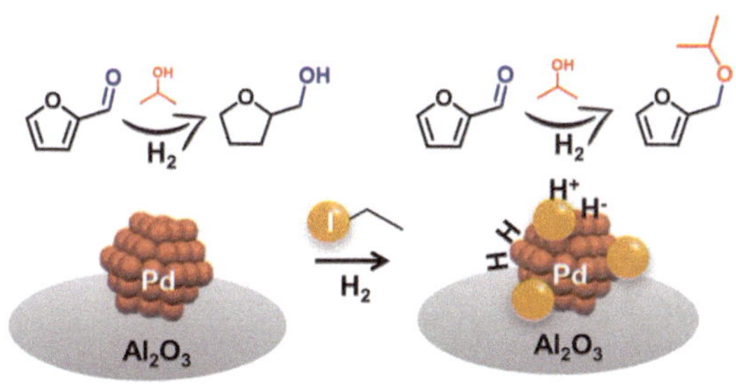

Figure 7.10 Reaction pathways during the reductive etherification of furfural (FF) in isopropyl alcohol over Pd/Al_2O_3 and I–Pd/Al_2O_3 catalysts. Reproduced from ref. 76 with permission from American Chemical Society, Copyright 2019.

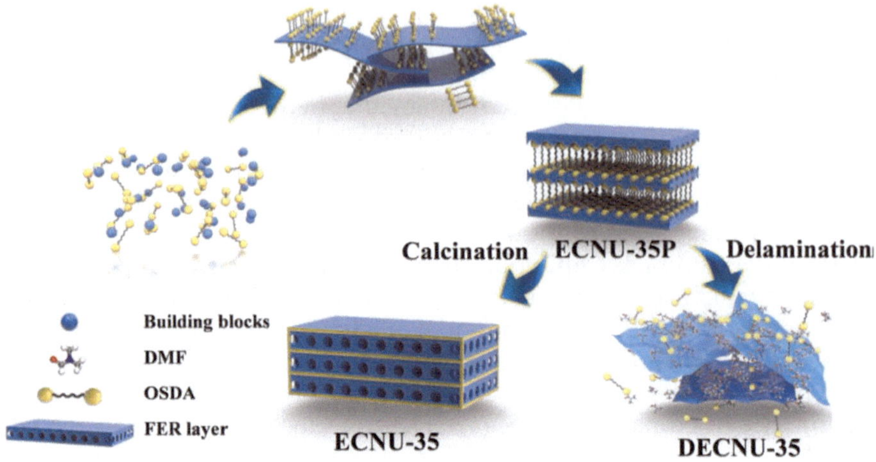

Figure 7.11 Graphic description for the formation and delamination of ECNU-35P. Reproduced from ref. 77 with permission from John Wiley & Sons, Copyright © 2022 Wiley-VCH GmbH.

sites and the acidic sites on Nb_2O_5 nanowires. The particle size of Pd is found to be critical for reductive etherification, due to size-dependent adsorption of –C=O and –C=C– groups on the corners, edges, and planes of Pd nanoparticles. Guo *et al.*[79] synthesized Pd/SiO_2 catalysts with different Pd particle sizes in the range 1–28 nm (see Figure 7.12) through various methods. After comparing the catalytic activity of Pd particles with different sizes, the authors found that smaller Pd particles were beneficial to the hydrogenation of the furan ring and C=O, and bigger Pd particles were conducive to the formation of furanic ethers. With this in mind, they proposed an organic aminosilane-aided strategy to immobilize Pd nanoparticles of 5–8 nm along with the rational tuning of the acidity of HZSM-5 by organosilane. The prepared Pd(TPE)/HZSM-5 catalyst exhibited 84% selectivity to EFE with furfural conversion of 88%, which was higher than the traditional acid-catalysed etherification route. Yang *et al.*[80] recently found that the incorporation of Si onto Al_2O_3 also weakened the metal–support interaction between Pd and SiO_2–Al_2O_3, which resulted in larger Pd particles, consequently resulting in a lower capacity for H_2 adsorption and activation. This effectively avoids the over-hydrogenation of the intermediate product furfuryl alcohol, which is conducive to the formation of furanic ethers.

7.4.2 Non-noble Metals

According to Pizzi and co-workers,[71] the non-noble metals showed low activity as well as very low selectivity for ether. However, from the viewpoint of practical application, it is desirable to develop non-noble metal catalysts for this purpose. The disadvantage of using non-noble metals in hydrogenation is always associated with their very low activity in H_2 dissociation, which

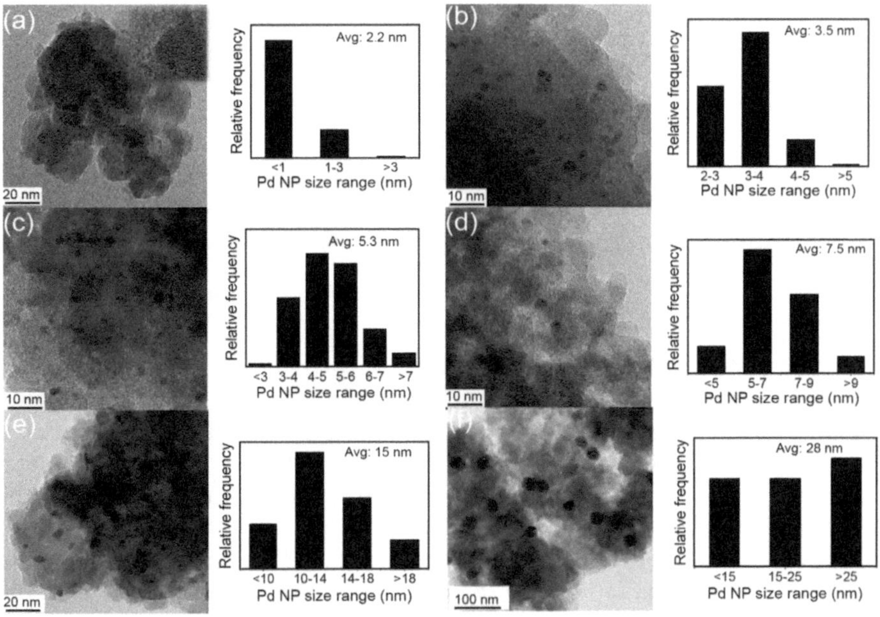

Figure 7.12 Typical TEM images of Pd nanoparticles supported on siliceous materials prepared by various methods: (a) Pd-2.2, (b) Pd-3.5, (c) Pd-5.3, (d) Pd-7.5, (e) Pd-15, (f) Pd-28. Reproduced from ref. 79 with permission from KeAi Publishing, Copyright 2023.

requires a high reaction temperature. At this temperature, furanic compounds tend to polymerize. Therefore, a few efforts have been made to improve the catalytic efficiency of Ni,[81,82] Co,[83] and Cu[84–87] catalysts in combination with acidic catalysts.

7.4.2.1 *Ni*

The first example of reductive etherification was achieved by using a supported Ni catalyst in 1927.[67] Since then, very few reports have been made for Ni catalysts. Recently, Tong *et al.*[81] developed a Ni/C catalyst from a 2D Ni-based metal–organic framework (MOF). This catalyst can efficiently promote the transformation of the –CHO of vanillin, in methanol and in the presence of H_2, to a C–O ether bond or a methyl group (–CH_3) *via* the reductive etherification and subsequent hydrogenolysis of the ether bond. Song *et al.*[82] synthesized a series of Ni-loaded ZSM-5 zeolite catalysts with various Si/Al ratios. They applied them in the reductive etherification of furfural into the furanic ether in alcohols. The results indicated that the variation of the Si/Al ratio of ZSM-5 will not only influence the amount of Brønsted acid sites which catalyse the etherification of furfuryl alcohol but also affect the reducibility of Ni species. The catalytic results suggested that zeolite with Si/Al = 50 (Z50) was the best choice. When Ni is supported on Z50, the surface is dominated by bulk NiO species, and only a few Ni–O–Al species can be observed, which makes the Ni component in Ni/Z50 more easily reduced than the Ni component in other catalysts. Furfural conversion with Ni/Z50 with 15 wt% Ni content can reach 91%; however, when further increasing Ni content to 20 wt% the furfural conversion decreased to 41%, which could be due to the larger size of the Ni particles. Under optimum conditions, the yield of etherification products exceeded 80% over the Ni/Z50 catalyst, and the yield of hydrogenation products was always less than 10%.

7.4.2.2 Co

Reports on the catalytic performance of Co in reductive etherification are very few. Li *et al.*[83] found that cobalt catalysts could also exhibit high catalytic activity for reductive etherification of HMF to BMMF under mild conditions. The authors investigated the effects of different metals and metal oxides on product distribution. The results indicated that metal oxides cannot achieve the conversion of HMF to BHMF, and for the metal catalysts only Co-400 exhibited higher etherification catalytic activity. Through analysis of the characterization and reaction results, the author found that the surface of Co-400 simultaneously contains Co^0 and $Co^{2+/3+}$ (see Figure 7.13), and the coexistence of Co^0 and $Co^{2+/3+}$ was more conducive to the electron transfer, which endowed the catalyst with not only hydrogenation activity but also with strong etherification activity. The authors also found that the Co catalyst showed strong adsorption of C=O double bonds and repellency of the C=C double bonds on the furan rings, which avoided the formation of excessive hydrogenation by-products. Using a Co-400 catalyst at 140 °C, 2 MPa H_2 and a reaction time of 1 h, the yield of etherified product BMMF was 98.5% and the conversion of HMF was almost 100%.

7.4.2.3 Cu

Copper, as one of the most common non-noble metal catalyst, has good selectivity toward the saturation of the carbonyl group. Moreover, Cu-based catalysts do not cleave the C–O bond and only show minor activity in C–C cleavage and ring hydrogenation, which gives Cu-based catalysts great potential for the etherification of 5-hydroxymethylfurfural and furfural. Cao and co-workers[84] performed a one-pot conversion of HMF into BMMF by using Cu/SiO_2 and HZSM-5 (Si/Al = 25) as catalysts simultaneously and obtained 68% yield of BMMF at 120 °C, 2.5 MPa H_2 and reaction time of 12 h. Later on, Wei *et al.*[85] used Cu as a metal centre and applied different zeolites as supports to prepare CuO-USY and Cu/ZSM-5 catalysts and then used them to direct the conversion of HMF to furanic ethers. Both works indicated that the degree of dispersion of copper species has a remarkable influence on the catalytic activity of Cu-based catalysts during the direct conversion of HMF to furanic ethers. In addition, the introduction of CuO resulted in the formation of Al–O–Cu(II) species, which reduces Brønsted acid sites

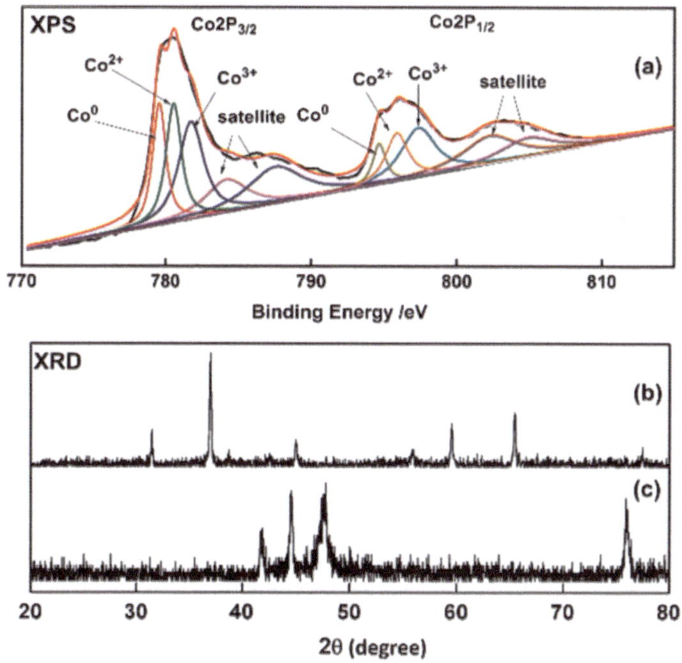

Figure 7.13 XPS and XRD patterns of catalysts: (a) XPS pattern of Co-400 catalyst; (b) XRD pattern of commercial Co_3O_4: (c) XRD pattern of Co-400 catalyst. Reproduced from ref. 83 with permission from the Royal Society of Chemistry.

accompanied by the generation of Lewis acid sites (see Figure 7.14). The introduction of Cu species gives the catalysts suitable acidity, which is beneficial to the etherification process of HMF or furfural.

Similarly, Hu *et al.*[86] reported the highly efficient conversion of HMF to 2,5-bis(ethoxymethyl)furan (BEMF) in a continuous-flow fixed-bed reactor. The *in situ* reduction process can efficiently avoid the secondary oxidation of Cu^0 species. HMF was first hydrogenated into BHMF by the Cu^0 species and then BHMF was etherified with ethanol over the acid sites to produce EMFA (ethoxymethylfurfuryl alcohol). The Cu/HSZ-AE (HSZ = hierarchically structured ZSM-5) catalyst prepared by deposition–precipitation using ammonia evaporation, with higher dispersion of copper and suitable acidity, exhibited better BEMF yield (79.7%) and stability than other catalysts in this work at 140 °C, 2.0 MPa H_2 and reaction time of 8 h. Very recently, Nuzhdin *et al.*[87] reported the application of Cu-doped porous metal oxides obtained from layered double hydroxides (LDHs) in combination with HZSM-5-Al_2O_3 for hydrogenation of furfural derivatives. The $CuAlO_x$ ensured the hydrogenation of furfural to FA, which subsequently underwent etherification with isopropanol in the presence of the HZSM-5-Al_2O_3 composite. The combination of $CuAlO_x$ with ZSM-5(Si/Al = 40)-Al_2O_3 material made it possible to

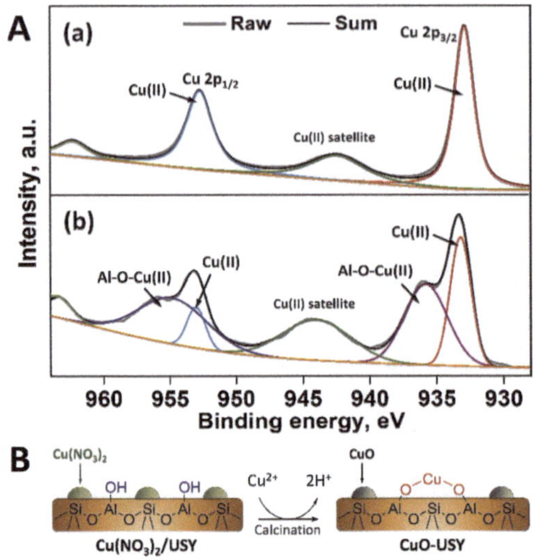

Figure 7.14 (A) XPS spectra of (a) pure CuO and (b) CuO(5)-USY; (B) Cu^{2+}/H^+ exchange on the surface of USY during in the calcination procedure. Reproduced from ref. 85 with permission from Elsevier, Copyright 2019.

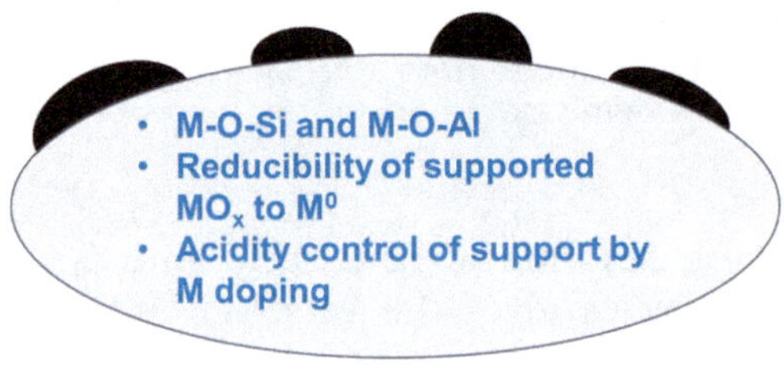

Figure 7.15 Schematic illustration of the design of supported transition metal oxides.

obtain IMF (2-(isopropoxymethyl)furan) with a yield of 94% at 20 °C, total pressure of 10 bar, liquid flow rate of 0.5 mL min^{-1} and H_2 flow rate of 60 mL min^{-1}.

In summary, compared with noble metals, transition metals show considerably lower activity in reductive etherification because they are inert toward H_2 activation. Moreover, when supported on commonly used supports, metal silicates or metal aluminates usually form and hinder the reduction of MO_x because of strong M–O–Al or M–O–Si bonds. As a result, relatively higher reaction temperatures are required for the hydrogenation of the C=O bond, while promoting the occurrence of side reactions including total hydrogenation and hydrogenolysis. The coordination of MO_x with the support in turn affects the acidity of the surface. Therefore, a moderate-to-weak interaction strength between MO_x and the support should be considered when designing supported transition metal catalysts (see Figure 7.15).

7.5 Catalytic Transfer Hydrogenation

Catalytic transfer hydrogenation (CTH) of carbonyl compounds was first observed in 1925 with alcohols used as a hydrogen source and aluminium alkoxide serving as a homogeneous catalyst, known as the Meerwein–Ponndorf–Verley (MPV) reaction after its discoverers.[88] The MPV reaction using a homogeneous aluminium catalyst has encountered several challenges for practical application.[89] First, it is difficult to control the selectivity for the target product. Second, the aluminium catalyst is moisture-sensitive and difficult to handle. Third, the separation and recovery of homogeneous catalysts remains a problematic task. The development of heterogeneous transition metal/metal oxide catalysts can efficiently tackle these problems, giving catalytic transfer hydrogenation several merits such as enhanced safety and easy handling by using safe hydrogen donors. Renewable and liquid-phase hydrogen sources such as alcohols, formic acid, formate and hydrosilane are commonly used as hydrogen donors. In this regard, catalytic transfer hydrogenation has attracted much attention, especially for the transformation of biomass-derived platform chemicals, which typically contain C=O and C=C functional groups.[90] There are two main possible pathways for CTH reactions: direct hydrogen transfer and metal hydride transfer (see Figure 7.16). In this chapter, we focus on etherification involving CTH with furfural as a reactant.

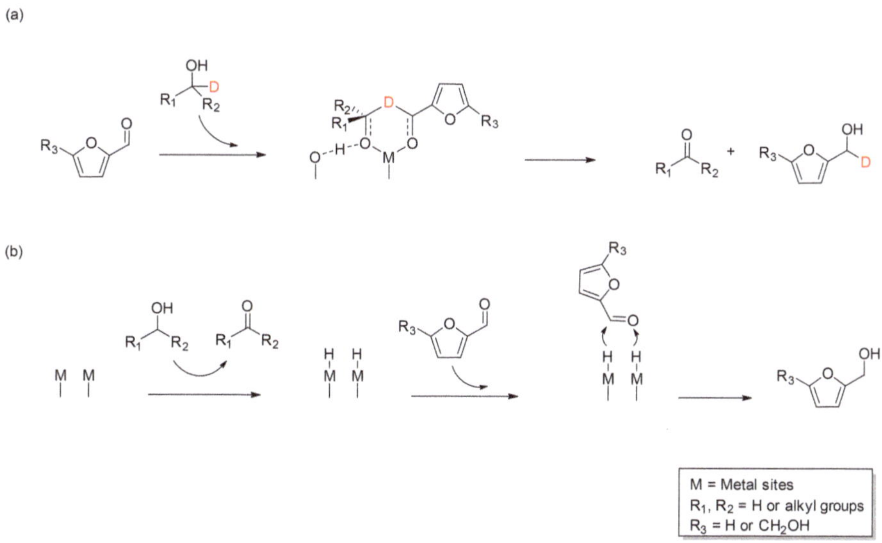

Figure 7.16 Two possible pathways of CTH reactions: (a) direct hydrogen transfer and (b) metal hydride route. Reproduced from ref. 90 with permission from the Royal Society of Chemistry.

The first report of cascade CTH and etherification was made by Corma and co-workers[91] in 2006 over Sn- and Zr-containing silicate molecular sieves. Starting from *p*-methoxybenzaldehyde and 2-butanol, ether can be effectively prepared through a cascade reaction that involves an MPV reduction of the aldehyde followed by etherification of the alcohol. It should be noted that this process proceeds smoothly without requiring the removal of water. Inspired by this elegant work, significant progress has been made in this field. We will summarize most of the recent studies in this section.

7.5.1 Al-, Zr-, Sn-, and Hf-based Zeolites

Zr(Sn, Hf)-based heteroatom zeolites have been widely investigated in transfer hydrogenation (Table 7.5). In general, secondary alcohols are commonly used in this process since the primary and tertiary alcohols do not show transfer hydrogenation activity. Vlachos *et al.*[92] systematically investigated the effects of reaction conditions, topology (Beta and MFI), framework metal (Al, Sn, and Zr), and alcohol structure on the cascade reaction (see Figure 7.17). Both the conversion of HMF and selectivity of the target compound (bis-ether) increased with the reaction temperature. Under the reaction conditions investigated, the selectivity of bis-ether could reach 87% at 453 K. Further increase of the temperature led to hydrogenolysis reaction. By carefully tuning the zeolite structure, the authors have proved that isolated Lewis acid sites in Sn-Beta are the active sites for both the MPV reaction and the etherification in the production of bis-ether. Si-Beta, having a zeolite beta structure without any active sites, was completely inactive for the reaction. The non-framework Sn showed rather low activity for both MPV

Table 7.5 The catalytic performances of Al(Zr, Sn) containing Beta(MFI) in cascade CTH of HMF in alcoholic compounds (ref. 92). Reaction conditions: 453 K, 6 h. The compound numbers refer to those in Figure 7.17.[a]

			Yield (%)						
Zeolite	Solvent	Conv. (%)	1 (furfural ether)	2	3 (bis-ether)	4	5	6	BHMF
EF-Sn-Beta	2-Propanol	<20	0.5	4.4	1.5	—	2	—	9.7
Si-Beta	2-Propanol	<10	—	—	—	—	—	—	3.9
Al-Beta	2-Propanol	98	67.2	—	—	—	—	—	—
Zr-Beta	2-Propanol	100	—	18.0	77.0	—	—	—	—
Sn-MFI	2-Propanol	85	—	28.1	36.6	—	—	—	10.0
Sn-Beta	2-Propanol	95	0.7	0.7	81.7	1.1	—	—	—
Sn-Beta	2-Butanol	85	0.8	—	72.2	4.0	—	—	—
Sn-Beta	1-Propanol	85	10.8	0.8	61.2	—	0.6	0.7	—
Sn-Beta	1-Butanol	88	13.2	0.8	59.0	—	0.7	4.4	—
Sn-Beta	EtOH	90	11.1	1.1	66.2	0.6	1.0	0.5	—

[a] 5-[(1-Methylethoxy)methyl]furfural (compound **1**), 5-[(1-methylethoxy)methyl]-2-furanmethanol (compound **2**), 2,5-bis[(1-methylethoxy)methyl]furan (compound **3**), 2-methyl-5-[(1-methylethoxy)methyl]furan (compound **4**), 5-[bis(1-methylethoxy)methyl]-2-furanmethanol (compound **5**), 2-[bis(1-methylethoxy)methyl]-5-[(1-methylethoxy)methyl]furan (compound **6**).

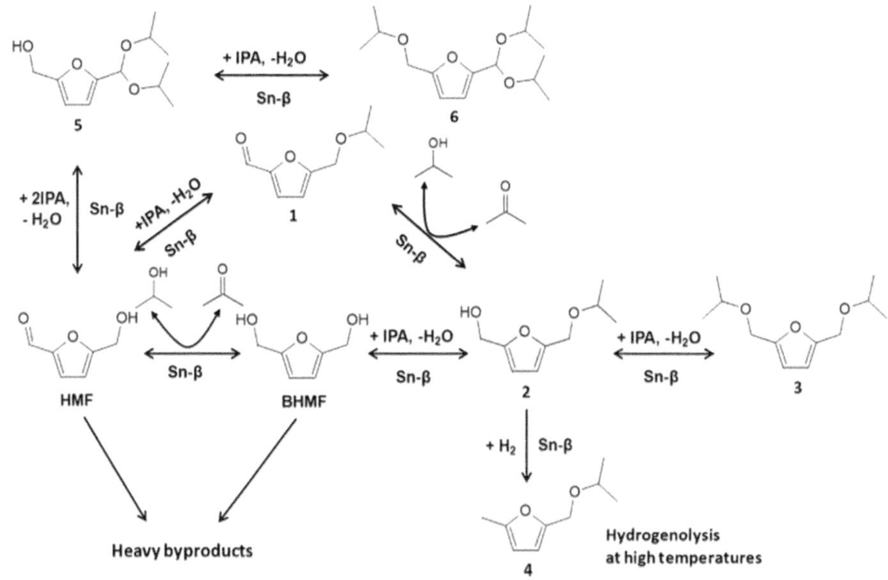

Figure 7.17 Reaction network of sequential CTH and etherification of HMF to 2,5-bis(alkoxymethyl)furan with 2-propanol and Sn-Beta catalyst. Compounds obtained are as follows: HMF, BHMF, 5-[(1-methylethoxy)methyl]furfural (compound **1**), 5-[(1-methylethoxy)methyl]-2-furanmethanol (compound **2**), 2,5-bis[(1-methylethoxy)methyl]furan (compound **3**), 2-methyl-5-[(1-methylethoxy)methyl]furan (compound **4**), 5-[bis(1-methylethoxy)methyl]-2-furanmethanol (compound **5**), and 2-[bis-(1-methylethoxy)methyl]-5-[(1-methylethoxy)methyl]furan (compound **6**). IPA = 2-propanol. Reproduced from ref. 92 with permission from John Wiley & Sons, Copyright © 2014 Wiley-VCH Verlag GmbH & Co. KGaA, Weinheim.

reaction and etherification. Al-Beta, having both Brønsted and Lewis acid sites, was able to catalyse the etherification of HMF while the MPV reaction could not proceed. Zr-Beta was also active for this cascade reaction, but a lower activity was noticed compared to Sn-Beta. It should be noted that Sn-MFI had a lower activity for HMF conversion and etherification, which is attributed to the slower diffusion of the reactants and products within the channels of the MFI structure. The authors also compared the activities of alcohols by using several alcohols such as 1-propanol, 1-butanol, 2-butanol, and ethanol. The results showed that transfer hydrogenation using primary alcohols was slower than that using secondary alcohols because the secondary alcohols tend to dehydrogenate more easily. Roman-Leshkov and co-workers[93] studied the long-term stability of the Lewis acid zeolites (Hf-Beta and Sn-Beta) and their deactivation mechanism in continuous-flow mode by using a fixed-bed plug flow reactor. 2-Butanol as a reactant increased the product selectivity towards the desired ether compound. For the reaction with Hf-Beta, the initial BBMF selectivity was 93% at a conversion of 91%. The HMF conversion remained constant throughout 96 h, while the selectivity for BBMF decreased in exponential decay along with the increase in BHMF selectivity. Almost complete recovery of the etherification activity of the Hf-Beta catalyst could be achieved by calcination at 823 K. The shift in product selectivity from BBMF to BHMF could be interpreted as the result of structural changes occurring at the Hf centres within the zeolite. Compared with Hf-Beta, Sn-Beta showed a significantly lower transfer-hydrogenation activity but a remarkably higher selectivity for etherification within 100 h.

The deactivation of Sn-Beta is likely due to the deposition of heavy by-products formed by the polymerization of furfural and furfuryl alcohol occurring under the reaction conditions. Hammond *et al.*[94] considered that the fast diffusion of furfuryl alcohol from Lewis acid sites to Brønsted acid sites would minimize the coking tendency. Therefore, bifunctional zeolites comprising Lewis and Brønsted acidic sites in close proximity in one matrix would be helpful for the cascade reaction. The conventional method of incorporating Sn(IV) into a partially dealuminated framework by solid-state incorporation was not successful due to the co-presence of residual Brønsted acid sites. They then proposed a solid-state incorporation method in which both Sn(IV) and Al(III) were introduced simultaneously with a suitable Al precursor (Al(III) acetylacetonate) and Sn precursor (Sn(II) acetate) onto the dealuminated sample. The thus-obtained bifunctional catalyst showed an order of magnitude higher turnover numbers than previously reported catalysts.

Besides Beta, aluminium phosphates also showed satisfactory activity in furfural conversion. Riisager and Fang[95] investigated the catalytic performance of aluminium phosphates (APO-5s) with different Al/P ratios (1, 1.25, and 1.5) for the transformation of furfural in alcohol. The results suggested that furfuryl acetals and furfuryl alcohol ethers were selectively produced depending on the acidity. Weakly acidic APO-5 (Al/P = 1) exhibited excellent catalytic performance with 96% acetal yield in ethylene glycol. More acidic APO-5 (Al/P = 1.5) catalysed the one-pot synthesis of ether with 55% yield in 2-propanol with unchanged performance in three successive reactions after intermediate calcination. The ether formation is likely associated with the direct hydrogenolysis of acetal by hydrogen transfer with 2-propanol as hydrogen donor. 1,4-Dioxane as solvent showed enhanced activity in etherification possibly due to the hydrogen bond formation between 1,4-doxane and the formed water.

7.5.2 Zr (Hf)-Based Oxides and Hydroxides

ZrO_2 (HfO_2) contains acid–base pairs on the surface and shows excellent catalytic transfer hydrogenation of the carbonyl group by the MPV reaction but only forms the corresponding alcohols. When combined with Brønsted acid sites, ZrO_2 may serve as a good candidate in CTH for the production of ethers. Ravasio *et al.*[96] found that the introduction of SiO_2 onto ZrO_2 alters the product selectivity from alcohol (FA) to ether (2-butyl furfuryl ether) in the case of CTH of furfural with 2-butanol. The product selectivity is strongly related to the dispersion of ZrO_2 on SiO_2. Under the reaction conditions investigated, pure ZrO_2 produced the alcohol in a quantitative yield in 4 h at 120 °C. Introducing SiO_2 amounts of 3 wt% and 10 wt%, the ether selectivity reached about 30–40% in the case of ZrSi3 and ZrSi10 catalysts. Further, increasing Zr loading on ZrSi oxides led to 95% ether selectivity for ZrSi30. The characterization results suggested that the high surface area and the homogeneity of the materials in terms of Zr and Si interdispersion led to the formation of a Zr–O–Si network comprising a nearly one-to-one ratio between Lewis and Brønsted acidity (see Figure 7.18). This process was also applied to other secondary alcohols such as 2-octanol, but not to primary alcohols such as 1-butanol. Acetal was observed as the major product along with a low amount of ether and unidentified compound when a primary alcohol was used. The authors also explored ZrSi30 for transfer hydrogenation and subsequent etherification of HMF. A high yield of about 96% of ether was noticed in this case, further broadening the applicability of this catalyst.

Besides amorphous SiO_2, mesoporous SiO_2 also shows its potential as a support. Valente *et al.*[97] found that Hf-containing TUD-1-type mesoporous silicate catalysts (Hf-TUD-1) showed satisfactory activity in CTH and etherification of furanic compounds. The Hf-TUD-1 was synthesized *via* a one-pot approach without using surfactants or expensive polymeric organic templates. It seems that Hf-TUD-1 with various Si/Hf ratios acted as essentially Lewis acid catalysts with relatively few weak Brønsted acid sites. Moreover, the mesoporous nature of the catalyst favours the diffusion of reactants, leading to a good yield of 2-alkoxymethylfuran ethers (AMFs) and bis(2-alkoxymethyl)furan ethers (BAMFs). Under the optimized reaction conditions,

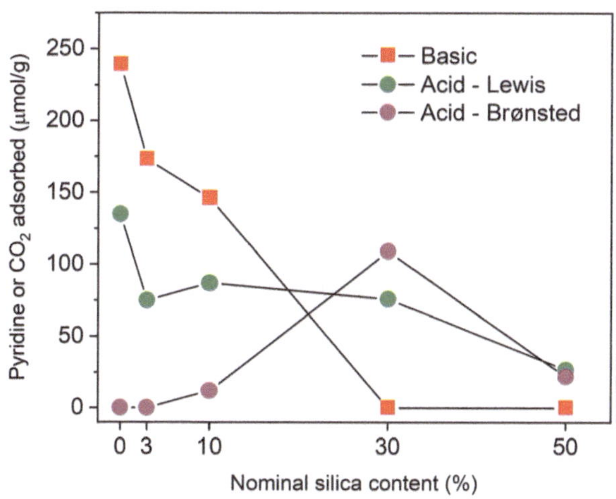

Figure 7.18 Acidic and basic sites of ($\mu mol\,g^{-1}$) Zr/Si catalysts with varying composition determined by pyridine and CO_2 adsorption, respectively. Adapted from ref. 96 with permission from the Royal Society of Chemistry.

Hf-TUD-1(75) gave BBMF from HMF with conversion of 96% and selectivity of 97% in 2-butanol at 150 °C after 21 h reaction.

Compared with ZrO_2 oxide, its hydroxide counterpart is more active in CTH. In this sense, a combination of $ZrO(OH)_2$ with an acidic component has been used in this cascade reaction. Rode and Shinde[98] succeeded in the etherification of HMF with 2-propanol over a combination of acidic Zr–montmorillonite (Zr–Mont), and the charge-transfer hydrogenation catalyst [$ZrO(OH)_2$]. They found that the OH group of $ZrO(OH)_2$ was important during the reaction process. The OH groups were proposed to act as a site for ligand exchange with 2-propanol to form zirconium(v) propoxide, while the Brønsted acid sites facilitated the etherification of HMF (see Figure 7.19). Another study by Rana and co-workers[99] also emphasized the application of $ZrO(OH)_2$ prepared by the precipitation method in etherification. By physically mixing $ZrO(OH)_2$ with Zr-HBeta, a one-pot synthesis of 2,5-bis(propoxymethyl)furan (BPMF) from HMF was achieved using 2-propanol as the hydrogen donor. In this study, the authors found that high selectivity toward BPMF was more likely related to the excess Lewis acid sites. Under the reaction conditions (140 °C, 0.25 g catalyst loading, 4 h reaction time, 1 g HMF), 61.6% HMF conversion and 91.23% BPMF selectivity were achieved. Despite its high catalytic performance and reusability in MPV reactions, the thermal stability of $ZrO(OH)_2$ remains an open question.

As mentioned above, a combination of Lewis acid and Brønsted acid is usually required for CTH and etherification. Different to this known acid-catalysed etherification, Li *et al.*[100] revealed a radical-coupling path for the construction of C–O bonds in the reduction–etherification of HMF by designing the unpaired-electron-defect-rich Cu/ZrO_2 catalyst in the absence of Brønsted acid sites (see Figure 7.20). As shown in Figure 7.20, the O–H bond of BHMF underwent cleavage at V_{Zr}–O^- to form an alkoxy radical intermediate while simultaneously the C–O bond of isopropyl alcohol dissociated at V_O to generate an alkyl radical intermediate, and then the two radicals underwent a radical cross-coupling to give the etherification product. By

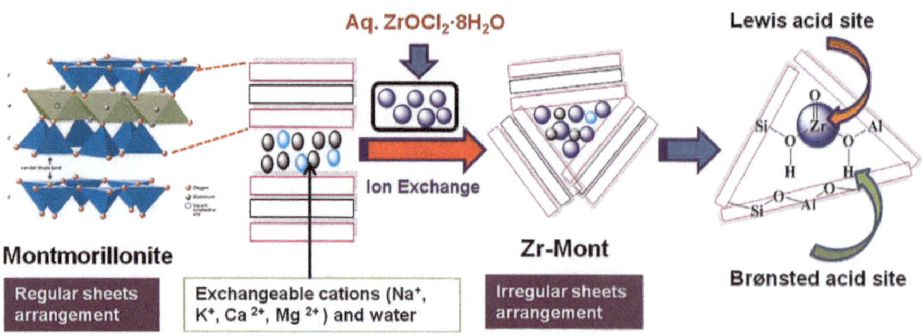

Figure 7.19 Schematic representation of the formation of Zr–Mont and the generation of Brønsted and Lewis acid sites. Adapted from ref. 98 with permission from John Wiley & Sons, Copyright © 2017 Wiley-VCH Verlag GmbH & Co. KGaA, Weinheim.

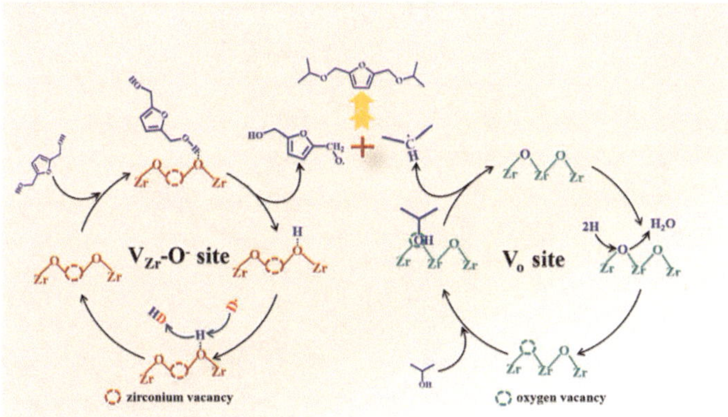

Figure 7.20 Etherification mechanism of BHMF and isopropanol promoted by V_{Zr}–O^- and V_O sites. Reproduced from ref. 100 with permission from American Chemical Society, Copyright 2022.

comparing the catalytic performance of different Cu/ZrO_2 catalysts, it was found that different catalysts showed little difference in the C=O hydrogenation of HMF to BHMF, but had a great effect on the subsequent degree of etherification of 2-hydroxymethyl-5-isopropoxymethylfuran (HPMF) to the desired BPMF. The density of V_{Zr}–O^- sites was the key factor in the etherification step. The rate of etherification shows a monotonic trend with the density of metal vacancies but a nonmonotonic trend with oxygen vacancies. With the assistance of V_{Zr}–O^- and V_O sites, 86.3% BPMF yield was achieved without introducing H_2 and external pressure.

7.5.3 Single Zr Atoms Coordinated with Organic Functional Groups and POSS

Similar to Zr atoms confined in a zeolite framework, Zr atoms stabilized by specific coordination environments have been explored in the cascade reaction. It is well known that Zr-based metal–organic frameworks show sufficient Lewis acidity and good MPV activity. To realize the cascade reaction, Brønsted acidity is required in such a framework. Song *et al.*[101] prepared sulfonic acid-functionalized zirconium-based coordination catalysts (Zr-BDC-Sx) by a convenient solvothermal self-assembly approach, in which 1,4-benzenedicarboxylic acid (BDC) and 2-sulfo-1,4-benzenedicarboxylic acid monosodium (BDC-SO_3Na) were used as dual ligands to regulate the acid–base properties of the catalysts (see Figure 7.21). Due to the superior cooperative actions of Lewis acid–base sites (Zr^{4+}–O^{2-}) and Brønsted acid sites (–SO_3H), Zr-BDC-S60 displayed the highest catalytic activity, affording 100.0% HMF conversion and 96.9% 2,5-bis(isopropoxymethyl)furan (BIPMF) yield in 3 h at 120 °C. Additionally, Zr-BDC-S60 was a robust catalyst and could be reused in at least three runs without an obvious loss in catalytic

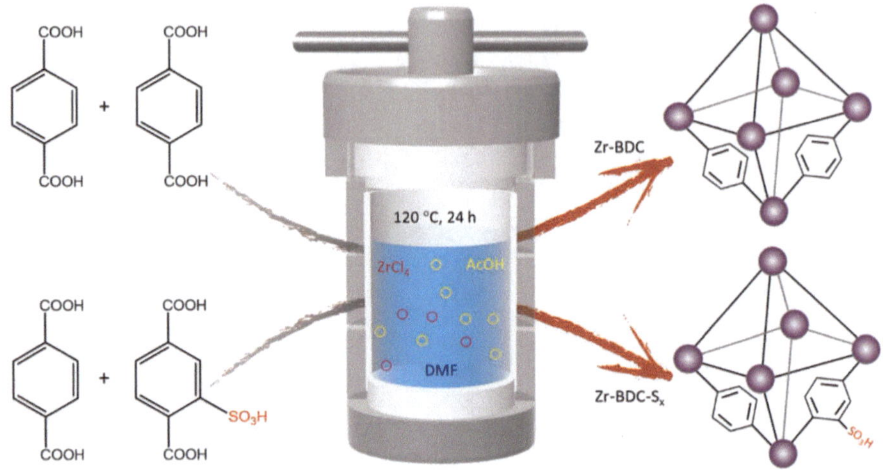

Figure 7.21 Schematic illustration of the preparation of sulfonic acid-functionalized zirconium-based coordination catalysts by the dual-ligands strategy. Adapted from ref. 101 with permission from the Royal Society of Chemistry.

activity. Next, they designed a series of zirconium–carbon coordination catalysts using a similar strategy starting with glucose and *p*-TSA.[102] First, a glucose and *p*-TSA mixture was thermally treated at 180 °C for 24 h. After milling and washing, a black foamy solid glucose-based carbonaceous acid (GC-SO_3H) material containing a large amount of carboxylate and sulfonic acid groups was obtained. Next, Zr(IV) was introduced onto the catalyst *via* coordination between Zr^{4+} and O^{2-} atoms of carboxylates and Ph-OH groups. By regulating the proportions of zirconium and organic ligand, Zr-GC-SO_3H-1.0, possessing a proper Lewis/Brønsted acid ratio as well as strong Lewis acid–base (Zr^{4+}–O^{2-}) and Brønsted acid (–SO_3H) sites, showed the optimum acid–base synergistic effects and displayed the highest catalytic performance. When the one-pot reductive etherification of HMF was performed in isopropanol, BIPMF yield could reach 95.8% in 6 h at 180 °C. In addition, Zr-GC-SO_3H-1.0 could be reused for five reaction cycles without an obvious loss in the yield of BIPMF.

Lin *et al.*[103] proposed a facile and firm assembly of Zr/2-methylimidazole (2-MeIM) coordination polymer over USY zeolite (USY@ZrCP) through N–Al bonds formed by the condensation of 2-MeIM and the hydroxyls on the USY surface at room temperature. USY@ZrCP possesses an abundance of strong Lewis acid sites with a suitable amount of Brønsted acid sites, which are mainly attributed to the highly dispersed Zr(IV) species and the hydroxyls over USY@ZrCP, respectively. According to Py-IR adsorption, ZrCP possessed medium Lewis acid sites (0.093 mmol g^{-1}) with negligible Brønsted acid sites, while USY zeolite had the most Brønsted acid sites (0.362 mmol g^{-1}) with negligible Lewis acid sites. In contrast, the amount of Lewis acid sites of

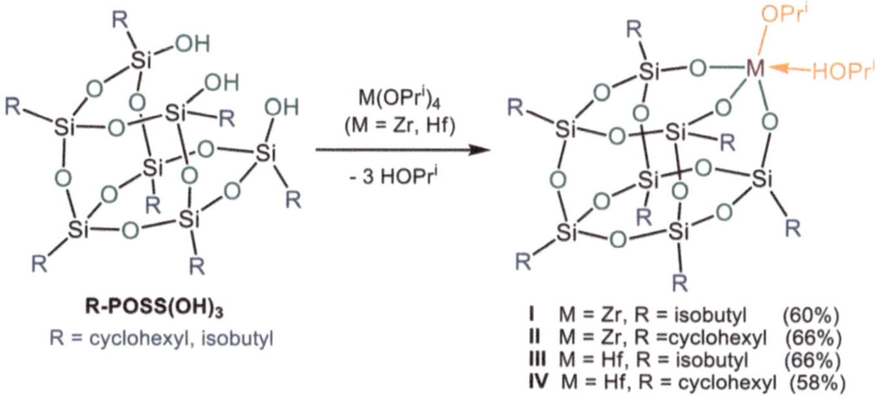

Figure 7.22 Synthesis of thermally-stable homogeneous zirconium and hafnium POSS complexes. Adapted from ref. 104 with permission from the Royal Society of Chemistry.

USY@ZrCP increased to 0.432 mmol g^{-1} at the cost of Brønsted acid sites (0.043 mmol g^{-1}). Therefore, this specific structure led to HMF conversion of 79% and BMF selectivity in the range of 80–90% depending on the amount of USY used under the reaction conditions investigated.

Early transition metal polyhedral oligosilsesquioxane (POSS) complexes of a tripodal coordination geometry have shown some potential as homogeneous catalysts in alkene epoxidations and polymerizations. The electron-withdrawing properties of the POSS ligand combined with its steric profile give rise to well-defined metal complexes that possess relatively high Lewis acidity while maintaining good chemical resistance, thus showing potential application in etherification. Krempner and coworkers[104] synthesized thermally-stable homogeneous zirconium and hafnium POSS complexes in good yields from the reactions of M(OPri)$_4$ (M = Zr, Hf) with R-POSS(OH)$_3$ (R = isobutyl, cyclohexyl) (see Figure 7.22). This type of catalyst was durable and robust, operating at temperatures of 100–160 °C for days without significant loss of catalytic activity for etherification of HMF into 2,5-bis(isopropoxymethyl)furan with an isolated yield of 69%. Moreover, such a catalyst can be prepared *in situ* by using a combination of M(Opri)$_4$ and R-POSS(OH)$_3$. In summary, Zr single atoms or clusters immobilized by an inorganic/organic matrix contain a large amount of unsaturated Zr–O bonds, which serve as Lewis acid sites. These sites can catalyse either the catalytic transfer hydrogenation of C=O bonds with an alcoholic solvent or, when combined with Brønsted acid sites such as Zr–OH, zeolites or sulfonic acid groups, a cascade reaction will that lead to the production of furanic ether.

7.6 Prospectives

Due to the depletion of fossil carbon resources, there is a rapidly growing interest in the search for renewable carbon resources, especially for use as fuel additives or surfactants. As an alternative to dimethyl ether or methyl *tert*-butyl ether, the applications of biomass-derived furanic ethers have been widely explored. Despite great success in recent decades, the industrial production of bioethers remains a challenging task. Several bottlenecks in catalyst design and reaction processes must be overcome for practical application. Firstly, catalysts with balanced acidity are required, which can selectively catalyse etherification while suppressing the decomposition/polymerisation of furan rings at a suitable acidity. Taking zeolites as an example, on the one hand the strong acidity generally leads to side reactions and on the other hand the porous structure can favour the polymerisation of intermediates. Therefore, the development of novel zeolites with weak Brønsted acidity and suitable topology will limit side reactions. Secondly, balanced metal/support interaction will be essential to control the selectivity toward the furan ring and C=O group, which requires a deep understanding of the location of metal nanoparticles in the matrix. A rational design of bifunctional catalysts with spatially distributed metal and acid sites is required. Thirdly, the one-pot transformation of lignocellulose materials instead of using FA/HMF as substrates is more favourable.

Abbreviations

HMF	5-Hydroxymethyfurfural
FAL	Furfural
LA	Levulinic acid
LE	Levulinate ester
PM	Particulate matter
IPA	Isopropyl alcohol
TBA	*tert*-Butyl alcohol
n-BuOH	Butyl alcohol
IBA	Isobutyl alcohol
MTBE	Methyl *tert*-butyl ether
DME	Dimethyl ether
FA	Furfuryl alcohol
HPAs	Heteropolyacids
POMs	Polyoxometalates
AMF	2-Alkoxymethyfuran
EMF	5-(Ethoxymethyl)furfural
BMF	5-(Butyloxymethyl)fufural
DME	Dimethyl ether
DEE	Diethyl ether
DMA	*N*,*N*-Dimethylacetamide
NMP	*N*-Methyl-2-pyrrolidinone
p-TSA	*p*-Toluenesulfonic acid
MFE	Methyl furfuryl ether
iPFE	Isopropyl furfuryl ether
EFE	Ethyl furfuryl ether

TMOF	Trimethyl orthoformate
TEOF	Triethyl orthoformate
BAMF	2,5-Bis(alkoxymethyl)furan
BBMF	2,5-Bis(*n*-butoxymethyl)furan
BMMF	2,5-Bis(methoxymethyl)furan
BHMF	2,5-Bis(hydroxymethyl)furan
DEOF	2-(Diethoxymethyl)furan
HPW	Phophotungstic acid
MZM	Mesoscopic zirconium-based mixed oxides
TPE	*N*-[3-(Trimethoxysilyl)propyl]ethylenediamine
BDC	1,4-Benzenedicarboxylic acid

Acknowledgements

This research was funded by the National Natural Science Foundation of China (22172050 and 21773067), the National Key R&D Program of China (2021YFA1501401), the Research Funds of Happiness Flower ECNU (2020ST2203), and the Open Research fund of Shanghai Key Laboratory of Green Chemistry and Chemical Processes.

References

1. R. G. Grim, Z. Huang, M. T. Guarnieri, J. R. Ferrell, III, L. Tao and J. A. Schaidle, *Energy Environ. Sci.*, 2020, **13**, 472.
2. Q. Fu, L. Liu and H. Wang, *Resour. Policy*, 2023, **86**, 104126.
3. S. C. Peter, *ACS Energy Lett.*, 2018, **3**, 1557.
4. M. He, Y. Sun and B. Han, *Angew. Chem., Int. Ed.*, 2022, **61**, e202112835.
5. M. He, K. Zhang, Y. Guan, Y. Sun and B. Han, *Nat. Sci. Rev.*, 2023, **10**, nwad046.
6. A. Favero, A. Daigneault and B. Sohngen, *Sci. Adv.*, 2020, **6**, eaay6792.
7. W. De Jong and R. Van Ommen, *Biomass as a Sustainable Energy Source for the Future: Fundamentals of Conversion Processes*, John Wiley & Sons, Inc., Hoboken, New Jersey, 2014.
8. L. Petridis and J. C. Smith, *Nat. Rev. Chem.*, 2018, **2**, 382.
9. A. A. Modenbach and S. E. Nokes, *Biomass Bioenergy*, 2013, **56**, 526.
10. J. J. Bozell and G. R. Petersen, *Green Chem.*, 2010, **12**, 539.
11. N. S. Bhat, S. S. Mal and S. Dutta, *Mol. Catal.*, 2021, **505**, 111484.
12. E. Cousin, K. Namhaed, Y. Peres, P. Cognet, M. Delmas, H. Hermansyah, M. Gozan, P. A. Alaba and M. K. Aroua, *Sci. Total Environ.*, 2022, **847**, 157599.
13. K. J. Yong, T. Y. Wu, C. B. T. L. Lee, Z. J. Lee, Q. Liu, J. M. Jahim, Q. Zhou and L. Zhang, *Biomass Bioenergy*, 2022, **161**, 106458.
14. J. He, Q. Qiang, S. Liu, K. Song, X. Zhou, J. Guo, B. Zhang and C. Li, *Fuel*, 2021, **306**, 121765.
15. X. Yue and Y. Queneau, *ChemSusChem*, 2022, **15**, e202102660.

16. A. A. Turkin, E. V. Makshina and B. F. Sels, *ChemSusChem*, 2022, **15**, e202200412.
17. X. Li, P. Jia and T. Wang, *ACS Catal.*, 2016, **6**, 7621.
18. A. Jaswal, P. P. Singh and T. Mondal, *Green Chem.*, 2022, **24**, 510.
19. K. Yan, G. Wu, T. Lafleur and C. Jarvis, *Renewable Sustainable Energy Rev.*, 2014, **38**, 663.
20. S. Dutta, V. Madav, G. Joshi, N. Naik and S. Kumar, *Fuel*, 2023, **347**, 128437.
21. K. Kohse-Hoinghaus, P. Obwald, T. A. Cool, T. Kasper, N. Hansen, F. Qi, C. K. Westbrook and P. R. Westmoreland, *Angew. Chem., Int. Ed.*, 2010, **49**, 3572.
22. C. K. Westbrook, *Annu. Rev. Phys. Chem.*, 2013, **64**, 201.
23. S. Y. No, *Application of Liquid Biofuels to Internal Combustion Engines*, Springer Nature Singapore Pte Ltd, Singapore, 2019.
24. B. Dogan, D. Erol, H. Yaman and E. Kodanli, *Appl. Therm. Eng.*, 2017, **120**, 433.
25. A. S. Mohammed, S. M. Atnaw, A. V. Ramaya and G. Alemayehu, *J. Energy Inst.*, 2023, **108**, 101227.
26. A. K. Agarwal and H. Valera, *Potential and Challenges of Low Carbon Fuels for Sustainable Transport*, Springer Nature Singapore Pte Ltd, 2022
27. J. E. Rorrer, A. T. Bell and D. Toste, *ChemSusChem*, 2019, **12**, 2835.
28. O. I. Awad, R. Mamat, O. M. Ali, N. A. C. Sidik, T. Yusaf and K. Kadirgama, *Renewable Sustainable Energy Rev.*, 2018, **82**, 2586.
29. R. J. Haan and J.-P. Lange (Shell), *WO Pat.*, WO-2009/77606, 2009.
30. T. A. Natsir and S. Shimazu, *Fuel Process. Technol.*, 2020, **200**, 106308.
31. C. J. Donahue, T. D'Amico and J. A. Exline, *J. Chem. Educ.*, 2002, **79**, 724.
32. Inamuddin, T. A. Rangreez and A. M. Asiri, *Applications of Ion Exchange Materials in Chemical and Food Industries*, Springer Nature Switzerland AG, Gewerbestrasse 11, 6330 Cham, Switzerland, 2019.
33. M. Balakrishnan, E. R. Sacia and A. T. Bell, *Green Chem.*, 2012, **14**, 1626.
34. H. Nguyen, N. Xiao, S. Daniels, N. Marcella, J. Timoshenko, A. Frenkel and D. G. Vlachos, *ACS Catal.*, 2017, **7**, 7363.
35. M. I. Alam, S. De, S. Dutta and B. Saha, *RSC Adv.*, 2012, **2**, 6890.
36. P. A. Russo, M. M. Antunes, P. Neves, P. V. Wiper, E. Fazio, F. Neri, F. Barreca, L. Mafra, M. Pillinger, N. Pinna and A. A. Valente, *J. Mater. Chem. A*, 2014, **2**, 11813.
37. Y. Xiang, S. Wen, Y. Tian, K. Zhao, D. Guo, F. Cheng, Q. Xu, X. Liu and D. Yin, *RSC Adv.*, 2021, **11**, 3585.
38. Y. Xiang, K. Zhao, S. Zhou, W. Zhao, Z. Zeng, X. Zhu and X. Liu, *Sustainable Energy Fuels*, 2022, **6**, 2368.
39. Q. Cao, J. Guan, G. Peng, T. Hou, J. Zhou and X. Mu, *Catal. Commun.*, 2015, **58**, 76.
40. M. Paniagua, J. A. Melero, J. Iglesias, G. Morales, B. Hernandez and C. Lopez-Aguado, *Appl. Catal., A*, 2017, **537**, 74.
41. S. S. Wang and G. Y. Yang, *Chem. Rev.*, 2015, **115**, 4893.

42. J. Zhong, J. Perez-Ramirez and N. Yan, *Green Chem.*, 2021, **23**, 18.
43. L. Lian, H. Zhang, S. An, W. Chen and Y. F. Song, *Sci. China: Chem.*, 2021, **64**, 1117.
44. N. L. Mulik, P. S. Niphadkar and V. V. Bokade, *Res. Chem. Intermed.*, 2020, **46**, 2309.
45. Y. Ren, B. Liu, Z. Zhang and J. Lin, *J. Ind. Eng. Chem.*, 2015, **21**, 1127.
46. F. Yang, J. Tang, R. Ou, Z. Guo, S. Gao, Y. Wang, X. Wang, L. Chen and A. Yuan, *Appl. Catal., B*, 2019, **256**, 117786.
47. L. Bing, Z. Zhang and K. Deng, *Ind. Eng. Chem. Res.*, 2012, **51**, 15331.
48. P. Che, F. Lu, J. Zhang, Y. Huang, X. Nie, J. Gao and J. Xu, *Bioresour. Technol.*, 2012, **119**, 433.
49. A. Liu, Z. Zhang, Z. Fang, B. Liu and K. Huang, *J. Ind. Eng. Chem.*, 2014, **20**, 1977.
50. O. M. Portilla-Zuniga, J. J. Martinez, M. Casella, D. I. Lick, A. G. Sathicq, R. Luque and G. P. Romanelli, *Mol. Catal.*, 2020, **494**, 111125.
51. S. Wang, Z. Zhang, B. Liu and J. Li, *Catal. Sci. Technol.*, 2013, **3**, 2104.
52. W. Zhang, Z. Wang, J. Huang and Y. Jiang, *Engery Fuels*, 2021, **35**, 9209.
53. P. Neves, P. A. Russo, A. Fernandes, M. M. Antunes, J. Farinha, M. Pillinger, M. F. Riberio, J. E. Castaheiro and A. A. Valente, *Appl. Catal., A*, 2014, **487**, 148.
54. C. R. Patil and C. V. Rode, *ChemistrySelect*, 2018, **3**, 12504.
55. P. Lanzafame, D. M. Temi, S. Perathoner, G. Centi, A. Macario, A. Aloise and G. Giordano, *Catal. Today*, 2011, **175**, 435.
56. J. Xiong, L. Li, X. Lu, L. Ye, Z. Yu, W. Li, M. Zhang, Y. Li, J. Tao, R. Zhang and Y. Qiao, *Biomass Bioenergy*, 2024, **182**, 107117.
57. H. Hafizi, G. Walker, J. Iqbal, J. J. Leahy and M. N. Collins, *Mol. Catal.*, 2020, **496**, 111176.
58. C. Chizallet, C. Bouchy, K. Larmier and G. Perngruber, *Chem. Rev.*, 2023, **123**, 6107.
59. J. G. Goodwin Jr., S. Natesakhawat, A. A. Nikolopoulous and S. Y. Kim, *Catal. Rev.*, 2002, **44**, 287.
60. D. R. Chaffey, T. E. Davies, S. H. Taylor and A. E. Graham, *ACS Sustainable Chem. Eng.*, 2018, **6**, 4996.
61. D. Hu, H. Hualei, H. Jin, P. Zhang, Y. Hu, S. Ying, X. Li, Y. Yang, J. Zhang and L. Wang, *Appl. Catal., A*, 2020, **590**, 117338.
62. K. Barbera, P. Lanzafame, S. Perathoner, G. Centi, M. Migliori, A. Aloise and G. Giordano, *New J. Chem.*, 2016, **40**, 4300.
63. P. Neves, M. M. Antunes, P. A. Russo, J. P. Abrantes, S. Lima, A. Fernandes, M. Pillinger, S. M. Rocha, M. F. Ribeiro and A. A. Valente, *Green Chem.*, 2013, **15**, 3367.
64. T. A. Natsir, T. Hara, N. Ichikuni and S. Shimazu, *ACS Appl. Energy Mater.*, 2018, **1**, 2460.
65. H. Li, J. Yu, K. Du, W. Li, L. Ding, W. Chen, S. Xie, Y. Zhang and Y. Tang, *Angew. Chem., Int. Ed.*, 2024, e202405092.
66. F. Sigmund and G. Marchart, *Monatsh. Chem.*, 1927, **48**, 267.

67. F. Sigmund and R. Uchann, *Monatsh. Chem.*, 1929, **51**, 234.
68. W. L. Howard and J. H. Brown Jr, *J. Org. Chem.*, 1961, **26**, 1026.
69. A. M. Ajeebi, H. S. Alghamdi, M. A. Sanhoob, M. Aktary, M. A. Aziz, A. S. Alzahrani, A. Q. Alsayoud and M. N. Shaikh, *J. Ind. Eng. Chem.*, 2024, **137**, 572.
70. M. Verzele, M. Acke and M. Anteunis, *J. Chem. Soc.*, 1963, 5598.
71. R. Pizzi, R.-J. van Putten, H. Brust, S. Perahoner, G. Centi and J. C. van der Waal, *Catalysts*, 2015, **5**, 2244.
72. Z. Sun, M. Chen, K. Wang, C. Chen, J. Fei, W. Guo, C. Zhu, H. He, Y. Liu and Y. Cao, *Green Chem.*, 2023, **25**, 10381.
73. V. Bethmont, F. Fache and M. Lemaire, *Tetrahedron Lett.*, 1995, **36**, 4235.
74. Y. Wang, Q. Cui, Y. Guan and P. Wu, *Green Chem.*, 2018, **20**, 2110.
75. L. Long, Y. Wang, H. Wu, T. Xue, P. Wu and Y. Guan, *RSC Adv.*, 2019, **9**, 25345.
76. D. Wu, W. Y. Hernandez, S. Zhang, E. I. Vovk, X. Zhou, Y. Yang, A. Y. Khodakov and V. V. Ordamsky, *ACS Catal.*, 2019, **9**, 2940.
77. J. Wang, Y. Fan, X. Guo, Q. Gu, J. Jiang, Y. Guan, X. He, Y. Ma, H. Xu and P. Wu, *ChemCatChem*, 2022, **14**, e202200535.
78. Z. Zhang, X. Ma, C. Yue, X. Wei and B. Liu, *React. Kinet., Mech. Catal.*, 2022, **135**, 1541.
79. X. Guo, H. Wu, P. Wu, M. He and Y. Guan, *Green Energy. Environ.*, 2023, **8**, 519.
80. K. Yang, R. Wang, D. Xu, X. Ma, D. Ding, M. Zhang, L. Zhang, Z. Bai, Z. Zhu and H. Lv, *J. Catal.*, 2023, **425**, 170.
81. X. Tong, P. Guo, S. Liao, S. Xue and H. Zhang, *Green Chem.*, 2019, **21**, 5828.
82. M. Song, C. Qiu, P. Ma, J. Zhong, Z. Zhang, W. Fang, W. Song, J. Fan and W. Lai, *Renewable Energy.*, 2023, **212**, 468.
83. X. Li, K. Zhang, S. Chen, C. Li, F. Li, H. Xu and Y. Fu, *Green Chem.*, 2018, **20**, 1095.
84. Q. Cao, W. Liang, J. Guan, L. Wang, Q. Qu, X. Zhang, X. Wang and X. Mu, *Appl. Catal., A*, 2014, **481**, 49.
85. J. Wei, T. Wang, X. Cao, H. Liu, X. Tang, Y. Sun, X. Zeng, T. Lei, S. Liu and L. Lin, *Appl. Catal., B*, 2019, **258**, 117793.
86. H. Hu, T. Xue, Z. Zhang, J. Gan, L. Chen, J. Zhang, F. Qu, W. Cai and L. Wang, *ChemCatChem*, 2021, **13**, 3461.
87. A. L. Nuzhdin, Y. Wang, E. N. Vlasova, M. V. Bukhtiyarova, I. G. Danilova, S. A. Yashnik, A. A. Pochatar, L. Xiao, W. Wu and G. A. Bukhtiyarova, *Fuel*, 2024, **356**, 129622.
88. J. F. Minambres and J. Cejka, *Catal. Rev.*, 2023, **66**, 2111.
89. B. Taleb, R. Jahjah, D. Cornu, M. Bechelany, M. A. Ajami, G. Kataya, A. Hijazi and M. H. El-Dakdouki, *Molecules*, 2023, **28**, 7541.
90. W. Fang and A. Risager, *Green Chem.*, 2021, **23**, 670.
91. A. Corma and M. Renz, *Angew. Chem., Int. Ed.*, 2007, **46**, 298.
92. J. Jae, E. Mahmoud, R. F. Lobo and D. G. Vlachos, *ChemCatChem*, 2014, **6**, 508.

93. J. D. Lewis, S. van de Vyver, A. J. Crisci, W. R. Gunther, V. K. Michaelis, R. G. Griffin and Y. Roman-Leshkov, *ChemSusChem*, 2014, **7**, 2255.
94. D. Padovan, A. Al-Nayili and C. Hammond, *Green Chem.*, 2017, **19**, 2846.
95. W. Fang and A. Riisager, *Appl. Catal., B*, 2021, **298**, 120575.
96. F. Zaccheria, F. Bossola, N. Scotti, C. Evangelisti, V. Dal Santo and N. Ravasio, *Catal. Sci. Technol.*, 2020, **10**, 7502.
97. M. M. Antunes, A. F. Silva, C. D. Bernardino, A. Fernandes, F. Riberio and A. A. Valente, *Molecules*, 2021, **26**, 7203.
98. S. Shinde and C. Rode, *ChemSusChem*, 2017, **10**, 4090.
99. S. P. Rana and P. H. Rana, *New J. Chem.*, 2023, **47**, 891.
100. X. Li, H. Bai, Q. Wang, Y. Zhao, J. Feng and D. Li, *ACS Catal.*, 2022, **12**, 7357.
101. A. He, Q. Gu, X. Shen, J. Zheng, L. Hu, X. Wang, Y. Jiang, Z. Wu, J. Xu and J. Song, *Green Chem.*, 2023, **25**, 2349.
102. X. Shen, J. Zheng, L. Hu, Q. Gu, J. Li, K. Chen, Y. Jiang, X. Wang, Z. Wu and J. Song, *Green Chem.*, 2023, **25**, 4090.
103. J. Wei, T. Wang, H. Liu, Y. Liu, X. Tang, Y. Sun, X. Zeng, T. Lei, S. Liu and L. Lin, *J. Catal.*, 2020, **389**, 87.
104. S. Garg, D. K. Unruh and C. Krempner, *Catal. Sci. Technol.*, 2021, **11**, 211.

CHAPTER 8

Biorefining of Second-generation Liquid Biomass into Renewable Diesel

T. S. MARLIZA,*[a] S. MOHD RAZALI,[b] N. ASIKIN-MIJAN,[c] M. H. A. FADLI[c] AND FARIS A. J. ALDOGHACHI[d]

[a] Department of Science and Technology, Universiti Putra Malaysia Bintulu Campus, 97008 Bintulu, Sarawak, Malaysia; [b] Preparatory Centre for Science and Technology, Universiti Malaysia Sabah, 88400 Kota Kinabalu, Sabah, Malaysia; [c] Department of Chemical Sciences, Faculty of Science and Technology, Universiti Kebangsaan Malaysia, 43600 UKM Bangi, Selangor, Malaysia; [d] Chemistry Department, Faculty of Science, University of Basrah, 61004, Basrah, Iraq
*Email: t_marliza@upm.edu.my

8.1 Introduction

Second-generation liquid biomass provides a huge step forward in renewable energy and sustainability efforts. Unlike first-generation liquid biomass, which primarily refers to a fuel made from an edible oil or food biomass such as from food crops like corn and sugarcane, second-generation liquid biomass refers to a fuel produced from non-food biomass sources such as agricultural and forestry residues and organic municipal waste. The shift from first-generation to second-generation liquid biomass reflects a broader trend towards sustainable bioenergy production. Second-generation biofuels are seen as having greater potential to contribute the reductions in greenhouse-gas (GHG) emissions and towards sustainable development goals by utilizing waste materials and

Green Chemistry Series No. 89
Biorefining of Renewable Diesel: From Production to Market
Edited by Mohammad Aslam, Sanjeev Mishra and Jorge Aburto

Published by the Royal Society of Chemistry, www.rsc.org

non-food crops. These feedstocks are particularly valuable since they do not compete with food production, which addresses concerns about food security and land use. Second-generation liquid biomass also has the potential to improve the environment by reusing waste materials and lowering greenhouse-gas emissions.

The development of advanced biofuels, biopower, and bio-based products from second-generation biomass feedstocks is still being researched. Technologies such as biochemical and thermochemical conversion processes can efficiently convert these feedstocks into biofuels like ethanol and biodiesel, as well as other valuable chemicals and materials. Second-generation biomass feedstocks play an essential role in diversifying energy sources, reducing greenhouse-gas emissions, and promoting sustainable development, making them a promising avenue for achieving a more environmentally friendly energy future.

8.2 Second-generation Biomass Feedstock

Biomass is organic material derived from plants and animals that can be utilized to generate renewable and sustainable energy. It is divided into two categories: primary biomass which refers to materials that are directly obtained from natural ecosystems such as wood, agricultural products, and grasses, and secondary biomass, which consists of organic waste materials produced by human activities such as food waste and animal manure. Both types of biomass can be converted into various forms of energy through different processes, including the production of biofuels, biopower and bioproducts.

It is important to develop advanced biofuels and bioenergy technologies aimed at reducing dependency on fossil fuels and addressing climate change. Utilization of second-generation biomass as feedstocks are expected to lower lifecycle greenhouse-gas emissions compared to the traditional fossil fuels.

8.2.1 Primary Liquid Biomass

The primary category of liquid biomass is distinguished based on their origin and composition. Biofuel produced from feedstocks derived from energy crops which do not compete directly with food and feed crops, *i.e.* wood and agricultural products such as forestry residue, wood chips, saw dust, wood pallets, corn and sugarcane, can be used as sources of heat and electricity. Biomass can be burned directly for heat and used in biomass power plants by conversion to liquid and gaseous fuels through various processes. Aside from that, lignocellulosic biomass also can be used to produce cellulosic ethanol and used as feedstocks for biochemical processes. Lignocellulosic biomass is a plant biomass that is composed of cellulose, hemicellulose, and lignin. The sources of this biomass including tree trunks, wheat straw and rice husks.

There are plenty of studies on the potential of primary liquid biomass. Recently Luo *et al.* studied an efficient production of second-generation biofuels (bioethanol and biobutanol) from renewable lignocellulosic biomass. In their study, a ternary deep eutectic solvent was synthesized with choline chloride, succinic acid, and glycerol at a molar ratio of 1 : 0.5 : 0.5. The results indicated that 46.6–49.2 $g L^{-1}$ ethanol could be obtained using *S. cerevisiae* under either separate hydrolysis and fermentation strategies or simultaneous saccharification and fermentation strategies. Their study provides a sustainable and green pretreatment technology to deconstruct lignocellulose and possesses universality to produce various low-carbon bioalcohols by microbial fermentation.[1]

Another research by Deshmukh and Pathan used *Bambusa tulda*, which is a type of Indian bamboo, as raw material for the extraction of bioethanol. Their study covers both the ethanol extraction from bamboo and the effects of various mixtures on spark-ignited engine performance. Four separate blends were developed, and the engine performance of bioethanol 15 (BE 15) was found to be the greatest among all the gasoline blends tested. This invention offers a bioethanol-production process that is good for environmental sustainability.[2]

8.2.2 Secondary Liquid Biomass

Secondary liquid biomass is a category of fuel produced from biomass derived from waste products. Animal manure produced from livestock such as cows, pigs and chickens and human activities waste, *i.e.* food waste and sewage sludge from wastewater, can be used to produce biogas through anaerobic digestion. In the Netherlands, manure biogas plants were developed by the Host-Bioenergy company with manure digesters that can efficiently process various types of manure, providing a sustainable method of producing renewable energy for the agricultural and farming sectors. The plant produces biomethane from manure *via* anaerobic digestion and biogas upgrading, and this upgrading generates up to 40 $N m^3 h^{-1}$ of biomethane with an efficiency of 99.5% and with utilization of 15 000 ton manure per year.[3]

Current research by Afifi *et al.* aimed at optimizing an economical procedure to produce biogas and bio-ethanol from different organic wastes such as sewage sludge and/or cattle dung and/or poultry manure. Each waste type was mixed with the starter, $CaCO_3$ and water, then loaded in a fermenter and kept for 35 days at 35 °C for anaerobic digestion. The results indicated that the digester containing sewage sludge produced the greatest biogas volume of 27.45 L_b (litres of biogas) per digester per day, and their findings show that these wastes can be promising for bio-energy production, being economical and environment friendly.[4]

8.3 Second-generation Biomass Technology

In the 19th century, before fossil fuels were widely used as a source of energy, biomass was the main source of energy. However, this biomass produces

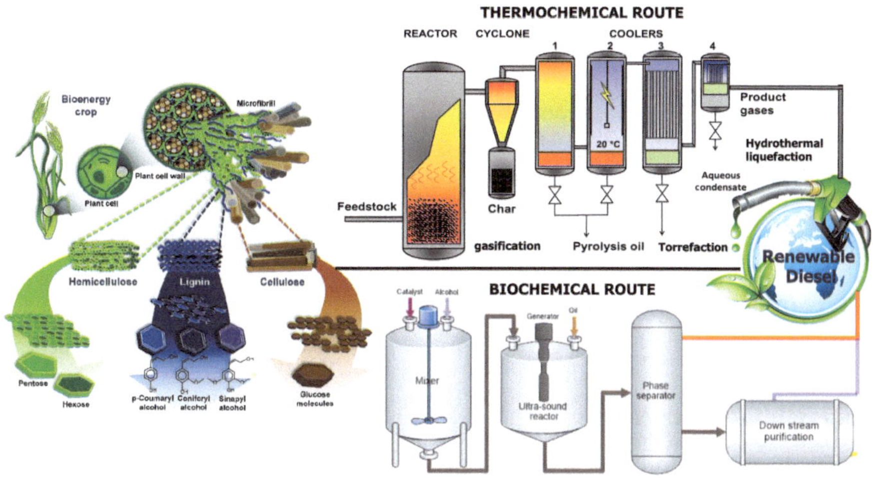

Figure 8.1 Second-generation liquid biomass technology into renewable diesel.

much less energy than fossil fuel. Therefore, fossil fuels are used as the source of energy due to the high energy demands with increasing industry. However, increased energy demands are expected up to a point where fossil fuels cannot accommodate future economic growth. Thus, major changes need to be made in the use of energy resources because biomass is one of several types of sources of sustainable and renewable energy.[5,6]

The production of renewable diesel from biomass can generally be divided into two categories, namely thermochemical and biochemical (Figure 8.1).[7] Thermochemical routes have three subdivisions: pyrolysis, gasification, and supercritical water gasification (SCWG). Meanwhile, biochemical routes are comparable to those used to produce ethanol and have the potential to yield renewable diesel. Microorganisms such as bacteria and yeast metabolize plant sugars during fermentation to make ethanol, and a similar process may be used to generate sustainable fuel from biomass.[8] The thermochemical methods have the benefit of increased overall efficiency (thermal to renewable diesel) and reduced production costs.[9] On the other hand, depending on dry biomass weight,[10] the yield of renewable diesel that may be generated from biomass is quite low. Biological production processes for renewable diesel are found to be more environmentally friendly and less energy intensive compared to thermochemical processes.[11]

8.3.1 Thermochemical Routes

The thermochemical pathway employed in second-generation renewable-diesel synthesis involves the conversion of biomass into extended hydrocarbon chains tailored for utilization in aviation and maritime contexts.[12] It is possible to heat carbon-based materials to high temperatures with or without the presence of steam, oxygen, or air. A mixture of gases, including

H_2, CO, CO_2, CH_4, and other hydrocarbons, as well as water, are produced by these thermochemical processes. Several fuels, such as ethanol, synthetic diesel, synthetic petrol and jet fuel, may be created from the products by fermentation or chemical synthesis.

Determinants impacting the selection of these pathways encompass the appropriateness of the final product for specific applications, energy content, and production expenses.[9] This paradigm assumes critical significance in the transition from conventional fossil fuels to sustainable alternatives, mitigating concerns regarding energy balance, greenhouse-gas emissions, and competition for land use typically associated with first-generation biofuels.[13]

8.3.1.1 *Gasification*

Gasification technology assumes a pivotal role in the synthesis of renewable diesel by facilitating the conversion of biomass into hydrocarbons suitable for transportation applications.[14] Gasification, a thermochemical process, entails the conversion of biomass into syngas, subsequently subject to catalytic transformation into hydrocarbon fuels such as renewable diesel.[14] This technology enables the generation of a diverse array of low-grade fuel derivatives, amenable to further refinement *via* hydrotreating and hydrocracking processes to yield "drop-in" transportation fuels like renewable diesel.[15] Conventional feedstocks like coal and crude oil have well-established gasification processes. Waste wood, energy crops, black liquor, and agricultural and forest leftovers are among the materials that may be gasified using second-generation technology.[16]

The initial syngas produced during gasification harbors impurities such as tars, sulfur, and chloride compounds, necessitating purification steps to attain the desired fuel-grade composition. Gasification presents a viable avenue to produce renewable natural gas, specialized chemicals, and renewable diesel from sustainable biomass feedstocks, thereby fostering solutions with reduced carbon intensity and diminishing reliance on conventional fossil fuels.[17] The product usually obtained is syngas, which may be utilized for further synthesis to produce various Fischer–Tropsch (FT) products such as diesel fuel, bio-methanol, bio-DME (dimethyl ether), petrol by catalytic conversion of dimethyl ether or biomethane (synthetic natural gas).[18] Moreover, syngas may be utilized to produce heat and to power gas turbines or gas motors to provide both mechanical and electrical power.

Chan *et al.* examined the viability of desulfurization and catalytic tar reforming at high temperatures to purify syngas from municipal solid waste gasification.[19] A multi-stage system purifies synthetic gas generated by a downdraft fixed-bed gasifier. The system consists of a fluidized-bed catalytic tar reformer, a particle filter, and a fixed-bed reactor for dichlorination and ultimately desulfurization, with operating temperatures cascading downward throughout. They synthesized a novel nano-structured nickel catalyst supported on alumina and a regenerable Ni–Zn desulfurization sorbent loaded onto a honeycomb (cordierite mullite). Complementary sampling and analysis procedures are used to measure contaminants and identify their distribution at various phases. The experimental and thermodynamic modeling findings are compared to establish the integrated system's kinetic constraints.[19]

8.3.1.2 Pyrolysis

The process of pyrolysis is a well-known method for breaking down organic materials without oxygen at high temperatures. Forestry and agricultural leftovers, wood debris, and energy crops may all be utilized as feedstock in second-generation biofuel applications to create products like bio-oil for fuel oil uses.[20] Bio-oil often requires extensive further processing before it can be used as a refinery feedstock to substitute crude oil.

Recent investigations into pyrolysis technology for renewable-diesel production have concentrated on optimizing efficiency and sustainability. One study underscored the significance of judicious biomass selection to maximize bio-oil yields, advocating for biomass varieties abundant in cellulose and low in moisture content. Furthermore, research has explored advancements in catalyst development tailored for biomass pyrolysis, aiming to bolster gas production on an industrial scale.[21] Economic evaluations of large-scale pyrolysis facilities have been recommended to position this technology competitively *vis-à-vis* fossil fuels. Moreover, environmental assessments have elucidated the benefits of pyrolysis-derived fuels, demonstrating substantial reductions in greenhouse-gas emissions relative to conventional gasoline. Employing lifecycle analysis (LCA) has facilitated the tracking of energy fluxes, accentuating the favorable energy balance and diminished CO_2 emissions associated with biomass-derived pyrolysis fuels.[20] Collectively, recent inquiries underscore the promise of pyrolysis technology in efficiently and sustainably producing renewable diesel, while concurrently mitigating environmental impacts compared to conventional fossil fuels.

Asikin-Mijan *et al.* reported the pyrolysis of *Jatropha curcas* oil (JCO), which underwent deoxygenation using NiO–Fe_2O_3 and NiO–ZnO catalysts supported on multi-walled carbon nanotubes (MWCNTs).[22] It was discovered that high Fe and Zn doses were unsuccessful in pyrolysis, with the maximum activity recorded on the $(NiO)_{20}(Fe_2O_3)_5$/MWCNT catalyst. According to structure–activity correlations, n-$(C_{15}+C_{17})$ selectivity and catalytic activity are greatly enhanced by minimal metal loading, weak and medium acidic sites, and strong basic sites. In a comparison of carbon nanostructures and micron-sized carbon-supported NiO–Fe_2O_3, the deoxygenated green diesel resulting from NiO–Fe_2O_3/MWCNT-catalyzed deoxygenation had the lowest oxygen concentration and the greatest heating value. As a result, it demonstrated the relevance of carbon nanostructures as catalytic supports for enhancing diesel quality.

8.3.1.3 Torrefaction

Torrefaction technology plays a critical part in the thermochemical approach to sustainable fuel generation. It includes heating biomass in a low-oxygen environment to improve its characteristics, making it more suitable for conversion into advanced biofuels like renewable diesel.[23] Torrefaction is a kind of pyrolysis that takes place at temperatures ranging from 200 °C to 320 °C.[24] Torrefaction improves the efficiency and quality of the following conversion operations by converting biomass into a denser and more stable form, allowing production of sustainable diesel fuels. This technology emerges as a cornerstone for promoting sustainable development and decarbonizing energy production, providing a viable path to lowering greenhouse-gas emissions and pushing for the use of renewable resources in the transportation sector.[25]

Recent research on torrefaction technology for renewable-diesel production underscores its potential in refining biomass to generate solid fuel suitable for diverse applications.[26] Torrefaction stands out as a promising thermochemical pre-treatment method capable of enhancing biomass characteristics, rendering it a viable resource for renewable-diesel synthesis. This process entails subjecting biomass to elevated temperatures in an oxygen-depleted environment, yielding a product characterized by heightened energy density and diminished moisture content, crucial attributes for efficient conversion into advanced biofuels like renewable diesel.[27] Despite encountering challenges such as elevated energy input requirements and the potential loss of valuable components during torrefaction, ongoing investigations are directed toward optimizing this technology to foster a more sustainable and renewable energy landscape.

Gan *et al.* produced solid fuel by torrefying jatropha seed kernels at 200 °C, 250 °C, and 300 °C for 15 min, 30 min, 45 min, and 60 min while maintaining particle sizes of 0.5–1.0 nm and 1.0–2.0 mm.[28] Torrefaction temperature had a greater impact on torrefaction performance than holding time. Following torrefaction at 300 °C for 60 min with particle sizes of 0.5–1.0 mm, the higher heating value (HHV) enhancement factor rose to 1.243. The large particle size resulted in minimal enhancement in HHV and good solid yield by slowing the diffusion rate of torrefaction vapor through interior pores. Research on the torrefaction severity index demonstrates that an increase in HHV is closely correlated with weight loss, which lowers the overall energy content of biochar.

8.3.1.4 Hydrothermal Liquefaction

Hydrothermal liquefaction (HTL) technology emerges as a promising avenue for renewable-diesel synthesis, leveraging the conversion of moisture-laden biomass into crude-like oil under moderate temperature and elevated pressure conditions.[29,30] Hydrothermal liquefaction is a wet material processing method that is akin to pyrolysis. The procedure is normally carried out at moderate temperatures of up to 400 °C and pressures higher than atmospheric. Hydrothermal liquefaction is a promising method for manufacturing fuel and chemical feedstock due to its capacity to handle a wide range of materials.[29] This thermochemical process entails the conversion of carbon and hydrogen constituents from organic matter such as biomass into hydrophobic compounds, tailored for heavy-duty engine, marine, and rail applications, or further refinement into transportation fuels like diesel, gasoline, or jet fuels.[31]

HTL presents distinct advantages over alternative biofuel methodologies, manifesting in the generation of products boasting heightened energy density, necessitating abbreviated reaction durations, and accommodating a broader spectrum of input materials.[30] The resultant biocrude derived from HTL holds the potential for refinement into renewable fuels like green diesel or sustainable aviation fuel (SAF). Corporations are actively advancing the scalability of HTL technologies for commercial deployment, fostering collaborations to explore diverse feedstock options encompassing lignocellulosic biomass, end-of-life plastics, and municipal solid waste for biocrude generation.[29] Notwithstanding challenges such as the transition to continuous operational frameworks and the optimization of operating temperatures for optimal output quality, HTL remains a pivotal technology in renewable-diesel production, offering diminished environmental impact relative to conventional fossil fuels.

The techno-economic analysis (TEA) and lifecycle analysis of algal biofuel production in two stages, HTL and hydrodeoxygenation (HDO), were reported by Masoumi and Dalai.[32] At 200 dry metric tonnes per day, microalgae were converted to biocrude oil *via* HTL in the methanol–water system, which was then catalytically upgraded to create biofuels. Based on HTL data, the greatest biocrude oil production of 57.8 wt% was achieved with a microalgae to solvent mass ratio of 1 : 5 and a methanol–water mass ratio of 3 : 1 ratio. The biocrude oil contained 14.5 wt% oxygen and had an HHV of 33.4 MJ per kg of biocrude oil, which needed to be upgraded before it could be used as a transportation fuel. HDO was used to improve the quality of biocrude oil by removing 3.1 wt% oxygen.

8.3.2 Biochemical Routes

The biochemical pathway for second-generation renewable-diesel synthesis encompasses the enzymatic conversion of complex sugar polymers inherent in biomass, such as cellulose and hemicellulose, into monomeric sugars.[33] Pre-treatment is commonly used in biochemical operations to speed up the hydrolysis process that separates lignin, hemicellulose, and cellulose. Once these materials have been separated, the cellulose fractions may undergo fermentation mediated by microorganisms to yield fuels and chemicals. This process represents a fundamental aspect of second-generation biofuel production, which prioritizes the utilization of non-edible lignocellulosic biomass sources, including but not limited to vegetable grasses, forest residues, and agricultural byproducts.[34]

Second-generation biofuels, exemplified by lignocellulosic ethanol and Fischer–Tropsch fuels, present notable advantages over fossil fuels, including heightened energy output, expanded feedstock versatility, mitigated land competition, and reduced environmental footprints.[35] The biochemical pathway stands as a principal conduit for the production of second-generation biofuels derived from lignocellulosic feedstocks, complementing the thermochemical route that yields an assortment of elongated hydrocarbon chains tailored for aviation and maritime applications. These routes collectively represent pivotal avenues for advancing the renewable energy landscape, aligning with sustainability imperatives, and striving towards a more diversified and environmentally benign energy portfolio.[36]

In the biochemical conversion process, biomass is converted to ethanol by sequential steps of pretreatment (to reduce the recalcitrance of biomass), hydrolysis (conversion of sugar polymers to monomers) and fermentation (sugars to ethanol). In Kumar's dissertation, he investigated the potential of cellulosic ethanol production from grass straw, assessed the techno-economic viability and environmental impacts of the bioethanol production and developed a stochastic molecular model for modelling cellulose hydrolysis. The study revealed that ethanol production from grass straw provides environmental benefits compared to the use of gasoline, with 57.43–112.67% reduction in fossil energy use to produce 10 000 MJ of fuel. The GHG emissions during the lifecycle of ethanol production were estimated in the range of (−131) to (−555.4) kg CO_2 eq. per 10 000 MJ of fuel. It was observed that the assumptions and allocation procedures used during the analysis had a significant effect on the LCA results.[37]

8.4 Conversion of Second-generation Liquid Biomass into Renewable Diesel

Renewable diesel is a mixture of diesel-like hydrocarbons in range of C_{13}–C_{24}. Recently, the majority of renewable diesel has been produced *via* a catalytic reaction involving hydroprocessing and/or decarboxylation/decarbonylation of triglycerides from various agricultural feedstocks.[38,39] In the work conducted by Larnaudie *et al.*, renewable diesel was produced through the hydroprocessing of triglycerides, which can be derived from animal, plant or vegetable oils.[40] However, the renewable diesel produced using these feedstocks may result in a shortage of edible oils, impacting food prices, while utilizing animal fats, used frying oils, and waste oils as feedstock could mitigate production costs, yet these resources are limited and may not meet the growing demand for renewable fuels.[41,42] Hence, the biological conversion of the second generation of feedstock known as lignocellulose biomass emerges as an alternative approach to this particular process in order to reduce the dependency on the fossil fuels.

Notably, lignocellulosic biomass represents a promising option as feedstock for renewable-diesel production, considering its output/input energy ratio, availability, and low cost.[42] The bio-oil currently produced *via* lignocellulosic hydroprocessing approach will further react with hydrogen, forming hydrogenation-derived renewable diesel (HDRD). The HDRD produced mainly contains components such as alkanes, aromatic compounds, and alkyl side chains (larger molecule alkyl groups: it can be a linear, branched or cyclic compound).[43,44] The chemical structures of HDRD components will determine the physical properties of the HDRD. Notably, due to effective isomerization and hydrocracking processes in producing HDRD, it offers a lower cloud point and higher cetane number compared to commercial diesel. Hence, HDRD may become an ideal fuel for blending with fossil fuel.[44] Maize, wheat, rice, and sugarcane are the four agricultural crops with the highest production and are responsible for generating the majority of lignocellulosic biomass in the global agriculture sector, and are considered the ideal feedstocks for production of HDRD. Other than these sources of biomass, crop residues and agriculture waste, forest waste, and perennial plants have also been used as HDRD feedstocks.[45] Figure 8.2 depicts the Scopus network visualization map of lignocellulosic biomass for renewable-diesel production. Agricultural waste emerges as the primary feedstock, dominated by wheat straw, maize, switchgrass, sugarcane, and municipal

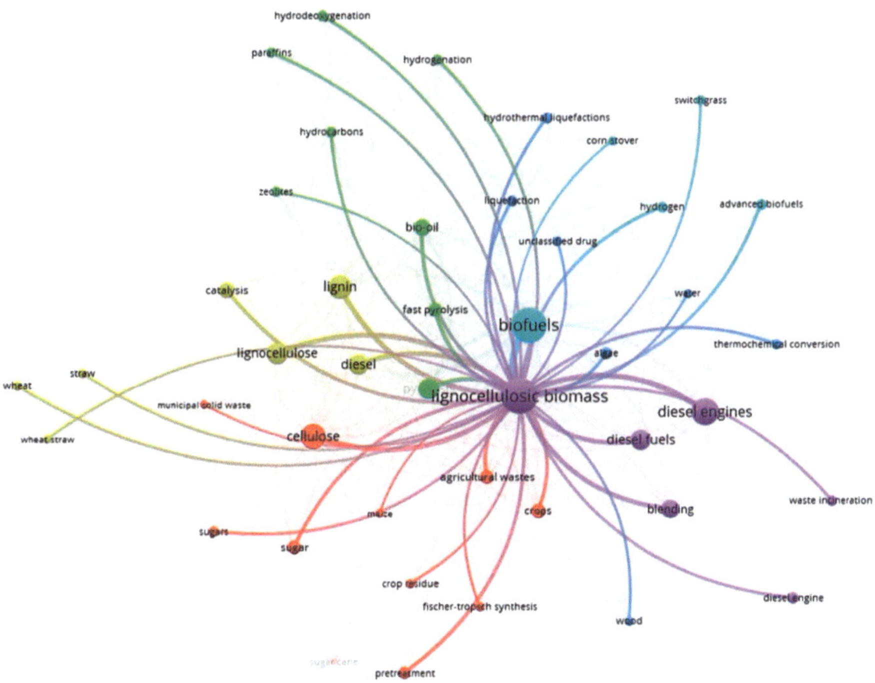

Figure 8.2 Scopus network visualization map of studies of lignocellulosic process into diesel.

solid waste. Intriguingly, algae feedstock has also been employed for HDRD production. The process for production of HDRD from lignocellulose is typically *via* hydrothermal conversion including pyrolysis, HTL and Fischer–Tropsch. Indeed, the process also needs the presence of hydrogen to promote both HDO and hydrogenation reactions. HDO and hydrogenation reactions are crucial steps in both HTL and FT of lignocellulosic processes since they can facilitate oxygen removal, making the resulting biofuels cleaner, more stable, and higher quality. By reducing harmful compounds and improving fuel properties like energy density and combustion efficiency, these reactions play a key role in producing HDRD that closely resembles traditional fossil fuels.

Additionally, catalysts such as zeolites are used to speed up these reactions, making the conversion process more efficient overall.[46] Catalysts are needed to enhance the bio-oil yield and augment the amount of aromatic hydrocarbons. Several studies have demonstrated that catalytic pyrolysis, compared to non-catalytic methods, enhances bio-oil yield and enriches it with desirable petrochemicals. According to research conducted by Christoforou *et al.*, pyrolysis of olive-mill solid waste with catalysts results in improved bio-oil yields and enhanced selectivity towards aliphatic and aromatic hydrocarbons, which indicated that catalytic pyrolysis offers promising avenues to produce bio-oils rich in valuable petrochemicals.[47] To improve the poor microwave absorption of switchgrass (SG) and speed up the heating rate of the catalytic pyrolysis process, K_3PO_4 (tripotassium phosphate) was combined with SG as a microwave absorber and catalyst, thus improving the properties of the biochar and gas and producing phenolic-rich bio-oil.[48] The samples of pre-mixed SG catalyst were put into the reactor, and N_2 was added to establish an anoxic environment for almost 30 min at a rate of 1.5 $L\,min^{-1}$. Based on the information provided, K_3PO_4 enhanced the selectivity for phenolic synthesis; for KP-300 and KP-400, the phenolic and aromatic compounds respectively increased by 385% and 303%.

However, the bio-oil obtained through pyrolysis cannot be directly used as a transportation fuel due to its high viscosity, high oxygen content (~39%), high corrosiveness due to high acidity, and thermal instability.[49–51] Therefore, to replace partially or fully transportation fuel with bio-oil, it is necessary to improve the quality of bio-oil by upgrading through HDO or decarboxylation/decarbonylation (DO; a catalytic thermal process for oxygenated species that removes CO_2/CO and H_2O without the presence of hydrogen).[52] Indeed, both HDO and DO methods are crucial in converting lignocellulosic biomass-based intermediates like bio-oil into sustainable renewable-diesel fuels.[53] Benzu *et al.* and Mohammad *et al.* reported that hydrogen will facilitate the breaking of the carbon–oxygen bond *via* the HDO reaction for oxygen elimination.[54,55] During direct HDO, the C=O double bond is broken first, which produces alcohol.[56] Oxygen is then eliminated as CO_2 during the decarboxylation process by directly attacking the C–C single bond.[57] Under minimal working circumstances (such as low temperature and pressure), catalytic HDO might be carried out, providing effective fuel generation.[58,59] According to Hachemi *et al.*, HDO of fatty acids was conducted in a semi-batch reactor operating at 300 °C and 30 bar H_2.[60] The information provided stated that the principal by-products generated were aliphatic hydrocarbons, specifically *n*-heptadecane and *n*-octadecane, which were produced in varying proportions using various catalysts. Similarly, in the work of Nur Azreena *et al.*, the catalyst La(10)HZSM5(90) had the highest overall HDO activity, producing 97% hydrocarbons with 92% diesel selectivity.[61]

Other than lignocellulosic biomass, second-generation feedstocks also include non-edible oil, waste cooking oil (WCO) and animal fats that can be used as renewable-diesel feedstocks. There are vast numbers of non-edible oil plants in nature; these include neem (*A. indica*), jatropha (*J. curcas*), karanja (*P. pinnata*), tobacco seed (*N. tabacum* L.), rice bran, mahua (*M. indica*), rubber plant (*H. brasiliensis*), castor, linseed, ceiba, sterculia and microalgae.[62] Indeed, utilization of jatropha oil is widely reported for production of renewable diesel, which mainly involved HDO and DO processes.[63,64] The most common process for transformation of these feedstocks is *via* HDO and DO. Indeed, these reactions need the presence of a catalyst, which is mainly a noble metal or metal oxide-based catalyst such as Pd, Pt, Ni, Mo, Co, Cu, or W. Due to cost constraints of noble metal catalysts, many researchers have utilized metal oxide-based catalysts. Indeed, Ni showed the most potential as a metal oxide. Recently, Adzahar *et al.* successfully deoxygenized ceiba oil into renewable diesel at 350 °C using Ni supported on red mud (Ni/RM3) with 86% hydrocarbon yield.[65] A high yield of diesel fractions (89%) with product selectivity toward *n*-($C_{15}+C_{17}$) (79%) was observed by Asikin-Mijan *et al.* using NiO–Fe_2O_3 supported on multi-walled carbon nanotubes (MWCNT) from jatropha oil.[66]

Notably, compared with monometallic catalysts, bimetallic catalysts showed effective HDO and DO activities. This was shown in Gamal *et al.*'s study of palm fatty acid distillate (PFAD) DO over CoMo/AC.[67] Based on the study results, the bimetallic catalyst $Co_{10}Mo_{10}$/AC exhibited excellent catalytic performance with 92% yield of hydrocarbon components (C_8–C_{20}) and 89% selectivity for *n*-($C_{15}+C_{17}$), with a total acid number of 24 mg KOH per mg. The high effectiveness of $Co_{10}Mo_{10}$/AC is due to the super acid–base sites that result from synergism between the CoMo and AC support. Wan Khalit *et al.* studied the DO of WCO using $Ni_{20}Zn_{10}$/AC and the result showed superior DO reaction by yielding 86% hydrocarbons and 79% *n*-($C_{15}+C_{17}$) selectivity.[68] Similarly effective was DO of WCO reported by Abdulkareem-Alsultan *et al.* using Ag_2O_3–La_2O_3/AC nano-catalyst.[69] Notably, Wan Khalit *et al.*'s studies investigated the fuel properties of pure renewable diesel (G100) and it was revealed that the renewable diesel exhibited outstanding fuel properties (acid value, kinematic viscosity, flash point, cetane index, and calorific value) in comparison with ultra-low sulfur diesel (ULSD), suggesting that G100 can be used in vehicle engines without modification and has great potential for commercialization. A recent report studied DO of chicken fat oil over binary metal pairs (Ni–Mg, Ni–Mn, Ni–Cu, Ni–Ce) supported on MWCNTs.[70] Indeed, oxygen-free rich diesel-range hydrocarbons were produced efficiently by $Ni_{10}Mg_{15}$/MWCNT and $Ni_{10}Mn_5$/MWCNT catalysts with a yield of >84% hydrocarbons, with n-($C_{15}+C_{17}$) selectivity of >85%. The heating value of the obtained renewable diesel complied with the ultra-low sulfur diesel standard. In the case of HDO, several studies have reported conversion of fatty acids into renewable diesel; nevertheless, the majority of reactions focus on model compounds such as palmitic acid, oleic acid, stearic acid, guaiacol, phenol, *etc.*[49] Similar to DO catalysts, many of the effective catalysts are mixed-metal oxide catalysts like NiMo, Ni_2P, Pd–Fe, and Pt–Re, which are supported on materials with high surface area and high porosity such as zeolites, Al_2O_3, SBA-15, MCM-41, and carbon supports. Our recent studies have reported that a renewable-diesel blend produced lower CO_2 and SO_2 emissions than petroleum diesel as a result of the rich oxygen-free fuel content.[71,72] The obtained fuel properties of pure renewable diesel and blended renewable diesel are in compliance with ASTM D6751, ASTM D240-17, and EN 14214 standards. Indeed, based on these findings, it is shown that blended renewable diesel is a clean fuel for the environment and a promising alternative fuel for internal combustion engines. Table 8.1 presents a summary of previous studies on non-edible feedstocks and their synthesis methods.

8.5 Current Situation

Second-generation biomass materials are seen as promising feedstocks for producing renewable diesel. However, several related sustainability issues require careful planning, technological innovation, and strong policy frameworks to ensure that the benefits of these feedstocks are fully utilized. One of the existing issues in second-generation biomass is indirect land-use changes to expand cultivation of biomass crops. Natural ecosystems such as forest are converted into agricultural land and this may result in biodiversity loss. Furthermore, energy efficiency also is a concern, where the process of converting second-generation biomass into renewable diesel requires high energy. Poor waste management may also contribute to diminish the sustainability of the process. In terms of the socioeconomic impact, biomass cultivation can provide economic opportunities for rural areas; however, it also can lead to land grabbing with consequences of social inequality in rural development. Besides, the production of renewable diesel from second-generation biomass may not always be economically viable without subsidies or incentives from the government. Full support from the government and local authorities will assure that the production of renewable diesel from second-generation biomass is entirely implemented.

The post-pandemic world has shown one of the most catastrophic disruptions in demand in bioenergy history. With the outbreak of COVID-19, the world experienced a huge economic crisis, with a steep decline in fuel demand and price. The crisis has generally shocked the renewable energy sector and caused a shutdown of many factories in China that accounted for 50% of the global wind-power supply chain. Seyed *et al.* analysed the global

Table 8.1 Summary of studies on non-edible feedstocks and their synthesis methods.

Biomass feedstock	Reaction	Conditions	Product	Ref.
Dried olive husk and olive kernels	Catalytic fast pyrolysis	Temperature: 500 °C, ZMS-5 zeolite catalyst, N_2 gas	Bio-oil yield of 53.62%	47
Switchgrass (lignocellulosic biomass)	Microwave catalytic pyrolysis	Temperature: 300 °C, K_3PO_4 catalyst, 1.5 L min^{-1} N_2 gas	Bio-oil (selectivity for phenol and alkylphenol increased by 385% and 420%)	48
PFAD	Catalytic deoxygenation	Time: 2 h, temperature: 350 °C, catalyst loading: 3 wt%	93% (C_8–C_{20}) and 90% ($C_{15}+C_{17}$)	67
Waste cooking oil (WCO)	Catalytic deoxygenation	Continuous N_2 gas flow, temperature: 350 °C, catalyst loading: 3 wt%	86% hydrocarbons with 79% selectivity for *n*-($C_{15}+C_{17}$)	68
Ceiba oil	Catalytic deoxygenation	Semi-batch reactor, time: 2 h, temperature: 350 °C, catalyst loading: 3 wt%	Saturated diesel *n*-($C_{15}+C_{17}$)	65
Chicken fat oil (CFO)	Catalytic deoxygenation	Semi-batch reactor, time: 2 h, temperature: 350 °C, catalyst loading: 3 wt%	Hydrocarbon yields of >75% with *n*-($C_{15}+C_{17}$) selectivity of >81%	70
PFAD	Hydrodeoxygenation	Stainless-steel autoclave, time: 2 h, temperature: 400 °C, pressure: 5 MPa of H_2	Superior diesel (C_{13}–C_{18}) with 93% selectivity	61
Tall oil fatty acids (pulp)	Hydrodeoxygenation	Semi-batch reactor, time: 6 h, temperature: 300 °C, pressure: 30 bar	C_{17} and C_{18} with >75% conversion	60
Jatropha curcas oil	Pyrolytic deoxygenation	Semi-batch reactor, time: 1 h, N_2 gas, catalyst loading: 3 wt%	79% *n*-($C_{15}+C_{17}$) selectivity	66
Waste cooking oil	Pyrolytic deoxygenation	Semi-batch reactor, time: 120 min, catalyst loading: 1 wt%, 350 °C	~89% yield and excellent *n*-($C_{15}+C_{17}$) selectivity (93%)	69

challenges and pressing issues of the biomass supply chain under the COVID-19 crisis. The findings revealed a significant worldwide decline in woody residue and palm oil biomass production, supply, export, and demand. At the same time, oil prices and energy production were adversely impacted. It is observed that electricity generation from biomass may decelerate due to supply-chain disruptions and logistical issues in delivering solid biomass to large power facilities.[73]

Recently, Pattnaik *et al.* also listed some challenges for commercialization of biorefineries, including technology advancement, scale-up challenges, lack of investment and biomass availability.[74] The complex structure of lignocellulosic biomass is a primary obstacle to the efficient use of biomass. It is necessary to have pre-treatment for lignin removal to eliminate the complex nature of lignocellulosic biomass to ease further reaction processes.

8.6 Conclusion

Second-generation liquid biomass is a promising enhancement in renewable-diesel production, since it utilizes non-food feedstocks, reducing competition with food resources and waste to promote environmental sustainability. The numerous conversion technologies, including thermochemical methods such as gasification, pyrolysis, and hydrothermal liquefaction, are emphasized for their efficiency in creating hydrocarbons suited for diesel applications. Biochemical approaches, on the other hand, highlight the enzymatic conversion of biomass into biofuels. They provide effective paths for converting this biomass into renewable diesel, minimizing lifecycle greenhouse-gas emissions compared to fossil fuels. The environmental benefits of this strategy are consistent with global sustainability goals, since they reduce carbon intensity and reliance on traditional energy sources. However, the long-term viability of second-generation biomass is dependent on continued research and development to optimize conversion processes, increase economic feasibility, and improve overall technology. Exploration into the effectiveness of various biomass types, catalyst advancement, and process optimization are critical for progressing the technology while rendering it competitive with fossil fuels. In conclusion, as technologies grow more advanced and more cost-effective, they may play an important part in moving to a more ecologically friendly energy future. Persistent advancements in technology improvements are critical to realizing the entirety of its potential in the energy industry.

References

1. H. Luo, T. Zhou, R. Zhang, Q. Yang, X. You, S. Wang, J. Wang, F. Xie and R. Yang, *Ind. Crops Prod.*, 2024, **22015**, 119271.
2. M. Deshmukh and A. Pathan, *Renewable Sustainable Energy Rev.*, 2024, **192**, 114270.

3. Manure Biogas Plants, https://www.host-bioenergy.com/solutions/biogas-plants/manure-biogas-plants/ (accessed August 2024).
4. M. M. I. Afifi, A. A. El-Shehawy and F. A. A. Ali, *Pol. J. Chem. Technol.*, 2024, **26**(2), 48.
5. A. Tanksale, J. N. Beltramini and G. M. Lu, *Renewable Sustainable Energy Rev.*, 2010, **14**, 166.
6. C. E. Wyman, B. E. Dale, R. T. Elander, M. Holtzapple, M. R. Ladisch and Y. Y. Lee, *Bioresour. Technol.*, 2005, **96**, 1959–1966.
7. P. Kumar and L. Fiori, *Energy Convers. Manage: X*, 2024, **23**, 100659.
8. A. K. Rai, N. H. Al Makishah, Z. Wen, G. Gupta, S. Pandit and R. Prasad, *Fermentatio*, 2022, **8**(4), 161.
9. A. G. Patel, N. K. Maheshwari, P. K. Vijayan and R. K. Sinha, in *Sixteenth Annual Conference of Indian Nuclear Society (INSAC-2005)*, ed. S. Ganesan, Indian Nuclear Society, India, 2005, ch. 8.
10. A. Demirbaş, *Fuel*, 2001, **80**, 1885.
11. D. Das and T. N. Veziroğlu, *Int. J. Hydrogen Energy*, 2001, **26**, 13.
12. J. Zhang and X. Zhang, in *Biomass, Biopolymer-Based Materials, and Bioenergy: Construction, Biomedical, and other Industrial Applications*, ed. D. Verma, E. Fortunati, S. Jain and X. Zhang, Elsevier, Netherland, 2019, p. 327.
13. S. Ganguli, A. Somani, R. K. Motkuri and C. Bloyd, *India Alternative Fuel Infrastructure: The Potential for Second-generation Biofuel Technology*, United States, 2018. https://www.osti.gov/servlets/purl/1530891 (accessed August 2024).
14. M. Shahabuddin, M. T. Alam, B. B. Krishna, T. Bhaskar and G. Perkins, *Bioresour. Technol.*, 2020, **312**, 123596.
15. S. R. Chia, S. Nomanbhay, M. Y. Ong, A. H. B. in Shamsuddin, K. W. Chew and P. L. Show, *Fuel*, 2022, **314**, 123137.
16. E4Tech, Review of Technologies for Gasification of Biomass and Wastes, 2009. http://wiki.gekgasifier.com/f/Review+of+Biomass+Gasification+Technologies. NNFCC. Jun09.pdf (accessed August 2024).
17. S. Jones, D. Elliott, C. Kinchin, C. Valkenburg, J. Holladay, S. Czernik and C. D. Walton Stevens, *Production of Gasoline and Diesel from Biomass via Fast Pyrolysis, Hydrotreating, and Hydrocracking*, United State, 2009. https://www.osti.gov/servlets/purl/1218327 (accessed August 2024).
18. IRENA And Methanol Institute, *Innovation Outlook: Renewable Methanol*, International Renewable Energy Agency, Abu Dhabi, 2021.
19. W. P. Chan, A. Veksha, J. Lei, W. D. a Oh, X. Dou, A. Giannis, G. Lisak and T. T. Lim, *Appl. Energy*, 2019, **237**, 227.
20. S. Spatari, V. Larnaudie, I. Mannoh, M. C. Wheeler, N. A. Macken, C. A. Mullen and A. A. Boateng, *Renewable Energy*, 2020, **162**, 371.
21. M. I. Jahirul, M. G. Rasul, A. A. Chowdhury and N. Ashwath, *Energies*, 2012, **5**, 4952.
22. N. Asikin-Mijan, N. A. Rosman, G. AbdulKareem-Alsultan, M. S. Mastuli, H. V. Lee, N. Nabihah-Fauzi, I. M. Lokman, F. A. Alharthi, A. A. Alghamdi, A. A. Aisyahi and Y. H. Taufiq-Yap, *Process Saf. Environ. Prot.*, 2020, **142**, 336.

23. T. O. Olugbade and O. T. Ojo, *Bioenergy Res.*, 2020, **13**, 999.
24. M. Patel, A. O. Oyedun, A. Kumar and R. Gupta, *Fuel Process. Technol.*, 2019, **191**, 79.
25. T. A. Mamvura and G. Danha, *Heliyon*, 2020, **6**(3), e30531.
26. P. N. Y. Yek, Y. W. Cheng, R. K. Liew, W. A. Wan Mahari, H. C. Ong, W. H. Chen, W. Peng, Y. K. Park, C. Sonne, S. H. Kong, M. Tabatabaei, M. Aghbashlo and S. S. Lam, *Renewable Sustainable Energy Rev.*, 2021, **151**, 111645.
27. L. D. B. Pestaño and W. I. José, in *Gasification for Low-grade Feedstock*, ed. Y. Yun, InTechOpen Limited, United Kingdom, 2018, p. 43.
28. Y. Y. Gan, H. C. Ong, T. C. Ling, W. H. Chen and C. T. Chong, *Energy*, 2019, **170**, 367.
29. H. Shahbeik, H. Kazemi Shariat Panahi, M. Dehhaghi, G. J. Guillemin, A. Fallahi, H. Hosseinzadeh-Bandbafha, H. Amiri, M. Rehan, D. Raikwar, H. Latine, B. Pandalone, B. Khoshnevisan, C. Sonne, L. Vaccaro, A. S. Nizami, V. K. Gupta, S. S. Lam, J. Pan, R. Luque, B. Sels, W. Peng, M. Tabatabaei and M. Aghbashlo, *Renewable Sustainable Energy Rev.*, 2024, **189**, 11396.
30. L. Grande, I. Pedroarena, S. A. Korili and A. Gil, *Materials*, 2021, **14**(18), 5286.
31. B. B. Marangon, J. S. Castro, P. P. Assemany, E. A. Couto and M. L. Calijuri, *Renewable Sustainable Energy Rev.*, 2022, **155**, 111910.
32. S. Masoumi and A. K. Dalai, *Biomass Bioenergy*, 2021, **151**, 106168.
33. V. Balan, *ISRN Biotechnol.*, 2014, **2014**, 463074.
34. G. Kumar, J. Dharmaraja, S. Arvindnarayan, S. Shoban, P. Bakonyi, G. D. Saratale, N. Nemestóthy, K. Bélafi-Bakó, J. J. Yoon and S. H. Kim, *Fuel*, 2019, **251**, 352.
35. J. Y. Lee, S. E. Lee and D. W. Lee, *Crit. Rev. Environ. Sci. Technol.*, 2022, **52**, 2453.
36. V. A. Marulanda, C. D. B. Gutierrez and C. A. C. Alzate, in *Advanced Bioprocessing for Alternative Fuels, Biobased Chemicals, and Bioproducts: Technologies and Approaches for Scale-Up and Commercialization*, ed. M. Hosseini, Elsevier, Netherland, 2019, p. 59.
37. Kumar D., PhD thesis, Oregon State University, 2015.
38. C. H. Ko, S. H. Park, J. K. Jeon, D. J. Suh, K. E. Jeong and Y. K. Park, *Korean J. Chem. Eng.*, 2012, **29**, 1657.
39. M. H. Hassan and M. A. Kalam, *Procedia Eng.*, 2013, **56**, 39.
40. V. Larnaudie, M. Bule, K. Y. San, P. V. Vadlani, J. Mosby, S. Elangovan, M. Karanjikar and S. Spatari, *Fuel*, 2020, **279**, 118429.
41. A. Yousuf, *Waste Manage.*, 2012, **32**, 2061.
42. T. Koizumi, *Renewable Sustainable Energy Rev.*, 2015, **52**, 829.
43. P. Wang and X. Lü, in *Advances in 2nd Generation of Bioethanol Production*, ed. X. Lu, Elsevier, Netherland, 2021, p. 1.
44. D. Tuli and S. Kasture, in *Advanced Biofuel Technologies: Present Status, Challenges and Future Prospects*, ed. D. Tuli, S. Kasture and A. Kuila, Elesevier, Netherland, 2022, p. 119.

45. A. Favero, A. Daigneault and B. Sohngen, *Sci. Adv.*, 2020, **6**(13), eaay6792.
46. Y. L. Ding, H. Q. Wang, M. Xiang, P. Yu, R. Q. Li and Q. P. Ke, *Front. Chem.*, 2020, **8**, 564465.
47. E. A. Christoforou, P. A. Fokaides, S. W. Banks, D. Nowakowski, A. V. Bridgwater, S. Stefanidis, K. G. Kalogiannis, E. F. Iliopoulou and A. A. Lappas, *Waste Biomass Valorization*, 2018, **9**, 301.
48. B. A. Mohamed, M. Bilal, E. S. Salama, S. Periyasamy, I. M. R. Fattah, R. Ruan, M. K. Awasthi and L. Leng, *J. Cleaner Prod.*, 2022, **366**, 132668.
49. M. X. Y. Ravindran, N. Asikin-Mijan, G. AbdulKareem-Alsultan, H. C. Ong, M. M. Nurfarhana, H. V. Lee, T. A. Kurniawan, D. Derawi, S. F. M. Yusoff, I. M. Lokman and Y. H. Taufiq-Yap, *J. Environ. Chem. Eng.*, 2024, **12**, 112330.
50. P. Prasertpong, C. Jaroenkhasemmeesuk, N. Tippayawong and Y. Thanmongkhon, *Chiang Mai Univ. J. Nat. Sci.*, 2017, **16**, 135.
51. D. Martins-Vieira, J. C. Missau, J. Anshu, K. Siakpebru, O. K. Thengane, S. K. Morais, A. R. C. Tanabe, E. H. Bertuol, D. A. Review, D. Lachos-Perez, J. Cláudio Martins-Vieira, J. Missau, K. Anshu, O. K. Siakpebru, S. K. Thengane, A. Rita, C. Morais, E. H. Tanabe and D. Assumpção Bertuol, *Analytica*, 2023, **4**, 182.
52. M. Patel and A. Kumar, *Renewable Sustainable Energy Rev.*, 2016, **58**, 1293.
53. M. Zhang, Y. Hu, H. Wang, H. Li, X. Han, Y. Zeng and C. C. Xu, *Mol. Catal.*, 2021, **504**, 111438.
54. S. Bendezú, R. Cid, J. L. G. Fierro and A. L. Agudo, *Appl. Catal., A*, 2000, **197**(1), 47.
55. M. Mohammad, T. Kandaramath Hari, Z. Yaakob, Y. Chandra Sharma and K. Sopian, *Renewable Sustainable Energy Rev.*, 2013, **22**, 121.
56. N. Arun, R. V. Sharma and A. K. Dalai, *Renewable Sustainable Energy Rev.*, 2015, **48**, 240.
57. C. Dupont, R. Lemeur, A. Daudin and P. Raybaud, *J. Catal.*, 2011, **279**, 276.
58. M. M. Ambursa, J. C. Juan, Y. Yahaya, Y. H. Taufiq-Yap, Y. C. Lin and H. V. Lee, *Renewable Sustainable Energy Rev.*, 2021, **138**, 110667.
59. T. M. H. Dabros, M. Z. Stummann, M. Høj, P. A. Jensen, J. D. Grunwaldt, J. Gabrielsen, P. M. Mortensen and A. D. Jensen, *Prog. Energy Combust. Sci.*, 2018, **68**, 268.
60. I. Hachemi, N. Kumar, P. Mäki-Arvela, J. Roine, M. Peurla, J. Hemming, J. Salonen and D. Y. Murzin, *J. Catal.*, 2017, **347**, 205.
61. I. Nur Azreena, H. L. N. Lau, N. Asikin-Mijan, M. A. Hassan, S. M. Izham, M. Safa Gamal, W. Nor Adira Wan Khalit, M. Arumugam, E. Kennedy, M. Stockenhuber and Y. H. Taufiq-Yap, *J. Anal. Appl. Pyrolysis*, 2022, **161**, 105406.
62. E. F. Aransiola, E. Betiku, D. Ikhuomoregbe and T. Ojum, *Afr. J. Biotechnol.*, 2012, **11**(22), 6178.
63. S. Thongkumkoon, W. Kiatkittipong, U. W. Hartley, N. Laosiripojana and P. Daorattanachai, *Renewable Energy*, 2019, **140**, 111.

64. K. W. Jeon, J. O. Shim, J. W. Cho, W. J. Jang, H. S. Na, H. M. Kim, Y. L. Lee, B. H. Jeon, J. W. Bae and H. S. Roh, *Fuel*, 2019, **236**, 928.
65. N. A. Adzahar, N. Asikin-Mijan, M. I. Saiman, G. A. Alsultan, M. S. Mastuli, M. R. Shamsuddin and Y. H. Taufiq-Yap, *RSC Adv.*, 2022, **12**, 16903.
66. N. Asikin-Mijan, N. A. Rosman, G. AbdulKareem-Alsultan, M. S. Mastuli, H. V. Lee, N. Nabihah-Fauzi, I. M. Lokman, F. A. Alharthi, A. A. Alghamdi, A. A. Aisyahi and Y. H. Taufiq-Yap, *Process Saf. Environ. Prot.*, 2020, **142**, 336.
67. M. Safa-Gamal, N. Asikin-Mijan, W. N. A. Wan Khalit, M. Arumugam, M. I. Saiman and Y. H. Taufiq-Yap, *Fuel Process. Technol.*, 2020, **208**, 106519.
68. W. N. A. Wan Khalit, N. Asikin-Mijan, T. S. Marliza, M. Safa-Gamal, M. R. Shamsuddin, I. N. Azreena, M. I. Saiman and Y. H. Taufiq-Yap, *J. Anal. Appl. Pyrolysis*, 2022, **164**, 105505.
69. G. Abdulkareem-Alsultan, N. Asikin-Mijan, N. Mansir, H. V. Lee, Z. Zainal, A. Islam and Y. H. Taufiq-Yap, *J. Anal. Appl. Pyrolysis*, 2019, **137**, 171.
70. N. Aliana-Nasharuddin, N. Asikin-Mijan, G. Abdulkareem-Alsultan, M. I. Saiman, F. A. Alharthi, A. A. Alghamdi and Y. H. Taufiq-Yap, *RSC Adv.*, 2020, **10**, 626.
71. Using a Renewable Fuel to Cut GHG Emissions, https://www.neste.us/neste-my-renewable-diesel/product-information/biofuels (accessed August 2024).
72. A. G. Alsultan, N. Asikin Mijan, N. Mansir, S. Z. Razali, R. Yunus and Y. H. Taufiq-Yap, *ACS Omega*, 2021, **6**, 408.
73. M. Z. Seyed, N. Shiwakoti and P. Stasinopoulos, *Clean. Logist. Supply Chain.*, 2022, **4**, 100049.
74. Challenges for Commercialization of Second-Generation Biorefineries, https://encyclopedia.pub/entry/55134 (accessed August 2024).

CHAPTER 9

Biorefining of Third-generation Algal Biomass into Green Fuels and High-value Products

IFTIKHAR HUSSAIN, KHALIL UL REHMAN, ISHFAQ MAJEED MIR, RAQEEBA AZIZ AND MOHAMMAD ASLAM*

Biofuels Laboratory, Department of Chemistry, National Institute of Technology Srinagar (J&K), India
*Email: maslam@nitsri.ac.in

9.1 Introduction

According to UN forecasts, the world population is expected to reach 8.6 billion in 2030, 9.8 billion in 2050 and 11.2 billion in 2100, with developing nations mostly in Asia and Africa accounting for the majority of this rise. The increase in the world's population has resulted in a significant rise in energy usage, especially in the fields of transportation and industry.[1] The most abundant source of primary energy worldwide and the primary fuel source is crude oil. An important part of the world economy is the crude oil sector. Crude oil reserves are an estimate of the amount of oil that has not yet been extracted from the earth in a given nation. Petroleum refineries process it to create a variety of fuels for transportation, including aviation fuel, diesel, gasoline, and other petroleum-based goods. The top ten producers and consumers of crude oil in the world are shown in Figures 9.1 and 9.2, respectively.[2] The majority of the proven oil reserves worldwide are found in OPEC members (79.5%), with Venezuela at the top of the list. A rise in the

Green Chemistry Series No. 89
Biorefining of Renewable Diesel: From Production to Market
Edited by Mohammad Aslam, Sanjeev Mishra and Jorge Aburto

Published by the Royal Society of Chemistry, www.rsc.org

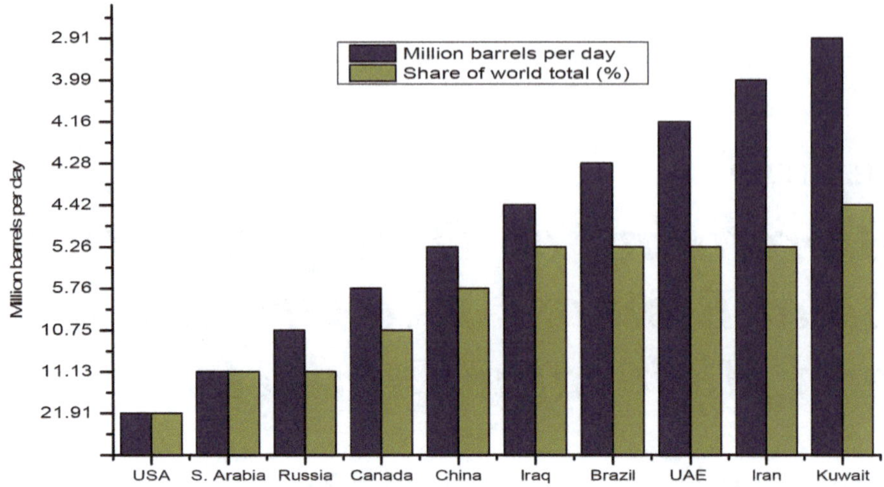

Figure 9.1 Top ten producers of crude oil in the world (EIA, USA).

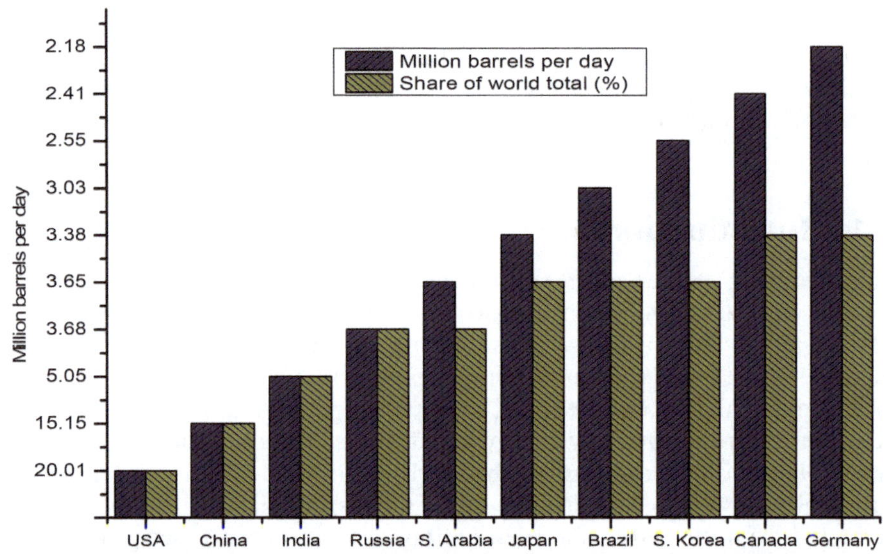

Figure 9.2 Top ten consumers of crude oil in the world (EIA, USA).

price of crude oil may cause inflation, unstable economies, and more rivalry for the few remaining oil supplies. Furthermore, the primary energy source at the moment is crude petroleum, which is non-renewable and contributes significantly to greenhouse-gas (GHG) emissions, global warming, and climate change. Finding a sustainable substitute for crude oil is therefore crucial.[3,4]

Biofuels made from renewable resources are unquestionably a practical solution for meeting future energy needs while lowering carbon emissions in this risky situation.[5] Algal biomass appears to be a potential renewable energy source that can be used for the manufacture of green fuels associated with significantly lower GHG emissions, especially in view of the greater carbon emissions linked to finite petroleum reserves. Algal biomass is now an excellent feedstock for the synthesis of green hydrocarbons and value-added products, despite its extremely low commercial value. Algae, including macro- and microalgae, are utilized as a base feedstock in third-generation (3-G) biorefineries and as a practical means of initiating the production of biorenewables. Algae may be utilized to create a wide range of products because they are high in lipids, proteins, and carbohydrates. Algae biorefineries typically use a combination of biochemical, chemical, and thermochemical processes to extract and transform these components into a variety of products, such as biofuels (like biodiesel and bioethanol), biochemicals (like pigments, antioxidants, and nutraceuticals), animal feed, fertilizer, nutraceuticals, and even pharmaceuticals.[6]

A biorefinery is a refinery that converts biomass into fuels, energy and other valuable products (Figures 9.3 and 9.4). The steps involved in converting biomass take place in a biorefinery plant, which gathers together the facilities, equipment configurations, and technologies required for converting a variety of biomass into chemicals, fuels, and electricity. "A biorefinery is a facility that integrates biomass conversion processes and equipment to produce fuels, power, and value-added chemicals from biomass", states the National Renewable Energy Laboratory (NREL), USA. The definition of biorefining given by IEA Bioenergy Task 42 is among the most thorough: "The sustainable processing of biomass into a spectrum of bio-based products (food, feed, chemicals, and materials) and bioenergy (biofuels, power and/or heat)". The biorefinery concept is similar to a traditional petroleum refinery that produces multiple fuels and products from petroleum. The objective of a biorefinery is to convert biomass into marketable fuels, chemicals, or compounds. Modern biorefineries or advanced biorefineries aim to explore the utilization of lignocellulosic biomass, algal biomass into energy, fuels, and value-added chemicals.[7]

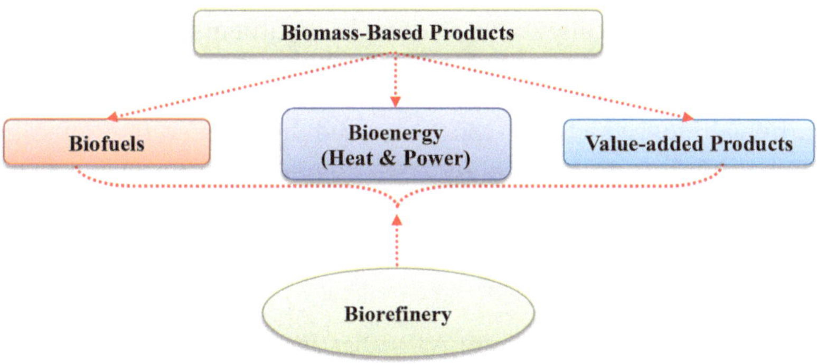

Figure 9.3 Biorefinery concept.

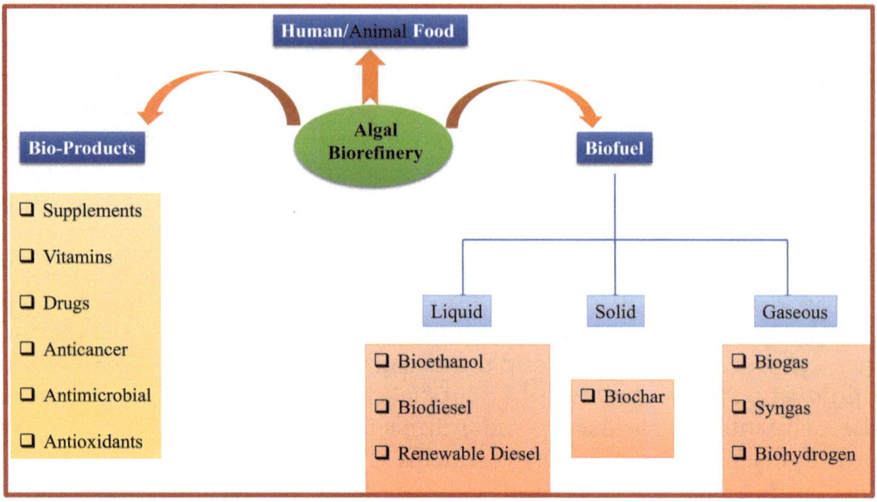

Figure 9.4 An overview of algal biomass-based biorefining.

The most effective way to generate revenue in a biorefinery is by recovering multiple products through various processing routes *viz* thermo-chemical routes, chemical routes and biochemical routes. Considering the importance of 3-G algal biorefining, the present chapter covers the conversion of algal biomass into green fuels and its contribution to sustainability by reducing the emission of greenhouse gases.

9.2 Algae

Algae are photosynthetic organisms that thrive in aquatic environments, producing energy from sunlight through photosynthesis similarly to plants. These organisms are incredibly diverse, and can be found in various forms of water bodies, including fresh water, salt water, brackish water, and even in waste-water bodies. These microorganisms are photosynthetic and utilize water, carbon dioxide, and sunlight to produce carbohydrates, proteins, lipids, and other biologically active chemical constituents. They exist either as single cells or in the form of colonies or filaments.[8] Algae only need CO_2, nutrients, H_2O and sunlight to grow and produce large amounts of lipids and carbohydrates which can be processed into different biofuels.[9] According to the size and morphological characteristics, algal species are usually classified into microalgae and microalgae.

9.2.1 Microalgae

Microalgae are single-celled organisms that live in marine and freshwater habitats, particularly in benthic and littoral zones. They are photosynthetic organisms containing chlorophyll as their primary photosynthetic pigment.

Due to their simpler cellular structure, microalgae are more effective than macroalgae in converting solar radiation into biomass *via* photosynthesis. Microalgae are easily cultivated in a photoreactor because they are phototrophic – they need solar energy for growth. Microalgae are classified into several groups based on their pigment composition, including Chlorophyceae, Phaeophyceae, Pyrrophyceae, Chrysophyceae, Bacillariophyceae, and Rhodophyceae. They play an important part in aquatic photosynthesis and produce enormous amounts of bioactive ingredients that have applications in the pharmaceutical, food, cosmetics, and bioenergy industries and in phycoremediation. These chemical compounds include important fatty acids like eicosapentaenoic acid and docosahexaenoic acid, as well as pigments like carotenoids (*e.g.* fucoxanthin, β-carotene, lutein, and astaxanthin). These compounds are in great demand to meet the nutritional needs and manage acute and chronic diseases of the growing population of the world. Microalgae are also prized in the cosmetic and biofuel industries because of their high pigment and fatty-acid content. They can be used for the generation of a range of biofuels.[10] Comparatively speaking, macroalgae grow more slowly than microalgae and have a lower lipid content. High concentrations of proteins, lipids, and complex carbohydrates found in microalgae can be used as feedstocks to produce a range of bioproducts including biofuel.

9.2.2 Macroalgae

Macroalgae, commonly referred to as seaweeds, are multicellular and macroscopic autotrophic organisms found in marine and freshwater environments. They are distinguished from microalgae by their larger size and complex structure. They are taxonomically classified into three main groups based on the colour of their thallus (the main body of the algae), *viz.* Chlorophyta, Rhodophyta, and Phaeophyta.[11]

Algae are seen as a feasible alternative energy source to first-generation (1-G) and second-generation (2-G) feedstocks because of their rapid growth rate, low harvest time, high lipid content, greater CO_2 absorption, low land utilization and the fact that they do not compete with food crops for arable land, avoiding the food *vs.* fuel debate. The algal biomass can be converted into a variety of sustainable biofuels including bioethanol, biodiesel, renewable diesel, biogas, and bio-hydrogen by different conversion methods *via* thermochemical and biochemical processes.[12]

9.2.3 Biochemical Composition of Algae

Algal biomass is composed of three main components: proteins, carbohydrates, and fats. The chemical compositions of algae species are given in Table 9.1. Important chemicals like vitamins and pigments are also present in lesser amounts. The chemical makeup of different types of algae influences how suitable they are for particular applications, such as food supplements or biofuels. The composition of the algal biomass can be greatly

Table 9.1 Biochemical composition of some algae species (% dry weight basis).[13–16]

S. no	Algal species	Composition (% dry weight)		
		Protein	Carbohydrates	Lipids
1	*Acutodesmus obliquus*	30–45	14–18	14–17
2	*Scenedesmus rotundus*	33–35	19–44	29–35
3	*Porphyridium* spp.	42.9	32.1	17.34
4	*Chlorella vulgaris*	51–58	12–17	14–22
5	*Dunaliella salina*	57	32	6
6	*Nannochloropsis oculata*	63	15	11
7	*Odontella aurita*	39–61	14–18	22–38
8	*Spirulina platensis*	46–63	4–9	8–14

impacted by various environmental factors, including pH, temperature, light, nutrients, population density, and CO_2 and nitrogen levels.

9.3 Algae-based Biorefineries (3-G Biorefineries)

Algae-based biorefineries (3-G biorefineries) have advanced the transformation of biomass into fuels, cosmetics, synthetic chemicals, food, feed, pharmaceuticals and high-value products (Figure 9.5). Algae (microalgae and macroalgae) serve as a feasible starting material for the production of biorenewables. Algae, being rich in lipids, proteins, and carbohydrates, can be used to produce a variety of products. To extract and transform these components into a variety of products, such as biofuels, biochemical, animal feed, fertilizer, nutraceuticals and even pharmaceuticals, algae biorefineries usually use a combination of biochemical, chemical, and thermochemical processes.

Microalgae generate a wide range of compounds such as carotenoids, docosahexaenoic acid, eicosapentaenoic acid, bioactive substances, natural pigments, colours, polysaccharides, and compounds that help prevent cancer. Additionally, algal biomass can supply protein for dietary supplements and animal feed, along with carbohydrates, antioxidants, pigments, essential omega-3 and omega-6 fatty acids, vitamins, and various industrial compounds. Omega-6 fatty acids improve health by lowering the risk of arthritis, hypertension, and cardiovascular disease. Due to increased health consciousness, products like astaxanthin, β-carotene, and omega-6 fatty acids are in great demand in society.[17,18] Carotenoids, natural isoprenoids found in algae, have been identified for their anti-inflammatory, anti-aging, UV-radiation protection, immune system stimulation, and anticancer activities.[19] Phycobiliproteins have anti-inflammatory, neuroprotective, antiviral, anticancer, antidiabetic and antioxidant properties, making them a unique substance in the medical industry.[20] Microalgae are also a prolific source of many cosmetic ingredients. Furthermore, glycerol, a by-product of the algal biorefinery process, is widely used in the pharmaceutical and cosmetic industries. Phycocyanin, a blue pigment found in *Spirulina*, is commonly used

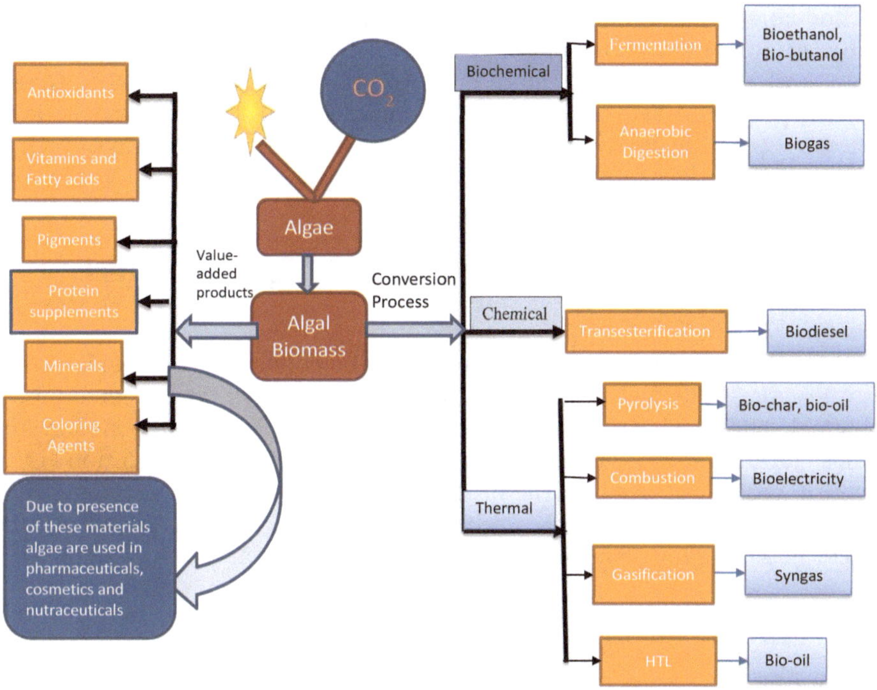

Figure 9.5 Systematic scheme of 3-G algal biorefinery.

as a colorant within the food and cosmetics industries. Lycopene, a carotenoid, is considered a powerful antioxidant that may neutralize oxygen-derived free radicals. When used as a sunblock agent, its antioxidant effectiveness exceeds that of lutein, tocopherol, and β-carotene.[21]

Algal biomass can also be used to produce biofuels. Anaerobic digestion of algal biomass yields bio-methane.[22] Fermentation of starch present in algal biomass yields bioethanol.[23] Photo-fermentation, dark fermentation, and bio-photosynthesis are all ways to make bio-hydrogen.[24] Biodiesel production can occur through traditional chemical catalysts, such as acid or alkali transesterification, or through enzyme-mediated transesterification. Enzymes, as natural catalysts, offer environmentally friendly conditions, making enzyme-based processes a promising alternative to address the limitations of chemical synthesis. Currently, about 95% of global biodiesel production relies on edible oils, while recycled oils and animal fats contribute around 10% and 6%, respectively. In the transesterification reaction, the triglycerides in algal lipids react with alcohol and a catalyst to produces fatty acid methyl esters (FAME) and glycerol.[25] Hydrothermal liquefaction (HTL), gasification, pyrolysis, and torrefaction are the thermal methods employed for the generation of liquid transportation fuels. However when it comes to biofuel generation, hydrothermal liquefaction outperforms gasification, pyrolysis, and torrefaction. This method converts a wider spectrum of biomass feedstocks into high-quality biofuels more efficiently, operates at moderate temperatures and pressures, and can handle wet material without substantial drying.[26] Green diesel is produced by deoxygenation of feedstocks containing triglycerides and fatty acids. This process typically involves three pathways: hydrodeoxygenation, decarboxylation, and decarbonylation.[27]

9.4 Conversion Technologies for Algal Biomass into Bio-oil

There are different techniques employed for the conversion of algal biomass into bio-oil. These techniques are classified into three types, namely biochemical, chemical, and thermal techniques. Biochemical technology includes fermentation, anaerobic digestion, *etc.* Transesterification is a chemical technique used for the production of biodiesel. Currently, thermal techniques are gaining much attention and mainly include pyrolysis, torrefaction, gasification and hydrothermal liquefaction and discussed below.

9.4.1 Thermochemical Conversion

The process of thermally breaking down the organic molecules present in biomass to produce biofuels is known as thermochemical conversion. The aim of thermochemical conversion is to reduce the amount of undesirable by-products produced. To convert algal biomass into useful products, thermochemical conversion methods such as HTL, pyrolysis, torrefaction, and gasification are frequently conducted. Algal biomass can be thermochemically converted into solid, liquid, and gaseous biofuels. Solid biofuels, such as biochar, are produced by torrefaction, while liquid biofuels, such as bio-oil, are produced by pyrolysis and HTL. Gaseous biofuels, such as syngas or synthetic gas (primarily composed of hydrogen and carbon monoxide), are produced by gasification.[28] Bio-oil is the main liquid biofuel that may be produced from algal biomass by direct thermochemical conversion. Compared to other fuels, bio-oil has numerous environmental benefits since it is a renewable resource. For instance, it is carbon neutral, has low sulfur and nitrogen content, and also releases fewer SO_x and NO_x pollutants into the atmosphere when burnt.[29] Furthermore, it is found that microalgal bio-oil has a higher heating value than cellulosic bio-oil. HTL and pyrolysis are the main thermochemical conversion processes employed for synthesizing bio-oil.

9.4.2 Hydrothermal Liquefaction for Bio-crude Oil Production

HTL is emerging as a viable method for sustainable energy conversion. One of its primary benefits is its capacity to process a diverse range of biomass feedstocks, including wet waste materials. This flexibility not only reduces feedstock pre-processing costs, but also has a smaller environmental impact. HTL is a thermochemical conversion technique that takes place in an aqueous environment at temperatures ranging from 250 °C to 400 °C and pressures of 2–25 MPa. This process yields a variety of important products (biocrude, biofuel, solids (biochar), and gases).[30] HTL takes place in subcritical water. In its subcritical phase, water has quite unique properties and acts as both a reactant and a catalyst in the reaction. At the subcritical point, water loses viscosity and organic molecules become more soluble. Water-insoluble compounds are now soluble in water as the dielectric constant of water drops in its subcritical phase. Water has a dielectric constant of 20 at 350 °C and 20 MPa, compared to 78 at 25 °C and 0.1 MPa. This also has an impact on the solubility of ionic compounds, because water becomes less polar near the subcritical point.[31]

Table 9.2 HTL study of various algae species.[35–40]

S. no	Algae species	Operating parameters	Catalysts	Bio-oil yield (wt%)
1	*Chlamydomonas nivalis* and *Nannochloropsis gaditana*	*C. nivalis*: 300 °C, 60 min, water/algae ratio of 4 *N. gaditana*: 300 °C, 30 min, water/algae ratio of 9	Cu-MOR, Cu-ZSM-5, Cu-SSZ13	18.8% (*C. nivalis*), 34.0% (*N. gaditana*)
2	*Kappaphycus alvarezii* (KA) and *Eucheuma denticulatum* (ED)	$T = 350$ °C	5 wt% neutral and alkali catalysts, such as Na_2CO_3, K_2CO_3, $CaCO_3$, Na_2SO_4, NaOH, and KOH	26.7% for Na_2CO_3 treatment of KA, 18.5% for KOH treatment of ED
3	*Fucus vesiculosus*	$T = 300$ °C, residence time = 20 min	Hβ zeolite	27.6%
4	*Chlorella vulgaris*	$T = 287$ °C, reaction time = 40 min, microalgae/water ratio = 0.07	Ni–Mo/Al_2O_3	56.21%
5	*Salicornia bigelovii* Torr. (SL) *Cyanidioschyzon merolae* (CM)	$T = 300$ °C, reaction time = 30 min	—	7.63% for SL, 34.63% for CM
7	*Nannochloropsis* sp.	$T = 300$ °C, $t = 30$ min	Ni/γ-Al_2O_3, Ni–Cu/γ-Al_2O_3, Ni–Re/γ-Al_2O_3, Ni–Cu–Re/γ-Al_2O_3 (10% Ni, 5% Cu, 2.5% Re)	38% with Ni/γ-Al_2O_3 53% with Ni–Re/γ-Al_2O_3 58% with Ni–Cu–Re/γ-Al_2O_3

HTL is a promising technology for converting algal biomass into bio-crude oil due to its ability to transform high moisture-content (>90%) algae without the requirement for drying, resulting in significant energy savings. HTL of various algal species are depicted in Table 9.2. To optimize biocrude production and quality in HTL, operational parameters such as temperature, pressure, and residence time must be tailored to the particular feedstocks employed. It is also worth noting that bio-crude produced *via* HTL is of inferior quality to petroleum-based fuels due to undesirable qualities such as high acidity, high viscosity, and low thermal stability. These qualities can cause engine damage and the release of pollutants such as NO_x and SO_x gases. Therefore, upgrading is necessary to satisfy the requisite criteria. There are several techniques for upgrading bio-oil, such as catalytic pyrolysis, emulsification, hydrodeoxygenation, molecular distillation, hydrogenation, catalytic cracking, esterification, and supercritical fluids.[32] Catalytic hydroprocessing has proven to be an effective method for upgrading bio-oil due to its excellent activity in removing heteroatoms, resulting in improved properties that are comparable to petroleum fuel. Various catalysts have been explored for this purpose, including NiMo and CoMo sulfide catalysts,[33] Pt, Pd, Ni and Ru.[34]

9.4.3 Pyrolysis for Bio-oil Production

Pyrolysis is an efficient technique for transforming the organic compounds in algal biomass into liquid biofuels. It operates at temperatures ranging from 400 °C to 700 °C in an oxygen-free atmosphere (anoxic conditions), producing syngas, charcoal, and bio-oil. Pyrolysis involves complex processes such as rearrangement reactions, polymerization, fragmentation, decarboxylation, and dehydration. The lipids, carbohydrates, and proteins in microalgae make them suitable for pyrolysis in energy production. As pyrolysis is energy- and yield-efficient, it is considered a cost-effective method for converting algae biomass into fuel.[41] Pyrolysis of algal biomass species is depicted in Table 9.3. Depending on the residence duration, reaction temperature and heating rate, pyrolysis can be classified as slow pyrolysis and rapid pyrolysis. Slow pyrolysis

Table 9.3 Pyrolysis of algal biomass species.[43–48]

Pyrolysis	Algae species	Conditions	Bio-oil yield (wt%)
Fast pyrolysis	*C. vulgaris*	550 °C	47.7
	Nannochloropsis gaditana	480 °C	20–31
Slow pyrolysis	*Spirulina platensis*	556 °C	66.0
	Dunaliella salina	500–550 °C	55.4
Catalytic pyrolysis	*Gracilaria gracilis*	500 °C; 30 min; argon gas flow rate: 30 mL min^{-1}; particle size: 0.15 mm	38.65
	Chlorella vulgaris	300–600 °C; algae-to-catalyst ratio: 5 : 1, 1 : 2, 1 : 1, 2 : 1; catalyst: 5% Ni on zeolite	10.4

involves heating the organic materials at relatively low temperatures (300–500 °C) over long periods in the absence of oxygen. It has a longer vapour residence time (5–60 min) and lower heating rates (5–80 °C min^{-1}). Low heating rates and extended residence periods (450–550 s) enable a further repolymerization phase, increasing solid biofuel production. On the other hand, fast pyrolysis involves short residence time, high temperature (500–800 °C) and heat transfer as well as extremely high heating rates, followed by the rapid cooling of pyrolysis vapour to give bio-crude oil.[42]

9.5 Green Diesel

In the United States, the Department of Energy (DOE), in collaboration with the Internal Revenue Service and the Environmental Protection Agency (EPA), has produced a definition for renewable diesel. This term describes biomass-based diesel fuels that fulfil ASTM D975 requirements but are not mono-alkyl esters. Green diesel, also known as third-generation super-cetane diesel or renewable diesel (cetane no. 90), is made up of hydrocarbons ranging from C_{15} to C_{18}. Green diesel has a high cetane number, resulting in a shorter ignition delay, faster combustion, improved engine performance, and lower fuel usage. Green diesel, which is currently considered the most favorable biofuel, outperforms petroleum-based diesel and biodiesels in terms of heating value, energy density, cetane number, and quality. It is highly stable, has minimal oxygen concentration, and burns cleaner than petroleum diesel due to its non-corrosive properties.[49]

9.5.1 Feedstocks for Green-diesel Production

Feedstocks used for the generation of green diesel are classified into 1-G, 2-G, 3-G, and 4-G as shown in Figure 9.6. These feedstocks consist of one or

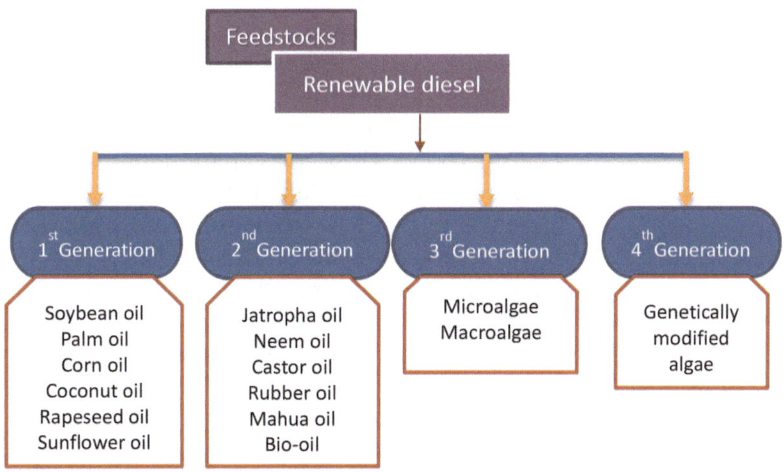

Figure 9.6 Classification of feedstocks for the production of green diesel.

Table 9.4 Key components of the feedstocks.[51–57]

Key components	Feedstock
Triglycerides	Vegetable oils (rapeseed oil, sunflower oil, olive oil, palm oil, karanja oil, rubber seed oil, *etc.*), animal fats (chicken fat), waste cooking oil, algae oil, palm stearin
Free fatty acids	Waste from refining process (palm fatty acid, soybean, canola and sunflower deodorizer distillates), triglyceride hydrolysis, steam stripping of a biodiesel feedstock
Fatty acid esters	Methyl esters of fatty acids (FAME)/biodiesel, product from reaction of fatty acids and alcohol

more of the following substances: triglycerides, fatty acid esters, and free fatty acid compounds. Table 9.4 highlights the key components of these feedstocks. Due to the drawbacks faced by first- and second-generation feedstocks, third-generation feedstocks (algae) are the best choice for green-diesel production. These are better than first- and second-generation feedstocks since they do not compete with food resources, grow on non-arable land, and have better growth rate and productivity, high lipid content, greater CO_2 absorption, low harvest time and emit less greenhouse gases. The high lipid content make them a suitable feedstock for green-diesel production. The algal biomass can be converted into a variety of sustainable biofuels including bioethanol, renewable diesel, biogas, and bio-hydrogen by different conversion methods *via* thermochemical and biochemical processes.[50] Some species of algae widely used for biofuel generation are *Chlorella vulgaris, Nannochloropsis* sp., *Tetraselmis* sp., *Dunaliella* sp. *etc.*

9.5.2 Green-diesel Production *Via* Deoxygenation Process

Green diesel is created through the catalytic deoxygenation (DO) of biomass containing triglycerides and fatty acids. DO is a thermal process that produces alkanes (hydrocarbons) from oily biomass by the removal of hydrogen, oxygen, and carbon atoms from the fatty biomass by releasing CO_2, CO, and H_2O.[58] The DO process involves heating of biomass in the presence of hydrogen and a heterogeneous metal catalyst, which converts the triglycerides and fatty acids into hydrocarbons. The deoxygenation reaction occurs at 200–400 °C and 1–10 MPa over metal catalysts. This process effectively removes oxygen from the biomass, releasing it in the form of by-products like water (H_2O), carbon monoxide (CO), and carbon dioxide (CO_2).[59]

The initial step of deoxygenation is the hydrogenation of fatty esters or fatty acids containing double bonds, due to the reactivity of double bond being thermodynamically higher than the acid group. This step is followed by the cracking of hydrogenated triglycerides to produce free fatty acids. Cracking is a potential mechanism for removing oxygenated species from fatty acid molecules and also for reducing the fatty acid chain through C–C bond breaking. Cracking often includes thermal cleavage or catalytic destruction of triglycerides.[60] The cracked product contains light alkanes, alkenes, and fatty acids. Afterwards, free

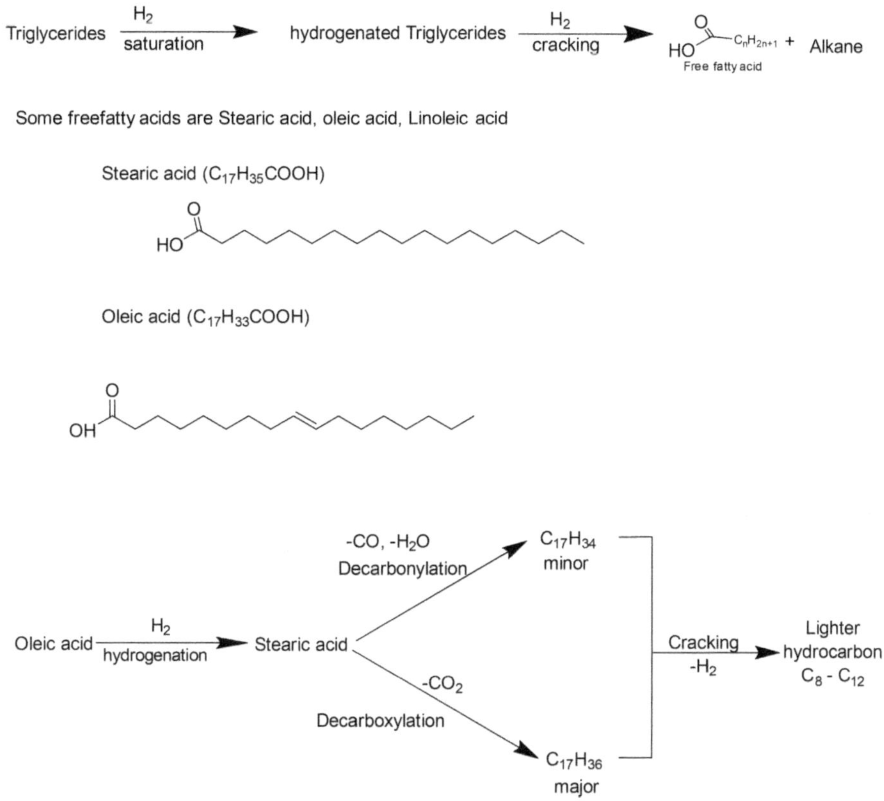

Figure 9.7 Reactions in the production of green diesel.

fatty acids are converted to alkanes through the hydrodeoxygenation, decarboxylation and decarbonylation pathways[61] as shown in Figure 9.7.

9.5.3 Catalysts for Green-diesel Production

The deoxygenation (DO) process requires substantial energy, so catalysts are preferred under green chemistry principles to lower the energy demand. Effective DO catalysts should have strong acidity, small particle size with a large metal surface area, good dispersion, and high heat stability, with the ability to split H_2 during hydrodeoxygenation. These properties depend on the active metal location, concentration, quantity, and support. An ideal catalyst can produce renewable diesel with high cetane number and low coke formation. Developing catalysts with high stability, selectivity, and activity is key for efficient DO reactions, often using solid catalysts with an active phase on a support material. Figure 9.8 shows the catalyst employed for the generation of renewable diesel, including active metals and different supports. Catalyst properties depend on the support, the active phase, and the

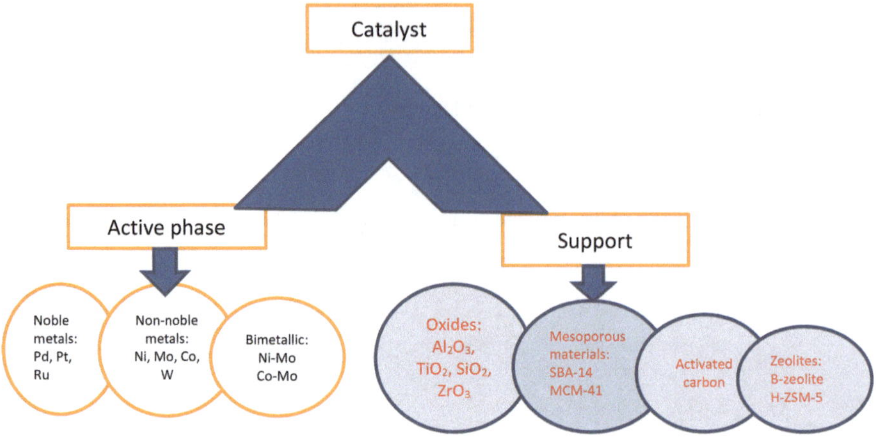

Figure 9.8 Catalysts employed for the generation of green diesel, including active metals and different supports.

interactions with reagents and products. However, catalyst selection should also consider factors like regenerability, regeneration costs, and toxicity.

Supported metal catalysts play a crucial role in deoxygenation (DO) by providing active metal sites that enable hydrogen splitting, insertion, and carbon–carbon (C–C) bond cleavage. The choice of metals for these catalysts, typically forming the active phase in solid catalysts, includes noble metals (*e.g.* Pt, Pd, Rh, Ru) and non-noble metals (*e.g.* Ni, Mo, Cu, Co, Zn). Noble metals like palladium and platinum are favored for their high activity and selectivity, promoting decarbonylation (DCO) and decarboxylation (DCO_2) with lower hydrogen consumption. However, their high cost and susceptibility to issues like water contamination and deactivation reduce their desirability. To address these limitations, non-noble metals are being explored, with nickel-based catalysts gaining popularity due to their affordability and stability, and the relatively low reduction temperature of Ni oxides. However poor dispersion and aggregation of Ni metal can reduce its catalytic efficiency, resulting in fewer active sites. To enhance performance, other metals such as Cu, Sn, Mo, and Zn are often added. For instance, adding copper to nickel increases dispersion and electron density, enhancing catalytic activity and hydrogenation. Tungsten further improves the catalyst's physicochemical properties by increasing surface area, Ni dispersion, and acidity regulation. Nickel and nickel-based bimetallic catalysts are highly effective and economical for hydrodeoxygenation (HDO) due to their strong ability to break down oxygen-containing molecules. Combining nickel with metals like molybdenum or cobalt enhances catalyst stability, selectivity, and efficiency, while also reducing coke formation.[62–66]

9.5.4 Catalyst Supports

Supports used in the deoxygenation process greatly influence catalyst activity. Key considerations for support selection include interaction with the active phase, surface area, pore size and structure, metal distribution, and stability. Common supports in hydrodeoxygenation (HDO) include mesoporous materials, activated carbon, zeolites, and metal oxides like γ-Al_2O_3, SiO_2, Nb_2O_5, TiO_2, ZrO_2, and CeO_2.[67]

Zeolites, with their strong cracking ability, porous structure, and acidity, have been promising catalysts for green-diesel production. Ni/zeolite Y enabled 91.5% conversion in microalgae biodiesel deoxygenation, while zeolites achieved 84.8% conversion and 78.2% selectivity in palm fatty acid deoxygenation. Their acidity, influenced by the Si/Al ratio, is key to their catalytic activity.[68]

Mesoporous silica, formed by condensing sodium silicate or silicon alkoxide around a surfactant template, is influenced by factors like silicon supply, pH, and aging conditions, affecting its surface area, pore size, and wall thickness. Known for thermal stability, ease of surface modification, and large surface area, mesoporous silica types like SBA-15, SBA-16, and MCM-41 are favored for their moderate acidity and high pore diameter, making them suitable for processing large molecules in fatty acid-based reactions.[69]

Mixed-metal oxides such as TiO_2, Al_2O_3, SiO_2, ZrO_2, and CeO_2 are interesting supports for green-diesel production *via* DO. Ooi *et al.*[70] found that a mixed-metal oxide comprising Al_2O_3–TiO_2 improves catalyst stability, activation, surface area, reducibility, metal dispersion, and acidic centers compared to individual oxides. The outstanding performance of mesoporous Al_2O_3–TiO_2 is credited to its acidity, mesoporosity, and large surface area. The findings suggest that this catalyst holds great potential for the production of hydrocarbon-like biofuels.

Activated carbon serves as an effective support, offering a high surface area (*e.g.* 584 $m^2\ g^{-1}$ for Ni/AC) and essential acid and basic sites, aiding C–O bond activation and C–C bond cleavage. This support enables efficient decarbonylation, converting 90% of waste cooking oil into 89% C_{15} and C_{17} hydrocarbons, while also reducing catalyst deactivation from coke formation by adsorbing hydrocarbons during deoxygenation.[71] On the other hand, the deoxygenation of methyl laurate over Co/C resulted in a very low conversion of 6% because of the high strong adsorption capacity of C, which barely released the product.[72]

9.6 Benefits of 3-G Algal Biorefineries

The circular economy (CE) offers benefits for the environment, the economy, and society. The CE is closely align with the Sustainable Development Goals by reducing consumption and optimizing the utilization of raw materials, water, and electricity. Using algae, both micro- and macroalgae, helps to promote environmental sustainability by lowering greenhouse-gas emissions and improving water and wastewater treatment. Economically, it promotes the development of biofuels create jobs and benefits society by providing food, cosmetics, pharmaceuticals, fertilizers, and animal feed.

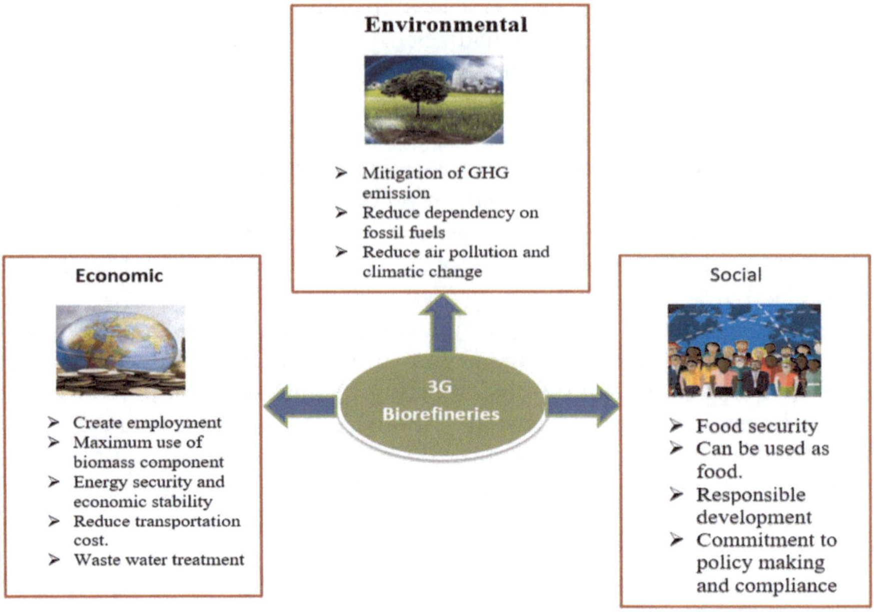

Figure 9.9 Environmental, economic, and social benefits of algal biorefineries towards sustainable technology development.

Taking into consideration the SDG goals 7 and 13, it is crucial to focus on 3-G biorefineries. When evaluating any process, it is important to consider the environmental, social, and economic impact of biorefineries across the entire bio-based product chain. The three main pillars of the sustainability of algal feedstock for biofuel and biochemical production are depicted in Figure 9.9.

Algae, as autotrophs, play a vital role in carbon fixation, with microalgae capable of sequestering approximately 0.5 gigatons of CO_2 annually. This process reduces greenhouse gases and produces biomass for biofuels and other products. Certain green algae, such as *Chlorella*, thrive in high CO_2 concentrations, making them effective for carbon sequestration.[73]

Algae play a crucial role in wastewater treatment by removing contaminants and nutrients, making the water suitable for discharge or reuse. Various microalgae species, such as *Scenedesmus*, *Chlorella*, *Euglena*, *Oscillatoria*, *Chlamydomonas*, and *Ankistrodesmus*, thrive in wastewater and effectively absorb excess nitrogen and phosphorus, helping to address eutrophication issues.[74] Certain algae species can effectively absorb heavy metals and toxins from wastewater, thereby eliminating these contaminants. Green microalgae such as *Chlorella miniata*, *C. vulgaris*, *C. reinhardtii*, and *Sphaeroplea* have been successfully employed in this process. This method, known as phycoremediation, utilizes algae's natural ability to bind and accumulate heavy metals through biosorption and bioaccumulation mechanisms.[75]

Dyes are increasingly used across various industries, but their entry into ecosystems leads to pollution and environmental risks due to their chemical stability and resistance to biodegradation. They can cause acute and chronic toxicity in aquatic life, diminishing biodiversity and disrupting ecosystem function, while also posing significant health risks to humans. Traditional methods of dye removal from wastewater often fall short, making phyco-remediation—a microalgae-based approach—a promising alternative for removing pollutants. This method is environmentally friendly, efficient, easy to implement, and cost-effective, and utilizes readily available resources.[76]

Algal biorefineries can generate various valuable products, including biofuels, biochemicals, nutraceuticals, pharmaceuticals, and cosmetics, with certain algae also serving as nutrient-rich food sources. However, only a few microalgae have received GRAS (Generally Recognized as Safe) status from the FDA, such as *Arthrospira platensis*, *Chlamydomonas reinhardtii*, and *Chlorella vulgaris*. These algae can contain significant protein levels, with *Arthrospira platensis* having up to 70% protein in its biomass, while other GRAS algae average around 40%, surpassing many plant and animal protein sources.[77] Algae are also loaded with vitamins and minerals, such as *Dunaliella tertiolecta*, which has been shown to contain high levels of vitamins A, B1, B9, and E. Algal groups like Rhodophyceae, Pheophyceae, and Chlorophyceae have strong cytotoxic, apoptotic, antiproliferative, and anti-angiogenic effects on cancer cells, including leukaemia. Green algae (*Ulva fasciata*), red algae (*Gracilaria corticata*), and brown algae (*Sargassum ilicifolium*) were tested for their apoptotic, antiproliferative, and cell-cycle inhibitory effects on human cancer cell lines (HT-29, HepG2, HeLa, MCF-7, and MDA-MB-231). All of the extracts were antiproliferative against all cancer cell lines in a dose-dependent manner, with *G. corticata* methanol extract (GCME) showing the strongest inhibitory effect against the MCF-7 cell line.[78] *Gracilaria tenuistipitata*, a species of red algae, possesses notable anti-cancer potential. Studies have shown that its methanolic extract can suppress the growth of Ca9-22 oral cancer cells by triggering DNA damage, promoting oxidative stress, and inducing cell death.[79] Additionally, a methanol extract from *Plocamium telfairiae* triggered caspase-dependent cell death in HT-29 colon cancer cells.[80]

9.7 Techno-economic Feasibility and Life-cycle Assessment of 3-G Biorefinery

The evaluation of bio-crude processes using techno-economic analysis (TEA) and life-cycle assessment (LCA) is critical, as shown in Figure 9.10. These studies provide important insights into costs, feasibility, and environmental implications, which are critical for guiding research and commercialization projects.[81] Masoumi and Dalai conducted a techno-economic analysis of microalgae-based biofuel generation, predominantly diesel, by hydrothermal liquefaction and hydroprocessing. Their studies revealed that the cost of feedstock has a significant impact on the selling price of biofuels. A 50% reduction in feedstock costs resulted in an approximate 27% drop in

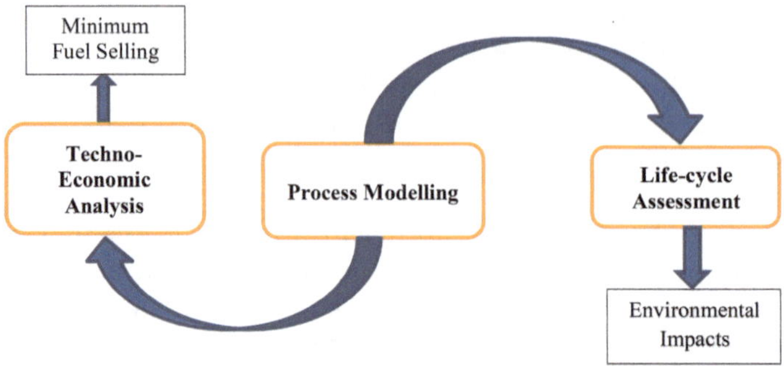

Figure 9.10 TEA and LCA of a 3-G algal biorefinery.

the fuel's minimum selling price. Their LCA study showed a considerable decrease in GHG emissions, with an estimated reduction of 1.13 gCO_2-eq per MJ (grams CO_2 equivalent per megajoule). This is much lower than conventional petroleum-based fuel generation.[82] DeRose *et al.* conducted TEA and LCA of renewable fuel production from low-lipid algae. Their study determined minimum selling prices (MSP) of US$12.85 per gasoline gallon equivalent (GGE) and US$10.41 per GGE, along with global warming potentials of 111.2 gCO_2-eq per MJ and 0.2 gCO_2-eq per MJ for the biochemical and thermochemical pathways, respectively. The authors further suggest that increasing the production of high-value co-products could reduce the MSP to US$3.85 per GGE.[83] Meyer *et al.* investigated the economics of producing biofuels from lignocellulosic feedstocks like switch grass and maize stover, as well as upgrading bio-oil by hydrodeoxygenation. The study found that capital expenses (30–40%) and feedstock prices (30%) make up the majority of the minimum fuel selling price (MFSP) for all biomass types. Blend feedstocks may be more cost-effective than pure feedstocks, thereby improving the overall economy. The minimum fuel selling price for produced fuels varies based on biomass feedstock, ranging from US$3.7 per GGE to US$5 per GGE.[84] Sun *et al.* assessed the net energy ratio (NER) of several conversion methods for producing biofuel from microalgae. They found that using an anaerobic digestion system with hydrothermal pretreatment is more viable for industrial use and environmentally friendly. This is because it has a low NER of 0.71 and results in reduced GHG emissions, at −60.84 gCO_2-eq per MJ of biogas.[85] However, it is crucial to recognize that the results of TEA and LCA might differ greatly based on a number of variables, including the kind of feedstock, its makeup, and the operating conditions.

9.8 Future Perspectives and Challenges in Algal Biorefining

Algal biorefineries are expected to play an important role in the circular economy by utilizing waste streams such as CO_2 emissions from power plants as inputs for algae cultivation. This decreases waste and greenhouse-gas emissions while encouraging more sustainable industrial practices. These biorefineries contribute to the circular economy by reducing their carbon footprint and creating a closed-loop system that maximizes resource efficiency while minimizing environmental effect. The production of high-value co-products is an important strategy for increasing the commercial viability of algal biorefining. By extending its product line to include nutraceuticals, cosmetics, animal feed, and bioplastics in addition to biofuels, the business may generate several revenue streams while reducing its reliance on the unpredictable biofuel market. For example, algal biomass can be utilized to produce important chemicals such as carotenoids, omega-3 fatty acids, and phycobiliproteins, all of which are in high demand in the nutraceutical and pharmaceutical industries. Furthermore, after extracting these compounds, the residual biomass may be used to produce bioenergy, making the process more sustainable and cost-effective. This integrated method not only maximizes the utilization of algal biomass, but it also helps to create a more circular bioeconomy, addressing both environmental and economic concerns. Scaling up from laboratory- and pilot-scale operations to commercial-scale facilities is critical for the future of algal biorefining. Innovations in photobioreactor designs, open-pond systems, and hybrid-culture technologies will be critical to increasing yields and cutting production costs.

In addition, there are some challenges associated with 3-G algal biorefinery, as mentioned below:

1. High production costs: one of the most significant challenges in algal biorefining is the high cost associated with cultivation, harvesting, and processing. Although technological advances have reduced costs in recent years, further reductions are necessary to compete with conventional biofuels and petrochemical products.
2. Water and nutrient management: algae cultivation requires substantial amounts of water and nutrients, which can be challenging to manage sustainably, especially in large-scale operations. Developing closed-loop systems that recycle water and nutrients can help mitigate these issues but requires significant investment and technological development.
3. Regulatory and policy barriers: the commercialization of algal biorefining is often hindered by complex regulatory frameworks and a lack of clear policies supporting the industry. Navigating these regulatory landscapes requires coordinated efforts between industry stakeholders, policymakers, and regulatory bodies.
4. Market demand and infrastructure: establishing a stable market demand for algal-based products and developing the necessary infrastructure for distribution and consumption are ongoing challenges. The success of algal biorefining will depend on consumer acceptance, competitive pricing, and the creation of supportive supply chains.

9.9 Conclusion

Biorefining algal biomass into green diesel and value-added products provides a long-term answer to global energy and environmental issues. Algae, a rapidly farmed renewable feedstock, can be turned into green diesel using processes such as transesterification or hydrothermal liquefaction, producing a fuel compatible with internal combustion engines while lowering greenhouse-gas emissions and fossil-fuel dependence. Additionally, algal biomass can be fractionated to produce a range of valuable co-products, including proteins for animal feed, carbohydrates for bioethanol production, and pigments or antioxidants for pharmaceuticals, cosmetics, and nutraceuticals. Overall, the biorefining of algal biomass into green diesel and value-added products represents a promising pathway toward sustainable energy production, reduced environmental impact, and circular-economy development. However, challenges related to high production costs, technological barriers, and scalability need to be addressed through further research, innovation, and supportive policies to fully realize the potential of this technology. With continued research, innovation, and policy support, it has the potential to play a key role in the global transition to a more sustainable and circular economy.

Declarations

Mohammad Aslam is an editorial board member for this book titled *Biorefining of Renewable Diesel: From Production to Market* and was not involved in the editorial review of, or the decision to publish, this chapter. All authors declare that there are no competing interests.

Acknowledgements

The authors would like to thank to Anusandhan National Research Foundation (formerly known as Science and Engineering Research Board, Department of Science and Technology), Government of India, for their support of this research through CRG/2022/003994.

References

1. https://www.un.org/en/desa/world-population-projected-reach-98-billion-2050-and-112-billion-2100.
2. International Energy Agency (IEA). https://www.iea.org/ (assessed in 2023).
3. https://www.iea.org/reports/world-energy-outlook-2023/executive-summary.
4. B. Lin and R. Bai, Oil prices and economic policy uncertainty: Evidence from global, oil importers, and exporters' perspective, *Res. Int. Bus. Finance*, 2021, **56**, 101357.

5. D. Mignogna, M. Szabó, P. Ceci and P. Avino, Biomass energy and biofuels: perspective, potentials, and challenges in the energy transition, *Sustainability*, 2024, **16**(16), 7036.
6. L. Bhatia, R. K. Bachheti, V. K. Garlapati and A. K. Chandel, Third-generation biorefineries: a sustainable platform for food, clean energy, and nutraceuticals production, *Biomass Convers. Biorefin.*, 2022, **12**(9), 4215–4230.
7. A. Gupta, T. Sahni and S. Kumar, Biorefineries: An Analogue to Petroleum Refineries, in *Paddy Straw Waste for Biorefinery Applications*, Springer Nature Singapore, Singapore, 2024, pp. 1–29.
8. J. Singh and R. C. Saxena, An introduction to microalgae: diversity and significance, in *Handbook of marine microalgae*, 2015, pp. 11–24.
9. P. Prabakaran and S. Karthikeyan, Algae biofuel: A futuristic, sustainable, renewable and green fuel for IC engines, *Mater. Today: Proc.*, 2023, DOI: 10.1016/j.matpr.2023.03.579.
10. R. Shaikh, A. Rizvi, S. Pandit, N. Desai and R. Patil, Microalgae: Classification, bioactives, medicinal properties, industrial applications, and future prospectives, in *An Integration of Phycoremediation Processes in Wastewater Treatment*, Elsevier, 2022, pp. 451–486.
11. J. S. Sangha, S. Kelloway, A. T. Critchley and B. Prithiviraj, Seaweeds (macroalgae) and their extracts as contributors of plant productivity and quality: the current status of our understanding, *Adv. Bot. Res.*, 2014, **71**, 189–219.
12. O. M. Adeniyi, U. Azimov and A. Burluka, Algae biofuel: current status and future applications, *Renewable Sustainable Energy Rev.*, 2018, **90**, 316–335.
13. F. A. Ansari, S. K. Ratha, N. Renuka, L. Ramanna, S. K. Gupta, I. Rawat and F. Bux, Effect of microplastics on growth and biochemical composition of microalga Acutodesmus obliquus, *Algal Res.*, 2021, **56**, 102296.
14. R. Dixit, S. Singh and A. Singh, Effect of nitrogen deficiency on the physiology and biochemical composition of microalga Scenedesmus rotundus-MG910488, *J. Basic Microbiol.*, 2020, **60**(2), 158–172.
15. P. Ardiles, P. Cerezal-Mezquita, F. Salinas-Fuentes, D. Órdenes, G. Renato and M. C. Ruiz-Domínguez, Biochemical composition and phycoerythrin extraction from red microalgae: A comparative study using green extraction technologies, *Processes*, 2020, **8**(12), 1628.
16. A. K. Koyande, P. L. Show, R. Guo, B. Tang, C. Ogino and J. S. Chang, Bioprocessing of algal bio-refinery: a review on current advances and future perspectives, *Bioengineered*, 2019, **10**(1), 574–592.
17. M. Bhattacharya and S. Goswami, Microalgae–A green multi-product biorefinery for future industrial prospects, *Biocatal. Agric. Biotechnol.*, 2020, **25**, 101580.
18. F. Camacho, A. Macedo and F. Malcata, Potential industrial applications and commercialization of microalgae in the functional food and feed industries: A short review, *Mar. Drugs*, 2019, **17**(6), 312.
19. J. Ávila-Román, S. García-Gil, A. Rodríguez-Luna, V. Motilva and E. Talero, Anti-inflammatory and anticancer effects of microalgal carotenoids, *Mar. Drugs*, 2021, **19**(10), 531.

20. H. Chen, H. Qi and P. Xiong, Phycobiliproteins—A family of algae-derived biliproteins: productions, characterization and pharmaceutical potentials, *Mar. Drugs*, 2022, **20**(7), 450.
21. R. Pangestuti, I. T. Suryaningtyas, E. A. Siahaan and S. K. Kim, Cosmetics and cosmeceutical applications of microalgae pigments, in *Pigments from Microalgae Handbook*, 2020, pp. 611–633.
22. R. G. Saratale, G. Kumar, R. Banu, A. Xia, S. Periyasamy and G. D. Saratale, A critical review on anaerobic digestion of microalgae and macroalgae and co-digestion of biomass for enhanced methane generation, *Bioresour. Technol.*, 2018, **262**, 319–332.
23. S. Jain and S. Kumar, A comprehensive review of bioethanol production from diverse feedstocks: Current advancements and economic perspectives, *Energy*, 2024, 131130.
24. S. F. Ahmed, N. Rafa, M. Mofijur, I. A. Badruddin, A. Inayat, M. S. Ali, O. Farrok and T. M. Yunus Khan, Biohydrogen production from biomass sources: metabolic pathways and economic analysis, *Front. Energy Res.*, 2021, **9**, 753878.
25. G. M. Mathew, D. Raina, V. Narisetty, V. Kumar, S. Saran, A. Pugazhendi, R. Sindhu, A. Pandey and P. Binod, Recent advances in biodiesel production: Challenges and solutions, *Sci. Total Environ.*, 2021, **794**, 148751.
26. M. Elhassan, R. Abdullah, M. R. R. Kooh and Y. F. C. Chau, Hydrothermal liquefaction: a technological review on reactor design and operating parameters, *Bioresour. Technol. Rep.*, 2023, **21**, 101314.
27. Z. Rahmawati, L. Santoso, A. McCue, N. L. A. Jamari, S. Y. Ninglasari, T. Gunawan and H. Fansuri, Selectivity of reaction pathways for renewable diesel production towards biojet fuel applications, *RSC Adv.*, 2023, **13**(20), 13698–13714.
28. Y. Wang and J. J. Wu, Thermochemical conversion of biomass: Potential future prospects, *Renewable Sustainable Energy Rev.*, 2023, **187**, 113754.
29. O. Sarkar, M. Agarwal, A. N. Kumar and S. V. Mohan, Retrofitting hetrotrophically cultivated algae biomass as pyrolytic feedstock for biogas, bio-char and bio-oil production encompassing biorefinery, *Bioresour. Technol.*, 2015, **178**, 132–138.
30. E. Ocampo, V. V. Beltrán, D. Ocampo, E. A. Gómez and L. A. Ríos, Hydrothermal liquefaction process: review and trends, *Curr. Res. Green Sustainable Chem.*, 2023, 100382.
31. W. Wang, Y. Xu, X. Wang, B. Zhang, W. Tian and J. Zhang, Hydrothermal liquefaction of microalgae over transition metal supported TiO_2 catalyst, *Bioresour. Technol.*, 2018, **250**, 474–480.
32. H. A. Baloch, S. Nizamuddin, M. T. H. Siddiqui, S. Riaz, A. S. Jatoi, D. K. Dumbre, N. M. Mubarak, M. P. Srinivasan and G. J. Griffin, Recent advances in production and upgrading of bio-oil from biomass: A critical overview, *J. Environ. Chem. Eng.*, 2018, **6**(4), 5101–5118.
33. P. Biller, B. K. Sharma, B. Kunwar and A. B. Ross, Hydroprocessing of bio-crude from continuous hydrothermal liquefaction of microalgae, *Fuel*, 2015, **159**, 197–205.

34. Z. He, Y. Jiao, J. Wang and Y. Chen, Effects of M (Zr, Nb, Y) modifiers on the catalytic performance of Ni/Ce–Al_2O_3 bimetallic catalyst in steam reforming of n-decane, *J. Anal. Appl. Pyrolysis*, 2016, **122**, 142–150.
35. H. Yousefzadeh, E. Borhan, B. Z. Haznedaroglu, A. Uzun and C. Erkey, Hydrothermal liquefaction of Chlamydomonas nivalis and Nannochloropsis gaditana microalgae under different operating conditions over copper-exchanged zeolites, *Biochem. Eng. J.*, 2024, **205**, 109237.
36. J. Nallasivam, P. F. Prashanth, S. Harisankar, S. Nori, S. Suryanarayan, S. R. Chakravarthy and R. Vinu, Valorization of red macroalgae biomass *via* hydrothermal liquefaction using homogeneous catalysts, *Bioresour. Technol.*, 2022, **346**, 126515.
37. A. A. Jazie, J. Haydary, S. A. Abed and M. F. Al-Dawody, Hydrothermal liquefaction of Fucus vesiculosus algae catalyzed by Hβ zeolite catalyst for Biocrude oil production, *Algal Res.*, 2022, **61**, 102596.
38. M. R. Moazezi, H. Bayat, O. Tavakoli and A. Hallajisani, Hydrothermal liquefaction of Chlorella vulgaris and catalytic upgrading of product: Effect of process parameter on bio-oil yield and thermodynamics modeling, *Fuel*, 2022, **318**, 123595.
39. K. P. R. Dandamudi, K. M. Luboowa, M. Laideson, T. Murdock, M. Seger, J. McGowen, P. J. Lammers and S. Deng, Hydrothermal liquefaction of Cyanidioschyzon merolae and Salicornia bigelovii Torr.: The interaction effect on product distribution and chemistry, *Fuel*, 2020, **277**, 118146.
40. K. Pongsiriyakul, W. Kiatkittipong, S. Adhikari, J. W. Lim, S. S. Lam, K. Kiatkittipong, A. Dankeaw, P. Reubroycharoen, N. Laosiripojana, K. Faungnawakij and S. Assabumrungrat, Effective Cu/Re promoted Ni-supported γ-Al_2O_3 catalyst for upgrading algae bio-crude oil produced by hydrothermal liquefaction, *Fuel Process. Technol.*, 2021, **216**, 106670.
41. F. Li, S. C. Srivatsa and S. Bhattacharya, A review on catalytic pyrolysis of microalgae to high-quality bio-oil with low oxygeneous and nitrogenous compounds, *Renewable Sustainable Energy Rev.*, 2019, **108**, 481–497.
42. K. Brindhadevi, S. Anto, E. R. Rene, M. Sekar, T. Mathimani, N. T. L. Chi and A. Pugazhendhi, Effect of reaction temperature on the conversion of algal biomass to bio-oil and biochar through pyrolysis and hydrothermal liquefaction, *Fuel*, 2021, **285**, 119106.
43. F. Sotoudehniakarani, A. Alayat and A. G. McDonald, Characterization and comparison of pyrolysis products from fast pyrolysis of commercial Chlorella vulgaris and cultivated microalgae, *J. Anal. Appl. Pyrolysis*, 2019, **139**, 258–273.
44. N. Priharto, F. Ronsse, W. Prins, R. Carleer and H. J. Heeres, Experimental studies on a two-step fast pyrolysis-catalytic hydrotreatment process for hydrocarbons from microalgae (Nannochloropsis gaditana and Scenedesmus almeriensis), *Fuel Process. Technol.*, 2020, **206**, 106466.
45. K. C. Rocha, C. G. Alonso, W. G. O. Leal, E. L. Schultz, L. A. Andrade and I. C. Ostroski, Slow pyrolysis of Spirulina platensis for the production of nitrogenous compounds and potential routes for their separation, *Bioresour. Technol.*, 2020, **313**, 123709.

46. C. Yang, R. Li, B. Zhang, Q. Qiu, B. Wang, H. Yang, Y. Ding and C. Wang, Pyrolysis of microalgae: A critical review, *Fuel Process. Technol.*, 2019, **186**, 53–72.
47. O. Norouzi, A. Tavasoli, S. Jafarian and S. Esmailpour, Catalytic upgrading of bio-products derived from pyrolysis of red macroalgae Gracilaria gracilis with a promising novel micro/mesoporous catalyst, *Bioresour. Technol.*, 2017, **243**, 1–8.
48. N. H. Zainan, S. C. Srivatsa, F. Li and S. Bhattacharya, Quality of bio-oil from catalytic pyrolysis of microalgae Chlorella vulgaris, *Fuel*, 2018, **223**, 12–19.
49. M. Aslam, Transformation of 1-G and 2-G liquid biomass to green fuels using hydroprocessing technology: A promising technology for biorefinery development, *Biomass Bioenergy*, 2022, **163**, 106510.
50. N. Mishra, S. Mishra and R. Prasad, Current Status and Challenges of Microalgae as an Eco-Friendly Biofuel Feedstock: A Review, *Present Environ. Sustainable Dev.*, 2021, **15**(1), 179–189.
51. S. R. Yenumala, P. Kumar, S. K. Maity and D. Shee, Production of renewable diesel from karanja oil (Pongamia pinnata) using mesoporous NiMo-alumina composite catalysts, *Bioresour. Technol. Rep.*, 2019, 7, 100288.
52. M. Ameen, M. T. Azizan, A. Ramli, S. Yusup and B. Abdullah, The effect of metal loading over Ni/γ-Al_2O_3 and Mo/γ-Al_2O_3 catalysts on reaction routes of hydrodeoxygenation of rubber seed oil for renewable diesel production, *Catal. Today*, 2020, **355**, 51–64.
53. M. Pratiwi, O. Muraza, G. F. Neonufa, R. Purwadi, T. Prakoso and T. H. Soerawidjaja, Production of sustainable diesel *via* decarboxylation of palm stearin basic soaps, *Energy Fuels*, 2019, **33**(11), 11648–11654.
54. M. F. Kamaruzaman, Y. H. Taufiq-Yap and D. Derawi, Renewable diesel production from palm fatty acid distillate over SBA-15-supported nickel, cobalt, and nickel/cobalt catalysts, *Biomass Bioenergy*, 2020, **134**, 105476.
55. M. S. Gamal, N. Asikin-Mijan, M. Arumugam, U. Rashid and Y. H. Taufiq-Yap, Solvent-free catalytic deoxygenation of palm fatty acid distillate over cobalt and manganese supported on activated carbon originating from waste coconut shell, *J. Anal. Appl. Pyrolysis*, 2019, **144**, 104690.
56. R. G. Kukushkin, P. M. Yeletsky, C. T. Grassin, B. H. Chen, O. A. Bulavchenko, A. A. Saraev and V. A. Yakovlev, Deoxygenation of esters over sulfur-free Ni–W/Al_2O_3 catalysts for production of biofuel components, *Chem. Eng. J.*, 2020, **396**, 125202.
57. C. Ochoa-Hernández, J. M. Coronado and D. P. Serrano, Hydrotreating of methyl esters to produce renewable diesel over Co-and Ni-containing Zr-SBA-15 catalysts, *Catalysts*, 2020, **10**(2), 186.
58. N. A. A. Razak, N. A. Mijan, Y. H. Taufiq-Yap and D. Derawi, Production of renewable diesel *via* hydrogen-free and solventless deoxygenation reaction of waste cooking oil, *J. Cleaner Prod.*, 2022, **366**, 132971.

59. J. L. C. W. Pimenta, M. de Oliveira Camargo, R. B. Duarte, O. A. A. dos Santos and L. M. de Matos Jorge, Deoxygenation of vegetable oils for the production of renewable diesel: Improved aerogel based catalysts, *Fuel*, 2021, **290**, 119979.
60. M. Žula, M. Grilc and B. Likozar, Hydrocracking, hydrogenation and hydro-deoxygenation of fatty acids, esters and glycerides: Mechanisms, kinetics and transport phenomena, *Chem. Eng. J.*, 2022, **444**, 136564.
61. K. W. Cheah, M. J. Taylor, A. Osatiashtiani, S. K. Beaumont, D. J. Nowakowski, S. Yusup, A. V. Bridgwater and G. Kyriakou, Monometallic and bimetallic catalysts based on Pd, Cu and Ni for hydrogen transfer deoxygenation of a prototypical fatty acid to diesel range hydrocarbons, *Catal. Today*, 2020, **355**, 882–892.
62. A. Bjelić, M. Grilc, M. Huš and B. Likozar, Hydrogenation and hydro-deoxygenation of aromatic lignin monomers over Cu/C, Ni/C, Pd/C, Pt/C, Rh/C and Ru/C catalysts: Mechanisms, reaction micro-kinetic modelling and quantitative structure-activity relationships, *Chem. Eng. J.*, 2019, **359**, 305–320.
63. S. Lycourghiotis, E. Kordouli, C. Kordulis and K. Bourikas, Transformation of residual fatty raw materials into third generation renewable diesel over a nickel catalyst supported on mineral palygorskite, *Renewable Energy*, 2021, **180**, 773–786.
64. M. Gousi, E. Kordouli, K. Bourikas, E. Symianakis, S. Ladas, C. Kordulis and A. , Lycourghiotis, Renewable diesel production over nickel-alumina nanostructured catalysts promoted by copper, *Energies*, 2020, **13**(14), 3707.
65. X. Yu and C. T. Williams, Recent applications of nickel and nickel-based bimetallic catalysts for hydrodeoxygenation of biomass-derived oxygenates to fuels, *Catal. Sci. Technol.*, 2023, **13**(3), 802–825.
66. I. Vázquez-Garrido, A. López-Benítez, A. Guevara-Lara and G. Berhault, Synthesis of NiMo catalysts supported on Mn–Al_2O_3 for obtaining renewable diesel from waste soybean oil, *Catal. Today*, 2021, **365**, 327–340.
67. F. Wang, J. Jiang, K. Wang, Q. Zhai, H. Sun, P. Liu, J. Feng, H. Xia, J. Ye, Z. Li and F. Li, Activated carbon supported molybdenum and tungsten carbides for hydrotreatment of fatty acids into green diesel, *Fuel*, 2018, **228**, 103–111.
68. J. Cheng, Z. Zhang, X. Zhang, Z. Fan, J. Liu and J. Zhou, Continuous hydroprocessing of microalgae biodiesel to jet fuel range hydrocarbons promoted by Ni/hierarchical mesoporous Y zeolite catalyst, *Int. J. Hydrogen Energy*, 2019, **44**(23), 11765–11773.
69. J. A. Cecilia, R. Moreno Tost and M. Retuerto Millán, Mesoporous materials: from synthesis to applications, *Int. J. Mol. Sci.*, 2019, **20**(13), 3213.
70. X. Y. Ooi, L. E. Oi, M. Y. Choo, H. C. Ong, H. V. Lee, P. L. Show, Y. C. Lin and J. C. Juan, Efficient deoxygenation of triglycerides to hydrocarbon-biofuel over mesoporous Al_2O_3–TiO_2 catalyst, *Fuel Process. Technol.*, 2019, **194**, 106120.

71. W. N. A. W. Khalit, N. Asikin-Mijan, T. S. Marliza, M. S. Gamal, M. R. Shamsuddin, M. I. Saiman and Y. H. Taufiq-Yap, Catalytic deoxygenation of waste cooking oil utilizing nickel oxide catalysts over various supports to produce renewable diesel fuel, *Biomass Bioenergy*, 2021, **154**, 106248.
72. Z. Ding, T. Zhao, Q. Zhu, S. Liao, L. Ning, Y. Bi and H. Chen, Facile synthesis Cobalt catalysts for regulating the deoxygenation pathways in producing diesel-like hydrocarbon fuels, *Biomass Bioenergy*, 2020, **143**, 105879.
73. S. Tripathi, S. Choudhary, A. Meena and K. M. Poluri, Carbon capture, storage, and usage with microalgae: a review, *Environ. Chem. Lett.*, 2023, **21**(4), 2085–2128.
74. A. López-Sánchez, A. L. Silva-Gálvez, Ó. Aguilar-Juárez, C. Senés-Guerrero, D. A. Orozco-Nunnelly, D. Carrillo-Nieves and M. S. Gradilla-Hernández, Microalgae-based livestock wastewater treatment (MbWT) as a circular bioeconomy approach: Enhancement of biomass productivity, pollutant removal and high-value compound production, *J. Environ. Manage.*, 2022, **308**, 114612.
75. Z. Chen, A. I. Osman, D. W. Rooney, W. D. Oh and P. S. Yap, Remediation of heavy metals in polluted water by immobilized algae: current applications and future perspectives, *Sustainability*, 2023, **15**(6), 5128.
76. A. Ayele, D. Getachew, M. Kamaraj and A. Suresh, Phycoremediation of synthetic dyes: an effective and eco-friendly algal technology for the dye abatement, *J. Chem.*, 2021, 9923643.
77. N. J. Szabo, R. A. Matulka and T. Chan, Safety evaluation of whole algalin protein (wap) from Chlorella protothecoides, *Food Chem. Toxicol.*, 2013, **59**, 34–45.
78. A. Mridha and S. Paul, Algae as potential repository of anticancerous natural compounds, *Int. J. Phytomed.*, 2017, 181–194.
79. C. C. Yeh, J. I. Yang, J. C. Lee, C. N. Tseng, Y. C. Chan, Y. C. Hseu, J. Y. Tang, L. Y. Chuang, H. W. Huang, F. R. Chang and H. W. Chang, Anti-proliferative effect of methanolic extract of Gracilaria tenuistipitata on oral cancer cells involves apoptosis, DNA damage, and oxidative stress, *BMC Complementary Altern. Med.*, 2012, **12**, 1–9.
80. J. Y. Kim, M. Y. Yoon, M. R. Cha, J. H. Hwang, E. Park, S. U. Choi, H. R. Park and Y. I. Hwang, Methanolic extracts of Plocamium telfairiae induce cytotoxicity and caspase-dependent apoptosis in HT-29 human colon carcinoma cells, *J. Med. Food*, 2007, **10**(4), 587–593.
81. J. L. Zheng, Y. H. Zhu, M. Q. Zhu, G. T. Sun and R. C. Sun, Life-cycle assessment and techno-economic analysis of the utilization of bio-oil components for the production of three chemicals, *Green Chem.*, 2018, **20**(14), 3287–3301.
82. S. Masoumi and A. K. Dalai, Techno-economic and life cycle analysis of biofuel production *via* hydrothermal liquefaction of microalgae in a methanol–water system and catalytic hydrotreatment using hydrochar as a catalyst support, *Biomass Bioenergy*, 2021, **151**, 106168.

83. K. DeRose, C. DeMill, R. W. Davis and J. C. Quinn, Integrated techno economic and life cycle assessment of the conversion of high productivity, low lipid algae to renewable fuels, *Algal Res.*, 2019, **38**, 101412.
84. P. A. Meyer, L. J. Snowden-Swan, S. B. Jones, K. G. Rappé and D. S. Hartley, The effect of feedstock composition on fast pyrolysis and upgrading to transportation fuels: Techno-economic analysis and greenhouse gas life cycle analysis, *Fuel*, 2020, **259**, 116218.
85. C. H. Sun, Q. Fu, Q. Liao, A. Xia, Y. Huang, X. Zhu, A. Reungsang and H. X. Chang, Life-cycle assessment of biofuel production from microalgae *via* various bioenergy conversion systems, *Energy*, 2019, **171**, 1033–1045.

CHAPTER 10

Organic Waste as a Feedstock to Produce Renewable Diesel

SNIGDHENDUBALA PRADHAN, PRAKASH PARTHASARATHY, TAREQ AL-ANSARI AND GORDON MCKAY*

Division of Sustainable Development, College of Science and Engineering, Hamad Bin Khalifa University, Doha. Emails: spradhan@hbku.edu.qa; prakrock@gmail.com; talansari@hbku.edu.qa
*Email: gmckay@hbku.edu.qa

Highlights

- This chapter summarizes organic-waste generation and management challenges.
- A detailed analysis of organic-waste characterization.
- In depth analysis of renewable-diesel production technology from biomass.
- A comprehensive analysis of renewable-diesel production strategies and issues.
- An economic aspect of organic waste as a feedstock.

10.1 Introduction

Worldwide population growth and economic development continuously demand more energy every year. From 1990 to 2020, global energy consumption increased significantly by 69.22% due to the growing world population and is expected to reach about 660 quadrillion Btu in 2050.[1] Most of the total worldwide energy consumption, about 80%, results from the combustion of traditional fossil fuels such as coal, oil, and gas, which is projected to rise at an average annual rate of 0.9% in 2030. However, their use is not sustainable as

Green Chemistry Series No. 89
Biorefining of Renewable Diesel: From Production to Market
Edited by Mohammad Aslam, Sanjeev Mishra and Jorge Aburto

Published by the Royal Society of Chemistry, www.rsc.org

their resources are limited and supplies are being exhausted.[2] In Asian countries, energy consumption is higher due to the increasing population rate, while energy use in European countries has remained unchanged over the past three years.[2,3] Fossil fuels also contribute to global pollution and deteriorate human health. To sustainably meet the growing energy demands, alternative energy sources need to be explored and implemented. Fuel use in transportation is one of the vital energy sectors and plays an extremely important role in our daily life. According to a survey, the transportation sector is the world's largest consumer of fossil energy, which is almost 40% of the total energy sectors.[4] Gasoline, diesel, and jet fuel are the leading transportation fuels which are depleting rapidly due to rising population and growing demands for petroleum products. Consequently, the price of petroleum products is also increasing continuously. Additionally, the use of petroleum has adverse impacts on the Earth's environment, which indicates the necessity to reduce high fossil-fuel use in the transportation sector.[2]

Therefore, an alternative ecofriendly and sustainable energy source is needed to replace fossil fuels worldwide. Renewable energy sources are the world's fastest growing sector of energy, with a yearly consumption rate of 3.0% by 2030.[4] Fatty acid methyl ester (FAME) biodiesel or renewable diesel has been recognized as having considerable potential in the transportation sector for diesel-fuel vehicles.[5] FAME biodiesel consists of a mixture of fatty acid methyl esters, while renewable diesel is a mixture of hydrocarbons in the boiling-point range of diesel that is free of aromatic compounds. In recent years, there has been a big concern raised to develop bio-based fuel sources, *i.e.* bioethanol, biodiesel, microbial fuel cells, *etc.* Fluctuation in cost and limited availability of these renewables also remain evergreen concerns at the ecological, economic, and political levels.[6] Waste-to-energy, particularly from biomass, could be an excellent substitute in this regard. In almost all parts of the world, plant biomass has been used as a substantial fuel for cooking and heat generation. On the other hand, many developed countries are in the process of shifting to biomass-based fuels and energy systems for cooking, heat, transportation and electricity generation to avoid carbon dioxide (CO_2) gas emissions during the fossil fuels combustion.[7]

Biofuels including biodiesel, bioethanol, biogas, and sustainable hydrocarbon fuels, are broadly classified as first- to fourth-generation biofuels depending on the feedstock types and technologies used for the production.[8] First-generation biofuels, also known as conventional biofuels, were produced from food crops grown on farmlands. Second-generation biofuels are produced from different types of non-food biomass, while third-generation biofuels are produced from aquatic plant biomass such as algae. The fourth generation of biofuels are substantially advanced and emerging types of biofuels produced from engineered plants or microorganisms with higher yields. The two most frequently utilized biofuels are biodiesel and bioethanol. Considering reported studies, it has become a critical need to develop economical, eco-friendly, and sustainable processes for biomass-based energy production in a monetarily sound manner.[9]

Biodiesel is biodegradable and nontoxic, with lubricating properties that increase the life of diesel engines and can be produced sulfur-free. It increases the life of the catalytic converter, is miscible with hydrocarbons, and produces less pollution than petrodiesel.[10] Renewable diesel is also one of the best options to use safely – it is sustainable, cost-effective, and produced by eco-friendly processes. On Earth, biomass is the only source of renewable carbon and it is well-acknowledged that fossil fuels, such as coal, natural gas, and petroleum, originate from the decomposition of biomass over millions of years.[11] Biomass now provides more than 10% of energy worldwide. The European Renewable Energy Council (EREC) projected 50% of the world's energy would be replaced in the form of renewables by 2040.[12]

In this context, this chapter covers the major aspects of the role of organic-waste biomass in bioenergy production, in particular as renewable diesel. This chapter summarizes in detail the various properties of biomass and the benefits of applying renewable-diesel production. This chapter aims to elaborate the benefits and economic aspects of renewable diesel and its production techniques from biomass to fulfil the replacement of fossil fuels. This chapter also demonstrates the various strategies and future scope of organic-waste valorization that could encourage researchers, stakeholders, government agencies and policy makers for renewable-diesel production.

10.2 Organic Waste

10.2.1 Sources of Generation and Their Classification

Organic waste, also known as biodegradable waste, is produced mainly from plants and animals or industrial processes. Nowadays, a wide variety of potential organic wastes are generated worldwide with the growing population. Broadly, organic wastes are classified into three major categories: plant-based, animal-based and industrial (Figure 10.1).

Organic-waste generation is a big threat to the environment and world economy due to the emissions of carbon dioxide (CO_2) and methane (CH_4) in landfilling. The waste breaks down into CO_2 and CH_4 by water, microorganisms and bacteria through composting, aerobic digestion, anaerobic digestion, and insect farming. Recycling of organic waste is a sustainable solution for different types of biomass to resolve the bulk waste into different forms of bioenergy, such as chemical energy, electrical energy, thermal energy, mechanical energy, agriculture and pollution remediation, that will support reduction of greenhouse gas (GHG) emissions. All business places including restaurants, hotels, retail establishments and multi-family residences produce large quantities of organic waste and are motivated to recycle organic waste. Different types of organic waste generated from different sources are reported in Table 10.1.

Agriculture is the backbone of the world economy and food security; however, day by day, the increasing demand for food supply is generating a huge quantity of wastes from agricultural activities.[13,14] This waste has become a

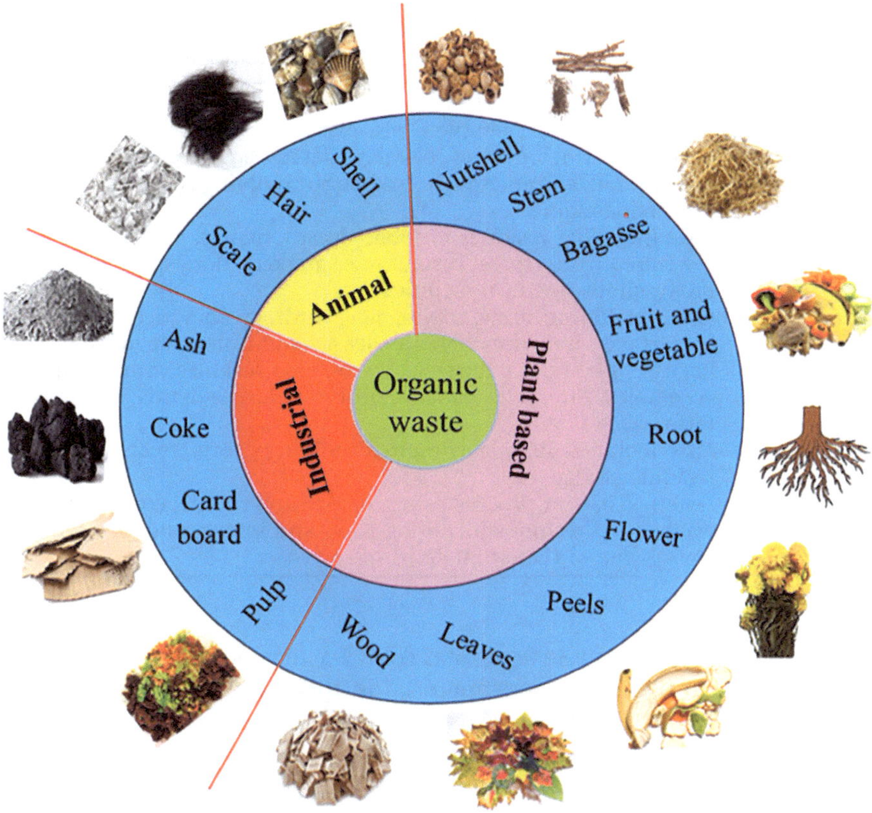

Figure 10.1 Circular diagram represents the classification of organic waste.

serious threat to environment. Agricultural wastes have become a significant source of pollution due to the lack of farmer's skills in handling and re-processing such wastes. As a result, due to the random burning of wastes, the release of GHGs has created a series of problems such as air pollution, ash runoff by rainwater and irrigation water, and heavily polluted soil and water. Likewise, many food industries are responsible for the generation of large quantities of peelings and scraps from fruits and vegetables by food processing.[15] Municipal wastes, such as the wastes of animals, vegetables, and fruits, are generated from different types of commercial businesses, households, and others. These wastes are created because of poor maintenance standards and hygiene of vegetable, pulp and fiber generated after the extraction of sugar and starch, filter sludge and coffee grounds.[16]

10.2.2 Global Perspectives of Organic-waste Statistics

All socioeconomic groups, rural and urban geographic locations, modern lifestyle, tourism, and economic activities have increased organic-waste

Table 10.1 Different sources of organic wastes.

Source	Types of organic waste
Residential	The waste is generated from dwellings, apartments, *etc.*, and consists of leftover food, vegetable peels, plastic, cloths, *etc.*
Institutional	Waste consisting of paper, plastics, glasses, *etc.* generated from educational, administrative, and public buildings (schools, colleges, universities, offices, *etc.*).
Commercial	Waste including residues of food, glasses, metals, ashes, *etc.*, generated from stores, restaurants, markets, hotels, motels, auto-repair shops, medical facilities, *etc.*
Agricultural	Wastes including straw, stover, cane trash, green agricultural waste, tree trunks, tree branches, tree leaves and fronds, sugarcane bagasse, rice husks, livestock waste, crop residues, fruits and vegetables, litter, *etc.*, generated from fields, orchards, vineyards, farms, *etc.*
Municipal	Waste includes dust, leafy matter, building debris, treatment-plant residual sludge, *etc.*
Industrial	Waste consists of process ashes, demolition rubble, construction waste, food residues including peels and scraps, and hazardous wastes due to industrial activities.

generation, which is about 50% of total solid waste generation.[17] Each year, an adult consumes 30–75 tons of material in their companies and houses, which results in the generation of organic waste in the form of food waste and other consumable products.[16] According to the World Bank, urban areas produce 2.01 billion tonnes of waste which includes 0.745 kg of waste per person per day.[18] According to Bhatia *et al.*,[16] in the USA 292.4 million tons of municipal solid waste (MSW) are generated each year, out of which 21.6% is from courtyard trimmings and 12.1% is food waste. Elfeki and Tkadlec[19] reported that in the village of Desoq, located in Kafr El Sheikh, Egypt, approximately 50.1% of the waste produced is organic, with nearly 30% consisting of food waste and yard trimmings. On the other hand, almost 998 million metric tonnes of agricultural waste are generated each year and 90% is disposed directly into landfills without auxiliary processing.[20] According to world statistical data analysis, in 2007 a total of 25% of all organic waste was disposed to landfills and incineration in Germany. In the USA, a total of 67% organic waste was directly discarded in 2008,[21] which generated a significant amount of GHGs and had a negative impact on the environment. The US urban forest generates more than 25 million tons of dried leaf waste and 20 million tons of dried tree waste per year, which is equivalent to more than 20 Mt of carbon (C) by mass.[22] Tadesse[23] reported that wood pulp was generating more than 195.79 million metric tonnes globally in 2022.

According to world statistics, 75% of food waste is accounted from total waste in Bangladesh, 50% of total waste in China, 40 to 60% of total waste in Malaysia, and 20 to 30% of total waste in European countries and 40% of total waste in USA.[20] According to the National Resources Defense Council (NDRC) statistical report for Asian countries, India and China have the highest contribution of food-waste production, accounting for approximately 1.34 billion tonnes per year. In India, approximately 68.7 million tonnes of food are wasted just from households annually for a consumption of 50 kg per person.[24] In Middle East countries, the majority of the produced waste is organic waste.[18] Qatar, a potential revenue country, generates 6.2 million metric tonnes of organic waste each year and it is predicted recycling of this waste could save 2 631 650 USD per year.[25] In many countries, organic-waste generation is highly based on income, as it directly reflects expenditure patterns. High-income countries generate relatively less organic waste at 36%, while lower middle-income and higher middle-income countries respectively generated 54% and 55% organic waste due to the lower economic value for management systems. In lower-income countries, high levels of organic waste generation and the rising costs of landfilling except in regions like Europe, Central Asia, and North America are driving efforts to recycle organic waste as part of a sustainable global solution. Almost 50% of the world organic biomass is believed to be lignocellulose and can be used as a feedstock for the production of biofuel due to the persistence in its qualitative properties.[26]

10.2.3 Characteristics of Organic-waste Biomass

The utilization of biomass from organic-waste, by converting it in an efficient and environmentally friendly way, is significant and desirable for renewable energy that entails socioeconomic benefits. Nowadays, biomass waste from various sources is generally treated by two major processing technologies: thermochemical and biological.[27] Lignocellulosic biomass is a promising raw material, containing approximately 50% of its composition in the form of carbon, and is considered an environmentally friendly and renewable resource.[28,29] Lignocellulose biomass comprised of mainly cellulose, hemicellulose, lignin, protein, and carbohydrate. Detailed properties of lignocellulose from different types of organic-waste biomass are reported in Table 10.2.[30–39] Lignocellulose can play an important role in various forms of energy products such as bioethanol, biodiesel, biogas and bio-H_2.[40] Cellulose has high crystallinity, whereas hemicellulose and lignin are amorphous.

Cellulose and hemicellulose present in organic-waste biomass are the raw feedstocks of second generation biofuels. Cellulose is the most abundant renewable organic compound in nature, with a high molecular weight of 300 000 and 500 000 g mol^{-1}, and can be symbolized as $(C_6H_{10}O_5)_n$. Cellulose consists of repeating units of β-D-glucopyranose monomer with a high polymerization degree as linear chains, which is beneficial for biofuel production. There is high availability of lignocellulosic biomass, but there is a lack of a straightforward process to convert lignocellulosic biomass into biofuel. The estimated global market of cellulose fiber grew from US$21.5 billion in 2017 to US$41.5 billion by 2025.[41]

The hemicellulose content of biomass, shown in Table 10.2, consists of several kinds of pentose and hexose monomer units, such as xylose,

Table 10.2 Dry composition of the organic fractions of various organic biomass sources.

Organic biomass	Cellulose (%)	Hemicellulose (%)	Lignin (%)	Protein (%)	Carbohydrates (%)	CH_4 yield/mL per g VS
Hardwood	40–55	15–30	18–25	0	0.51–58.6	130–180
Softwood	40–44	30–32	25–32	0	1.89–13.2	191–273
Eucalyptus	54.0	18.0	21.0	13.8	41	—
Pine wood pellet	39.5	22.1	37.1	—	1.52–3.01	—
Oak wood	43.2	21.9	35.4	—	2.7	—
Rubber wood	39.6	28.4	27.6	—	10.2	—
Cotton, flax	80–95	5–20	2.5–5.0	17–26	3.8	356
Giant reed	41.5	20.5	18.4	13.2	—	131
Maize straw	38.33 ± 0.8	29.76 ± 1.35	3.82 ± 0.5	4–9	7	272
Rice straw	28–36	23–28	12–14	2–7	20	203
Wheat straw	33–40	20–25	15–20	3.5	87	214–334
Cotton seed hairs	80–95	5–20	0	22	19.3	—
Barley straw	31–45	27–38	9–14	4.9	65	370
Groundnut shell	74	5.5	4.5	3.63	17–25.6	121.6
Sugarcane bagasse	40–45	30–35	20–30	1–3	22.3	161.8
Wheat sorghum bagasse	45	27	21	4.4–21.1	84.5	296
Bagasse	32–48	19–24	1–3	55.4	56.1	106.8–243.80
Oat straw	34.8	26.7	8.7	4	0.1	186.2 ± 10.5
Rice husk	40.3	12.5	25.4	2.9–3.6	37.1	145
Rice straw	35–50	20–35	10–25	3–6	20	135
Hemp stalk	52.0	25.0	17.0	11–38	2.6	190–219
Cotton gin	20.0	9.1	17.6	6–14	34–49	282.89–94.22
Grasses	25–40	25–50	10–30	0.22–59.5	44.5	
Lemongrass	45	5	10	2.98	25.3	403
Orchard grass	52.3	42.9	6.6	10–12	—	300

Meadow grass	28–40	15–38	12–18		6.5	376
Rye grass	6.5	27.9	42.38	20	2–3	431
Switchgrass	45	31.4	12	—	—	220
Coastal Bermuda grass	25	35.7	6.4	9–16	52.6	340
Leaves	15–20	80–85	—	1.4	2.9	—
Pineapple leaf	66.2	19.5	4.28	6	13	571
Bamboo leaves	34.1	25.6	35.0	12.92	2.3	—
Cashew leaves	72.5	3.5	12	3.01	22	140–210
Corn stover	38–40	18–28	7–21	12.06	15–18	258.8
Corncob	80	3.0	2.5	2.31	22	102
Sponge gourd	66.6	17.4	15.5	4	3	—
Banana peels	13.2	14.8	14	1.95 ± 0.14	11.82 ± 2.17	188–334
Birch	40.1 ± 0.6	17.5 ± 0.2	24.2 ± 0.1	17	42	365
Potato	4.1–20.2	2–14.3	15–20	18	15–25	321
Weet-Bix	9.4	2.0	8.6	3.6	69	—
Peapod	30–77.6	17.1	25–33.4	19.80 ± 1.65	4.8	256–357
Apple peel	47.49	27.77	24.72	12	90	320
Orange peel	14.2–69.1	0.2–9.02	1.0 ± 0.1	5.9 ± 0.3	64.2 ± 3.1	177
Tomato peel	33.1	10.5	7.8	0.88	4	561
Grape waste	17.5	6.9	51.7	0.66	95	110
Coconut husk	34	21	27	13.13	7	232.1
Cashew apple bagasse	20.6	10.2	35.3	1.09–1.13	50	506
Walnut shell	23.3	20.4	53.5	2.5	19	913
Durian	60.45	13.09	15.45	1.5	27	225
Lettuce	25	34	34.5	1.4	2.9	210
Mixed vegetable waste	40–53	9–12	10–17	15.30	42	—
Assorted fruit rind	—	—	—	—	98	180–732
Watermelon rind	20	23	10	17.23	6.4	581
Rambutan peel	24.28	11.62	35.34	1.95–2.91	22.73	—

Table 10.2 *(Continued)*

Organic biomass	Cellulose (%)	Hemicellulose (%)	Lignin (%)	Protein (%)	Carbohydrates (%)	CH_4 yield/mL per g VS
Olive stone flour	25	34	34.5	0.5–15	5–15	—
Mango peel	9.19	14.51	4.25	1.5–6.6	70	692
Carrot peel	28–51.6	6.7–12.3	17.5–32.1	1	9.6	310–482
Cabbage waste	17.32	9.12	5.94	16.1	8.2	440
Watermelon rind	20	23	10	17.23	6.4	581
Cauliflower waste	17.32	9.12	5.94	16.1	7.15	249.61
Banana peel	18–59	17–20	10–20	1.95 ± 0.14	11.82–75	227–294
Soyabean straw	44.2	5.9	19.2	2.5–11.9	11.6	223.2
Pistachio shell	30–55	20–32	12–38	20	19.21	—
Nut shell	25–30	19.3–29	30–40	2.4	14	121.6
Paper	85–99	0	0–15	0	0.5	326
Newspaper	40–55	25–40	18–30	0	1	210
Waste papers from chemical pulp	60–70	10–20	5–10	—	28.4–69.7	215
Primary wastewater solids	8–15	12.5–7.1	20–29	23.9	31	220–250
Swine manure	6.0–20.6	19–28	8.48	17.09	25	485
Cattle manure	4.7–23	8–20.45	11.48–30.4	7	30	37.5–270.6
Poultry manure	4.4–44.3	3.3–8.4	10–25.5	34.5	21	158.0–273.9

mannose, glucose, and galactose, in a branched, amorphous, and irregular arrangement with a lower degree of polymerization.[42–48] Therefore, hemicellulose is more easily degraded at lower temperatures and by using dilute acid or basic solution compared to cellulose. Hemicellulose, a combination of different types of polymerized monosaccharides, is the second-most abundant chemical component of dry wood mass, encompassing 25–35%. Hemicelluloses are chemically related to cellulose and generally heterogeneous, built up of different pentoses (C_5 sugars) and hexoses (C_6 sugars).

The third major component of biomass is lignin, which accounts for approximately 23–33% of the mass of softwoods and 16–25% of the mass of hardwoods. It is an amorphous cross-linked resin with no exact structure that is deposited by woody plants. Furthermore, the recalcitrant effect of lignin, the third component of lignocellulosic biomass, affects biofuel processes by requiring the introduction of a pre-treatment step to degrade lignocellulosic material.[49] The components of propyl groups, methoxy groups, hydroxy groups and benzene rings are joined together to form the chemical structure of lignin. Lignin has higher carbon content (approximately 61%) compared to cellulose (42%) and hemicellulose (40%) in lignocellulosic biomass.[50] Lignin consists of principally three kinds of phenyl-propane units in a three-dimensional branched polymer structure with aromatic rings. It has a wide range of degradation temperatures depending on the origin.[51]

Biomass materials have great potential to produce biofuel, particularly due to their microporous structure as in compounds like cellulose, hemicellulose and lignin.[52] Petrov *et al.*[53] carried out synthesis of porous carbon from several species of biomass that varied in the composition of cellulose and lignin. Biomass also contains other macromolecules, carbohydrates, lipids, and proteins. Renewable diesel can be produced by several different technological pathways.

Carbohydrate biomass resources are abundant, widely distributed and cheaper than lipid-containing biomass, and thus the production of biofuels from carbohydrate biomass is expected to be more sustainable than from lipid biomass. In addition, the limitations of using bioethanol as a direct substitute for petrol have inspired the development of chemical catalytic methods to produce more efficient biofuels from carbohydrate-rich biomass. The quantity of biogas produced increases with higher volatile solids (VS) content, reaching an optimum yield at 91.1% VS, corresponding to 2.88 kg. Volatile solids gave the maximum amount of biogas yield in terms of mL CH_4 per g organic waste.[54] The protein and carbohydrate content of different organic biomass sources are presented in Table 10.2.[55–57]

Biomass also contains inorganics, mainly potassium (K), sodium (Na), calcium (Ca), magnesium (Mg) and phosphorus (P), which end up in the ash. Organic extractives, including fats, waxes, alkaloids, proteins, phenolics, simple sugars, pectins, resins, terpenes, starches, glycosides *etc.*, are also essential for biofuel production. Biomass contains lignocellulosic complexes that hold water in the form of fibers, vessels and other anatomical parts.[58]

Proximate and ultimate analyses are two important characterization methods of organic waste for renewable-diesel production with either biochemical or thermochemical conversion. The proximate and ultimate analyses of different biomasses are reported in Table 10.3.[59–67] Proximate analysis is used to determine the weight percentage of moisture, volatile matter, fixed carbon and ash content of the biomass, whereas ultimate analysis is used to determine the weight percentages of chemical elements (carbon, hydrogen, nitrogen, oxygen and sulfur). The thermochemical conversion for energy production is directly linked to the proximate and ultimate analyses and the lignocellulose properties of the biomass. Lower moisture and ash content[68] with higher fixed carbon and volatile matter represent higher conversion in renewable-diesel production. The ultimate analysis signifies the fuel efficiency of the organic biomass.[69]

10.3 Organic-waste Valorization

10.3.1 Methods

Organic-waste management is a process where waste biomass is recycled to a useful product, such as fertilizer, renewable energy, *etc.*, through different recycling methods. Almost 44% of organic waste is food waste, which is produced naturally or artificially and releases methane (CH_4) gas by anaerobic decomposition when dumped into landfills. CH_4 gas has 20 times more potential than carbon dioxide (CO_2) in its effect on climate change and damage the environment. Organic-waste management by local governments and urban authorities greatly needs strategic decisions to improve collection, recycling, treatment and disposal approaches in a cost-effective manner for a sustainable global environment. There are two major types of management methods for organic waste: (1) the conventional method and (b) probiotic strategies for byproducts (Figure 10.2). Considering the economic feasibility, valorization of organic waste to a value-added product and application in different fields relieves pressure on waste disposal, encourages recycling, and mitigates adverse environmental impacts. Production of value-added products such as bio-oil, fertilizer, compost, biosorbent, and biochar from organic-waste biomass enhances the global economy by reducing chemical pollution, reducing GHG emissions, mitigating climate change, improving agricultural productivity and enhancing bioenergy processing. Bioproducts are used to produce electricity, generate heat, and in transportation.[70]

Many agricultural wastes, such as sugarcane bagasse, peanut hulls, maize stalks, potato peels, palm oil leaves, jatropha oil, sunflower waste, and others, are processed into bioethanol, biodiesel, biosorbent, biosurfactant and biogas. The increasing quantity of agricultural waste and its inappropriate treatment, especially in developing countries, have always been critical issues for environmental safety and human health.[71] Many research studies indicate all agricultural wastes have incredible potential benefits

Table 10.3 Proximate and ultimate analyses of different organic biomasses.

Organic biomass	Moisture (%)	Volatile matter (%)	Fixed carbon (%)	Ash (%)	C (%)	N (%)	H (%)	O (%)	S (%)
Food waste	6.0–10	74–93	1.5–13	3.6–5.5	47.0 ± 2.4	2.7 ± 0.4	7.3 ± 0.4	42.9 ± 3.3	0.2
Wood and woody biomass	5–63	30–80	6–26	1–8	49–57	<1	5–10	32–45	<1
Herbaceous biomass	4–48	41–77	9–35	1–19	42–58	1–3	3–9	34–49	<1
Animal and human waste biomass	3–9	43–62	12–13	23–34	57–61	6–12	7–8	21–25	1–2
Aquatic biomass	8–14	42–53	22–33	11–38	27–43	1–3	4–6	34–46	1–3
Crop residue	4.2–8.4	71.6–96.8	4.2–16	14.5–20	34.0–50	0.2–0.8	4.6–6.7	32.8–45.7	0.08–0.13
Agro-industrial waste	7.70–17.20	62.10–86.4	9.20–34.6	2.4–18.8	38.69–53.70	0.15–5.50	5.40–6.70	36.20–45.20	<1
Nut shell	11–14	74–78	20–25	1–2	48–51	0.2–05	4–6	41–44	0.01–0.03
Pruning waste	7–25	70–85	12–20	0.5–4	45–49	0.1–08	5–6	36–44	0.01–0.08
Dairy manure		79.55 ± 0.19	9.51 ± 0.24	10.94 ± 0.27	42.63 ± 0.01	2.01 ± 0.13	6.43 ± 0.08	39.70 ± 0.03	0.42 ± 0.02
Paper waste	5.95	78.55	7.93	7.57	41.43	1.01	6.87	49.83	0.86
Textile waste	6.85	82.37	10.61	0.17	41.19	0.01	6.97	51	0.84

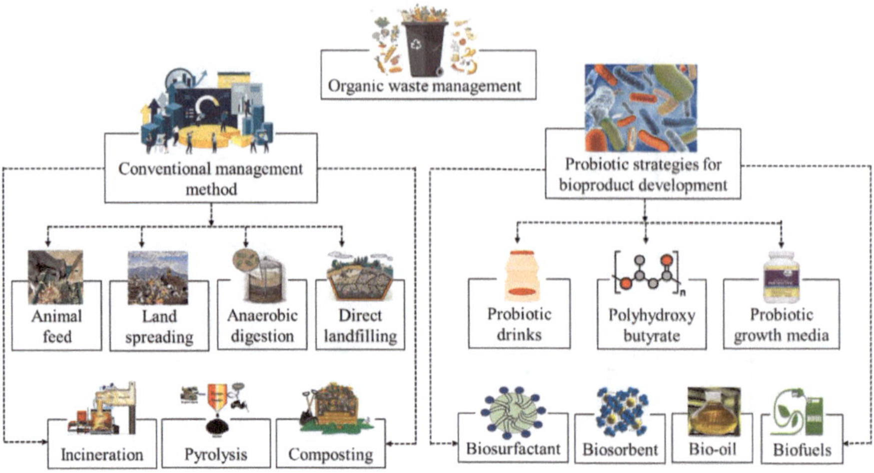

Figure 10.2 Different types of organic-waste management methods.

when recycled to produce bioenergy products to control environmental pollution[72] and mitigate GHG emissions. Therefore, utilizing all the agricultural biomass is needed to address the greatest challenges of climate change. Consequently, there is an urgent need to develop the most suitable methods to transform such agricultural wastes into economic and valuable materials.

Anaerobic digestion or fermentation into bioethanol production is one of the potential conversional methods to recycle industrial food waste into biogas. Food waste from households and industries is a major issue from liquid waste discharge from washing meat, washing fruit and vegetables, pre-cooked and boiled meats, poultry and fish, and operations of cleaning and managing wine making. This wastewater includes sugars, starches, fats, oils, and other dissolved organic matter, which can be used to produce biodiesel and renewable diesel. In the bakery industry, a lot of organic chemicals are used, which are later considered to be harmless as they are natural, but are a major challenge in disposal. This waste is degraded rapidly, and huge quantities of residue are not used in other applications or in the bioenergy sector. There are many management technologies, such as pyrolysis, gasification, digestion, incineration, composting, *etc.*, which are used for energy conversion.[73]

Incineration of food waste to produce thermal and can greatly reduce the volume of food waste.[74] But the major drawback is the high installation cost of incinerators, higher risk for long-term negative health impacts on the communities near the incineration plant, and the requirement for a consistent and enduring feed for optimal operation. The burning process in incinerators can pollute the environment if not properly operated and monitored. It can impact on human health for a long time by developing complications such as genetic disabilities and cancer.[74] After incineration, the end product "ash" also contains toxic compounds and heavy metals that may be potentially harmful to people during the recovery process.[75,76] Recycling of organic waste by composting is an ideal technique to improve soil fertility, crop production and reduce GHG emissions.[77] Though there are several benefits, the major drawback is high costs for waste collection and transportation, a large land requirement for composting, and the long processing period for the decomposition.[78]

Thermochemical pyrolysis is a sustainable, simple and convenient process for reducing more than 80% of organic waste by producing technoeconomic products such as biochar, biogas, bio-oil, *etc.*[79] The pyrolysis process has recently gained interest as it offers a shorter reaction time and allows increased energy recovery efficiency for various types of organic wastes.[80] Compared to incineration, pyrolysis processes thus represent a useful technique by minimizing environmentally toxic pollutants such as polybrominated diphenylethers.[81,82] Pyrolysis is generally undertaken in the absence of oxygen and is widely considered to comprise two types, namely (a) slow and (b) fast pyrolysis. These processes are categorized based on the temperature range, heating rate and residence times. Slow pyrolysis is mainly used for the production of biochar, while fast pyrolysis is used for bio-oil production.[83] Depending on the type of feedstock, 20–40% liquid oil, 20–35% solid char and 30–40% gaseous products are achieved by slow pyrolysis, but it is not favourable for high-quality bio-oil production.[84] The limitations of the slow pyrolysis process are high energy consumption at lower pyrolysis temperatures and long residence times.[85]

Organic-waste bioremediation is a valuable technology that mitigates pollutants from the environment through microbial activities, since biosurfactants are eco-friendly and effective compounds and are generated by some microorganisms. They can enhance the decay of petroleum hydrocarbons by increasing the bioavailability of contaminants for bioremediation by oleophilic microorganisms in the polluted aquatic environments. Thus, biosurfactants are a crucial factor in an optimized bioremediation process.[86] The objective of the present review is to evaluate the biosurfactant-producing microorganisms and their potentiality in bioremediation of petroleum hydrocarbons in aquatic environments.[87]

10.3.2 Opportunities

Government and industries are showing an increasing interest in waste management plans for the future, especially organic-waste management. Emerging markets and innovative uses for different types of organic waste have prompted stakeholders to re-examine how to exchange and utilize waste.[88] Stakeholders across the governmental, industrial, academic and non-governmental organization (NGO) sectors have encouraged investment in organic-waste management rather than organic-waste disposal.[89] When exploring potential waste streams, stakeholders have noted existing knowledge gaps and wanted to address them through research. Many researchers, project partners and stakeholders in organic-waste management have noted previous and ongoing work commissioned by government agencies over the past decade and desired to continue and mobilize. Stakeholders need diverse techniques for organic-waste management to meet diversion targets, environmental concerns over GHG emissions, and shortening lifespan of landfills.[90] Utilizing organic waste instead of landfilling has potential for environmental, economic, and social benefits. As the world regulates the impacts of climate change, stakeholders acknowledge that utilization of organic waste can be beneficial to the environment and people. Representatives from all sectors noted the benefits of bioenergy production from small scale to large scale, and also noted opportunities to process different types of organic waste into high-value products such as renewable diesel, biodiesel and biogas.[91] Energy demand in different sectors has a high impact on socioeconomic factors such as urbanization, population, industrialization, enhancement of technologies and net capital income and is an important component in energy planning and policy making. Before forecasting the future energy demand, policymakers need a complete knowledge of the growth and pattern of energy demand in different sectors, and this requires analysis of the energy demand.[92] Furthermore, through researchers and by stakeholders, organic waste is being managed in various innovative ways to explore environmental, economic, and social benefits.

10.3.3 Waste-management Crisis

Early research showed several significant challenges to adopt better organic-waste management practices for long-term purposes, including funding to coordinate ongoing curbside pickup, costs associated with retrofitting existing collection infrastructure (such as garbage trucks), manpower, and transfer-station processing.[93,94] Causes of the waste-management crisis include:

- Trucks with loaded waste became difficult to pass in heavy traffic.
- Lack of an alternative waste-disposal plan to landfills and exploitation.
- Insufficient government policies for long-term purposes.
- High charges by private contractors to manage organic waste.
- Traditionally centralized waste-management systems have very little sorting capabilities, and lack money to sustain the sorting process.
- Energy costs for incinerators of organic waste that is too wet to be processed and requires strict sorting before incineration.
- Lack of existing policies to support non-traditional uses and processing of organic waste.

10.4 Energy Demand and Impact of Fossil-fuel Energy

World energy demand is a global concern for the production, supply and consumption of energy related to the primary trading, innovation, automation, and economic growth, which are largely dependent on future energy use.[95] Energy is used from different sources for transportation, electricity and cooking. Details of energy sources, production and consumption are shown in Table 10.4. The energy demand is classified primarily into three sectors, namely (i) residential and commercial buildings, (ii) industry, and (iii) transportation. In 2022, 80% of energy was produced from fossil fuels.[96] The Gulf States and Russia are major energy exporters, with notable customers being the European Union and China, where domestically not enough energy is produced in order to satisfy energy demand. Energy consumption generally increases about 1–2% per year, except solar and wind energy, which averaged 20% per year in the 2010s.[96]

In 2021, nearly 36 billion tonnes of CO_2 was emitted from fossil-fuel consumption, out of 50 billion tonnes of total GHG emissions worldwide.[97] Many scenarios have been projected to reduce GHG emissions, typically to achieve net zero by 2050. The costs of developing various alternative energy sources to meet energy needs involves deciding how to effectively utilize available renewable energy resources to satisfy demand.[98] Energy efficiency advantages include decreased greenhouse-gas emissions and local air pollution, as well as fewer energy infrastructure expenditures, lower fossil-fuel reliance, better competitiveness, and improved customer welfare.[99] Overuse of fossil fuels is resulting in an increase of fossil-fuel prices as the supply is dominated by a few Middle Eastern countries.[100] According to the International Energy Agency (IEA) report, the energy demand will increase by 76% of in Asia by 2030.[101] Dependence on fossil-fuels imports poses a serious threat to the energy security and economic stability of all countries, as the world is expected to run out of petroleum in the next 50 years. Due to the limited reserves of fossil fuels and their negative impact on global climate and human health, much research is focused on replacement by alternative energy sources. Along with different energy resources such as solar, wind, hydroelectric, and nuclear energy, renewable diesel can help to reduce fossil-fuel dependence.

Table 10.4 Energy demand and consumption from various resources.

Energy sources	Demand (%)	Consumption (%)
Biomass	10	13
Coal	22	37
Gas	24	—
Oil	35	19
Hydroelectric	2	13
Nuclear	5	—
Other renewables	1.12	—
Traditional biomass	—	61
Wind	—	8
Biofuel	—	5

10.5 Renewable Diesel: A Key Source of Energy

Renewable diesel is a mixture of hydrocarbons that is chemically equivalent to petroleum diesel, and when used in its pure form is called R100. It can be co-processed with petroleum diesel, and can be effortlessly blended with petroleum diesel or biodiesel in various amounts. It can be transported in petroleum pipelines and traded at retail stations with or without blending. Blends of renewable diesel and petroleum diesel are labelled with an R followed by the volume percentage of renewable-diesel content. For example, a blend of 20% renewable diesel and 80% petroleum diesel is called R20. A blend of 20% biodiesel and 80% renewable diesel is called B20R80 to make a 100% biofuel. A blend of 20% biodiesel, 20% renewable diesel, and 60% petroleum diesel is called B20R20.[102] In 2021, the United States produced 815 million gallons of renewable diesel, which helped reduce petroleum diesel imports by 392 million gallons, out of a total national diesel consumption of 1163 million gallons.[102] California plays a major role in US imports of renewable diesel. The first generation of biodiesel was derived from soybeans, sunflowers, palm oil and corn, the second generation from jatropha, animal fats and waste cooking oil, and third-generation biodiesel was derived from algae. The first-generation sources (crops) are unable to meet demand because they account for 70–85% of total biodiesel production costs. The use of second- and third-generation sources is increasing due to lower production costs.[103] Algae-based renewable diesel is advantageous for the high growth rate of 15–25 tons per ha per year when compared to first- and second-generation sources, which have biomass production rates of 0.4 metric tonnes per ha per year for soybeans, 3.62 metric tonnes per ha per year for palm oil, and 4.14 metric tonnes per ha per year for jatropha.[104,105] The algae-based renewable diesel has a high energy content (35 800 kJ kg^{-1}, or 80% of the energy in petroleum),[106] is noncompetitive with food markets (as compared to first-generation sources) and can grow by utilizing wastewater nutrients.

Renewable diesel can be produced from any biomass feedstock, including those used in biodiesel production. It is a nontoxic, biodegradable energy source that emits fewer greenhouse gases than conventional diesel. Renewable diesel is also a nontoxic energy source and is gaining attention globally due to its direct applicability in preexisting engines without requiring modification, since its properties are similar to those of conventional diesel fuel. A major benefit of renewable diesel is that it is free from sulfur and polycyclic aromatic compounds.

10.5.1 Benefits of Renewable Diesel

Renewable diesel is 100% purified improves air quality by reducing emissions linked to heavy-duty traffic, including (a) 33% lower levels of fine particulate material, (b) 9% lower nitrogen oxides (NO_x), (c) 30% lower hydrocarbons (HC), (d) 24% lower carbon monoxide (CO), and (e) reduced levels of polyaromatic hydrocarbons (PAH).[107] Renewable diesel has high performance with a minimum cetane number of 70 compared to 50 for traditional diesel. It is fully compatible for use with the existing fuel distribution infrastructure without additional investments. Renewable diesel has no blending limit with conventional diesel. The fuel can be stored over long periods without deterioration in quality or water accumulation. The fuel is odorless, sulfur-free, oxygen-free, and aromatic-free and suitable for very cold weather conditions. Renewable diesel is chemically the same as petroleum diesel, but the refinery process makes it a cleaner fuel. Renewable diesel is an excellent choice for businesses to use as a supplemental fuel and it carries a premium price tag, which makes it a worthwhile investment.[108] Some of the benefits of renewable diesel include:

a. Improved vehicle performance: renewable diesel is refined to be a cleaner fuel that usually burns more cleanly in engines and helps to reduce the impact on vehicles. Since vehicles sustain less wear and tear over time, vehicle performance is improved and the need for costly repairs is reduced.
b. Compatibility of engines and infrastructure: renewable diesel meets the conventional petroleum ASTM D975 specification allowing it to be used in existing infrastructure and diesel engines.
c. Fewer emissions: renewable diesel reduces both CO_2 and NO_x emissions compared to petroleum diesel. According to the Low Carbon Fuel Standard Certified Carbon Intensities (LCFSCCI) of California, USA, renewable diesel reduces carbon intensity by an average of 65% in comparison to petroleum diesel.
d. More flexibility: renewable diesel can fully replace with diesel or be blended by any amount. Renewable diesel is a sustainable fuel that can be made from many sources of green waste biomass.
e. Reduced maintenance: renewable diesel can improve the performance of various engines by reducing wear and tear and increasing efficiency. As a result, renewable-diesel vehicles have reduced frequency of repair and help to save maintenance costs.
f. Cleaner burns: many businesses are turning to renewable diesel because it burns cleaner than standard petroleum diesel. Some mixtures have the potential to reduce emissions entirely, while other mixtures can be used to reduce carbon emissions significantly. Businesses are attracted to renewable diesel because of the increasing demand from consumers for companies to become more eco-friendly.
g. Improved storage: renewable diesel can be stored at low temperatures because the refining process improves its cold flow properties, making it resistant to freezing even in harsh winter conditions. This benefit makes renewable diesel more reliable for businesses that consistently operate in cold weather conditions and need fuel that can be stored and operated at low temperatures.
h. Limited odor: petroleum diesel is often associated with a strong smell, making some people feel sick or lightheaded from the pungent scent.

A minor benefit of renewable diesel is that it has no odor after its refinement process, making it easier to withstand being in the same room when it is burning in an engine.

10.6 Production of Renewable Diesel

Renewable diesel, also known as green diesel, is produced most often by hydrotreating and also through gasification, pyrolysis, and other biochemical and thermochemical technologies. Indonesia is the world's largest producer of biodiesel and renewable diesel, producing more than the combined amount of Brazil and the United States, which are second and third, respectively.[109] Renewable-diesel production facilities are increasingly converted from some parts of crude oil refineries or are complete conversions of refineries. Some are entirely new refinery facilities. All refining technologies have the potential to switch between processing crude oil and organic fats and oils, depending on economic conditions. Because crude-oil refining technology is used to produce renewable diesel, the capital costs are substantially higher for renewable-diesel production compared to FAME production.[102] Renewable diesel can be produced by different technological pathways, including (a) traditional hydrotreating, (b) biological sugar upgrading, (c) catalytic conversion of sugars, (d) gasification, (e) pyrolysis, and (f) hydrothermal processing.[110] Currently, commercial production facilities use the hydrotreating pathway in which fats, oils, and greases are the most common feedstocks.

The first step of making renewable diesel starts with hydrogenation or hydrotreating. The hydrotreating process parallels the process used to "crack" crude oil into gasoline, diesel, and other petroleum products in a crude-oil refinery.[111] Hydrotreating is a process that utilizes high-pressure hydrogen to remove oxygen from the fuel, which is then converted into water.[112] The double bonds in the fuel are saturated, removing the oxygen from fatty acid chains as either CO_2 or water, depending on the availability of hydrogen (H_2).

Renewable diesel is an ideal solution for businesses in cold environments that need reliable fuel to continue their operations (Zore *et al.*, 2021). Isomerization is an essential part of the renewable-diesel production process because it helps the fuel to be efficient in cold weather by preventing freezing. The process works by changing the long-chain *n*-paraffins into iso-paraffins of the same molecular weight, which can also be referred to as branching.[113] This process improves cold flow by slightly altering the boiling point, which reduces the risk of yield loss to lower boiling fractions. At this stage of renewable diesel production, the remaining hydrocarbon molecules are restructured without changing their atomic composition. Many businesses prefer renewable diesel because it performs well in cold climates-an important advantage in many regions around the world.

After the isomerization process, the product goes through fractionation, allowing the final product to be ready for distribution.[113] Fractionation helps create a product that can be stored until businesses purchase it. Renewable diesel can be stored in different temperatures, even extreme cold if done correctly. The final step of the production process is distribution of renewable diesel to businesses and consumers.[113] The petroleum logistics system allows distributors to deliver renewable diesel to their customers. This system ensures the smooth transportation of petroleum and petroleum-related products. Renewable diesel may sometimes be blended with other fuels for distribution. There are multiple stages in the distribution process, including storage and point of sale. Renewable diesel can be delivered through different methods, including pipelines, ships, trucks or trains.

10.7 Organic Waste as a Feedstock to Produce Renewable Diesel

Based on the production process and fuel properties, renewable fuels are classified into biodiesel, renewable diesel, and syn-diesel (Fischer–Tropsch diesel), as shown in Figure 10.3. The sustainability analysis of renewable fuel production processes includes multiple criteria, such as technical performance, fuel quality, environmental impact, and economic benefits. Most sustainable processes are associated with Fischer–Tropsch diesel from biomass and the catalytic hydrotreatment process to produce grade I (GD I) renewable diesel from vegetable oils. While algae make a major contribution in production of grade III (GD III) renewable diesel, it is the least favourable process because GD III has the highest impact on the water-scarcity footprint and unit production cost indicators.[114]

Organic-waste biomass resources are classified into first, second, and third generations of feedstock types for renewable fuel production.[115] Figure 10.3 shows a schematic diagram of selected renewable-fuel production processes

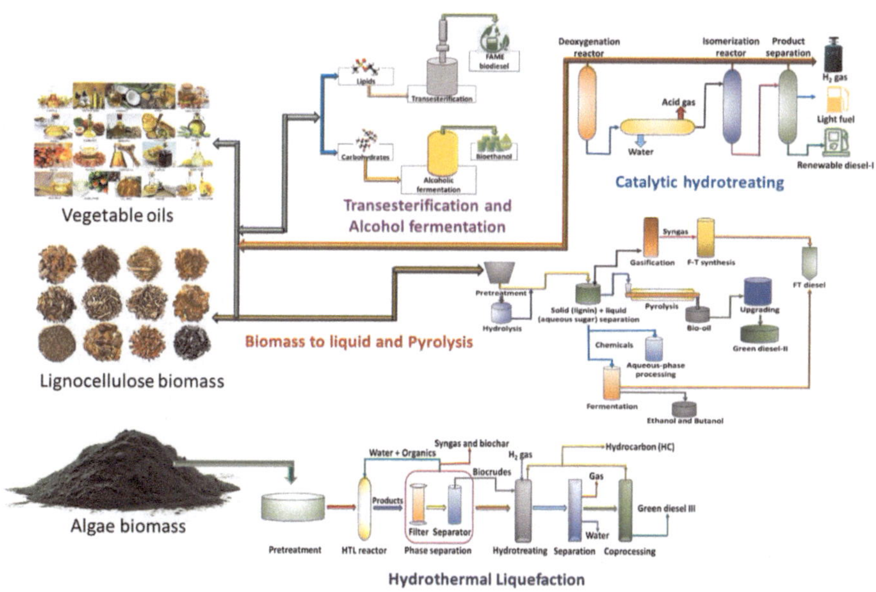

Figure 10.3 Flow diagram illustrating the conversion of three biomass sources-vegetable oils, lignocellulosic biomass, and algae biomass- into renewable fuels.

from organic-waste biomass. Renewable diesel I (GD I), FAME biodiesel and bioethanol are generated from vegetable oils and organic-waste biomass by transesterification, catalytic hydrotreating and alcohol fermentation methods. Renewable diesel II (GD II) and FT diesel are considered to be second-generation fuels, which utilize lignocellulosic organic biomass in biomass-to-liquid and pyrolysis processes.[114] Finally, renewable diesel III (GD III) is a third-generation biofuel produced from algae by the hydrothermal liquefication process.[116]

10.7.1 Transesterification Process

Transesterification is an organic reaction in which the alkoxy group of an ester is exchanged with an alcohol. Transesterification of vegetable oils is a mature process using triglycerides, methanol, and catalysts. Different sources of feedstocks contain triglycerides, which are available in soybean, canola, rapeseed, and palm oil *etc.* A catalyst is regularly used to accelerate the process and boost its yield. The catalyst might be a strong base (potassium hydroxide and sodium hydroxide) or strong acid (sulfuric acid). Since the transesterification reaction is reversible, excess alcohol is required to shift the equilibrium toward the product.[117] However, excess methanol can lead to the formation of emulsions or by-products that interfere with phase separation, making biodiesel purification more difficult.

The mechanism of biodiesel generation is usually assumed to consist of an initial mass-transfer-controlled process followed by a kinetically controlled process.[117] Using very effective homogeneous alkali catalysts, such as sodium or potassium hydroxides, carbonates or alkoxides, the great majority of commercial biodiesel is produced from vegetable oils. The performances of alkali-catalyzed, acid-catalyzed and non-catalyzed processes are used to determine the molar ratio of alcohol (methanol or ethanol) to oil (raw material) in the transesterification reaction.[117] The rate of the alkali-catalyzed process is considerably higher than the acid-catalyzed process. In general, the alkali-catalyzed method constitutes the vast majority of biodiesel manufacturing operations. In the transesterification process, a source of triglycerides, such as oil or fat, is used to react with methanol or ethanol in the presence of a catalyst. In general, the glycerol produced is considered to be a byproduct; however, the glycerol can be refined to a greater purity due to its commercial value in the pharmaceutical, food and cosmetics industry.[118] There is growing interest in simplifying and reducing the cost of the transesterification process by minimizing hazardous waste generation during the refining of biodiesel and glycerol.[119]

10.7.2 Catalytic Hydrotreatment of Vegetable Oils to Renewable Diesel I

Renewable diesel can be produced by different processes. One of the most favourable and common methods is catalytic hydroprocessing or hydrotreating. In catalytic hydrotreating, vegetable oil (liquid phase) is reacted with hydrogen (gas phase) in the presence of a metal catalyst (solid phase) to produce renewable diesel. Gaseous hydrogen is used to remove oxygen atoms from the triglyceride molecules in vegetable oil. The most frequently used reactors are trickle-bed reactors and slurry reactors. A trickle-bed reactor is a fixed-bed reactor, while a slurry reactor is a type of bubble column or stirred reactor. Hydrotreatment of vegetable oils in trickle-bed reactors takes place at a temperature of 300–350 °C and pressure of 50–100 bar, where triglyceride molecules are saturated through hydrogenation with renewable hydrogen to produce renewable diesel and propane.[120] Trickle-bed reactors have many advantages compared to slurry reactors, such as higher reaction conversion, lower operational or maintenance costs, lower catalyst loss, and lower separation cost.[121] In trickle-bed reactors, the first step is the production of fatty acids by the breakdown of triglycerides, and the second step is deoxygenation of the intermediate fatty acids to form diesel-range hydrocarbons by three key pathways, namely decarbonylation, decarboxylation, and hydrodeoxygenation.[122] The catalysts used during hydrotreatment are noble-metal catalysts (Pd or Pt) or sulfided transition-metal catalyst based on Ni or Mo.

10.7.3 Lignocellulosic Biomass Pyrolysis to Renewable Diesel II

Pyrolysis processes, which operate in an oxygen-free environment at 400–600 °C, have been extensively applied to produce pyrolytic bio-oil from lignocellulosic biomass.[84] The pyrolysis of organic waste is an emerging approach to produce renewable diesel and chemical products such as bio-oil, biochar, and syngas.[123] Pyrolysis of biomass is an environmentally friendly and low-cost process mainly used to decompose three primary components, cellulose, hemicellulose and lignin (Table 10.2), at different temperature ranges. Hemicellulose decomposes at 250–350 °C, cellulose decomposes at 325–400 °C, and lignin decomposes at 300–500 °C.[124] Biomass pyrolysis is most advantageous for end-product fixed carbon with ash, called biochar. The volatiles released from the bulk of biomass condensation produce the liquid product bio-oil. Bio-oil produced from pyrolysis has certain drawbacks, such as high acidity, lower heating value and high viscosity, and is unstable with petroleum products.[125]

10.7.4 Fischer–Tropsch Diesel

One of the most promising process for conversion of biomass to biofuel is called Fischer–Tropsch (FT) technology, in which the solid biomass is first converted to synthesis gas, containing mainly carbon monoxide and hydrogen, within a gasifier by fuel synthesis.[126] The gasification process is a thermochemical conversion technology that operates at 500–800 °C in the presence of a gasifying agent such as pure oxygen, steam, air or CO_2. Different types of gasifiers, such as fluidized bed, entrained flow or staged beds, are used to produce biofuel. After biomass gasification, the produced synthesis gas contains small amounts of undesired gas products such as sulfur, methane, ethane, propane, nitrogen compounds, halogens, volatile metals, tars, and particulates that might affect downstream processes.

10.7.5 Algae Hydrothermal Liquefaction for Renewable Diesel III

Algae can be cultivated in open ponds where the climate is conducive to algal growth or in closed-system photo-bioreactors where growth conditions, like temperature, pH and other factors, can be controlled and adjusted to adequate levels.[127] After cultivation, the algae biomass is harvested from the growth media with filtration techniques. Further dewatering is necessary to increase the biomass concentration to approximately 200 $g\,L^{-1}$ for the hydrothermal liquefaction (HTL) process.[128] This process takes place at temperatures of 250–400 °C and pressures of 50–150 bar with residence times of 3–60 min. Under these operating conditions, algal cells liquify, and water catalyzes the reactions among the biomass constituents. The gas fraction mainly consists of CO_2, while the aqueous fraction contains dissolved organic compounds (DOC) and nitrogen-containing species, such as ammonia or nitrates. The products resulting from liquefaction include biocrude, gaseous byproducts, an aqueous phase containing dissolved organic compounds, and solid residues such as char or coke.

10.8 Major Challenges in Renewable-Diesel Production

10.8.1 Feedstock Procurement

The increasing rate of renewable-diesel production has rapidly increased the demand for feedstocks. According to a survey, the price of cooking oil in 2021 increased by up to 50% compared to 2020, while the price of cattle and sheep fat products increased by 30%.[129] Renewable-diesel feedstocks, including oilseeds like soybeans and canola, have also expanded more than other sources such as wastes. The immediate result will be increased agricultural output and higher prices, including food prices. The rapid change of the renewable-diesel marketplace makes the feedstock market unstable. The marketplace for agricultural triglycerides and beef tallow, which serve as feedstocks for renewable diesel, is highly competitive. This is because production has decreased as farmers and cattle producers are not increasing their herds to supply more beef tallow. Additionally, renewable-diesel producers have little ability to drive increases in triglyceride production.

10.8.2 Wastewater Treatment

Renewable-diesel refining produces wastewater highly contaminated with residual oils. After filtration of contaminants, the resulting wastewater has a high chemical oxygen demand (COD) of 40 000–100 000 mg L^{-1} and total suspended solids (TSS) of 1000–10 000 mg L^{-1}, with emulsified oils at low pH level.[130] Therefore, a major challenge is wastewater-treatment facilities for these highly concentrated pollutants.

10.8.3 Waste Treatment

The distillation process in pretreatment generates spent bleaching clay that contains contaminated oil, water, and solids.[131] Evacuating those solids to a non-hazardous landfill can be dangerous because of self-ignition during transportation. Additionally, the residual oil will degrade to CH_4 in the landfill, increasing the refinery's GHG footprint. CH_4 emission has a carbon footprint 25 times higher than CO_2.

10.8.4 Finding the Right Place for Production

Determining the best region to develop infrastructure is complicated and it makes the most financial sense to produce renewable diesel in regions that offer tax credits.[132] Producers also have to choose a location with sufficient resources to handle the operation. Renewable diesel requires water for production, which is challenging in water-scarce areas.

10.8.5 Converting Facilities

Hydrotreating renewable diesel releases considerably more heat than with petroleum, so hydrotreaters need to be equipped with high liquid-product recycling capacities to absorb the heat as well as emergency depressurization systems.[133] Renewable-diesel production also requires more hydrogen than fossil diesel, so most refineries will require increased availability of hydrogen.

10.8.6 Reputation

Most major refining companies must earn the trust of the public and investors and demonstrate sustainability to successfully transition to renewable fuels. Many energy producers have fully adopted renewable diesel and are quickly ramping up new infrastructure construction, converting existing refineries, and aggressively planning to increase production of sustainable fuels. The challenge is to develop the new infrastructure, equipment, and distribution networks quickly, so that the market is not completely filled with former petroleum producers turned renewable-diesel producers.

10.9 Strategies to Enhance Production

Shareholders are willing to implement renewable diesel to reduce dependence on petroleum-based fuels and reduce climate change and fossil-fuel imports. There are several drawbacks in the biorefinery business in terms of

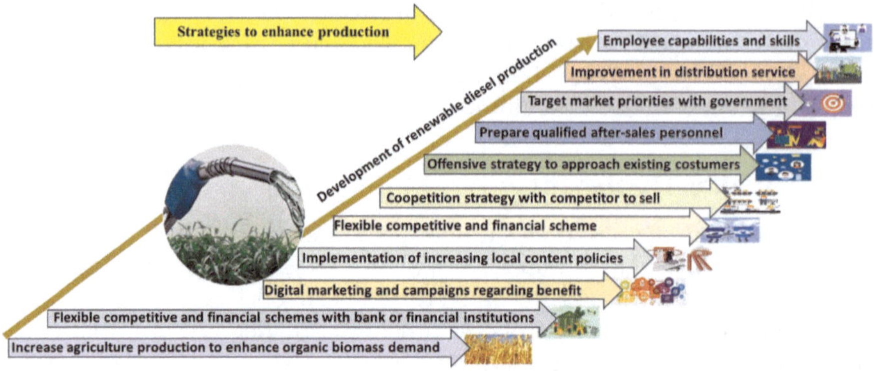

Figure 10.4 Flow diagram illustrating strategies to enhance renewable diesel production.

the renewable-diesel market, which has a higher price than fossil-fuel and crude oil. In accordance with those issues, biorefinery implementation will need an effective business strategy, especially in the marketing of renewable diesel to penetrate the domestic industrial fuel market. To develop a business strategy, a conceptual framework is necessary to have the right alternative strategies that are appropriate for the company, based on external and internal analysis. Based on the strategies shown in Figure 10.4, renewable-diesel production can be enhanced.

10.10 Economic Aspects of Renewable-diesel Production

Sustainable biofuel is an attractive alternative to fossil fuels because its production is not limited to specific locations with fossil fuel reserves and does not require the costly infrastructure associated with drilling and extraction. This allows for production in a wider range of geographic regions.[134] We can move towards a more sustainable future by providing crucial information for policymakers, investors, and industry leaders to understand the economic potential of renewable diesel (Figure 10.5). Renewable diesel has economic benefits if produced from waste oils by hydroprocessing and then converting the biologically derived oils into biodiesel. In the USA, renewable diesel generates sales revenues of \$4.60 per gallon as a commercial product, with lifecycle greenhouse gas emissions reduced by up to 75% compared to conventional diesel.

Renewable diesel production is highly sensitive to feedstock prices and conversion efficiency compared to other fuel production processes. In the USA, the biodiesel and renewable-diesel industry provided an economic boost of US\$23.2 billion while supporting 75 200 jobs and paying US\$3.6 billion in annual wages.[135] The farming, oilseed-processing, and fuel-production sectors gained the most economic benefits and employment

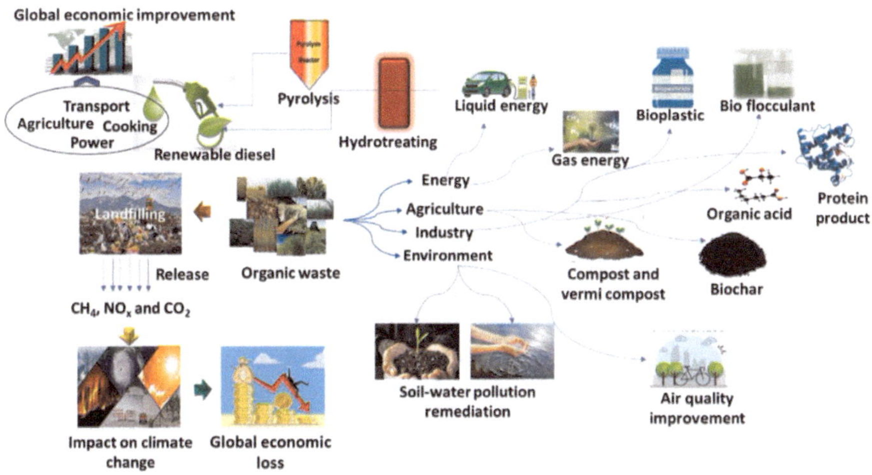

Figure 10.5 Flow diagram showing the economic and environmental benefits of renewable diesel production from organic waste.

opportunities through the production of renewable diesel. Renewable-diesel production also supports the waste-grease recycling and animal fat-rendering sectors. With the increasing demand for alternative fuels, the renewable-diesel industry is poised for further growth and expansion, bringing even more economic benefits and employment opportunities to communities across the country.

10.10.1 Case Study I

Patel and Kumar[133] and Patel *et al.*[136] developed a techno-economic model for assessment of the fast pyrolysis process for renewable diesel and gasoline from wood biomass. The scale factor of the lignocellulosic biomass-based renewable-diesel production was analyzed by sensitivity analysis to determine the economic benefits of renewable-diesel production. A plant capacity of 2000 tons per day (t/d) was used to analyze the production costs of renewable diesel and gasoline, utilizing merchant hydrogen and a sulfided nickel-molybdenum catalyst supported on alumina. The production costs of renewable diesel and gasoline were 1.09 USD per L and 1.04 USD per L, respectively, for a 2000 t per day capacity of biomass plant. To understand the variations in transportation costs, a wide range of production capacities has been considered, from 500 t per day to 5000 t per day. The optimum size of the production unit was found to be 3000 t per day for a renewable-diesel price of 1.007 USD per L with the minimum production cost. When the capacity is increased, raw-material delivery costs increase significantly; however, the capital cost per unit output decreases. The results indicate that the production cost of liquid fuels is most sensitive to the bio-oil yield, with the resulting production cost varying around 1.10 ± 0.038 per L.

10.10.2 Case Study II

Glisic *et al.*[88] reported that waste vegetable oils or lower-quality triglyceride-type oils could provide a sustainable renewable source for cost-efficient diesel fuel. Renewable diesel can be produced from a larger variety of feedstocks in comparison to biodiesel. The transesterification of an unsaturated feedstock provides a biodiesel blend of unsaturated FAMEs that needs special attention to the concentration of unsaturated fatty acids.[137] On the other hand, renewable diesel blends of saturated hydrocarbons can be produced through saturation of triglycerides with H_2.[138] Simple sugars (*e.g.* from sugarcane or sugar beet) or starches (*e.g.* from corn, wheat or sorghum) may be used as feedstocks for the production of renewable biofuels.[139] This study indicates that a large-capacity hydroprocessing unit located in a refinery, where most of the supporting process infrastructure is already available, is the most cost-effective way to process waste vegetable oils and similar low-cost oils, while the process-technology cost for FAME biodiesel is highly dependent on glycerol prices.

10.11 Conclusion

This chapter summarizes the impact on the environment and economy of biofuel production from waste biomass, reviewing the benefits, drawbacks, and future prospectives. This chapter provided a detailed update on the conventional processing methods, namely biochemical and thermochemical conversion technologies (*e.g.* hydrolysis, fermentation, pyrolysis, and gasification) for valorization of waste biomass to produce renewable diesel, biodiesel, and bio-oil *etc.* This review elaborated on the sustainable approach of recycling lignocellulosic biomass in a worldwide energy-production sector to achieve a bio-based economy. The global interest in processing organic-waste biomass to produce renewable diesel suggests an attractive economic future for the biorefinery industry. This study contributes an analysis of the importance of identifying the various properties of biomass needed to valorize a wide variety of urban, forestry, industrial, and agricultural organic waste as biofuel feedstocks. Renewable-diesel production from waste biomass has attracted significant attention as a promising alternative to fossil fuels. Organic waste as a renewable-diesel feedstock offers several advantages, such as reduction of greenhouse gas emissions, potential waste management solutions, and decreased reliance on non-renewable resources. However, this study encourages the need for a circular-economy approach in the large-scale economic production of renewable diesel from organic waste directed towards a "waste-to-energy" strategy.

Abbreviations

CO_2	Carbon dioxide
NO_x	Nitrogen oxides

GHG	Greenhouse gas
FAME	Fatty acid methyl ester
EREC	European Renewable Energy Council
MSW	Municipal solid waste
USA	United State of America
NRDC	Natural Resources Defense Council
VS	Volatile solid
K	Potassium
Na	Sodium
Ca	Calcium
Mg	Magnesium
P	Phosphorus
C	Carbon
N	Nitrogen
H	Hydrogen
O	Oxygen
S	Sulfur
IEA	International Energy Agency
HC	Hydrocarbons
PAH	Polyaromatic hydrocarbons
ASTM	American society for testing and materials
LCFSCCI	Low Carbon Fuel Standard Certified Carbon Intensities (California, USA)
FT	Fischer–Tropsch
GD	Grade
COD	Chemical oxygen demand
TSS	Total suspended solids

Acknowledgements

The authors would like to thank Qatar National Research Fund (QNRF) for their support of this research through NPRP-11S-0117-180328, the Supreme Committee for Delivery and Legacy (SCDL) and Hamad Bin Khalifa University (HBKU) and Qatar Foundation (QF).

References

1. A. Chanthakett, M. T. Arif, M. M. K. Khan and M. Subhani, Hydrogen production from municipal solid waste using gasification method, in *Hydrogen Energy Conversion and Management*, ed. M. M. K. Khan, A. K. Azad and A. M. T. Oo, 2024, ch. 4, pp. 103–131.
2. M. A. Azni, R. Md Khalid, U. A. Hasran and S. K. Kamarudin, Review of the effects of fossil fuels and the need for a hydrogen fuel cell policy in Malaysia, *Sustainability*, 2023, **15**, 4033.
3. S. R. Sharvini, Z. Z. Noor, C. S. Chong, L. C. Stringer and R. O. Yusuf, Energy consumption trends and their linkages with renewable energy

policies in East and Southeast Asian countries: Challenges and opportunities, *Sustainable Environ. Res.*, 2018, **28**, 257–266.
4. E. Lindstad, T. Ø. Ask, P. Cariou, G. S. Eskeland and A. Rialland, Wise use of renewable energy in transport, *Transp. Res., Part D: Transp. Environ.*, 2023, **119**, 103713.
5. P. Vignesh, A. R. Pradeep Kumar, N. Shankar Ganesh, V. Jayaseelan and K. Sudhakar, A review of conventional and renewable biodiesel production, *Chin. J. Chem. Eng.*, 2021, **40**, 1–17.
6. M. J. Burke and J. C. Stephens, Political power and renewable energy futures: A critical review, *Energy Res. Soc. Sci.*, 2018, **35**, 78–93.
7. M. Shahbaz, A. AlNouss, I. Ghiat, G. Mckay, H. Mackey, S. Elkhalifa and T. Al-Ansari, A comprehensive review of biomass based thermochemical conversion technologies integrated with CO_2 capture and utilisation within BECCS networks, *Resour., Conserv. Recycl.*, 2021, **173**, 105734.
8. P. Cavelius, S. Engelhart-Straub, N. Mehlmer, J. Lercher, D. Awad and T. Brück, The potential of biofuels from first to fourth generation, *PLoS Biol.*, 2023, **21**, e3002063.
9. H. Mukhtar, M. W. Mumtaz and S. Massa, Editorial: Waste to energy: Biomass-based energy systems, *Front. Bioeng. Biotechnol.*, 2022, **10**, 932981, https://www.frontiersin.org/articles/10.3389/fbioe.2022.932981 (accessed February 24, 2024).
10. L. G. Schumacher, J. V. Gerpen and B. Adams, Biodiesel Fuels, in *Encyclopedia Energy*, ed. C. J. Cleveland, Elsevier, New York, 2004, pp. 151–162.
11. J. C. Serrano-Ruiz, Biomass: A Renewable Source of Fuels, Chemicals and Carbon Materials, *Molecules*, 2020, **25**, 5217.
12. EREC, Renewables can meet 50 per cent of energy needs by 2040, claims report, CORDIS|European Commission (2023). https://cordis.europa.eu/article/id/22093-renewables-can-meet-50-per-cent-of-energy-needs-by-2040-claims-report (accessed February 24, 2024).
13. I. Ghiat, S. Pradhan, R. Govindan, H. R. Mackey, G. Mckay and T. Al-Ansari, Biochar: A sustainable approach of green waste management in agricultural practices under controlled microclimate, *Chem. Eng. Trans.*, 2022, **92**, 331–336.
14. K. Pawlak and M. Kołodziejczak, The role of agriculture in ensuring food security in developing countries: considerations in the context of the problem of sustainable food production, *Sustainability*, 2020, **12**, 5488.
15. S. Pradhan, M. Shahbaz, A. Abdelaal, T. Al-Ansari, H. R. Mackey and G. McKay, Optimization of process and properties of biochar from cabbage waste by response surface methodology, *Biomass Convers. Biorefin.*, 2020, **9**, 1–13.
16. L. Bhatia, H. Jha, T. Sarkar and P. K. Sarangi, Food waste utilization for reducing carbon footprints towards sustainable and cleaner environment: A Review, *Int. J. Environ. Res. Public Health*, 2023, **20**, 2318.

17. S. Syafrudin, J. Masjhoer and M. Maryono, Characterization and quantification of solid waste in rural regions, *Global J. Environ. Sci. Manage.*, 2023, **9**, 337–352.
18. S. Mariyam, L. Cochrane, S. Zuhara and G. McKay, Waste Management in Qatar: A systematic literature review and recommendations for system strengthening, *Sustainability*, 2022, **14**, 8991.
19. M. Elfeki and E. Tkadlec, Treatment of municipal organic solid waste in Egypt, *J. Mater. Environ. Sci.*, 2015, **6**, 756–764.
20. A. Sahoo, A. Dwivedi, P. Madheshiya, U. Kumar, R. K. Sharma and S. Tiwari, Insights into the management of food waste in developing countries: with special reference to India, *Environ. Sci. Pollut. Res. Int.*, 2023, 1–27.
21. V. Kumar, N. Sharma, M. Umesh, M. Selvaraj, B. M. Al-Shehri, P. Chakraborty, L. Duhan, S. Sharma, R. Pasrija, M. K. Awasthi, S. R. Lakkaboyana, R. Andler, A. Bhatnagar and S. S. Maitra, Emerging challenges for the agro-industrial food waste utilization: A review on food waste biorefinery, *Bioresour. Technol.*, 2022, **362**, 127790.
22. K. Lan, B. Zhang and Y. Yao, Circular utilization of urban tree waste contributes to the mitigation of climate change and eutrophication, *One Earth*, 2022, **5**, 944–957.
23. T. Karak, R. M. Bhagat and P. Bhattacharyya, Municipal solid waste generation, composition, and management: the world scenario, *Crit. Rev. Environ. Sci. Technol.*, 2012, **42**, 1509–1630.
24. P. Chaudhary, S. Garg, T. George, M. Shabin, S. Saha, S. Subodh and B. Sinha, Underreporting and open burning the two largest challenges for sustainable waste management in India, *Resour., Conserv. Recycl.*, 2021, **2021**, GC44D-08.
25. J. M. Alhumoud, Municipal solid waste recycling in the Gulf Co-operation Council states, *Resour., Conserv. Recycl.*, 2005, **45**, 142–158.
26. K. Sandesh Suresh, P. V. Suresh and T. G. Kudre, Prospective ecofuel feedstocks for sustainable production, in *Advances in Eco-Fuels for a Sustainable Environment*, ed. K. Azad, Woodhead Publishing, 2019, ch. 4, pp. 89–117.
27. S. Darmawan, N. Wistara, G. Pari, A. Maddu and W. Syafii, Characterization of lignocellulosic biomass as raw material for the production of porous carbon-based materials, *BioResources*, 2016, **11**, 3561–3574.
28. B. Hu, S.-H. Yu, K. Wang, L. Liu and X.-W. Xu, Functional carbonaceous materials from hydrothermal carbonization of biomass: an effective chemical process, *Dalton Trans.*, 2008, 5414–5423.
29. M. Sharon, M. Sharon, G. Kalita and B. Mukherjee, Hydrogen storage by carbon fibers synthesized by pyrolysis of cotton fibers, *Carbon Lett.*, 2011, **12**, 39–43.
30. B. Boneberg, G. Machado, D. Santos, F. Gomes, D. Faria, L. Gomes and F. Santos, Biorefinery of lignocellulosic biopolymers, *Revista Eletrônica Científica Da UERGS*, 2016, **2**, 79–100.

31. C. A. Cardona, J. A. Quintero and I. C. Paz, Production of bioethanol from sugarcane bagasse: Status and perspectives, *Bioresour. Technol.*, 2010, **101**, 4754–4766.
32. P. Ghanney, J. Kugbe and D. Anning, Role of Microbial Biomechanics in composting with special reference to lignocellulose biomass digestion, *Asian J. Biotechnol. Bioresour. Technol.*, 2021, 30–46.
33. D. Glushkov, G. Nyashina, A. Shvets, A. Pereira and A. Ramanathan, Current status of the pyrolysis and gasification mechanism of biomass, *Energies*, 2021, **14**, 7541.
34. P. Kaur, M. S. Taggar and J. Sidhu, Cellulolytic microorganisms: diversity and role in conversion of rice straw to bioethanol, *Cellul. Chem. Technol.*, 2020, **54**, 613–634.
35. S. Nayak and A. Mukherjee, Management of agricultural wastes using microbial agents, in *Waste Management: Challenges, Threats and Opportunities*, 2015, pp. 65–91.
36. K. O. Olatunji, N. Ahmed and O. Ogunkunle, Optimization of biogas yield from lignocellulosic materials with different pretreatment methods: A review, *Biotechnol. Biofuels*, 2021, **14**, 1–34.
37. Y. Sun and J. Cheng, Hydrolysis of lignocellulosic materials for ethanol production: a review, *Bioresour. Technol.*, 2002, **83**, 1–11.
38. F. Talebnia, D. Karakashev and I. Angelidaki, Production of bioethanol from wheat straw: An overview on pretreatment, hydrolysis and fermentation, *Bioresour. Technol.*, 2010, **101**, 4744–4753.
39. M. Tayyab, A. Noman, W. Islam, S. Waheed, Y. Arafat, F. Ali, M. Zaynab, S. Lin, H. Zhang and D. Khan, Bioethanol production from lignocellulosic biomass by environment-friendly pretreatment methods: A review, *Appl. Ecol. Environ. Res.*, 2017, **16**, 225–249.
40. A. Yousuf, D. Pirozzi and F. Sannino, Fundamentals of lignocellulosic biomass, in *Lignocellulosic Biomass to Liquid Biofuels*, ed. A. Yousuf, D. Pirozzi and F. Sannino, Academic Press, 2020, ch. 1, pp. 1–15.
41. M. Mujtaba, L. Fernandes Fraceto, M. Fazeli, S. Mukherjee, S. M. Savassa, G. Araujo de Medeiros, A. do Espírito Santo Pereira, S. D. Mancini, J. Lipponen and F. Vilaplana, Lignocellulosic biomass from agricultural waste to the circular economy: a review with focus on biofuels, biocomposites and bioplastics, *J. Cleaner Prod.*, 2023, **402**, 136815.
42. J. R. Ayala, G. Montero, M. A. Coronado, C. García, M. A. Curiel-Alvarez, J. A. León, C. A. Sagaste and D. G. Montes, Characterization of orange peel waste and valorization to obtain reducing sugars, *Molecules*, 2021, **26**, 1348.
43. A. Gowman, M. Picard, L. T. Lim, M. Misra and A. Mohanty, Fruit waste valorization for biodegradable biocomposite applications: A review, *BioResources*, 2019, **14**, 10047–10092.
44. J. Hernández Beltrán, I. Hernandez De Lira, M. Cruz-Santos, A. Saucedo-Luevanos, F. Hernández-Terán and N. Balagurusamy, Insight into pretreatment methods of lignocellulosic biomass to increase

biogas yield: current state, challenges, and opportunities, *Appl. Sci.*, 2019, **9**, 3721.

45. W. Holloway, C. Tasman-Jones and K. Maher, Towards an accurate measurement of dietary fibre, *N. Z. Med. J.*, 1977, **85**, 420–423.
46. M. Kim and D. F. Day, Composition of sugar cane, energy cane, and sweet sorghum suitable for ethanol production at Louisiana sugar mills, *J. Ind. Microbiol. Biotechnol.*, 2011, **38**, 803–807.
47. S. Kim, Evaluation of alkali-pretreated soybean straw for lignocellulosic bioethanol production, *Int. J. Polym. Sci.*, 2018, e5241748, DOI: 10.1155/2018/5241748.
48. A. M. Yahya, A. A. Adeleke, P. Nzerem, P. P. Ikubanni, S. Ayuba, H. A. Rasheed, A. Gimba, I. Okafor, J. A. Okolie and P. Paramasivam, Comprehensive characterization of some selected biomass for bioenergy production, *ACS Omega*, 2023, **8**, 43771–43791.
49. G. Velvizhi, K. Balakumar, N. P. Shetti, E. Ahmad, K. Kishore Pant and T. M. Aminabhavi, Integrated biorefinery processes for conversion of lignocellulosic biomass to value added materials: Paving a path towards circular economy, *Bioresour. Technol.*, 2022, **343**, 126151, DOI: 10.1016/j.biortech.2021.126151.
50. X. Cao, L. Zhong, X. Peng, S. Sun, S. Li, S. Liu and R. Sun, Comparative study of the pyrolysis of lignocellulose and its major components: Characterization and overall distribution of their biochars and volatiles, *Bioresour. Technol.*, 2014, **155**, 21–27.
51. H. Yang, R. Yan, H. Chen, C. Zheng, D. H. Lee and D. T. Liang, In-depth investigation of biomass pyrolysis based on three major components: Hemicellulose, cellulose and lignin, *Energy Fuels*, 2006, **20**, 388–393.
52. P. Simon and A. Burke, Nanostructured carbons: Double-layer capacitance and more, *Electrochem. Soc. Interface*, 2008, **17**, 38–43.
53. N. Petrov, T. Budinova, M. Razvigorova, V. Minkova, R. Zanzi and E. Bjoernbom, Preparation of activated carbons from cherry stones, apricot stones and grape seeds for removal of metal ions from water, 1999, pp. 46, https://www.osti.gov/etdeweb/biblio/20118137 (accessed February 15, 2024).
54. E. K. Orhorhoro, P. O. Ebunilo and G. E. Sadjere, Experimental determination of effect of total solid (TS) and volatile solid (VS) on biogas yield, *Am. J. Mod. Energy*, 2017, **3**, 131–135.
55. M. Chandran, A. Sinha and R. Rawat, Replacing controlled burning practice by Alternate methods of reducing fuel load in the Himalayan Long leaf Pine (Pinus roxburghii Sarg.) forests, In *5th international wildland fire conference*, South Africa, 2011, pp. 9–13.
56. J. Gao, D. Anderson and B. Levie, Saccharification of recalcitrant biomass and integration options for lignocellulosic sugars from Catchlight Energy's sugar process (CLE Sugar), *Biotechnol. Biofuels*, 2013, **6**, 10.
57. A. S. Nizami, M. Rehan, I. Ismail, T. Almeelbi and O. Ouda, Waste biorefinery in Makkah: A solution to convert waste produced during Hajj and Umrah seasons into wealth, 2015, pp. 90–104.

58. E. Apaydın Varol and U. Mutlu, TGA-FTIR Analysis of biomass samples based on the thermal decomposition behavior of hemicellulose, cellulose, and lignin, *Energies*, 2023, **16**, 3674.
59. Z. Akyurek, Sustainable valorization of animal manure and recycled polyester: Co-pyrolysis Synergy, *Sustainability*, 2019, **11**, 2280.
60. J. Asibor, P. Akhator and A. Obanor, Energy potential study of some tropical wood species from Nigeria, *Curr. J. App. Sci. Technol.*, 2019, **37**, 1–10.
61. D. Singh and S. Yadav, Evaluation of the physico-chemical development of kitchen food wastes through torrefaction – a biodiversity case study, *Biomass Convers. Biorefin.*, 2021, **11**, 1353–1362.
62. B. Kehinde, O. O. Agbede, I. I. Olateju, B. Samuel, O. M. Osuolale and F. Osuolale, Biomass gasification potential in Nigeria: A review, *LAUTECH J. Civil Environ. Stud.*, 2019, **3**, 83–97.
63. P. Muthu, G. Muniappan and R. B. Jeyakumar, Efficacious utilization of food waste for bioenergy generation through the anaerobic digestion method, *Processes*, 2023, **11**, 702.
64. W. T. Tsai, P. C. Huang and Y. Q. Lin, Characterization of biochars produced from dairy manure at high pyrolysis temperatures, *Agronomy*, 2019, **9**, 634.
65. S. Vassilev, D. Baxter, L. Andersen and C. Vassileva, An overview of the chemical composition of biomass, *Fuel*, 2010, **89**, 913–933.
66. M. Villarini, V. Marcantonio, A. Colantoni and E. Bocci, Sensitivity analysis of different parameters on the performance of a CHP internal combustion engine system fed by a biomass waste gasifier, *Energies*, 2019, **12**, 688.
67. C. Wambugu, E. Rene, J. van de Vossenberg, C. Dupont and E. van Hullebusch, Role of biochar in anaerobic digestion based biorefinery for food waste, *Front. Energy Res.*, 2019, 7, 14.
68. R. Garcia, Characterization of Spanish biomass wastes for energy use, *Bioresour. Technol.*, 2012, **103**, 249–258, https://www.academia.edu/23629434/Characterization of Spanish biomass_wastes for energy use (accessed February 20, 2024).
69. Y. D. Singh, P. Mahanta and U. Bora, Comprehensive characterization of lignocellulosic biomass through proximate, ultimate and compositional analysis for bioenergy production, *Renewable Energy*, 2017, **103**, 490–500.
70. I. M. Tijjani Usman, Y.-C. Ho, L. Baloo, M.-K. Lam and W. Sujarwo, A comprehensive review on the advances of bioproducts from biomass towards meeting net zero carbon emissions (NZCE), *Bioresour. Technol.*, 2022, **366**, 128167.
71. H. Elbasiouny, B. Elbanna, E. Al-Najoli, A. Alsherief, S. Negm, E. El-Nour, A. Nofal and S. Sharabash, Agricultural waste management for climate change mitigation: Some implications to Egypt, *Springer Water*, 2020, pp. 149–169.
72. G. Enaime and M. Lübken, Agricultural waste-based biochar for agronomic applications, *Appl. Sci.*, 2021, **11**, 8914.

73. P. Sharma, A. Bano, S. P. Singh, J. D. Atkinson, S. S. Lam, H. M. N. Iqbal and Y. W. Tong, Biotransformation of food waste into biogas and hydrogen fuel – A review, *Int. J. Hydrogen Energy*, 2024, **52**, 46–60.
74. A. Gao, Z. Tian, Z. Wang, R. Wennersten and Q. Sun, Comparison between the technologies for food waste treatment, *Energy Proc.*, 2017, **105**, 3915–3921.
75. S. Jacob, S. Nithianandam, S. Rastogi, S. Sakhuja and S. N. Sri Laxma Alankar, Handling and treatment strategies of biomedical wastes and biosolids contaminated with SARS-CoV-2 in waste environment, in *Environmental and Health Management of Novel Coronavirus Disease (COVID-19)*, ed. M. Hadi Dehghani, R. R. Karri and S. Roy, Academic Press, 2021, ch. 8, pp. 207–232.
76. Y. Kalmykova and K. Karlfeldt Fedje, Phosphorus recovery from municipal solid waste incineration fly ash, *Waste Manage.*, 2013, **33**, 1403–1410.
77. M. Kumar, Y. L. Ou and J. G. Lin, Co-composting of green waste and food waste at low C/N ratio, *Waste Manage.*, 2010, **30**, 602–609.
78. H. Jouhara, D. Czajczyńska, H. Ghazal, R. Krzyżyńska, L. Anguilano, A. J. Reynolds and N. Spencer, Municipal waste management systems for domestic use, *Energy*, 2017, **139**, 485–506.
79. S. Zuhara, S. Pradhan and G. McKay, Pyrolysis of biosolids with waste cardboard: effect of operating parameters, feedstock size and blending ratio, *Int. J. Environ. Sci. Technol.*, 2024, **21**, 617–630.
80. D. Czajczyńska, L. Anguilano, H. Ghazal, R. Krzyżyńska, A. J. Reynolds, N. Spencer and H. Jouhara, Potential of pyrolysis processes in the waste management sector, *Therm. Sci. Eng. Prog.*, 2017, **3**, 171–197.
81. D. Chen, L. Yin, H. Wang and P. He, Pyrolysis technologies for municipal solid waste: a review, *Waste Manage.*, 2014, **34**, 2466–2486.
82. S. Pradhan, A. H. Abdelaal, K. Mroue, T. Al-Ansari, H. R. Mackey and G. McKay, Biochar from vegetable wastes: agro-environmental characterization, *Biochar*, 2020, **2**, 439–453.
83. P. Parthasarathy, A. Fernandez, D. K. Singh, T. Al-Ansari, H. R. Mackey, R. Rodriguez, G. Mazza, J. V. Tirkey and G. McKay, Thermogravimetric analysis of camel dung, date stone, and their blend for pyrolytic, kinetic, and thermodynamic studies, *Cleaner Chem. Eng.*, 2022, **4**, 100072.
84. D. Aboelela, H. Saleh, A. Attia, Y. Elhenawy, T. Majozi and M. Bassyouni, Recent advances in biomass pyrolysis processes for bioenergy production: Optimization of operating conditions, *Sustainability*, 2023, **15**, 11238.
85. P. Parthasarathy, H. R. Mackey, S. Mariyam, S. Zuhara, T. Al-Ansari and G. McKay, Char products from bamboo waste pyrolysis and acid activation, *Front. Mater.*, 2021, 7, 482.
86. A. P. Karlapudi, T. C. Venkateswarulu, J. Tammineedi, L. Kanumuri, B. K. Ravuru, V. Ramu Dirisala and V. P. Kodali, Role of biosurfactants

in bioremediation of oil pollution-a review, *Petroleum*, 2018, **4**, 241–249.

87. M. Archacka, E. Celińska and W. Białas, Techno-economic analysis for probiotics preparation production using optimized corn flour medium and spray-drying protective blends, *Food Bioprod. Process.*, 2020, **123**, 354–366.
88. S. Glisic, J. Pajnik and A. Orlović, Process and techno-economic analysis of green diesel production from waste vegetable oil and the comparison with ester type biodiesel production, *Appl. Energy*, 2016, **170**, 176–185.
89. J. Koiwanit and V. Filimonau, Stakeholder collaboration for solid waste management in a small tourism island, *PLoS One*, 2023, **18**, e0288839.
90. H. B. Sharma, K. R. Vanapalli, B. Samal, V. R. S. Cheela, B. K. Dubey and J. Bhattacharya, Circular economy approach in solid waste management system to achieve UN-SDGs: Solutions for post-COVID recovery, *Sci. Total Environ.*, 2021, **800**, 149605.
91. D. Alfonso-Solar, C. Vargas-Salgado, C. Sánchez-Díaz and E. Hurtado-Pérez, Small-scale hybrid photovoltaic-biomass systems feasibility analysis for higher education buildings, *Sustainability*, 2020, **12**, 9300.
92. A. A. V. Julio, T. S. Milessi, E. A. Ocampo Batlle, E. E. Silva Lora, D. M. Yepes Maya and J. C. Escobar Palacio, techno-economic and environmental potential of renewable diesel as complementation for diesel and biodiesel in Brazil: A comprehensive review and perspectives, *J. Cleaner Prod.*, 2022, **371**, 133431.
93. M. A. Hossain, N. Ferdous and E. Ferdous, Crisis-driven disruptions in global waste management: Impacts, challenges and policy responses amid COVID-19, Russia-Ukraine war, climate change, and colossal food waste, *Environ. Challenges*, 2024, **14**, 100807.
94. A. A. Zorpas, J. Navarro-Pedreño, M. Jeguirim, G. Dimitriou, M. B. Almendro Candel, C. Argirusis, I. Vardopoulos, P. Loizia, G. Chatziparaskeva and I. Papamichael, Crisis in leadership *vs* waste management, *Euro-Mediterr. J. Environ. Integr.*, 2021, **6**, 80.
95. L. Jin, K. Duan and X. Tang, What Is the Relationship between Technological Innovation and Energy Consumption? Empirical Analysis Based on Provincial Panel Data from China, *Sustainability*, 2018, **10**, 145.
96. T. Ahmad and D. Zhang, A critical review of comparative global historical energy consumption and future demand: The story told so far, *Energy Rep.*, 2020, **6**, 1973–1991.
97. USEPA, Overview of greenhouse gases, 2022, https://www.epa.gov/ghgemissions/overview-greenhouse-gases (accessed February 1, 2024).
98. Y. Meng, Y. Yang, H. Chung, P. H. Lee and C. Shao, Enhancing sustainability and energy efficiency in smart factories: A review, *Sustainability*, 2018, **10**, 4779.
99. I. Vigna, R. Pernetti, W. Pasut and R. Lollini, New domain for promoting energy efficiency: energy flexible building cluster, *Sustainable Cities Soc.*, 2018, **38**, 526–533.

100. N. Benti, A. Aneseyee, C. Amente, T. Woldegiyorgis, G. Gurmesa, M. Bibiso, A. Asfaw, A. Woldesenbet, Y. Mekonnen, N. E. Benti, A. Aneseyee and C. Geffe, Biodiesel production in Ethiopia: Current status and future prospects, *Sci. Afr.*, 2023, **19**, e01531.
101. S. R. Chia, S. Nomanbhay, M. Y. Ong, A. H. B. Shamsuddin, K. W. Chew and P. L. Show, renewable diesel as fossil fuel substitution in Malaysia: A review, *Fuel*, 2022, **314**, 123137.
102. M. Gerveni, T. Hubbs and S. Irwin, Biodiesel and renewable diesel: What's the difference?, *Farmdoc Daily*, 2023, **13**, 22, https://farmdocdaily.illinois.edu/2023/02/biodiesel-and-renewable-diesel-whats-the-difference.html (accessed January 22, 2024).
103. T. S. B. Abd Manan, T. Khan, W. H. M. Wan Mohtar, A. Machmudah, D. Dutykh, S. Qazi, A. Ahmad and N. Wan Rasdi, Bioremediation of wastewater using algae for potential renewable bioenergy cogeneration, in *Algal Biotechnology*, ed. A. Ahmad, F. Banat and H. Taher, 2022, ch. 3, pp. 47–62.
104. M. Bošnjaković and N. Sinaga, The Perspective of Large-Scale Production of Algae Biodiesel, *Appl. Sci.*, 2020, **10**, 8181.
105. A. Das and S. L. Rokhum, Renewable diesel and biodiesel: a comparative analysis, in *Renewable Diesel*, ed. D. Kumar, B. Singh and S. K. Gupta, 2024, ch. 6, pp. 123–166.
106. N. Rashid, M. S. Ur Rehman, M. Sadiq, T. Mahmood and J.-I. Han, Current status, issues and developments in microalgae derived biodiesel production, *Renewable Sustainable Energy Rev.*, 2014, **40**, 760–778.
107. J. N. Paim, A. G. Santos, R. G. O. Araujo, M. M. Nascimento, J. B. De Andrade and L. L. N. Guarieiro, Emissions of PAHs, Nitro-PAHs and Quinones (Oxy-PAHs) Associated to PM1.0 and PM2.5 Emitted by a diesel engine fueled with diesel-biodiesel-ethanol blends, *Atmosphere*, 2023, **14**, 656.
108. J. Sheehan, V. Camobreco, J. Duffield, M. Graboski and H. Shapouri, An overview of biodiesel and petroleum diesel life cycles, 1998.
109. Statista, Global biofuel production by country 2022 | Statista, (2022). https://www.statista.com/statistics/274168/biofuel-production-in-leading-countries-in-oil-equivalent/ (accessed February 24, 2024).
110. S. Douvartzides, N. Charisiou, K. Papageridis and M. Goula, Green diesel: Biomass feedstocks, production technologies, catalytic research, fuel properties and performance in compression ignition internal combustion engines, *Energies*, 2019, **12**, 809.
111. M. A. Fahim, T. A. Alsahhaf and A. Elkilani, Hydroconversion, in *Fundamentals of Petroleum Refining*, ed. M. A. Fahim, T. A. Alsahhaf and A. Elkilani, Elsevier, Amsterdam, 2010, ch. 7, pp. 153–198.
112. I. Gandarias, P. L. Arias, I. Gandarias and P. L. Arias, Hydrotreating catalytic processes for oxygen removal in the upgrading of bio-oils and bio-chemicals, in *Liquid, Gaseous and Solid Biofuels – Conversion Techniques*, Intech Open, 2013, DOI: 10.5772/52581.
113. D. C. Elliott, Production of biofuels *via* bio-oil upgrading and refining, in *Handbook of Biofuels Production*, ed. R. Luque, C. S. K. Lin, K. Wilson and J. Clark, Woodhead Publishing, 2nd edn, 2016, ch. 19, pp. 595–613.

114. O. Awogbemi, D. V. V. Kallon, E. I. Onuh and V. S. Aigbodion, An overview of the classification, production and utilization of biofuels for internal combustion engine applications, *Energies*, 2021, **14**, 5687.
115. N. Asikin-Mijan, J. C. Juan, Y. H. Taufiq-Yap, H. C. Ong, Y.-C. Lin, G. AbdulKareem-Alsultan and H. V. Lee, Towards sustainable green diesel fuel production: Advancements and opportunities in acid-base catalyzed H2-free deoxygenation process, *Catal. Commun.*, 2023, **182**, 106741.
116. S. Khan, P. Das, M. Abdul Quadir, M. I. Thaher, C. Mahata, S. Sayadi and H. Al-Jabri, Microalgal feedstock for biofuel production: Recent advances, challenges, and future perspective, *Fermentation*, 2023, **9**, 281.
117. K. A. V. Miyuranga, U. S. P. R. Arachchige, T. M. M. Marso and G. Samarakoon, Biodiesel Production through the transesterification of waste cooking oil over typical heterogeneous base or acid catalysts, *Catalysts*, 2023, **13**, 546.
118. Y. Liu, B. Zhong and A. Lawal, Recovery and utilization of crude glycerol, a biodiesel byproduct, *RSC Adv.*, 2022, **12**, 27997–28008.
119. V. Gupta and K. Pal Singh, The impact of heterogeneous catalyst on biodiesel production; a review, *Mater. Today: Proc.*, 2023, **78**, 364–371.
120. Y. Muharam, U. Dianursanti and A. S. Wirya, Modelling of Jatropha oil hydrocracking in a trickle-bed reactor to produce green fuel, *Int. J. Chem. Eng.*, 2021, e7079459.
121. J. Tan, Y. N. Ji, W. S. Deng and Y. F. Su, Process intensification in gas/liquid/solid reaction in trickle bed reactors: A review, *Pet. Sci.*, 2021, **18**, 1203–1218.
122. M. Herskowitz, M. V. Landau, Y. Reizner and D. Berger, A commercially-viable, one-step process for production of green diesel from soybean oil on Pt/SAPO-11, *Fuel*, 2013, **111**, 157–164.
123. A. T. Hoang, H. C. Ong, I. M. R. Fattah, C. T. Chong, C. K. Cheng, R. Sakthivel and Y. S. Ok, Progress on the lignocellulosic biomass pyrolysis for biofuel production toward environmental sustainability, *Fuel Process. Technol.*, 2021, **223**, 106997.
124. D. Shen, R. Xiao, S. Gu, H. Zhang, D. Shen, R. Xiao, S. Gu and H. Zhang, The overview of thermal decomposition of cellulose in lignocellulosic biomass, in *Cellulose – Biomass Conversion*, Intech Open, 2013, DOI: 10.5772/51883.
125. H. Panchasara and N. Ashwath, Effects of pyrolysis bio-oils on fuel atomisation—A Review, *Energies*, 2021, **14**, 794.
126. J. Jenčík, V. Hönig, M. Obergruber, J. Hájek, A. Vráblík, R. Černý, D. Schlehöfer and T. Herink, Advanced biofuels based on Fischer–Tropsch synthesis for applications in diesel engines, *Materials*, 2021, **14**, 3077.
127. S. Abdur Razzak, K. Bahar, K. M. O. Islam, A. K. Haniffa, M. O. Faruque, S. M. Z. Hossain and M. M. Hossain, Microalgae cultivation in photobioreactors: Sustainable solutions for a greener future, *Green Chem. Eng.*, 2023, **4**, 418–439.
128. D. C. Elliott, P. Biller, A. B. Ross, A. J. Schmidt and S. B. Jones, Hydrothermal liquefaction of biomass: Developments from batch to continuous process, *Bioresour. Technol.*, 2015, **178**, 147–156.

129. D. W. Pethick, J. F. Hocquette, N. D. Scollan and F. R. Dunshea, Review: Improving the nutritional, sensory and market value of meat products from sheep and cattle, *Animal*, 2021, **15**, 100356.
130. I. Kabenge, A. Katimbo, N. Kiggundu and N. Banadda, Bioremediation technology potential for management of soil and water pollution from anticipated rapid industrialization and planned oil and gas sector in Uganda: A review, *J. Environ. Prot.*, 2017, **08**, 1393–1423.
131. S. M. Abdelbasir, A. I. Shehab and M. A. A. Khalek, Spent bleaching earth; recycling and utilization techniques: A review, *Resour., Conserv. Recycl. Adv.*, 2023, **17**, 200124.
132. S. A. Qadir, H. Al-Motairi, F. Tahir and L. Al-Fagih, Incentives and strategies for financing the renewable energy transition: A review, *Energy Rep.*, 2021, 7, 3590–3606.
133. M. Patel and A. Kumar, Production of renewable diesel through the hydroprocessing of lignocellulosic biomass-derived bio-oil: A review, *Renewable Sustainable Energy Rev.*, 2016, **58**, 1293–1307.
134. S. Mathur, H. Waswani, D. Singh and R. Ranjan, Alternative fuels for agriculture sustainability: carbon footprint and economic feasibility, *Agric. Eng.*, 2022, **4**, 993–1015.
135. M. Gerveni, T. Hubbs and S. Irwin, Overview of the Production Capacity of U.S. renewable diesel plants for 2023 and beyond, *Farmdoc Daily*, 2023, **13**, 57, https://farmdocdaily.illinois.edu/2023/03/overview-of-the-production-capacity-of-u-s-renewable-diesel-plants-for-2023-and-beyond.html (accessed February 24, 2024).
136. M. Patel, A. O. Oyedun, A. Kumar and R. Gupta, What is the production cost of renewable diesel from woody biomass and agricultural residue based on experimentation? A comparative assessment, *Fuel Process. Technol.*, 2019, **191**, 79–92.
137. A. H. Zacher, M. V. Olarte, D. M. Santosa, D. C. Elliott and S. B. Jones, A review and perspective of recent bio-oil hydrotreating research, *Green Chem.*, 2014, **16**, 491–515.
138. M. S. Graboski and R. L. McCormick, Combustion of fat and vegetable oil derived fuels in diesel engines, *Prog. Energy Combust. Sci.*, 1998, **24**, 125–164.
139. G. Santori, G. Di Nicola, M. Moglie and F. Polonara, A review analyzing the industrial biodiesel production practice starting from vegetable oil refining, *Appl. Energy*, 2012, **92**, 109–132.

CHAPTER 11

High-value Co-products from a Renewable Diesel-based Biorefinery Platform

DAYANA NASCIMENTO DARI,[a] LIDYA FERNANDES DA SILVA,[a] ANTÔNIO MAIRTON BEZERRA LIMA JÚNIOR,[a] VIVIANE DE CASTRO BIZERRA,[a] KAIANY MOREIRA DOS SANTOS,[a] PAULO GONÇALVES DE SOUSA JUNIOR,[b] PATRICK DA SILVA SOUSA,[b] ÉRICO CARLOS DE CASTRO[b] AND JOSÉ C. S. DOS SANTOS*[a]

[a] Instituto de Engenharias e Desenvolvimento Sustentável, Universidade da Integração Internacional da Lusofonia Afro-Brasileira, Campus das Auroras, Redenção, CEP 62790970, CE, Brasil; [b] Departamento de Engenharia Química, Universidade Federal do Ceará, Campus do Pici, Bloco 709, Fortaleza, CEP 60455760, CE, Brasil

*Email: jcs@unilab.edu.br

11.1 Introduction

Economic sectors such as industry and transportation depend closely on fossil fuels, including oil, natural gas, and coal.[1–3] This dependence can be attributed to the high energy yields and calorific value and the low production costs associated with these fuels and their derived inputs.[4–6] However, the excessive use of fossil-derived fuels and inputs has caused severe negative impacts on the planet, since burning fossil hydrocarbons results in the emission of greenhouse gases (GHG), contributing to atmospheric

Green Chemistry Series No. 89
Biorefining of Renewable Diesel: From Production to Market
Edited by Mohammad Aslam, Sanjeev Mishra and Jorge Aburto

Published by the Royal Society of Chemistry, www.rsc.org

air pollution and global warming.[7–9] Thus, fossil energy sources are unable to meet the growing global demand for energy, in addition to being associated with significant pollutant emissions, which makes the energy transition increasingly essential to mitigate this scenario.[10] The global climate emergency has created international agreements such as the 2015 Paris Agreement.[11,12] This pact seeks to drive the transition to renewable and sustainable energy sources, promoting their inclusion in energy matrices to limit global warming to 1.5 °C above pre-industrial levels.[13]

The fuel industry is essential to global economic development and is responsible for processing and marketing fossil-fuel derivatives and renewable biofuels.[4,14] In this regard, global energy consumption in 2015 was recorded at 544 284 GJ, of which 32.94% came from oil, 29.2% from coal, 23.85% from natural gas, 4.44% from nuclear energy, and 9.57% from renewable energy resources such as biofuels, solar, wind, and others.[15] Petroleum-derived fuels are mainly used in the light- and heavy-transport sectors.[16] In 2019 alone, there were approximately 1.4 billion vehicles in the world, and it is predicted that by 2035, the global vehicle fleet will reach the mark of 2 billion cars on the roads.[15] Therefore, biofuels are expected to be widely adopted, as they will cause less environmental damage throughout their life cycle.[17] Biofuels are environmentally friendly fuels capable of reducing CO_2 emissions by up to 65% compared to using fuels made from petroleum.[18] When burned, biofuels release fewer particles and produce less CO_2, sulfur compounds, and toxins than fossil fuels.[19] The term "biofuels" refers to fuels of animal or plant origin with low GHG emissions, with the main advantages being renewability and sustainability.[20,21] This is due to the possibility of partially or fully incorporating biofuels into petroleum diesel. Thus, the use of biofuels such as biodiesel, ethanol, and renewable diesel (or green diesel) has expanded throughout the world.[20]

From the source of raw materials to their conversion technologies to the final production of biofuel and its related byproducts, they are of interest for promoting a circular bioeconomy. Countries abundant in bio-based resources, including residual organic materials, have significant potential for a sustainable circular bioeconomy.[22–24] For example, Brazil is committed to investing in bioenergy and receives considerable support through programs to produce biofuels, with implementation in the energy and transportation sectors. In addition, "green diesel", which is composed of mainly hydrocarbon-based compounds and has properties similar to petroleum derivatives, has attracted the attention of the Brazilian transportation sector, especially Petrobras, the country's national oil company.[25]

Green diesel is synthesized by hydrotreating the triglycerides present in biomasses such as vegetable oils and animal fats by the application of hydrogen gas to the process and with a catalyst to convert the molecules into paraffinic hydrocarbons.[18,26,27] This biofuel can be mixed with petroleum diesel without any restrictions regarding fuel quality or even technical modification of the diesel engine.[28] In this line, one of the essential technologies used to produce green diesel is the integration of biorefineries into production processes.[19,21,29,30] This type of industry converts biomass into various products, aiming to optimize the use of resources and minimize the waste of natural resources, and therefore maximizing the benefits and profit derived from the generation of value-added co-products.[29,31] Biorefining is the process in which different energy outputs, such as fuel, heat, and electricity, are facilitated through various biomass conversion methods.[32]

The co-products generated by the green-diesel industry can be applied in numerous industrial areas, cover a diversification of production, reduce the generation of solid waste and effluents, reduce the exploitation of natural resources and promote a circular economy.[33] In addition, it is expected to reduce the waste of raw materials for other products and, in short, reduce new extractions of resources taken from nature.[34] Given this, this chapter explores the co-products generated through the biorefining of renewable diesel, addressing the biomass-conversion technologies in vogue, their industrial applications and their economic relevance in the current energy scenario.

11.2 Concept of Biorefineries Based on Renewable Diesel

11.2.1 Definition and Structure of a Biorefinery

Biorefineries are industries capable of converting renewable biomass into a wide range of products, such as fuels, chemicals, and energy. Green diesel's quality and renewable nature make it very prominent within the fuel spectrum. Biorefineries are thematically treated as hydrocarbon refineries, but they are entirely different because hydrocarbons are not the only raw materials: while conventional oil refineries work only with fossil hydrocarbons, biorefineries can work with raw materials such as vegetable oils, animal fats, and organic waste. This approach offers an alternative that is not only more sustainable but also much more economically viable.[4,35,36]

While conventional refineries have primarily used technologies such as catalytic cracking and steam reforming on crude oil – practical but energy-intensive and environmentally damaging approaches – biorefineries implement integrated processes that maximize the use of renewable resources and reduce emissions. Technological integration within biorefineries enables coordinated processing of various feedstocks and, as a result, increases production efficiency and end-product diversification.[35,37,38]

One of the leading technologies adopted in biorefineries is hydrogen catalysis hydrotreatment, essential for producing renewable diesel. In the reaction of oils or greases with hydrogen, the oxygen atoms are eliminated from the molecules under high temperatures and pressures, and a fuel chemically very similar to conventional diesel is finally obtained. However, renewable diesel is less polluting to the environment. Other biomass technologies are based on biomass gasification to synthesize gas through the

Fischer–Tropsch process or catalytic hydrothermolysis, allowing better use of byproducts and biomass.[4,20,35–40]

Another advantage of biorefineries is the valorization of byproducts resulting from the processes. Various industrial sectors widely adopt products such as glycerin (used in cosmetics and medicines) and biofertilizers (applied in agriculture). While the waste generated by oil refineries is often considered an environmental burden, integrating waste into a production cycle is what biorefineries try to accomplish. For example, solid waste and gases can be transformed into energy or valuable products, closing the loop and reducing operational costs.[37,39]

Furthermore, biorefineries can be very flexible in the feedstock used. Different biomass origins and technological routes can be mixed within the same unit without significant difficulties, representing an easy way to adapt to market and energy-demand changes. They can even become more resilient than conventional refineries, which are highly dependent on crude oil, a non-renewable resource with high price volatility and negative environmental impact.[20,35] In this way, biorefineries are essentially a sustainable and innovative alternative to oil refineries. By developing high-quality renewable diesel and utilizing the created byproducts, in other words, by providing added value to waste, they form an essential part of the shift towards a cleaner energy matrix.[4,20]

11.2.2 Feedstocks Used

Vegetable oils are widely used as feedstocks in biorefineries to produce renewable diesel. This is mainly due to their chemical composition, being rich in triglycerides that can be converted into renewable diesel.[41] Among the most common oils is soybean oil, which stands out for its wide availability and low cost. Palm oil, despite being highly efficient in terms of yield, faces challenges related to environmental issues such as deforestation and biodiversity loss. Alternatives such as sesame seed oil, coconut oil, and sunflower seed oil have been widely used because they are more sustainable and have composition profiles suitable for various industrial applications.[42,43]

Reusing waste oils and fats is a strategic solution that combines conservation and sustainability. Used cooking oil, for example, avoids the problem of improper disposal and offers a cost-effective alternative for producing renewable fuels and high-value chemicals.[44] Similarly, animal fats (a byproduct of the food industry) can be used to reduce costs and increase the sustainability of biorefineries.[45] In addition, oil-rich industrial waste, such as that generated in food processing, has the potential to be reused to minimize waste and promote a circular economy.

Lipids extracted from microalgae represent an unconventional and fruitful alternative for biorefinery platforms. This is because microalgae can be intensively cultivated on land that is not considered agricultural land while still being as efficient as needed in converting sunlight and carbon dioxide into high-lipid biomass.[21] The lipids produced are rich in triglycerides, crucial for producing renewable fuels and chemical co-products. Microalgae cultivation also has other environmental benefits, including mitigation of greenhouse gases and the promise of use in integrated wastewater-treatment systems.[46]

Although less conventional for renewable-diesel production, residual lignocellulosic biomass is an excellent resource for diversifying co-products in biorefineries. Agro-industrial residues, such as corn stover and sugarcane bagasse, are rich in compounds that are convertible into high-value chemical precursors, including lignin and cellulose.[47] Such materials increase the economic viability of biorefineries due to their low cost and the recycling of previously wasted resources.[29] Integration with residual biomass makes agricultural practices more sustainable by drastically reducing the burning of residues in the fields.

Unconventional feedstock sources have recently gained attention for their availability in specific market niches. Marine byproducts such as fish oil are found to be used in coastal regions to valorise waste from the fishing industry.[48] Equally viable are oils from alternative crops such as castor, camelina, and jatropha, which meet the need for new crops in areas where traditional oilseed climates cannot be established.[49,50] These non-food raw materials stand out for their ability to produce high-value chemicals that are highly resistant to adverse conditions, thus instilling much-needed sustainability in biorefineries

11.3 High-value Co-products: Types and Production

11.3.1 Description of the Main Co-products

It is possible to generate several high-value-added co-products from biorefineries based on renewable diesel or even environmentally friendly biorefineries, which have sustainable objectives and processes that are economically viable.[51,52] In a recent study, a microalgae biorefinery was used, which obtained the biomass of *Chlorella*, a freshwater green microalga, which was dried and pre-treated with a deep eutectic solvent (DES), which is an ecological solvent and minimizes the degradation of lipids present in the reaction, in addition to being assisted by ultrasound. This resulted in the effective extraction of lipids from the microalgae, considered high-quality lipids for the production of biodiesel with high cetane and calorific values. In addition to being considered an efficient and sustainable method for the extraction of lipids, this resulted in production of hydroponic biofertilizers.[53,54]

Several microalgae also produce fatty acids, especially polyunsaturated ones, such as docosahexaenoic acid (DHA) and eicosapentaenoic acid (EPA), which are omega fatty acids that are very beneficial to health. A recent study demonstrated the use of plant hormones as a treatment to increase the lipid content of microalgae, such as increasing the levels of omega fatty acids in *Chlorella* sp. with the plant hormone indole acetic acid (IAA).[55] Another study also identified that the transesterification reactions of *Aurantiochytrium* sp. KRS 101, during the process of using omega fatty acids, allowed the effective production of ethyl esters, which can be used in medicine, such as in the treatment of hypertriglyceridemia.[56] In addition, lipids extracted from residual oils or microalgae cultivated in biorefinery platforms can be used to produce nutritional supplements. Lipids are also integrated as high-value co-products in biodiesel production from transesterification reactions and extracted from lipid residues to make soap and lubricants.[57–59]

The biorefinery approach is growing for bioethanol and biodiesel production, especially using microalgae as a raw material, because alternative fuels from microalgae have become promising in scientific studies. In addition to lipids, microalgae also have cellulose and starch, which can be used to produce bioethanol and biodiesel. A study demonstrated the use of the microalga *Scenedesmus* sp., using the acid hydrolysis and fermentation process by *Saccharomyces cerevisiae*, for the production of glycerol dicarbonate, bioethanol, and fatty acid methyl esters. A yield of 93% was obtained for total sugar yield with the microalgae biomass pre-hydrated by acid hydrolysis, also producing 86% ethanol.[60–64]

Some high-value bioproducts were also found in the *Nicotiana glauca* plant, being a promising biomass for the search for high-value co-products, in which the leaves and stem of the plant were characterized and the metabolite profiles were evaluated by gas chromatography–mass spectrometry and liquid chromatography–mass spectrometry analyses. The dried samples of *N. glauca* were subjected to the application of some solvents to produce extracts enriched with metabolite content, which generated vegetable fatty acids, alkanes, fatty acid alcohols, and other components, as well as co-products such as fermentable oils and sugars, in addition to high levels of hydrocarbons. It was concluded that *N. glauca* is a renewable raw material and the metabolite profile can contribute to biorefining for the production of a wide range of high value-added co-products by optimizing the production processes.[65,66]

Scientific work using green biorefineries has also been explored. In a recent study, shrimp-shell raw material was used to extract α-chitin and some high-value co-products through the symbiotic microbial fermentation method of lactic acid, using *Lactobacillus plantarum* and *Streptococcus thermophilus*, in which mature coconut water was used as a low-cost nutrient source. The study demonstrated that the method was effective for producing lactic acid and protease, in addition to the raw material's excellent deproteinization and demineralization rates. The extracted chitin is similar to commercially sold α-chitin, with very similar structures. However, the chitin obtained has better quality based on the analyses of the work.[67–69]

Biorefinery platforms, especially those based on renewable diesel, make it possible to produce several high-value co-products in addition to the primary fuel, which allows for economic viability, sustainability, and efficiency to be added to the processes.[70–72] The co-products generated are derived from raw materials rich in lipids, including fatty residues, vegetable oils, or animal fats, with possible applications in various industrial sectors.[73–75]

Most high-value co-products can be obtained by transesterification methods, such as glycerin, which is widely used in food and cosmetics. Methyl and ethyl esters, omega-3 fatty acids, residual proteins, and lipid residues are some of the high-value co-products that can be obtained.[37,76,77] One of the most significant advantages of producing co-products from biorefineries is economic diversity, which allows for more substantial variation in profits from different products, and sustainability, which in addition to improving the efficiency of using raw materials also reduces waste during production.[78–80]

11.3.2 Production Mechanisms and Reactions Involved

The triglyceride hydrolysis process, widely applied in producing free fatty acids and glycerol, is crucial in cosmetics, food, and biodiesel.[81] This reaction involves the breaking of the triglyceride ester bonds through contact with water molecules under the influence of an acidic or basic catalyst.[82]

$$\text{triglyceride} + 3H_2O \xrightarrow{\text{catalyst}} 3 \times \text{fatty acids (RCOOH)} + \text{glycerol } (C_3H_8O_3)$$

The reaction process takes place in four fundamental phases. Initially, the catalyst activates the water molecule or the carbonyl group of the triglyceride. In an acidic environment, protonation of the oxygen of the carbonyl group makes the carbon more vulnerable to nucleophilic attack by water.[83] In a basic environment, the hydroxide ion (OH^-) acts directly on the carbon of the carbonyl group. In the second phase, an unstable tetrahedral intermediate appears, resulting from the attack of water on the carbon of the ester group. In phase three, the intermediate disintegrates, releasing a fatty acid molecule (RCOOH) and giving rise to a diglyceride.[84] Ultimately, the procedure is repeated until all three ester bonds are broken, giving rise to three fatty acids and one glycerol molecule.[83]

The reaction can be schematized as follows:

Initial structure (triglyceride): $CH_2(OOCR_1)$–$CH(OOCR_2)$–$CH_2(OOCR_3)$
Interaction with water (catalyzed): $CH_2(OOCR_1)$–... + $H_2O \rightarrow R_1COOH$ + diglyceride
Product: free fatty acids (RCOOH), the products from cleavage of each ester bond, and glycerol ($C_3H_8O_3$), the product remaining after removal of the three fatty acids.

This mechanism highlights the stepwise cleavage of ester bonds within the triglyceride structure, emphasizing the sequential formation of diglycerides, monoglycerides, and finally, free glycerol and fatty acids.[85]

The hydrocracking process for biomass is a sophisticated method that transforms renewable resources, such as lignocellulose, vegetable oils, and organic waste, into light hydrocarbons, such as propane, butane, and gasoline fractions, as well as valuable intermediates, such as diesel and kerosene.[86] This procedure is notable for its ability to generate renewable fuels with properties similar to fossil fuels but with a reduced environmental impact. Hydrocracking occurs under controlled high pressure and temperature conditions with hydrogen and suitable catalysts.[87] Complex biomass molecules undergo thermal and catalytic cracking throughout the process, breaking chemical bonds and forming smaller molecules.[88]

The intermediate products produced are subjected to hydrogenation, which saturates them with hydrogen, increasing the stability and energy content of the final products.[89] Additionally, the deoxygenation process occurs, which removes oxygen from the biomass in the form of water, carbon dioxide, or carbon monoxide, converting the biomass compounds into hydrocarbons.[90] The products obtained are categorized into light hydrocarbons, such as gas fractions (methane, ethane, propane, and butane), and intermediate hydrocarbons, such as kerosene, diesel, and intermediate oils used in refining processes and the formulation of chemical products.[91]

This method provides several benefits, such as transforming waste and renewable sources into high-value-added fuels, reducing net greenhouse-gas emissions due to the carbon neutrality of biomass, and manufacturing versatile fuels that can directly replace fossil fuels.[92] However, the hydrocracking process faces technical and financial obstacles, such as the high price of catalysts, the complexity of biomass as a feedstock, which complicates standardization, and the high need for energy resulting from the required pressure and temperature conditions [93] Typically, catalytic-hydrotreating reactors are used in biorefineries to produce renewable diesel and value-added co-products (see Figure 11.1).

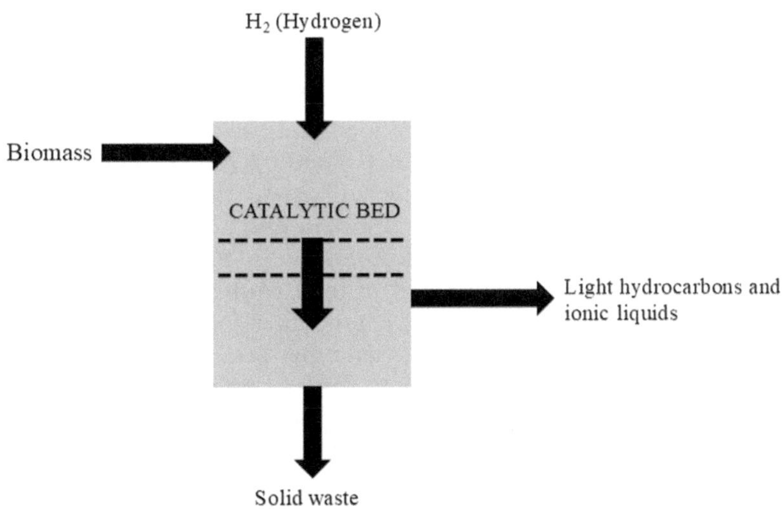

Figure 11.1 Illustration of a catalytic hydrotreatment reactor used in biorefineries to produce renewable diesel and value-added co-products.

The hydrotreatment reactor is made up of fundamental components that ensure the effectiveness of the procedure.[94] Biomass is introduced through a specific inlet, whether in a liquid or gaseous state. At the same time, compressed hydrogen is injected through another inlet to contribute to chemical reactions, such as hydrogenation and deoxygenation.[95] Inside the reactor is a catalytic bed formed of nickel, molybdenum, or cobalt, which helps break down complex biomass molecules and incorporate hyrogen.[96] In the reactor core, chemical changes occur under controlled conditions of high temperature and pressure in the reaction zone, transforming biomass into valuable hydrocarbons.[97,98]

11.4 Applications of Co-products

11.4.1 Industrial and Technological Use

The prospect of producing new materials through more sustainable methods has been gaining traction globally, driving increased demand for environmentally friendly methodologies.[99] In the biofuel sector, renewable diesel generates byproducts and industrial residues that, through chemical or enzymatic routes, can form molecules with high industrial interest.[100] One notable class of molecules in this context is fatty acids, which are byproducts of biorefineries and play a crucial role in manufacturing soaps and cosmetics.[101] Fatty acids are essential raw materials for these industrial sectors, as they are used in various applications such as lubricants and emulsifying agents, thus adding value to the process.[71] A biotechnological approach can be integrated into transforming these fatty acids into high-value industrial molecules.[74] Enzymes can produce glycerol esters, flavours, and biolubricants.[102–105] Some industrial processes have chosen biocatalysis as a production route for these new molecules due to increasing environmental awareness and incentives for ecological methodologies, as biocatalysis offers high selectivity and specificity, high reaction yields, and vigorous catalytic activity even under mild reaction conditions.[106]

Glycerol, another valuable byproduct obtained as an industrial residue from the biodiesel industry, is widely utilized by the pharmaceutical and synthetic industries.[107] It can serve as a solvent and excipient, a precursor for pharmaceuticals, a sweetener and humectant in food production, and a thermal bath medium in industrial synthesis equipment.[108] In the energy sector, glycerol is being explored as a precursor for fuel cells.[109] Its conversion into derivatives such as propylene glycol or acetic acid further expands the potential for obtaining new products, promoting the sustainability of the biorefinery and maximizing resource utilization.[110] Hydrocarbons from refining can be separated and purified to produce intermediate fuels, such as naphtha used in the petrochemical industry, and certain industrial solvents like hexane and toluene.[111] These products are in high demand across various sectors and complement the biorefinery value chain, providing greater flexibility and profitability to the system.[72]

Global efforts to fund and promote biorefinery byproducts have focused on strategies that foster sustainability, technological innovation, and circular-economy principles.[112] Many countries offer financial incentives, such as subsidies and tax credits, to companies investing in clean technologies, including biorefineries.[113] Examples include the Renewable Fuel Standard (RFS) in the United States and the National Biofuels Program (RenovaBio) in Brazil.[114,115] Environmental regulatory policies also play a key role in encouraging the development of technologies for utilizing these byproducts, such as setting targets for emission reductions and waste minimization.[116] Funding for research and development is another strategy to foster public–private partnerships and international programs.[117] Collaborations between governments, universities, and companies have been financing studies on innovative applications of byproducts derived from biorefineries.[118] This partnership increases attention to these byproducts, enabling sustainability certifications for products derived from biorefineries and encouraging the market to adopt certified ecological or sustainable products.[119,120]

11.4.2 Economic Value of Co-products

Refined glycerol and fatty acids stand out for their lower costs in biorefineries due to their generation as byproducts of already established processes such as biodiesel production.[121] These products are increasingly applied in industries such as cosmetics, pharmaceuticals, and food, reinforcing the economic feasibility of biorefineries. On the other hand, renewable diesel and light hydrocarbons incur higher costs, primarily due to substantial investments in technologies and processes like hydrogenation and purification. Nevertheless, regulatory pressures to reduce greenhouse-gas emissions and adopt renewable sources create a promising market for these products.[122]

Competitive advantages of biorefineries include incentives for sustainable policies to reduce environmental impacts, decrease reliance on petroleum, enhance market appeal for eco-friendly products, and integrate auxiliary products as revenue sources, contributing to growth of a circular economy. However, challenges persist, such as economic and technological factors, the need for more significant tax incentives to compete with the low costs of petroleum-derived products, uncertain regulatory demands for this market, and technological barriers to achieving high efficiency in these bioprocesses.[123]

A comparative analysis was made of the costs of producing byproducts using traditional and well-established methods and biorefineries, highlighting each approach's cost differences and the advantages and disadvantages (see Table 11.1).

Refined glycerol and fatty acids show significant cost reductions when produced in biorefineries, achieving savings of 20–35% and 25–40%, respectively, compared to traditional methods.[124] This cost efficiency is primarily due to their generation as byproducts of biodiesel production, which

Table 11.1 Production costs, advantages, and challenges of byproducts *via* biorefineries compared to traditional methods.

Product	Cost, traditional method	Cost, biorefinery	Cost difference	Advantages	Challenges	Ref.
Diesel fuel	US$0.50–0.70 per L (fossil)	US$0.80–1.20 per L (renewable)	+30% to +50%	Reduced GHG emissions, compatible with engines	High initial costs, dependency on subsidies	124–126
Refined glycerol	US$1.20–1.50 per kg (petroleum-based)	US$0.80–1.00 per kg (renewable byproduct)	−20% to −35%	Higher purity, lower energy costs	Limited scalability in some regions	127–129
Fatty acids	US$0.90–1.10 per kg (chemical extraction)	US$0.60–0.80 per kg (renewable byproduct)	−25% to −40%	Circular economy, lower environmental impact	Market dependence on regulations	91, 130 and 131
Hydrocarbons	US$0.30–0.50 per L (petroleum-based)	US$0.50–0.70 per L (renewable)	+20% to +40%	Sustainability, lower carbon emissions	Costs of catalysts and separation processes	132 and 133
Soaps	US$1.00–1.50 per kg (chemical fatty acids)	US$0.80–1.10 per kg (bio fatty acids)	−15% to −25%	Biodegradable products, better social acceptance	Volume depends on biodiesel demand	134–136
Cosmetics (base)	US$3.50–4.00 per kg	US$2.80–3.20 per kg	−10% to −20%	"Green" market appeal, innovative formulations	Standardization and production stability	137–139
Industrial solvents	US$0.40–0.60 per L (fossil-based)	US$0.50–0.75 per L (renewable)	+25% to +35%	Biodegradable, non-toxic, tax incentives	Low initial acceptance in industrial sectors	140–143

minimizes raw material and energy consumption costs. Additionally, these products have broad applicability in pharmaceutical, food, and cosmetic industries, where purity and sustainability add significant value.[125]

Other products, such as soaps and cosmetics, also benefit from cost reductions in biorefineries, with production costs 10–25% lower than conventional methods. This competitiveness arises from using raw materials derived from integrated processes that foster a circular economy while appealing to environmentally conscious consumers through their "green" profile.[126]

However, products like renewable diesel and light hydrocarbons incur significantly higher costs than their fossil-based counterparts. For instance, renewable diesel is 30–50% more expensive due to the complexity of its production, which includes hydrogenation and high-cost catalysts. Similarly, light hydrocarbons, which require advanced purification, exhibit an additional cost of 20–40%. These products face challenges competing within markets dominated by petroleum derivatives, where prices are traditionally lower. Nonetheless, their advantages in reducing greenhouse-gas emissions and complying with environmental regulations make them attractive in more sustainable markets.

Industrial solvents also reflect this trend, being 25–35% more expensive when produced in biorefineries.[127] Despite the higher costs, they offer benefits such as biodegradability and lower toxicity, which can outweigh their price in applications prioritising environmental safety.[128]

While production costs can pose a barrier to the competitiveness of biorefined products, their sustainability, purity, and environmental appeal advantages remain significant differentiators.[129] Products like glycerol and fatty acids demonstrate the potential of biorefineries to integrate processes and add value to byproducts, reducing waste and costs.[130] Conversely, products like renewable diesel and light hydrocarbons continue to face technological and market challenges, but hold promising prospects with technological advancements and the strengthening of supportive public policies.[131]

Therefore, the table highlights that biorefineries, despite their challenges, offer a sustainable and competitive alternative for producing various products. With ongoing technological developments and the scaling up of production, greater cost convergence between renewable and traditional methods is anticipated, strengthening the market for biorefined products and contributing to a low-carbon economy.

11.5 Economic Analysis and Feasibility

11.5.1 Comparison with Conventional Methods

Assessing co-product viability in renewable-diesel biorefineries is becoming increasingly important as we aim for more sustainable alternatives to traditional petrochemical methods.[113,114,132] In addition to contributing significantly to the reduction of greenhouse-gas emissions, the production of renewable diesel in a biorefinery makes it possible to explore opportunities for generating high-value-added co-products (fuels, biochemicals, and compounds for sustainable materials).[57] However, the obstacle lies in the economic viability of these processes and their competitiveness with conventional methods established in the market, which have lower costs due to the large scale and time in the petrochemical industry.[133–135] The rising cost of fossil fuels has driven the creation of ecologically sustainable solutions and consequently makes the economic analysis of biorefineries attractive.[136–138]

The production of renewable diesel integrated with the generation of high-value co-products offers a more sustainable approach, but initially with higher costs when compared to conventional methods.[139–141] A broader array of products, encompassing renewable diesel alongside fatty acids, alcohols, and even biopolymers, can be generated by biorefineries that utilize renewable resources such as vegetable oils, biomass, and organic waste.[142–144] Diversification through the development of multiple co-products can help reduce financial risks, provide increased resilience against market fluctuations, and offer more sustainable alternatives to conventional petrochemical processes, which heavily rely on crude oil and its price volatility.[145–147]

Manufacturing renewable diesel alongside co-products offers several benefits, including a reduction in biorefinery operating expenses.[148–150] Process efficiency is improved by utilizing the same infrastructure to produce multiple products leading to lower fixed and variable costs associated with renewable diesel production.[151–153] Byproducts like glycerin and fatty acids from diesel production can help reduce costs and enhance competitiveness in petrochemical enterprises.[154–156] Additionally, biofuels can function as an energy source in biorefineries, reducing reliance on external energy sources and strengthening the process's efficiency, economics, and environmental sustainability.[157–159]

Because of this, it is essential to highlight the importance of the regulatory context aligned with incentive policies to promote the energy transition or its economic viability.[160–162] The competitive advantage of biorefineries over conventional petrochemical processes may be significantly increased in many countries by implementing tax incentives and subsidies for producing renewable fuels and sustainable co-products.[163–165] Additionally, the economic viability of these co-products may be fueled by consumer demand and increased awareness of environmentally friendly products, making them not merely a technological solution but also a commercial opportunity with substantial development potential in the years to come.[166–168]

11.5.2 Created Value Chains

The circular economy has proven to be an effective business model for integrating the generation of high-value-added co-products in renewable-diesel biorefineries.[32,169,170] This business model has several potentialities, since it proposes maximizing the value of resources throughout the life cycle, including the reuse of waste from various industrial sectors and the recovery of materials and energy.[112,159,171] Therefore, biorefineries can adopt practices such as the "biorefining cascade", which aims to take advantage of each stage of the process to produce various products with value for the industry and therefore significantly reduce waste. Agricultural waste can be used to make biofuels.[172–174] During this process, byproducts are obtained with the potential to be transformed into organic fertilizers or valuable compounds for the chemical industry.[175–177] In this way, connected value chains are created.

Another advantageous aspect to be highlighted is that by using byproducts as inputs for new production processes, dependence on non-renewable raw materials is reduced, in addition to reducing greenhouse-gas emissions and waste.[178–180] In this way, it becomes sustainable to promote the integration in biorefineries of co-products derived during the synthesis of renewable diesel, as it increases the level of synergy between environmental and economic issues, thus aligning global sustainability objectives.[181–183] This can draw investors keen on sustainable technologies and consumers who prioritize eco-friendly products. As a result, cultivating strong value chains improves the durability of the business model, making biorefineries less susceptible to price variations in the fuel market.[184,185]

Given the positive aspects reported, it is essential to mention that the success of integrating the production of co-products of industrial interest during the production of renewable diesel will depend on incentive policies aligned with market awareness.[186–188] Therefore, it is necessary to implement policies to promote the bioeconomy and the circular economy and subsidies to encourage research and development of alternative technologies or tax incentives for responsible companies investing in cleaner technologies.[189–191]

11.6 Reactor Technologies and Configurations

Bioreactors are considered the core of bioprocesses, as they are essential elements in every biological process.[192] In addition to providing the optimal conditions to satisfy the requirements of the biological system, the bioreactor must also ensure a high yield of the bioprocess.[192,193] Biorefineries require the development and design of suitable bioreactor systems.[193,194] There are different reactor technologies and configurations.[195] Various reactor sets are combined to meet the biorefinery operating loads.[196]

11.6.1 Descriptions of Reactors

The type of reactor used is one of the many variables that can affect green-diesel production.[197–199] Continuous and batch reactors are the two reactors mainly applied in hydrothermal processes. Although continuous reactors require higher catalyst stability, their constant feeding and discharging make them more suitable for industrial applications using hydrothermal decarboxylation of oils (HTDO).[198]

Compared with other types of reactors, the most commonly used reactor for conducting exploratory studies to optimize reaction conditions and generate kinetic data quickly and economically is the batch reactor.[197,200] Although batch reactors have less complicated process control and simple setup, making them better for bench-scale studies, higher energy consumption and longer reaction times are required as each batch operation involves cooling, heating, and cleaning steps.[198] In hydrodeoxygenation, the dominance of coal at longer reaction times may generate potential long-term operational problems in continuous reactors.[201] Semi-batch and continuous reactors ensure high catalytic activity (by avoiding the poisoning of the catalysts by CO adsorption) and shift the balance of the reaction towards the products.[197,202]

In batch reactors, all chemical species have an identical residence time. On the other hand, continuous (flow) reactors, due to the constant introduction of raw material and the removal of products from these reactors, generate the phenomenon known as residence time distribution (RTD). This parameter must be optimized to improve renewable-diesel yield.[203] Reaction time is a crucial factor in the production of green diesel in batch processes. Lighter products are generated with increasing reaction time due to cracking reactions.[199] Higher hydrogen pressure in batch reactions produces more desirable products than in continuous reactions [204] The batch-stirred tank is the most commonly used in green-diesel production.[199]

Continuous fixed-bed reactors can be used to improve biofuel production.[205] Especially in industry, the more straightforward process control and maintenance of fixed-bed reactors make them more favourable than continuously stirred tank reactors.[206] Greater control over the chemistry and composition of the product can be achieved by using a fixed-bed reactor rather than a fluidized bed due to the catalyst flexibility. While one of the main advantages of a fluidized-bed system is the continuous and rapid regeneration of the catalyst, catalyst loss and attrition can occur. It is impossible to eliminate the propensity of catalysts for attrition, even with modified preparation methods.[207]

11.6.2 Process Diagrams

Conventional biomass biorefineries only partially consume their substrate to produce biofuels. Integrated biorefineries for the co-production of biofuels and high-value products have emerged as a way to achieve the complete conversion of biomass through various processing technologies.[32,208,209] García-Velásquez *et al.* (2018) proposed a biorefinery diagram focused on bioenergy production with different levels of integration in energy-driven biorefineries using fresh palm fruit bunches as feedstocks.[210] It can sustainably obtain an extensive range of products using raw materials in integrated processes.[211] According to Mongkhonsiri *et al.*,[212] integrated palm-oil biorefinery-network alternatives can provide high economic potential and less environmental impact than no integration.

Generally speaking, biorefineries process by thermochemical and biochemical routes, and in most cases these two need to be integrated at some point.[213] Integrated biorefineries pave a path toward a circular economy by closing the loop towards processes with effective waste management.[214]

11.7 Sustainability and Environmental Impacts

11.7.1 Waste Reduction

Biorefineries not only produce biofuels but also have the potential to enhance the value of co-products, which contributes to the reduction of waste disposal, in addition to reducing the exploration of new natural resources.[32,215] Producing renewable diesel from biomass can incidentally generate byproducts, such as glycerol, fatty acids, biogas, and fertilizers for crops, which can be reused by different industries.[22,26,216] These byproducts help to add value to waste, integrate new production chains (such as the use of glycerol in the chemical industry), and minimize disposal into the environment.[125,216]

A diverse range of raw materials is used in biorefineries, with advantages for those not competing with food crops, such as agricultural, food, and domestic or industrial waste. Thus, different biomasses have been studied to synthesise biofuels.[217–219] For example, Romero-Izquierdo *et al.* (2022) investigated the conversion of *Jatropha curcas* fruit in a biorefinery to produce biofuels and chemicals. The results indicated that the biorefinery could generate biofuels, including aviation biofuel, glycerol, fertilizer, green diesel, naphtha, and electricity.[217] The intensified sequence to separate aviation biofuel resulted in less than 1% energy savings, and 90% of the electricity consumption was supplied internally, increasing the sustainability of renewable aviation fuel production.[217] In another study, Karan *et al.* (2023) simulated a biorefinery for renewable diesel. They obtained multiple products, such as Cu-chlorophyllins, carotenoids, and proteins, from residual microalgae biomass, being able to replace their commercial fossil or synthetic analogues.[218] On the other hand, Karmee (2018) highlighted in his study that wastes such as spent coffee grounds can be converted into biofuels and a series of value-added products, such as food additives, biosorbents, activated carbons, polyols, polyurethane foams, phenolic antioxidants, biocomposites, and nutraceutical compounds.[219] In addition, biological treatment of organic waste, such as composting, can be a practical approach for valorising spent coffee grounds.[219]

11.7.2 Life-cycle Analysis

Life-cycle assessment (LCA) is a method used to analyze the environmental impact of a product, considering all stages of its life cycle, including the extraction of raw materials, manufacturing, consumption, and final disposal.[220] The LCA of biorefineries is essential for measuring their productive potential for a circular economy.

Biorefineries are strong candidates for reducing GHG emissions and ensuring a steady supply of renewable energy from converting sustainable feedstocks. It is estimated that 8–10% of global GHG emissions come from food waste, while the increasing demand for limited land and water resources threatens current biodiversity.[221] Therefore, it is essential to assess how waste-based biorefineries will benefit the environment and their competitiveness in the market through an LCA of their bioproducts.

In the study by Wong *et al.* (2016), an LCA was conducted for the conversion of lignocellulosic biomass into renewable diesel by hydrogenation processes (HDRD) to estimate the well-to-wheel GHG emissions and the fossil energy consumption in the treatment of biomass and its conversion to HDRD.[118] The GHG emissions of the feedstocks used for the study were 35.4–42.3 g CO_2-eq per MJ.[118] Okeke *et al.* (2020) performed an LCA on the production of drop-in renewable diesel from biogas generated by the anaerobic digestion of miscanthus. The analysis covered the cultivation, harvesting, and transportation of the feedstock to the biorefinery, the conversion of biomass to Fischer–Tropsch diesel, and the combustion of renewable diesel in a passenger vehicle. The results indicated that renewable diesel derived from miscanthus could reduce GHG emissions by up to 73% compared to fossil diesel.[222] Furthermore, researchers Cai *et al.* (2018) explored the concepts of integrated biorefineries (IBR) aimed at co-producing hydrocarbon fuels and value-added bioproducts to improve economic viability, increase the efficiency of biomass-based resource utilization, and thereby maximize potential in greenhouse-gas emission reductions.[72]

11.8 Conclusion

In summary, this chapter presents an overview of how biofuel biorefineries are a promising route to strengthening the bioeconomy and transitioning to a more sustainable economic development model. From this perspective, this chapter addressed the generation of value-added co-products in renewable-diesel biorefineries. It highlighted that, despite advances in the area, there are still challenges to be overcome, especially process optimization, cost reduction, technological barriers, and specific regulations. Various feedstocks, such as vegetable oils, animal fats, residual biomass, and non-food crops, have been reported in the production of renewable diesel, and the co-products generated, in addition to biofuel, have applications in different industries. Hydrogen-powered catalytic hydrotreating reactors are the emerging method for processing biomass and converting renewable-diesel biofuel. It is possible to conclude that co-products have enormous potential to add value to bioprocesses, promoting a circular economy and contributing to an economic forecast for biorefineries. To achieve its full potential, it is essential to intensify research and development efforts, prioritizing the characterization of new materials, the diversification of industrial applications, and improving the efficiency of associated processes. In addition, collaboration between scientific academia, industries, and government authorities emerges as a strategic element, capable of catalyzing technological advances and fostering the creation of regulations and incentives that boost the market for renewable bioproducts.

Acknowledgements

This study is supported in part by financial support from the following Brazilian agencies for scientific and technological development: Fundação Cearense de Apoio ao Desenvolvimento Científico e Tecnológico (FUNCAP) (PNE-0112-00048.01.00/16, PS1-0186-00216.01.00/21, PS1-00186-00255.01.00/21), Conselho Nacional de Desenvolvimento Científico e Tecnológico (CNPq) (311062/2019-9, 308452/2022-4), and Coordenação de Aperfeiçoamento de Ensino Superior (CAPES) (Finance Code 001, PROEX 23038.000509/2020-82).

References

1. T. Li, X. G. Yue, M. Qin and D. Norena-Chavez, Towards Paris Climate Agreement Goals: The Essential Role of Green Finance and Green Technology, *Energy Econ.*, 2024, **129**, 107273.
2. Z. Ruihui, Y. Xinmei and H. Yu, Cleaner Production and Total Factor Productivity of Polluting Enterprises, *J Cleaner Prod.*, 2023, **423**, 138827.
3. R. K. Srivastava, N. P. Shetti, K. R. Reddy and T. M. Aminabhavi, Biofuels, Biodiesel and Biohydrogen Production Using Bioprocesses. A Review, *Environ. Chem. Lett.*, 2020, 1049–1072.
4. R. C. de Oliveira and A. P. Junger, Use of Fossil Fuels in Brazil and Their Environmental Consequences, *Res., Soc. Dev.*, 2020, **9**(9), e466997537.
5. R. Murphy, What Is Undermining Climate Change Mitigation? How Fossil-Fuelled Practices Challenge Low-Carbon Transitions, *Energy Res. Soc. Sci.*, 2024, **108**, 103390.
6. A. Plantinga and B. Scholtens, The Finance Perspective on Fossil Fuel Divestment, *Curr. Opin. Environ. Sustainability*, 2024, **66**, 101394.
7. Y. Xie, K. S. Khoo, K. W. Chew, V. V. Devadas, S. J. Phang, H. R. Lim, S. Rajendran and P. L. Show, Advancement of Renewable Energy Technologies *via* Artificial and Microalgae Photosynthesis, *Bioresour. Technol.*, 2022, **363**, 127830.
8. S. Sahota, S. Kumar and L. Lombardi, Biohythane, Biogas, and Biohydrogen Production from Food Waste: Recent Advancements, Technical Bottlenecks, and Prospects, *Energies*, 2024, **17**(3), 666.
9. V. L. Trinh and C. K. Chung, Renewable Energy for SDG-7 and Sustainable Electrical Production, Integration, Industrial Application, and Globalization: Review, *Clean Eng. Technol.*, 2023, **15**, 100657.
10. A. M. Saleh, V. István, M. A. Khan, M. Waseem and A. N. Ali Ahmed, Power System Stability in the Era of Energy Transition: Importance,

Opportunities, Challenges, and Future Directions, *Energy Convers. Manage.: X*, 2024, **24**, 100820.
11. B. P. Sandaka and J. Kumar, Alternative Vehicular Fuels for Environmental Decarbonization: A Critical Review of Challenges in Using Electricity, Hydrogen, and Biofuels as a Sustainable Vehicular Fuel, *Chem. Eng. J. Adv.*, 2023, DOI: 10.1016/j.ceja.2022.100442.
12. D. N. Dari, I. S. Freitas, F. I. da S. Aires, R. L. F. Melo, K. M. dos Santos, P. da Silva Sousa, P. Gonçalves de Sousa Junior, A. Luthierre Gama Cavalcante, F. S. Neto, J. L. da Silva, É. Carlos de Castro, V. Santos Alexandre, A. M. Ana, J. Serpa, F. de, M. C. M. de Souza and J. C. S. dos Santos, An Updated Review of Recent Applications and Perspectives of Hydrogen Production from Biomass by Fermentation: A Comprehensive Analysis, *Biomass*, 2024, 132–163.
13. V. Del Duca, C. Ponsiglione, S. Primario and S. Strazzullo, Towards Economic, Environmental, and Societal Sustainable World: Reviewing the Interplay of Methodologies, Variables, and Impacts in Energy Transition Models, *J Cleaner Prod.*, 2024, **479**, 144074.
14. O. C. Anika, S. G. Nnabuife, A. Bello, E. R. Okoroafor, B. Kuang and R. Villa, Prospects of Low and Zero-Carbon Renewable Fuels in 1.5-Degree Net Zero Emission Actualisation by 2050: A Critical Review, *Carbon Capture Sci. Technol.*, 2022, **5**, 100072.
15. N. Yilmaz and A. Atmanli, Sustainable Alternative Fuels in Aviation, *Energy*, 2017, **140**, 1378–1386.
16. F. Liu, M. Shafique and X. Luo, Literature Review on Life Cycle Assessment of Transportation Alternative Fuels, *Environ. Technol. Innovation*, 2023, **32**, 103343.
17. A. Arias, C. E. Nika, V. Vasilaki, G. Feijoo, M. T. Moreira and E. Katsou, Assessing the Future Prospects of Emerging Technologies for Shipping and Aviation Biofuels: A Critical Review, *Renewable Sustainable Energy Rev.*, 2024, **197**, 114427.
18. K. W. Jeon, J. H. Gong, M. J. Kim, J. O. Shim, W. J. Jang and H. S. Roh, Review on the Production of Renewable Biofuel: Solvent-Free Deoxygenation, *Renewable Sustainable Energy Rev.*, 2024, **195**, 114325.
19. Z. Yang, K. Shah, C. Pilon-McCullough, R. Faragher, P. Azmi, B. Hollebone, B. Fieldhouse, C. Yang, D. Dey, P. Lambert and V. Beaulac, Characterization of Renewable Diesel, Petroleum Diesel and Renewable Diesel/Biodiesel/Petroleum Diesel Blends, *Renewable Energy*, 2024, **224**, 120151.
20. S. Borschiver, C. Realpe, M. A. Gimenes Couto and K. Coelho, Prospecção Tecnológica de Combustível Renovável Para Aviação: Estudo de Caso Do Diesel Verde, *Cadernos de Prospecção*, 2017, **10**(2), 263.
21. L. Brennan and P. Owende, Biofuels from Microalgae—A Review of Technologies for Production, Processing, and Extractions of Biofuels and Co-Products, *Renewable Sustainable Energy Rev.*, 2010, **14**(2), 557–577.

22. A. Amin, Review of Diesel Production from Renewable Resources: Catalysis, Process Kinetics and Technologies, *Ain Shams Eng. J.*, 2019, **10**(4), 821–839.
23. J. Sadhukhan, E. Martinez-Hernandez, M. A. Amezcua Allieri, J. A. Zermeño Eguía-Lis, A. Castillo, D. Dominguillo, E. Torres-García and J. Aburto, Strategic Navigation of World-Leading Biorefineries and Mexico's Policy Landscape: A Gateway to a Sustainable Circular Bioeconomy, *J Cleaner Prod.*, 2024, **434**, 140386.
24. S. R. Chia, S. Nomanbhay, M. Y. Ong, A. H. B. Shamsuddin, K. W. Chew and P. L. Show, Renewable Diesel as Fossil Fuel Substitution in Malaysia: A Review, *Fuel*, 2022, **314**, 123137.
25. T. dos S. Gonçalves, C. E. D. Oro, J. H. C. Wancura, M. S. N. dos Santos, A. Junges, R. M. Dallago and M. V. Tres, Challenges for Energy Guidelines in Crop-Based Liquid Biofuels Development in Brazil, *Next Sustainability*, 2023, **2**, 100002.
26. N. Pérez-Rangel, C. Coronado and J. Ancheyta, Approaches to Conditioning of Vegetable Oil Feedstock for Hydrotreating to Produce Renewable Diesel, *Fuel*, 2025, **383**, 133897.
27. S. Dey, A. Sreenivasulu, G. T. N. Veerendra, K. V. Rao and P. S. S. A. Babu, Renewable Energy Present Status and Future Potentials in India: An Overview, *Innovation Green Dev.*, 2022, **1**(1), 100006.
28. L. Pirjola, H. Kuuluvainen, H. Timonen, S. Saarikoski, K. Teinilä, L. Salo, A. Datta, P. Simonen, P. Karjalainen, K. Kulmala and T. Rönkkö, Potential of Renewable Fuel to Reduce Diesel Exhaust Particle Emissions, *Appl. Energy*, 2019, **254**, 113636.
29. J. C. Alvim, F. A. L. S. Alvim, V. H. G. Sales, P. V. G. Sales, E. M. de Oliveira and A. C. R. da Costa, Biorefineries: Concepts, Classification, Raw Materials and Products, *J. Bioenergy Food Sci.*, 2014, **1**, 61–77.
30. A. Demirbas, Biorefineries: Current Activities and Future Developments, *Energy Convers. Manage.*, 2009, **50**(11), 2782–2801.
31. N. T. Graham, N. Gakkhar, A. D. Singh, M. Evans, T. Stelmach, S. Durga, R. Godara, B. Gajera, M. Wise and A. K. Sarma, Integrated Analysis of Increased Bioenergy Futures in India, *Energy Policy*, 2022, **168**, 113125.
32. S. F. Ahmed, M. Kabir, A. Mehjabin, F. T. Z. Oishi, S. Ahmed, S. Mannan, M. Mofijur, F. Almomani, I. A. Badruddin and S. Kamangar, Waste Biorefinery to Produce Renewable Energy: Bioconversion Process and Circular Bioeconomy, *Energy Rep.*, 2023, **10**, 3073–3091.
33. P. V. Almeida, L. M. Gando-Ferreira and M. J. Quina, Biorefinery Perspective for Industrial Potato Peel Management: Technology Readiness Level and Economic Assessment, *J. Environ. Chem. Eng.*, 2023, **11**(3), 110049.
34. H. Long and G. Feng, Does National ESG Performance Curb Greenhouse Gas Emissions?, *Innovation Green Dev.*, 2024, **3**(3), 100138.
35. G. Chiatti, O. Chiavola and F. Palmieri, Impact on Combustion and Emissions of Jet Fuel as Additive in Diesel Engine Fueled with Blends of

Petrol Diesel, Renewable Diesel and Waste Cooking Oil Biodiesel, *Energies*, 2019, **12**(13), DOI: 10.3390/en12132488.
36. P. S. Nigam and A. Singh, Production of Liquid Biofuels from Renewable Resources, *Prog. Energy Combust. Sci.*, 2011, **37**(1), 52–68.
37. F. G. Calvo-Flores and F. J. Martin-Martinez, Biorefineries: Achievements and Challenges for a Bio-Based Economy, *Front. Chem.*, 2022, DOI: 10.3389/fchem.2022.973417.
38. H. Caliskan, I. Yildiz and K. Mori, Production and Assessment of New Biofuels from Waste Cooking Oils as Sustainable Bioenergy Sources, *Energies*, 2023, **16**(1), DOI: 10.3390/en16010463.
39. A. B. Kołtuniewicz and K. Dabkowska, Biorefineries - Factories of the Future, *Chem. Process Eng. – Inz. Chem. Procesowa*, 2016, **37**(1), 109–119.
40. S. K. Hoekman, A. Broch, C. Robbins, E. Ceniceros and M. Natarajan, Review of Biodiesel Composition, Properties, and Specifications, *Renewable Sustainable Energy Rev.*, 2012, **16**(1), 143–169.
41. J. Chaves Alvim, F. Aparecida Lima Silva Alvim, V. Hugo Gomes Sales, P. Victor Gomes Sales, E. Maria de Oliveira and A. Claudia Rodrigues da Costa, Biorefineries: Concepts, Classification, Raw Materials and Products, *J. Bioenergy Food Sci.*, 2014, **1**(3), 77402–77970.
42. E. Chizoo, M. F. Okechukwu, O. O. Dominic, A. A. Kingsley, A. S. Chimamkpam, M. C. Mariagorretti and E. P. Chinonso, Renewable Diesel Synthesis from Sesame Indicum (Bene) Seed Oil Using Novel Heterogeneous Biocatalyst Derived from the Chrysophyllum Albidium Seed Coat, *Heliyon*, 2023, **9**(11), DOI: 10.1016/j.heliyon.2023.e22006.
43. H. Yerrennagoudaru, K. Manjunatha, A. Raza, B. R. Kantharaj, A. Mujahed and K. Irshad, Analysis and Comparison of Performance and Emissions of Compression Ignition Engine Fuelled with Diesel and Different Bio-Fuels Blended with Methanol, *Mater. Today: Proc.*, 2018, **5**(2), 5175–5185.
44. A. Kumar, H. S. Pali and M. Kumar, Evaluation of Waste Plastic and Waste Cooking Oil as a Potential Alternative Fuel in Diesel Engine, *Next Energy*, 2024, **5**, 100181.
45. B. Riazi, J. M. Mosby, B. Millet and S. Spatari, Renewable Diesel from Oils and Animal Fat Waste: Implications of Feedstock, Technology, Co-Products and ILUC on Life Cycle GWP, *Resour., Conserv. Recycl.*, 2020, **161**, 104944.
46. A. Demirbas and M. Fatih Demirbas, Importance of Algae Oil as a Source of Biodiesel, *Energy Convers. Manage.*, 2011, **52**(1), 163–170.
47. M. Patel and A. Kumar, Production of Renewable Diesel through the Hydroprocessing of Lignocellulosic Biomass-Derived Bio-Oil: A Review, *Renewable Sustainable Energy Rev.*, 2016, **58**, 1293–1307.
48. S. K. Nayak, A. T. Hoang, B. Nayak and P. C. Mishra, Influence of Fish Oil and Waste Cooking Oil as Post Mixed Binary Biodiesel Blends on Performance Improvement and Emission Reduction in Diesel Engine, *Fuel*, 2021, **289**, 119948.

49. B. A. Oni and D. Oluwatosin, Emission Characteristics and Performance of Neem Seed (Azadirachta Indica) and Camelina (Camelina Sativa) Based Biodiesel in Diesel Engine, *Renewable Energy*, 2020, **149**, 725–734.
50. M. Khalaf, T. Xuan, W. A Abdel-Fadeel, H. M. M. Mustafa, S. Abdelhady and M. F. C. Esmail, A Comparative Study of Diesel Engine Fueled by Jatropha and Castor Biodiesel: Performance, Emissions, and Sustainability Assessment, *Process Saf. Environ. Prot.*, 2024, **188**, 453–466.
51. A. Bayu, A. Rachman, D. R. Noerdjito, M. Y. Putra and W. B. Widayatno, High-Value Chemicals from Marine Diatoms: A Biorefinery Approach, *IOP Conf. Ser.: Earth Environ. Sci.*, 2020, **460**(1), 012012.
52. A. Niño-Villalobos, J. Puello-Yarce, Á. D. González-Delgado, K. A. Ojeda and E. Sánchez-Tuirán, Biodiesel and Hydrogen Production in a Combined Palm and Jatropha Biomass Biorefinery: Simulation, Techno-Economic, and Environmental Evaluation, *ACS Omega*, 2020, **5**(13), 7074–7084.
53. A. Wichaphian, N. Sriket, S. Sensupa, J. Pekkoh, W. Pathom-aree, Y. Chromkaew, N. Suwannarach, J. Kumla, B. Cheirsilp and S. Srinuanpan, Value-Added Green Biorefinery Co-Products from Ultrasonically Assisted DES-Pretreated Chlorella Biomass, *Ultrason. Sonochem.*, 2023, **100**, 106628.
54. A. K. Singh, R. K. Srivastava, P. Pal, S. Mandal, U. K. Sahoo, A. Prakash, K. Sridhar, M. Sharma, P. K. Sarangi and B. S. Inbaraj, Microalgal Biorefinery as a Sustainable and Cost-Effective Platform for Co-Production of High-Value-Added Products/Metabolites: An Insight into Emerging Trends, Challenges, and Opportunities, *Biocatal. Agric. Biotechnol.*, 2024, **58**, 103192.
55. R. Sivaramakrishnan and A. Incharoensakdi, Plant Hormone Induced Enrichment of Chlorella Sp. Omega-3 Fatty Acids, *Biotechnol. Biofuels*, 2020, **13**(1), 7.
56. M. Sung and J.-I. Han, Ultrasound-Assisted in-Situ Transesterification of Wet Aurantiochytrium Sp. KRS 101 Using Potassium Carbonate, *Bioresour. Technol.*, 2018, **261**, 117–121.
57. K. W. Chew, J. Y. Yap, P. L. Show, N. H. Suan, J. C. Juan, T. C. Ling, D. J. Lee and J. S. Chang, Microalgae Biorefinery: High Value Products Perspectives, *Bioresour. Technol.*, 2017, **229**, 53–62.
58. R. A. Quevedo-Amador, B. P. Escalera-Velasco, A. M. R. Arias, H. E. Reynel-Ávila, J. C. Moreno-Piraján, L. Giraldo and A. Bonilla-Petriciolet, Application of Waste Biomass for the Production of Biofuels and Catalysts: A Review, *Clean Technol. Environ. Policy*, 2024, **26**(4), 943–997.
59. M. Narayanan, Biorefinery Products from Algal Biomass by Advanced Biotechnological and Hydrothermal Liquefaction Approaches, *Discover Appl. Sci.*, 2024, **6**(4), 146.
60. R. Sivaramakrishnan and A. Incharoensakdi, Utilization of Microalgae Feedstock for Concomitant Production of Bioethanol and Biodiesel, *Fuel*, 2018, **217**, 458–466.

61. L. E. González, G. C. Díaz, D. A. G. Aranda, Y. R. Cruz and M. M. Fortes, Biodiesel Production Based in Microalgae: A Biorefinery Approach, *Nat. Sci.*, 2015, **07**(07), 358–369.
62. A. S. Zaky, Introducing a Marine Biorefinery System for the Integrated Production of Biofuels, High-Value-Chemicals, and Co-Products: A Path Forward to a Sustainable Future, *Processes*, 2021, **9**(10), 1841.
63. E. S. Okeke, O. Ejeromedoghene, C. O. Okoye, T. P. C. Ezeorba, R. Nyaruaba, C. K. Ikechukwu, A. Oladipo and J. I. Orege, Microalgae Biorefinery: An Integrated Route for the Sustainable Production of High-Value-Added Products, *Energy Convers. Manage.: X*, 2022, **16**, 100323.
64. J. Roles, J. Yarnold, K. Hussey and B. Hankamer, Techno-Economic Evaluation of Microalgae High-Density Liquid Fuel Production at 12 International Locations, *Biotechnol. Biofuels*, 2021, **14**(1), 133.
65. N. Carreno-Quintero, T. Tohge, R. Van Acker, L. S. McKee, Q. Zhou, A. Bolze, X. Xing, M. Özparpucu, M. Rüggeberg, T. Piofczyk, Y. Koram, V. Bulone, W. Boerjan, A. R. Fernie and P. D. Fraser, Non-Targeted Discovery of High-Value Bio-Products in Nicotiana Glauca L: A Potential Renewable Plant Feedstock, *Bioresour. Bioprocess*, 2024, **11**(1), 12.
66. R. J. Stoklosa, Deriving Biofuels and Value-Added Co-products from Sorghum Bicolor: Prospects in Biorefinery Applications and Product Development, *ACS Symp. Ser.*, 2020, 43–62.
67. J. Ruangwicha, B. Cheirsilp and W. Suyotha, Green Biorefinery of Shrimp Shell Waste for α-Chitin and High-Value Co-Products through Successive Fermentation by Co-Lactic Acid Bacteria and Proteolytic Fungus, *Bioresour. Technol.*, 2024, **393**, 130106.
68. G. Di Blasio, R. Ianniello and C. Beatrice, Hydrotreated Vegetable Oil as Enabler for High-Efficient and Ultra-Low Emission Vehicles in the View of 2030 Targets, *Fuel*, 2022, **310**, 122206.
69. E. Monroe, J. S. Carlson, R. Dhaoui, M. Sarwar, P. T. Benavides, J. Zhu, C. S. McEnally, L. Pfefferle, A. George, N. Sizemore and R. W. Davis, Application of Alkoxyalkanoates (AOAs) as Renewable Diesel Blendstocks from Chemical Coupling of High-Yield Fermentation Products, *Energy Fuels*, 2023, **37**(3), 2091–2099.
70. K. N. Yogalakshmi, T. M. Mohamed Usman, S. Kavitha, S. Saloni, S. Thakur, S. Adish Kumar and J. Rajesh Banu, Lignocellulosic Biorefinery Technologies: A Perception into Recent Advances in Biomass Fractionation, Biorefineries, Economic Hurdles and Market Outlook, *Fermentation*, 2023, **9**(3), 238.
71. V. Larnaudie, M. Bule, K.-Y. San, P. V. Vadlani, J. Mosby, S. Elangovan, M. Karanjikar and S. Spatari, Life Cycle Environmental and Cost Evaluation of Renewable Diesel Production, *Fuel*, 2020, **279**, 118429.
72. H. Cai, J. Han, M. Wang, R. Davis, M. Biddy and E. Tan, Life-cycle Analysis of Integrated Biorefineries with Co-production of Biofuels and Bio-based Chemicals: Co-product Handling Methods and Implications, *Biofuels, Bioprod. Biorefin.*, 2018, **12**(5), 815–833.

73. C. Li, W. Yan, B. Wang, T. Xu, C. S. K. Lin, X. Zou and C. Du, Integrated Biorefineries for the Co-Production of Biofuels and High-Value Products, in *Handbook of Biofuels Production*, Elsevier, 2023, pp. 513–541.
74. J. S. Saral, A. R. Satheesh and P. Ranganathan, Economic and Environmental Analysis of Algal Biorefinery for the Production of Renewable Fuels and Co-Product, *Energy Convers. Manage.: X*, 2022, **14**, 100189.
75. D. Deb, N. Mallick and P. B. S. Bhadoria, Engineering Culture Medium for Enhanced Carbohydrate Accumulation in Anabaena Variabilis to Stimulate Production of Bioethanol and Other High-Value Co-Products under Cyanobacterial Refinery Approach, *Renewable Energy*, 2021, **163**, 1786–1801.
76. W. R. Gibbons and S. R. Hughes, Integrated Biorefineries with Engineered Microbes and High-Value Co-Products for Profitable Biofuels Production, *In Vitro Cell. Dev. Biol.: Plant*, 2009, **45**(3), 218–228.
77. A. K. Patel, R. R. Singhania, C.-D. Dong, P. K. Obulisami and S. J. Sim, Mixotrophic Biorefinery: A Promising Algal Platform for Sustainable Biofuels and High Value Coproducts, *Renewable Sustainable Energy Rev.*, 2021, **152**, 111669.
78. R. Reshmy, T. A. P. Paulose, E. Philip, D. Thomas, A. Madhavan, R. Sirohi, P. Binod, M. Kumar Awasthi, A. Pandey and R. Sindhu, Updates on High Value Products from Cellulosic Biorefinery, *Fuel*, 2022, **308**, 122056.
79. S. Wan, V. M. Truong-Trieu, T. Ward, J. K. Whalen and I. Altosaar, Advances in the Use of Genetically Modified Plant Biomass for Biodiesel Generation, *Biofuels, Bioprod. Biorefin.*, 2017, **11**(4), 749–764.
80. R. Sivaramakrishnan, S. Suresh, S. Kanwal, G. Ramadoss, B. Ramprakash and A. Incharoensakdi, Microalgal Biorefinery Concepts' Developments for Biofuel and Bioproducts: Current Perspective and Bottlenecks, *Int. J. Mol. Sci.*, 2022, **23**(5), 2623.
81. J. Prameswari, W. Widayat, L. Buchori and H. Hadiyanto, Novel Iron Sand-Derived α-Fe_2O_3/CaO_2 Bifunctional Catalyst for Waste Cooking Oil-Based Biodiesel Production, *Environ. Sci. Pollut. Res.*, 2022, **30**(44), 98832–98847.
82. F. O. Nitbani, P. J. P. Tjitda, B. A. Nurohmah and H. E. Wogo, Preparation of Fatty Acid and Monoglyceride from Vegetable Oil, *J. Oleo Sci.*, 2020, **69**(4), 277–295.
83. F. Li, A. Xia, X. Guo, W. Zhang, Y. Huang, X. Zhu, X. Zhu and Q. Liao, Continuous Hydrocarbon Fuels Production by Photoenzymatic Decarboxylation of Free Fatty Acids from Waste Oils, *J. Environ. Chem. Eng.*, 2023, **11**(5), 110748.
84. M. Li, J. Chen, Y. Huang, M. Li, X. Lin and T. Qiu, Reusable and Efficient Heterogeneous Catalysts for Biodiesel Production from Free Fatty Acids and Oils: Self-Solidifying Hybrid Ionic Liquids, *Energy*, 2020, **211**, 118631.

85. P. Zhang, D. Chen, Y. Tian, H. Li, T. Gong, J. Luo, J. Ruan, T. Gong and Z. Zhang, Comparison of Three In-Situ Gels Composed of Different Oil Types, *Int. J. Pharm.*, 2020, **587**, 119707.
86. I. Edeh, T. Overton and S. Bowra, Renewable Diesel Production by Hydrothermal Decarboxylation of Fatty Acids over Platinum on Carbon Catalyst, *Biofuels*, 2021, **12**(8), 945–952.
87. B. F. Honorato de Oliveira, L. F. de França, N. C. Fernandes Corrêa, N. F. da P. Ribeiro and M. Velasquez, Renewable Diesel Production from Palm Fatty Acids Distillate (PFAD) via Deoxygenation Reactions, *Catalysts*, 2021, **11**(9), 1088.
88. K. B. Baharudin, N. Abdullah, Y. H. Taufiq-Yap and D. Derawi, Renewable Diesel via Solventless and Hydrogen-Free Catalytic Deoxygenation of Palm Fatty Acid Distillate, *J Cleaner Prod.*, 2020, **274**, 122850.
89. H. Xu, L. Ou, Y. Li, T. R. Hawkins and M. Wang, Life Cycle Greenhouse Gas Emissions of Biodiesel and Renewable Diesel Production in the United States, *Environ. Sci. Technol.*, 2022, **56**(12), 7512–7521.
90. S. L. Douvartzides, N. D. Charisiou, K. N. Papageridis and M. A. Goula, Green Diesel: Biomass Feedstocks, Production Technologies, Catalytic Research, Fuel Properties and Performance in Compression Ignition Internal Combustion Engines, *Energies*, 2019, **12**(5), 809.
91. S. N. Gebremariam and J. M. Marchetti, Economics of Biodiesel Production: Review, *Energy Convers. Manage.*, 2018, **168**, 74–84.
92. I. Nur Azreena, H. L. N. Lau, N. Asikin-Mijan, M. A. Hassan, S. M. Izham, M. Safa Gamal, W. Nor Adira Wan Khalit, M. Arumugam, E. Kennedy, M. Stockenhuber and Y. H. Taufiq-Yap, Hydrodeoxygenation of Fatty Acid over La-Modified HZSM5 for Premium Quality Renewable Diesel Production, *J. Anal. Appl. Pyrolysis*, 2022, **161**, 105406.
93. W. Pérez, J. Marín, J. del Río, J. Peña and L. Rios, Upgrading of Palm Oil Renewable Diesel through Hydroisomerization and Formulation of an Optimal Blend, *Fuel*, 2017, **209**, 442–448.
94. H. Wang, K. Rogers, H. Zhang, G. Li, J. Pu, H. Zheng, H. Lin, Y. Zheng and S. Ng, The Effects of Catalyst Support and Temperature on the Hydrotreating of Waste Cooking Oil (WCO) over CoMo Sulfided Catalysts, *Catalysts*, 2019, **9**(8), 689.
95. C. Ochoa-Hernández, J. M. Coronado and D. P. Serrano, Hydrotreating of Methyl Esters to Produce Green Diesel over Co- and Ni-Containing Zr-SBA-15 Catalysts, *Catalysts*, 2020, **10**(2), 186.
96. Q. Liu, H. Zuo, Q. Zhang, T. Wang and L. Ma, Hydrodeoxygenation of Palm Oil to Hydrocarbon Fuels over Ni/SAPO-11 Catalysts, *Chin. J. Catal.*, 2014, **35**(5), 748–756.
97. A. Tirado, J. Ancheyta and F. Trejo, Kinetic and Reactor Modeling of Catalytic Hydrotreatment of Vegetable Oils, *Energy Fuels*, 2018, **32**(7), 7245–7261.
98. F. Alonso, J. Ancheyta, G. Centeno, G. Marroquín, P. Rayo and R. Silva-Rodrigo, Effect of Reactor Configuration on the Hydrotreating of Atmospheric Residue, *Energy Fuels*, 2019, **33**(2), 1649–1658.

99. L. D'Avino, R. Dainelli, L. Lazzeri and P. Spugnoli, The Role of Co-Products in Biorefinery Sustainability: Energy Allocation *versus* Substitution Method in Rapeseed and Carinata Biodiesel Chains, *J Cleaner Prod.*, 2015, **94**, 108–115.
100. R. K. Devappa, S. K. Rakshit and R. F. H. Dekker, Forest Biorefinery: Potential of Poplar Phytochemicals as Value-Added Co-Products, *Biotechnol. Adv.*, 2015, **33**(6), 681–716.
101. R. Lupitskyy, C. Staff and J. Satyavolu, Towards Integrated Biorefinery from Dried Distillers Grains: Evaluation of Feed Application for Co-Products, *Biomass Bioenergy*, 2015, **72**, 251–255.
102. P. G. de Sousa Junior, A. G. do Nascimento Camara, A. R. T. Oliveira, G. de Castro Lima, G. V. Lima, L. P. Coutinho, J. Y. Nunes Holanda Alexandre, L. F. Serafim, M. C. de Mattos, N. V. de Kássio Monteiro and J. C. Sousa dos Santos, Optimization and Theoretical Analysis of Lipase-Catalyzed Enzymatic Esterification of Glycerol for Efficient Glycerides Synthesis, *Biochem. Eng. J.*, 2023, **198**, 109033.
103. J. Y. N. H. Alexandre, F. T. T. Cavalcante, L. M. Freitas, A. P. Castro, P. T. Borges, P. G. de Sousa Junior, M. N. R. Filho, A. A. S. Lopes, A. M. da Fonseca, D. Lomonaco, M. A. de Sousa Rios and J. C. Sousa dos Santos, A Theoretical and Experimental Study for Enzymatic Biodiesel Production from Babassu Oil (Orbignya Sp.) Using Eversa Lipase, *Catalysts*, 2022, **12**(11), 1322.
104. K. M. dos Santos, J. de França Serpa, V. de Castro Bizerra, R. L. F. Melo, P. G. de Sousa Junior, V. Santos Alexandre, A. M. da Fonseca, P. B. A. Fechine, D. Lomonaco, J. C. Sousa dos Santos and M. C. Martins de Souza, Enhanced Biodiesel Production with Eversa Transform 2.0 Lipase on Magnetic Nanoparticles, *Langmuir*, 2024, DOI: 10.1021/acs.langmuir.4c02542.
105. F. Simão Neto, P. G. de Sousa Junior, C. J. A. da Silva Filho, L. Pinheiro Coutinho, R. L. F. Melo, J. Rocha-Martin, M. A. de S. Rios, A. A. Sanders Lopes, N. de K. V. Monteiro, M. C. de Mattos, L. F. Serafim and J. C. Sousa dos Santos, Green Enzymatic Synthesis of Geranyl Butyrate: Process Optimization and Mechanistic Insights, *ACS Omega*, 2024, DOI: 10.1021/acsomega.3c08483.
106. P. Kalita, B. Basumatary, P. Saikia, B. Das and S. Basumatary, Biodiesel as Renewable Biofuel Produced *via* Enzyme-Based Catalyzed Transesterification, *Energy Nexus*, 2022, **6**, 100087.
107. P. J. Westfall and T. S. Gardner, Industrial Fermentation of Renewable Diesel Fuels, *Curr. Opin. Biotechnol.*, 2011, **22**(3), 344–350.
108. T. Manchanda, R. Tyagi and D. K. Sharma, Comparison of Fuel Characteristics of Green (Renewable) Diesel with Biodiesel Obtainable from Algal Oil and Vegetable Oil, *Energy Sources, Part A*, 2018, **40**(1), 54–59.
109. L. Ferrari, F. Tuler, E. Promancio, L. Gusé, D. G. Touza, C. Casas and R. A. Comelli, Glycerol as Raw Material to an Argentinian Biorefinery, *Catal. Today*, 2022, **394–396**, 247–255.

110. J. R. M. Almeida, L. C. L. Fávaro and B. F. Quirino, Biodiesel Biorefinery: Opportunities and Challenges for Microbial Production of Fuels and Chemicals from Glycerol Waste, *Biotechnol. Biofuels*, 2012, 5(1), 48.
111. C. L. Gargalo, A. Carvalho, K. V. Gernaey and G. Sin, Supply Chain Optimization of Integrated Glycerol Biorefinery: GlyThink Model Development and Application, *Ind. Eng. Chem. Res.*, 2017, **56**(23), 6711–6727.
112. G. V. Salvador, V. M. M. Salim and F. S. Toniolo, Sustainability Assessment of a Decentralized Green Diesel Production in Small-scale Biorefineries, *Biofuels, Bioprod. Biorefin.*, 2022, **16**(6), 1527–1550.
113. E. G. Infante, F. T. F. Gomide, A. R. Secchi, L. F. Leite, A. M. de S. Antunes and A. R. G. Caranton, Diesel Production from Lignocellulosic Residues: Trends, Challenges and Opportunities, *Biofuels, Bioprod. Biorefin.*, 2024, **18**(5), 1711–1738.
114. A. A. V. Julio, T. S. Milessi, E. A. Ocampo Batlle, E. E. Silva Lora, D. M. Yepes Maya and J. C. Escobar Palacio, Techno-Economic and Environmental Potential of Renewable Diesel as Complementation for Diesel and Biodiesel in Brazil: A Comprehensive Review and Perspectives, *J Cleaner Prod.*, 2022, **371**, 133431.
115. L. Ou, S. Li, L. Tao, S. Phillips, T. Hawkins, A. Singh, L. Snowden-Swan and H. Cai, Techno-Economic Analysis and Life-Cycle Analysis of Renewable Diesel Fuels Produced with Waste Feedstocks, *ACS Sustainable Chem. Eng.*, 2022, **10**(1), 382–393.
116. S. Spatari, A. Stadel, P. R. Adler, S. Kar, W. J. Parton, K. B. Hicks, A. J. McAloon and P. L. Gurian, The Role of Biorefinery Co-Products, Market Proximity and Feedstock Environmental Footprint in Meeting Biofuel Policy Goals for Winter Barley-to-Ethanol, *Energies*, 2020, **13**(9), 2236.
117. E. Scopel and C. A. Rezende, Biorefinery On-Demand: Modulating Pretreatments to Recover Lignin, Hemicellulose, and Extractives as Co-Products during Ethanol Production, *Ind. Crops Prod.*, 2021, **163**, 113336.
118. A. Wong, H. Zhang and A. Kumar, Life Cycle Assessment of Renewable Diesel Production from Lignocellulosic Biomass, *Int. J. Life Cycle Assess.*, 2016, **21**(10), 1404–1424.
119. W. Kisiga, M. Chetty and S. Rathilal, Environmental Impact Assessment of Alternative Technologies for Production of Biofuels from Spent Coffee Grounds, *Energy Sci. Eng.*, 2024, **12**(11), 4823–4842.
120. D. E. Leiva-Candia, S. Tsakona, N. Kopsahelis, I. L. García, S. Papanikolaou, M. P. Dorado and A. A. Koutinas, Biorefining of By-Product Streams from Sunflower-Based Biodiesel Production Plants for Integrated Synthesis of Microbial Oil and Value-Added Co-Products, *Bioresour. Technol.*, 2015, **190**, 57–65.
121. L. S. Al-Saadi, V. C. Eze and A. P. Harvey, Techno-Economic Analysis of Processes for Biodiesel Production with Integrated Co-Production of

Higher Added Value Products from Glycerol, *Biofuels*, 2022, **13**(4), 489–496.
122. M. Anitha, S. K. Kamarudin and N. T. Kofli, The Potential of Glycerol as a Value-Added Commodity, *Chem. Eng. J.*, 2016, **295**, 119–130.
123. J. Calero, D. Luna, E. D. Sancho, C. Luna, F. M. Bautista, A. A. Romero, A. Posadillo, J. Berbel and C. Verdugo-Escamilla, An Overview on Glycerol-Free Processes for the Production of Renewable Liquid Biofuels, Applicable in Diesel Engines, *Renewable Sustainable Energy Rev.*, 2015, **42**, 1437–1452.
124. D. T. Johnson and K. A. Taconi, The Glycerin Glut: Options for the Value-added Conversion of Crude Glycerol Resulting from Biodiesel Production, *Environ. Prog.*, 2007, **26**(4), 338–348.
125. J. Kaur, A. K. Sarma, M. K. Jha and P. Gera, Valorisation of Crude Glycerol to Value-Added Products: Perspectives of Process Technology, Economics and Environmental Issues, *Biotechnol. Rep.*, 2020, **27**, e00487.
126. M. H. Moklis, S. Cheng and J. S. Cross, Current and Future Trends for Crude Glycerol Upgrading to High Value-Added Products, *Sustainability*, 2023, **15**(4), 2979.
127. S. Octave and D. Thomas, Biorefinery: Toward an Industrial Metabolism, *Biochimie*, 2009, **91**(6), 659–664.
128. L. Soh and M. J. Eckelman, Green Solvents in Biomass Processing, *ACS Sustainable Chem. Eng.*, 2016, **4**(11), 5821–5837.
129. S. De Meester, C. Callewaert, E. De Mol, H. Van Langenhove and J. Dewulf, The Resource Footprint of Biobased Products: A Key Issue in the Sustainable Development of Biorefineries, *Biofuels, Bioprod. Biorefin.*, 2011, **5**(5), 570–580.
130. L. Goswami, R. Kayalvizhi, P. K. Dikshit, K. C. Sherpa, S. Roy, A. Kushwaha, B. S. Kim, R. Banerjee, S. Jacob and R. C. Rajak, A Critical Review on Prospects of Bio-Refinery Products from Second and Third Generation Biomasses, *Chem. Eng. J.*, 2022, **448**, 137677.
131. M. K. Shahid, A. Batool, A. Kashif, M. H. Nawaz, M. Aslam, N. Iqbal and Y. Choi, Biofuels and Biorefineries: Development, Application and Future Perspectives Emphasizing the Environmental and Economic Aspects, *J. Environ. Manage.*, 2021, **297**, 113268.
132. M. Tsagkari, A. Kokossis and J. Dubois, A Method for Quick Capital Cost Estimation of Biorefineries beyond the State of the Art, *Biofuels, Bioprod. Biorefin.*, 2020, **14**(5), 1061–1088.
133. G. Mongkhonsiri, P. Charoensuppanimit, A. Anantpinijwatna, R. Gani and S. Assabumrungrat, Process Development of Sustainable Biorefinery System Integrated into the Existing Pulping Process, *J Cleaner Prod.*, 2020, **255**, 120278.
134. J. C. Solarte-Toro and C. A. Cardona Alzate, Sustainability of Biorefineries: Challenges and Perspectives, *Energies*, 2023, **16**(9), 3786.
135. M. A. Mehmood, C.-G. Liu and A. Reungsang, Editorial: Special Issue "Sustainable Biorefinery Approaches", *Biomass Convers. Biorefin.*, 2021, **11**(3), 741–742.

136. A. Akbari and P. I. Barton, Integrating Genome-Scale and Superstructure Optimization Models in Techno-Economic Studies of Biorefineries, *Processes*, 2019, 7(5), 286.
137. S. Frodeson, J. Berghel, M. Ståhl, K. Granström, C. Romlin and A. Thelander, The Potential for a Pellet Plant to Become a Biorefinery, *Processes*, 2019, 7(4), 233.
138. N. Dlangamandla, S. K. O. Ntwampe, J. O. Angadam, E. F. Itoba-Tombo, B. S. Chidi and L. Mekuto, Integrated Hydrolysis of Mixed Agro-Waste for a Second Generation Biorefinery Using Nepenthes Mirabilis Pod Digestive Fluids, *Processes*, 2019, 7(2), 64.
139. R. El-Araby, Biofuel Production: Exploring Renewable Energy Solutions for a Greener Future, *Biotechnol. Biofuels Bioprod.*, 2024, **17**(1), 129.
140. A. Suhara, Karyadi, S. G. Herawan, A. Tirta, M. Idris, M. F. Roslan, N. R. Putra, A. L. Hananto and I. Veza, Biodiesel Sustainability: Review of Progress and Challenges of Biodiesel as Sustainable Biofuel, *Clean Technol.*, 2024, **6**(3), 886–906.
141. J. V. L. Ruatpuia, G. Halder, M. Vanlalchhandama, F. Lalsangpuii, R. Boddula, N. Al-Qahtani, S. Niju, T. Mathimani and S. L. Rokhum, Jatropha Curcas Oil a Potential Feedstock for Biodiesel Production: A Critical Review, *Fuel*, 2024, **370**, 131829.
142. P. M. Yeletsky, R. G. Kukushkin, V. A. Yakovlev and B. H. Chen, Recent Advances in One-Stage Conversion of Lipid-Based Biomass-Derived Oils into Fuel Components – Aromatics and Isomerized Alkanes, *Fuel*, 2020, **278**, 118255.
143. S. Riaz, K. Y. Rhee and S. J. Park, Polyhydroxyalkanoates (PHAs): Biopolymers for Biofuel and Biorefineries, *Polymers*, 2021, **13**(2), 253.
144. J.-H. Kim, M. Kim, G. Park, E. Kim, H. Song, S. Jung, Y.-K. Park, Y. F. Tsang, J. Lee and E. E. Kwon, Chemicals and Fuels from Lipid-Containing Biomass: A Comprehensive Exploration, *Biotechnol. Adv.*, 2024, **75**, 108418.
145. C. G. Lopresto, Sustainable Biodiesel Production from Waste Cooking Oils for Energetically Independent Small Communities: An Overview, *Int. J. Environ. Sci. Technol.*, 2024, DOI: 10.1007/s13762-024-05779-2.
146. Z. Teimouri, S. Nanda, N. Abatzoglou and A. K. Dalai, Application of Activated Carbon in Renewable Energy Conversion and Storage Systems: A Review, *Environ. Chem. Lett.*, 2024, **22**(3), 1073–1092.
147. S. N. Omer, P. Saravanan, P. Kumar, R. R. Kannan, M. Rajasimman and V. Shanmugam, Insights into Renewable Biohydrogen Production from Algal Biomass: Technical Hurdles and Economic Analysis, *Biomass Convers. Biorefin.*, 2024, DOI: 10.1007/s13399-023-05263-w.
148. J. C. Solarte-Toro, M. Laghezza, S. Fiore, F. Berruti, K. Moustakas and C. A. Cardona Alzate, Review of the Impact of Socio-Economic Conditions on the Development and Implementation of Biorefineries, *Fuel*, 2022, **328**, 125169.
149. A. S. Nizami, M. Rehan, M. Waqas, M. Naqvi, O. K. M. Ouda, K. Shahzad, R. Miandad, M. Z. Khan, M. Syamsiro, I. M. I. Ismail and

D. Pant, Waste Biorefineries: Enabling Circular Economies in Developing Countries, *Bioresour. Technol.*, 2017, **241**, 1101–1117.
150. A. T. Ubando, C. B. Felix and W.-H. Chen, Biorefineries in Circular Bioeconomy: A Comprehensive Review, *Bioresour. Technol.*, 2020, **299**, 122585.
151. K. Dhamodharan, S. Ahlawat, M. Kaushal and K. Rajendran, Economics and Cost Analysis of Waste Biorefineries, in *Refining Biomass Residues for Sustainable Energy and Bioproducts*, Elsevier, 2020, pp. 545–565.
152. J. Louw, E. S. Dogbe, B. Yang and J. F. Görgens, Prioritisation of Biomass-Derived Products for Biorefineries Based on Economic Feasibility: A Review on the Comparability of Techno-Economic Assessment Results, *Renewable Sustainable Energy Rev.*, 2023, **188**, 113840.
153. J. C. Solarte-Toro, C. A. Rueda-Duran, M. Ortiz-Sanchez and C. A. Cardona Alzate, A Comprehensive Review on the Economic Assessment of Biorefineries: The First Step towards Sustainable Biomass Conversion, *Bioresour. Technol. Rep.*, 2021, **15**, 100776.
154. B. Kumar and P. Verma, Biomass-Based Biorefineries: An Important Architype towards a Circular Economy, *Fuel*, 2021, **288**, 119622.
155. S. Wijeyekoon, I. Suckling, M. Fahmy, P. Hall and P. Bennett, Techno-economic Analysis of Tannin and Briquette Co-production from Bark Waste: A Case Study Quantifying Symbiosis Benefits in Biorefinery, *Biofuels, Bioprod. Biorefin.*, 2021, **15**(5), 1332–1344.
156. G. L. Arruda, M. T. F. R. Raymundo, M. M. Cruz-Santos, V. P. Shibukawa, F. M. Jofre, C. A. Prado, S. S. da Silva, S. I. Mussatto and J. C. Santos, Lignocellulosic Materials Valorization in Second Generation Biorefineries: An Opportunity to Produce Fungal Biopigments, *Crit. Rev. Biotechnol.*, 2024, 1–20.
157. A. Arias, S. Estévez-Rivadulla, R. Rebolledo-Leiva, G. Feijoo, S. González-García and M. T. Moreira, Boosting the Transition to Biorefineries in Compliance with Sustainability and Circularity Criteria, *J. Environ. Chem. Eng.*, 2024, **12**(5), 113361.
158. S.-H. Lee, S. H. Park and H. Park, Assessing the Feasibility of Biorefineries for a Sustainable Citrus Waste Management in Korea, *Molecules*, 2024, **29**(7), 1589.
159. A. P. Clasen, F. Agostinho, C. M. V. B. Almeida, G. Liu and B. F. Giannetti, Advancing towards Circular Economy: Environmental Benefits of an Innovative Biorefinery for Municipal Solid Waste Management, *Sustainable Prod. Consumption*, 2024, **46**, 571–581.
160. S. T. Bryant, K. Straker and C. Wrigley, The Discourses of Power – Governmental Approaches to Business Models in the Renewable Energy Transition, *Energy Policy*, 2019, **130**, 41–59.
161. O. A. Adelekan, B. S. Ilugbusi, O. Adisa, O. C. Obi, K. F. Awonuga, O. F. Asuzu and N. L. Ndubuisi, Energy transition policies: a global review of shifts towards renewable sources, *Eng. Sci. Technol. J.*, 2024, **5**(2), 272–287.

162. C. Drago and A. Gatto, Policy, Regulation Effectiveness, and Sustainability in the Energy Sector: A Worldwide Interval-Based Composite Indicator, *Energy Policy*, 2022, **167**, 112889.
163. Y. Zou and M. Wang, Does Environmental Regulation Improve Energy Transition Performance in China?, *Environ. Impact Assess. Rev.*, 2024, **104**, 107335.
164. S. Bakhsh, W. Zhang, K. Ali and J. Oláh, Strategy towards Sustainable Energy Transition: The Effect of Environmental Governance, Economic Complexity and Geopolitics, *Energy Strategy Rev.*, 2024, **52**, 101330.
165. M. Aydin, T. Degirmenci, A. Erdem, Y. Sogut and N. Demirtas, From Public Policy towards the Green Energy Transition: Do Economic Freedom, Economic Globalization, Environmental Policy Stringency, and Material Productivity Matter?, *Energy*, 2024, **311**, 133404.
166. M. J. Watson, P. G. Machado, A. V. da Silva, Y. Saltar, C. O. Ribeiro, C. A. O. Nascimento and A. W. Dowling, Sustainable Aviation Fuel Technologies, Costs, Emissions, Policies, and Markets: A Critical Review, *J Cleaner Prod.*, 2024, **449**, 141472.
167. T. Mogany, V. Bhola and F. Bux, Algal-Based Bioplastics: Global Trends in Applied Research, Technologies, and Commercialization, *Environ. Sci. Pollut. Res.*, 2024, **31**(26), 38022–38044.
168. A. D. Pranta and M. T. Rahaman, Extraction of Eco-Friendly Natural Dyes and Biomordants for Textile Coloration: A Critical Review, *Nano-Struct. Nano-Objects*, 2024, **39**, 101243.
169. B. Long, F. Zhang, S. Y. Dai, M. Foston, Y. J. Tang and J. S. Yuan, Engineering Strategies to Optimize Lignocellulosic Biorefineries, *Nat. Rev. Bioeng.*, 2024, DOI: 10.1038/s44222-024-00247-5.
170. Y. Zhang, M. Goldberg, E. Tan and P. A. Meyer, Estimation of Economic Impacts of Cellulosic Biofuel Production: A Comparative Analysis of Three Biofuel Pathways, *Biofuels, Bioprod. Biorefin.*, 2016, **10**(3), 281–298.
171. A. V. Monclaro, H. A. R. Gomes, G. C. Duarte, L. R. de Souza Moreira and E. X. F. Filho, Unveiling the Biomass Valorization: The Microbial Diversity in Promoting a Sustainable Socio-Economy, *Bioenergy Res.*, 2024, **17**(3), 1355–1374.
172. M. A. Abdullah, M. S. Nazir, H. A. Hussein, S. M. U. Shah, N. Azra, R. Iftikhar, M. S. Iqbal, Z. Qamar, Z. Ahmad, M. Afzaal, A. D. Om, M. I. Shaharah, A. E. Rak and Y.-T. Hung, New Perspectives on Biomass Conversion and Circular Economy Based on Integrated Algal-Oil Palm Biorefinery Framework for Sustainable Energy and Bioproducts Co-Generation, *Ind. Crops Prod.*, 2024, **213**, 118452.
173. V. Kaushik and R. Kataria, Non-Conventional Strategies for Pretreatment of Lignocellulosic Biomass for Production of Value-Added Products: A Sustainable and Circular Economy Approach, *Biomass Convers. Biorefin.*, 2024, DOI: 10.1007/s13399-024-06388-2.
174. D. Chettri, D. Boro, M. Chirania and A. K. Verma, Integrating Biochar Production in Biorefineries: Towards a Sustainable Future and Circular Economy, *Biofuels, Bioprod. Biorefin.*, 2024, **18**(6), 2156–2176.

175. M. Eqbalpour, A. Andooz, E. Kowsari, S. Ramakrishna, Z. A. Cheshmeh and A. Chinnappan, A Comprehensive Review on How Ionic Liquids Enhance the Pyrolysis of Cellulose, Lignin, and Lignocellulose toward a Circular Economy, *WIREs Energy Environ.*, 2023, **12**(4), DOI: 10.1002/wene.473.
176. P. Miller, A. Sultana and A. Kumar, Optimum Scale of Feedstock Processing for Renewable Diesel Production, *Biofuels, Bioprod. Biorefin.*, 2012, **6**(2), 188–204.
177. A. K. Chandel, V. K. Garlapati, S. P. Jeevan Kumar, M. Hans, A. K. Singh and S. Kumar, The Role of Renewable Chemicals and Biofuels in Building a Bioeconomy, *Biofuels, Bioprod. Biorefin.*, 2020, **14**(4), 830–844.
178. M. V. L. Chhandama, P. K. Rai and Lalawmpuii, Coupling Bioremediation and Biorefinery Prospects of Microalgae for Circular Economy, *Bioresour. Technol. Rep.*, 2023, **22**, 101479.
179. S. Talekar, K. Ekanayake, B. Holland and C. Barrow, Food Waste Biorefinery towards Circular Economy in Australia, *Bioresour. Technol.*, 2023, **388**, 129761.
180. O. S. Dahunsi, Introduction to Biorefinery and Circular Bioeconomy, in *Biorefinery*, Elsevier, 2025, pp. 1–13.
181. L. A. Pfaltzgraff and J. H. Clark, Green Chemistry, Biorefineries and Second Generation Strategies for Re-Use of Waste: An Overview, in *Advances in Biorefineries*, Elsevier, 2014, pp. 3–33.
182. M. K. Awasthi, S. Harirchi, T. Sar, V. VS, K. Rajendran, R. Gómez-García, C. Hellwig, P. Binod, R. Sindhu, A. Madhavan, A. N. A. Kumar, V. Kumar, D. Kumar, Z. Zhang and M. J. Taherzadeh, Myco-Biorefinery Approaches for Food Waste Valorization: Present Status and Future Prospects, *Bioresour. Technol.*, 2022, **360**, 127592.
183. A. ElMekawy, L. Diels, H. De Wever and D. Pant, Valorization of Cereal Based Biorefinery Byproducts: Reality and Expectations, *Environ. Sci. Technol.*, 2013, **47**(16), 9014–9027.
184. M. I. Rashid, E. Benhelal and M. M. Abbas, A Critical Analysis on Transformation of Renewable Energy to Green Chemicals: Opportunities and Challenges, *ChemBioEng Rev.*, 2024, **11**(2), 363–385.
185. V. Venugopal, A. Sasidharan and T. Rustad, Green Chemistry to Valorize Seafood Side Streams: An Ecofriendly Roadmap toward Sustainability, *J. Agric. Food Chem.*, 2023, **71**(46), 17494–17509.
186. C. H. Geissler, J. Ryu and C. T. Maravelias, The Future of Biofuels in the United States Transportation Sector, *Renewable Sustainable Energy Rev.*, 2024, **192**, 114276.
187. R. A. Usmani, A. S. Mohammad and S. S. Ansari, Comprehensive Biofuel Policy Analysis Framework: A Novel Approach Evaluating the Policy Influences, *Renewable Sustainable Energy Rev.*, 2023, **183**, 113403.
188. S. A. Qadir, H. Al-Motairi, F. Tahir and L. Al-Fagih, Incentives and Strategies for Financing the Renewable Energy Transition: A Review, *Energy Rep.*, 2021, **7**, 3590–3606.

189. V. H. de Mello Santos, T. L. R. Campos, M. Espuny and O. J. de Oliveira, Towards a Green Industry through Cleaner Production Development, *Environ. Sci. Pollut. Res.*, 2022, **29**(1), 349–370.
190. D. Ferraz and A. Pyka, Circular Economy, Bioeconomy, and Sustainable Development Goals: A Systematic Literature Review, *Environ. Sci. Pollut. Res.*, 2023, DOI: 10.1007/s11356-023-29632-0.
191. T. M. W. Mak, X. Xiong, D. C. W. Tsang, I. K. M. Yu and C. S. Poon, Sustainable Food Waste Management towards Circular Bioeconomy: Policy Review, Limitations and Opportunities, *Bioresour. Technol.*, 2020, **297**, 122497.
192. M. S. Pino, R. M. Rodríguez-Jasso, M. Michelin, A. C. Flores-Gallegos, R. Morales-Rodriguez, J. A. Teixeira and H. A. Ruiz, Bioreactor Design for Enzymatic Hydrolysis of Biomass under the Biorefinery Concept, *Chem. Eng. J.*, 2018, **347**, 119–136.
193. Z. Zarei, P. Malekshahi, M. H. Morowvat and A. P. Trzcinski, A Review of Bioreactor Configurations for Hydrogen Production by Cyanobacteria and Microalgae, *Int. J. Hydrogen Energy*, 2024, **49**, 472–495.
194. H. N. Murthy, K. S. Joseph, K. Y. Paek and S. Y. Park, Bioreactor Configurations for Adventitious Root Culture: Recent Advances toward the Commercial Production of Specialized Metabolites, *Crit. Rev. Biotechnol.*, 2024, **44**(5), 837–859.
195. H. A. Ruiz, M. Conrad, S. N. Sun, A. Sanchez, G. J. M. Rocha, A. Romaní, E. Castro, A. Torres, R. M. Rodríguez-Jasso, L. P. Andrade, I. Smirnova, R. C. Sun and A. S. Meyer, Engineering Aspects of Hydrothermal Pretreatment: From Batch to Continuous Operation, Scale-up and Pilot Reactor under Biorefinery Concept, *Bioresour. Technol.*, 2020, **299**, 122685.
196. O. Paladino and M. Neviani, Interchangeable Modular Design and Operation of Photo-Bioreactors for Chlorella Vulgaris Cultivation towards a Zero-Waste Biorefinery, *Enzyme Microb. Technol.*, 2024, **173**, 110371.
197. G. Di Vito Nolfi, K. Gallucci and L. Rossi, Green Diesel Production by Catalytic Hydrodeoxygenation of Vegetables Oils, *Int. J. Environ. Res. Public Health*, 2021, **18**(24), 13041.
198. X. Yao, T. J. Strathmann, Y. Li, L. E. Cronmiller, H. Ma and J. Zhang, Catalytic Hydrothermal Deoxygenation of Lipids and Fatty Acids to Diesel-like Hydrocarbons: A Review, *Green Chem.*, 2021, **23**(3), 1114–1129.
199. S. Lucantonio, A. Di Giuliano, L. Rossi and K. Gallucci, Green Diesel Production *via* Deoxygenation Process: A Review, *Energies*, 2023, **16**(2), 1–44.
200. A. K. Noriega, A. Tirado, C. Méndez, G. Marroquín and J. Ancheyta, Hydrodeoxygenation of Vegetable Oil in Batch Reactor: Experimental Considerations, *Chin. J. Chem. Eng.*, 2020, **28**(6), 1670–1683.
201. S. Alkhoori, M. Khaleel, L. F. Vega and K. Polychronopoulou, Deoxygenation of Vegetable Oils and Fatty Acids: How Can We Steer the

Reaction Selectivity towards Diesel Range Hydrocarbons?, *J. Ind. Eng. Chem.*, 2023, 36–61.

202. E. Santillan-Jimenez and M. Crocker, Catalytic Deoxygenation of Fatty Acids and Their Derivatives to Hydrocarbon Fuels *via* Decarboxylation/Decarbonylation, *J. Chem. Technol. Biotechnol.*, 2012, **87**(8), 1041–1050.
203. M. Tabatabaei, M. Aghbashlo, M. Dehhaghi, H. K. S. Panahi, A. Mollahosseini, M. Hosseini and M. M. Soufiyan, Reactor Technologies for Biodiesel Production and Processing: A Review, *Prog. Energy Combust. Sci.*, 2019, **74**, 239–303.
204. S. K. Kim, S. Brand, H. S. Lee, Y. Kim and J. Kim, Production of Renewable Diesel by Hydrotreatment of Soybean Oil: Effect of Reaction Parameters, *Chem. Eng. J.*, 2013, **228**, 114–123.
205. A. Kaewchada, N. Akkarawatkhoosith, D. Bunpim, T. Bangjang, C. Ngamcharussrivichai and A. Jaree, Production of Bio-Hydrogenated Diesel from Palm Oil Using Rh/HZSM-5 in a Continuous Mini Fixed-Bed Reactor, *Chem. Eng. Process. - Process Intensif.*, 2021, **168**, DOI: 10.1016/j.cep.2021.108586.
206. A. Omidkar, H. Xu, Z. Li, K. Haddadian and H. Song, Techno-Economic and Life Cycle Assessment of Renewable Diesel Production *via* Methane-Assisted Catalytic Waste Cooking Oil Upgrading, *J Cleaner Prod.*, 2023, **414**, 137512.
207. A. Dutta, J. A. Schaidle, D. Humbird, F. G. Baddour and A. Sahir, Conceptual Process Design and Techno-Economic Assessment of Ex Situ Catalytic Fast Pyrolysis of Biomass: A Fixed Bed Reactor Implementation Scenario for Future Feasibility, *Top. Catal.*, 2016, **59**(1), 2–18.
208. A. Pandey and Y. C. Sharma, Advancements in Biomass Valorization in Integrated Biorefinery Systems, *Biofuels, Bioprod. Biorefin.*, 2024, **18**(6), 2078–2090.
209. J. Granacher, R. Castro-Amoedo, J. Schnidrig and F. Maréchal, Closing the Balance - on the Role of Integrating Biorefineries in the Future Energy System, *Sustainable Energy Fuels*, 2023, 7(19), 4839–4854.
210. C. A. García-Velásquez, V. Aristizábal-Marulanda and C. A. Cardona, Analysis of Bioenergy Production at Different Levels of Integration in Energy-Driven Biorefineries, *Clean Technol. Environ. Policy*, 2018, **20**(7), 1599–1613.
211. V. Aristizábal M, C. A. García V and C. A. Cardona A, Integrated Production of Different Types of Bioenergy from Oil Palm Through Biorefinery Concept, *Waste Biomass Valorization*, 2016, 7(4), 737–745.
212. G. Mongkhonsiri, N. Thongchul, A. Arpornwichanop, J. H. Lee, R. Gani and S. Assabumrungrat, A Systematic Design of Integrated Palm-Oil Biorefinery Networks: Identifying Sustainable Solutions, *Sustainable Prod. Consumption*, 2023, **42**, 138–157.
213. A. C. Kokossis, M. Tsakalova and K. Pyrgakis, Design of Integrated Biorefineries, *Comput. Chem. Eng.*, 2014, **81**, 40–56.
214. G. Velvizhi, K. Balakumar, N. P. Shetti, E. Ahmad, K. Kishore Pant and T. M. Aminabhavi, Integrated Biorefinery Processes for Conversion of

Lignocellulosic Biomass to Value Added Materials: Paving a Path towards Circular Economy, *Bioresour. Technol.*, 2022, **343**, 126151.
215. T. P. C. Ezeorba, E. S. Okeke, M. H. Mayel, C. O. Nwuche and T. C. Ezike, Recent Advances in Biotechnological Valorization of Agro-Food Wastes (AFW): Optimizing Integrated Approaches for Sustainable Biorefinery and Circular Bioeconomy, *Bioresour. Technol. Rep.*, 2024, **26**, 101823.
216. L. S. dos Santos, H. Gasparetto and N. P. G. Salau, Technological Maturity and Future Perspectives for Green Diesel Production in Brazil, *Cleaner Chem. Eng.*, 2024, **10**, 100127.
217. A. G. Romero-Izquierdo, C. Gutiérrez-Antonio, F. I. Gómez-Castro and S. Hernández, Synthesis and Intensification of a Biorefinery to Produce Renewable Aviation Fuel, Biofuels, Bioenergy and Chemical Products from Jatropha Curcas Fruit, *IET Renew. Power Gen.*, 2022, **16**(14), 2988–3008.
218. H. Karan, J. Roles, B. Hankamer and I. L. Ross, Targeting Greens and Yellows: A Solar Biorefinery Analysis for the Microalgae-Based Co-Production of Pigments, Proteins, and Fuel, *Algal Res.*, 2023, **74**, 103187.
219. S. K. Karmee, A Spent Coffee Grounds Based Biorefinery for the Production of Biofuels, Biopolymers, Antioxidants and Biocomposites, *Waste Manage.*, 2018, **72**, 240–254.
220. R. Fu, L. Kang, C. Zhang and Q. Fei, Application and Progress of Techno-Economic Analysis and Life Cycle Assessment in Biomanufacturing of Fuels and Chemicals, *Green Chem. Eng.*, 2023, **4**(2), 189–198.
221. J. Mandal, R. Mitra, V. K. Gupta, N. Subramanian, Y. Kayikci and M. K. Tiwari, Optimal Allocation of Near-Expiry Food in a Retailer-Foodbank Supply Network with Economic and Environmental Considerations: An Aggregator's Perspective, *J Cleaner Prod.*, 2021, **318**, 128481.
222. I. J. Okeke, K. Sahoo, N. Kaliyan and S. Mani, Life Cycle Assessment of Renewable Diesel Production via Anaerobic Digestion and Fischer-Tropsch Synthesis from Miscanthus Grown in Strip-Mined Soils, *J Cleaner Prod.*, 2020, **249**, DOI: 10.1016/j.jclepro.2019.119358.

CHAPTER 12

Blending, Storage, and Distribution of Renewable Diesel: Production to Market

JORGE ABURTO* AND JORGE GARDUÑO

Energy Efficiency and Biofuels Division, Instituto Mexicano del Petróleo, Mexico City 07730, Mexico
*Email: jaburto@yahoo.com

12.1 Introduction

The warming of the world due to steadily increasing CO_2 levels in the atmosphere is menacing ecosystems, human activities and every kind of life. Actions for mitigation and adaptation are required to limit temperature increase to 1.5 °C above pre-industrial levels. A mitigation action to achieve this limit is the reduction of greenhouse gas (GHG) emissions before 2025 at the latest and a decline of 43% by 2030, as agreed at the Paris COP 21.[1] Countries that have signed this agreement must implement national determined contributions (NDCs) to reduce GHGs and frame multinational efforts towards long-term reductions in GHG emissions. The transport sector accounts for 15% of global GHG emissions,[2] and one strategic action for mitigation considers its decarbonization using alternative fuels as biofuels and synthetic fuels.

Recently, green or renewable diesel has become a viable alternative to blend with petroleum diesel, helping to decarbonize the transport sector using compression fuel engines, also known as diesel engines. Renewable diesel is recognized as a drop-in fuel that can be produced in standalone facilities but also by co-processing in petroleum refineries. Since its

Green Chemistry Series No. 89
Biorefining of Renewable Diesel: From Production to Market
Edited by Mohammad Aslam, Sanjeev Mishra and Jorge Aburto

Published by the Royal Society of Chemistry, www.rsc.org

chemical composition is mainly C_{12} to C_{18} linear hydrocarbons, renewable diesel may be blended with petroleum diesel if it complies with the national and international standards. In this chapter, we present the standards and considerations for blending of renewable and petroleum diesel, as well as for storage and distribution from refineries to retail gas stations.

12.2 Blending

Diesel fuel is a hydrocarbon-containing middle distillate produced from the refining of petroleum and is intended for use in compression ignition engines. Several grades have been defined in terms of sulfur content from 15 ppm to 5000 ppm (S15, S500 and S5000).[3] The specifications for blends of diesel with biodiesel (fatty acid alkyl esters) are considered in ASTM D6751 and are out of the scope of this work.[4] On the other hand, renewable diesel is mainly composed of linear hydrocarbons or paraffins with a chain-length distribution of C_{12} to C_{18}, as are also found in the middle distillate fraction known as fossil diesel.

Because of this similarity in composition, renewable diesel is considered to be a drop-in fuel and, if the blends comply with the standards ASTM D975 and EN 590 (Table 12.1), can be transported, stored and used in the same way as its fossil counterpart without special handling, maintenance and safety.[5] Nevertheless, the Mexican standard NOM-016-CRE-2016 defines the quality criteria of petroleum-based fuels such as gasoline, diesel and jet fuel in Mexico, which only allows the use of additives for lubricity and conductivity, and does not allow the addition or blending with any other substance not defined in the standard.[6]

We can now briefly outline the refining process for fossil diesel and/or the hydrotreatment co-processing of biomass oils, tallow and greases to produce renewable diesel in petroleum refineries (Figure 12.1a);[7] its transport through tanker cars, pipelines or tankers (Figure 12.1b–d) to storage; the blending of renewable diesel from standalone plants in petroleum product terminals (Figure 12.1e and f); and finally the fuel distribution through tank cars to retail gas stations (Figure 12.1g and h).

Petroleum diesel must be blended with renewable diesel at the refinery or at the petroleum product terminal before distribution through pipelines, trains, barges and tanker trucks to local petroleum product terminals and finally to retail service stations. The international standards do not establish a range or limit for blending, but the quality specifications must be met (Table 12.1). For instance, the maximum sulfur content of the blend must be 15 ppm and, since renewable diesel may content a smaller amount of sulfur, the blending may allow the production of a petroleum diesel at less stringent hydrotreatment operational conditions with less operating expense. Since sulfur-containing organic molecules act as a natural lubricant and conductivity agent, ultra-low sulfur diesel and its blends with renewable diesel require the addition of such additives before retail.

Table 12.1 International and national standards for petroleum diesel and data for renewable diesel.[4–6,8]

Property	Unit	ASTM D975	EN 590	NOM-016-CRE-2016	Renewable diesel 1	Renewable diesel 2	Renewable diesel 3
Kinematic viscosity at 40 °C	$mm^2 s^{-1}$	1.3–24.0	2–4.5	1.9–4.1	3.94	2–4	2.92
Sulfur content	ppm, $mg\,kg^{-1}$	15	10	15	3.2	<5	7
Density at 15 °C	$kg\,m^{-3}$	—	820–845	—	—	770–790	829
Cloud point (max.)	°C	—	<10 (class 0)	To inform	18	<34	−22.1
Cetane number	Dimensionless	30–40	51 (min.)	45	—	70–90	58.3
Aromaticity	vol%	35	11 (max.)	35	0.6	0	—
Flash point	°C	38–55	>55	45–60	138	>59	—
Lubricity (max.)	Microns	520	460	520	—	—	—
Conductivity	$pS\,m^{-1}$	25	—	25	—	—	—

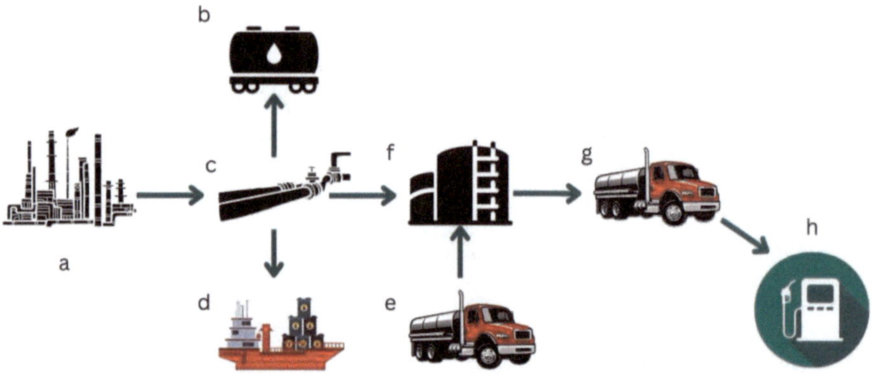

Figure 12.1 Schematic representation of fossil and renewable-diesel production, transportation, storage and distribution.

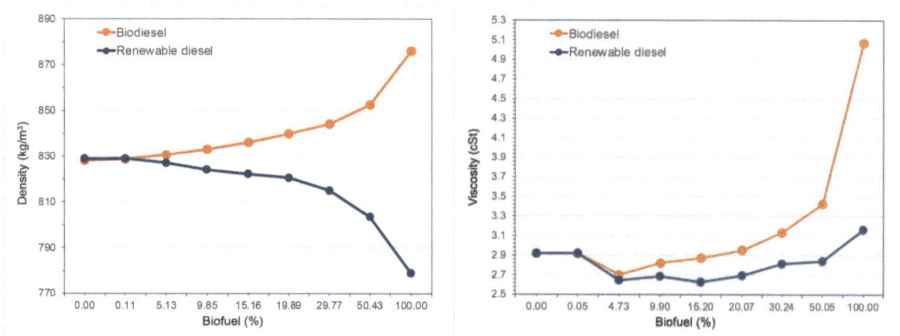

Figure 12.2 Behavior of density (left) and viscosity (right) of blends of petroleum diesel with biodiesel and renewable diesel.[10] Data from ref. 10.

Diesel engine performance depends on some key physicochemical properties of the blends like density, viscosity and cloud point, which change with the season and environmental temperature. We observed that the gradual addition of biodiesel, *i.e.* fatty acid methyl esters, to petroleum diesel increases every mentioned property (Figure 12.2), which may become a serious issue in winter by plugging hoses and fuel injection systems. On the other hand, the addition of renewable diesel diminished density, cloud point and cold filter plugging point, but does not have a significant impact on viscosity because of its similar composition to petroleum diesel. Therefore, the risk of operational issues would be mitigated with biodiesel. A more recent work uses the mixing rule to accurately predict density and viscosity according to Kay and Grumberg–Nissan expressions, respectively.[9]

Also, the use of biodiesel causes an increase in cloud point as well as in cold filter plugging point that may be also an issue during winter, but renewable diesel barely affects these properties.[10]

Hence, blending of fossil fuel with renewable diesel must comply with international and national standards to assure a good performance of the diesel engine, avoiding operational issues and guaranteeing a useful drop-in fuel with a recommended blend $\leq$50% v/v.

12.3 Storage

Renewable diesel is produced by different strategies: (1) in a standalone plant through different technologies such as hydrotreatment, hydrocracking, pyrolysis and gas-to-liquids, also known as Fischer–Tropsch; or (2) by co-processing with petroleum or its middle distillates through hydrotreatment or hydrocracking, which are current processes in petroleum refineries. Then, storage must consider two cases: the neat renewable fuel and the established blend of fossil and renewable diesel. They are both considered as drop-in fuels and storage must be considered in the state-of-the-art and must follow established standards (Table 12.2).

Table 12.2 National and international standards and codes for storage and distribution of fuels.[6,11–16]

Logistic operation	Standard	Scope
Storage	API 650 2012	Establishes minimum requirements for material, design, fabrication, erection, and testing for vertical, cylindrical, aboveground, closed- and open-top, welded storage tanks in various sizes and capacities for internal pressures approximating atmospheric pressure.
	UL 142 2018	Includes requirements that cover steel primary, secondary and diked type atmospheric storage tanks intended for noncorrosive, stable, flammable and combustible liquids that have a specific gravity not exceeding 1.0 in aboveground applications.
	UL 58 2018	Provides requirements for horizontal atmospheric steel underground tanks intended for storing flammable and combustible liquids.
	NOM-006-ASEA-2017	Defines technical specifications and criteria for industrial safety, operational safety and environmental protection for the design, construction, pre-start-up, operation, maintenance, closure and dismantling of onshore oil and petroleum storage facilities.
	NOM-001-SEDE-2012	All electrical installations inside reception, storage and delivery area of fuels.
Distribution	NOM-005-ASEA-2016	Establishes the specifications, parameters and technical requirements for industrial safety, operational safety and environmental protection that must be met in the design, construction, operation and maintenance of service stations for the storage and sale of diesel and gasoline.

In the case of new installations, it is necessary that the design step of the storage section considers a risk analysis and a consequence analysis to identify, analyze, evaluate, prioritize, and follow up on the recommendations resulting from them and mitigate the risks related to the project's own activities, using nationally and internationally accepted methodologies. For the elements and equipment that make up the installation, appropriate distribution must be observed for the products that are stored, complying with the minimum spacing established among storage and other facilities.[16] Moreover, electrical installations like equipment, materials and accessories must comply with the classification of electric areas.[14] The storage tanks may be vertical or horizontal (superficial or underground) and the area must consider at least the following items:

a. Control and measurement instrumentation in tanks
b. Land system
c. Lightning rod/lightning arrester
d. Containment dikes
e. Drains
f. Electrical installation
g. Gas and fire detection and alarm system
h. Fire system
i. Attack fronts
j. Roads and accesses.

The design of vertical tanks with fixed roof, internal floating roof and external floating roof must comply with the API 650 code,[11] while horizontal tanks must be designed and certified according to the current UL 142 and UL 58 codes.[12,13] The storage area must have reception and delivery areas that consider the following:

Reception:
a. Duct reception system
b. Measurement system
c. Unloading system for tank cars and/or tank vessels.

Delivery:
a. Pipeline delivery system
b. Pumping equipment
c. Measurement system
d. Loading system for tank cars and/or tank vessels.

12.4 Distribution

The blended renewable/fossil diesel leaves the refinery or the petroleum product terminal to smaller product terminals or retail gas stations for delivery to the consumer. Such distribution may be done by different transport systems.[17]

12.4.1 Pipeline

The connections of the delivery and measurement systems through a pipeline must be located within the property boundaries of the facility's terrestrial storage and must have containment systems for possible spills. The system must have the ability to communicate bidirectionally with the measurement systems and the controls for transfer of custody and with the facilities of the pipeline transportation system.

12.4.2 Tank Car

The area for product delivery by tank car must be made up of at least pipes, valves, filters, and the measurement and control system, which includes primary measuring instruments, transmitters and flow computer or other similar equipment for temperature, density and flow, a double block and bleed valve as well as, connections, pipes and hoses, which must be designed under national and international regulations currently applicable and be compatible with the product to be handled. All loading delivery positions must at a minimum consider loading arms and/or hoses, filtration and pumping systems, valves, connections, security system pipes and/or hoses, devices for removing air, permissive ground and safety systems.

12.4.3 Delivery by Tanker

Loading delivery arms and/or hoses must be designed in accordance with the Oil Companies International Design and Construction Specification for Marine Cargo Areas Marine Forum (OCIMF) in force or any other equivalent or higher.[17] To stop the flow in case of rupture, check valves must be installed in the product-loading lines, as well as isolation or cut-offs at the base of the equipment where the custody transfer operation is carried out, or near the approach to the dock, which may include pressure-activated valves that close automatically in case of a leak and non-return valves or other ways to prevent backflow in loading lines that are dedicated to unloading service.

12.4.4 Distribution to the Diesel Engine User

The retail gas station is the last step to deliver the diesel blend to the user of diesel engines and may have superficial or underground storage tanks with the following accessories: overfill valve, submersible pump, inventory control system, electronic annular leak detection, purge device, vapor recovery, man entrance, normal vent, emergency ventilation and emergency venting in secondary tank. The latter two accessories are mandatory for non-confined superficial tanks.[15]

A delivery system from the storage tanks to the fuel-dispatch zone is also needed, which must have a pump with an operational flow of no more than 50 liters per minute per dispatch hose for security reasons. It must be installed inside a hermetic container made of fiberglass, high-density polyethylene or other materials with UL or ULC certification, which guarantees the containment and handling of combustibles, with a wall thickness of at least 5 mm.

12.5 Concluding Remarks

The world's transport sector must undertake an energy transition from a high dependence on the fossil-fuel economy to versatile and diverse low- or negative-carbon energy sources to drastically reduce GHG emissions. The use of biofuels and synthetic fuels known as renewable diesel may contribute to this goal through blending with petroleum diesel, if national and international standards are fulfilled and some of them amended to allow the use of such alternative fuels. With the myriads of feedstocks and technologies to produce renewable diesel, its chemical composition being mostly based on linear hydrocarbons or paraffins makes it a viable fuel or drop-in fuel to blend with petroleum diesel in existing facilities of the downstream petroleum infrastructure, petroleum product terminals and retail gas stations that employ current transport systems such as tankers, pipelines and tank cars to transport fuel to the final users of compression diesel engines.

Acknowledgements

The authors acknowledge the Instituto Mexicano del Petróleo for its support.

References

1. United Nations, Paris Agreement, Paris: United Nations; 2015. Available from: https://unfccc.int/sites/default/files/english_paris_agreement.pdf.
2. Environmental Protection Agency (EPA), Global Emissions and Removals by Gas. USA: EPA. Available from: https://www.epa.gov/ghgemissions/global-greenhouse-gas-overview.
3. ASTM D975-17, Standard specifications for diesel fuel oils. 2017.
4. ASTM D6751-20, Specification for Biodiesel Fuel Blend Stock (B100) for Middle Distillate Fuels, 2020.
5. European standard, EN 590, Automotive fuels - Diesel - Requirements and test methods, 2009.
6. Norma Oficial Mexicana, NOM-016-CRE-2016, Especificaciones de calidad de los petrolíferos, 2016.
7. Agencia de Seguridad, Energía y Ambiente (ASEA), Cadena de valor de los hidrocarburos, México: ASEA. Available from: http://104.209.210.233/gobmx/repositorio/DIAGNOSTICO_DE_ACCESIBILIDAD/UNR/Cadena_de_valor_Modificacion_202112.pdf.

8. J. Aburto and M. A. Amezcua-Allieri, Biodiesel and Green Diesel Fuels: A Techno-Economic Analysis, in *Green Diesel: An Alternative to Biodiesel and Petrodiesel*, ed. M. Aslam, S. Shivaji Maktedar and A. K. Sarma, Advances in Sustainability Science and Technology, Springer, Singapore, 2022. Available from: DOI: 10.1007/978-981-19-2235-0_11.
9. L. F. Ramírez-Verduzco and M. J. Hernández-Sánchez, Blends of Green Diesel (Synthetized from Palm Oil) and Petroleum, *Bioenergy Res.*, 20212021, **14**, 1002–1013.
10. J. Rodríguez-Fernández, J. J. Hernández, A. Calle-Asensio, A. Ramos and J. Barba, Selection of blends of diesel fuel and advanced biofuels based on their physical and thermochemical properties, *Energies*, 2019, **12**(11), 2034.
11. American Petroleum Institute (API), API 650 Welded Tanks for Oil Storage, 2012.
12. UL 142 2018, Standard for Steel Aboveground Tanks for Flammable and Combustible Liquids.
13. UL 58 2018, Standard for Steel Underground Tanks for Flammable and Combustible Liquids.
14. Norma Oficial Mexicana. NOM-001-SEDE-2012. Instalaciones eléctricas (Utilización).
15. Norma Oficial Mexicana, NOM-005-ASEA-2016, Especificaciones y criterios técnicos de seguridad industrial, seguridad operativa y protección al medio ambiente para el diseño, construcción, mantenimeinto, cierre y desmantelamiento de las instalaciones terrestres de almacenamiento de petrolíferos y petróleo, excepto gas licuado de petróleo, 2016.
16. Norma Oficial Mexicana, NOM-006-ASEA-2017, Diseño, construcción, operacióh y mantenimiento de estaciones servicio para el almacenamiento y expendio de diesel y gasolina, 2017.
17. Oil Companies International Marine Forum (OCIMF), Guidelines for the Handling, Storage, Use, Maintenance and Testing of STS Hoses, London: OCIMF. Available from: https://www.ocimf.org/publications/information-papers/guidelines-for-the-handling,-storage,-use,-maintenance-and-testing-of-sts-hoses.

CHAPTER 13

Renewable-diesel Industry: A Real-world Scenario

PAULO GONÇALVES DE SOUSA JUNIOR,[a] KAIANY MOREIRA DOS SANTOS,[a] PATRICK DA SILVA SOUSA,[b] FRANCISCO SIMÃO NETO[b] AND JOSÉ CLEITON SOUSA DOS SANTOS*[a]

[a] Instituto de Engenharias e Desenvolvimento Sustentável, Universidade da Integração Internacional da Lusofonia Afro-Brasileira, Campus das Auroras, Redenção, CEP 62790970, CE, Brasil; [b] Departamento de Engenharia Química, Universidade Federal do Ceará, Campus do Pici, Bloco 709, Fortaleza, CEP 60455760, CE, Brasil
*Email: jcs@unilab.edu.br

13.1 Introduction

The global demand for more sustainable energy solutions has driven significant shifts in the world's energy matrix, emphasizing replacing fossil fuels with renewable alternatives.[1] In this context, renewable diesel emerges as a critical component, offering a cleaner and more efficient energy source that combines high technical performance with reduced greenhouse-gas (GHG) emissions.[2] Unlike conventional biodiesel, which is produced *via* transesterification, renewable diesel is obtained through hydrotreatment or pyrolysis processes, resulting in a product chemically similar to fossil diesel but derived from renewable feedstocks such as vegetable oils, animal fats, and organic waste.[3]

The energy transition required to mitigate the impacts of climate change demands an integrated approach that considers the technical, economic,

Green Chemistry Series No. 89
Biorefining of Renewable Diesel: From Production to Market
Edited by Mohammad Aslam, Sanjeev Mishra and Jorge Aburto

Published by the Royal Society of Chemistry, www.rsc.org

and environmental feasibility of emerging technologies.[4,5] Renewable diesel stands out in this context for it can directly replace fossil diesel in conventional engines without requiring modifications, leveraging existing infrastructure.[6] Furthermore, it contributes to decarbonizing of strategic sectors such as heavy transport, aviation, and maritime industries, which face significant challenges in adopting electrification or hydrogen technologies on a large scale.[7] Moreover, renewable diesel presents several advantages for its implementation. Figure 13.1 highlights some of them, from the economic aspects to the environmental impact of this energy transition.

This chapter aims to provide a comprehensive analysis of the role of renewable diesel in the global energy landscape. To achieve this, real-world case studies will be used to illustrate the successful production and use of renewable diesel in various countries, alongside corporate and governmental initiatives driving its implementation. Additionally, the chapter will discuss the technical, economic, and logistical challenges of its large-scale adoption, including feedstock costs, competition with subsidized fossil fuels, and regulatory barriers. Finally, it seeks to contextualize renewable diesel as a transitional and complementary solution in pursuing a more sustainable energy future. It highlights its potential to contribute to the Sustainable Development Goals (SDGs) and global climate neutrality commitments.

By addressing these aspects, this chapter aims not only to clarify the opportunities presented by renewable diesel but also to foster critical and informed discussions on the strategies required to overcome adoption challenges, solidifying its relevance within the context of the global energy transition.

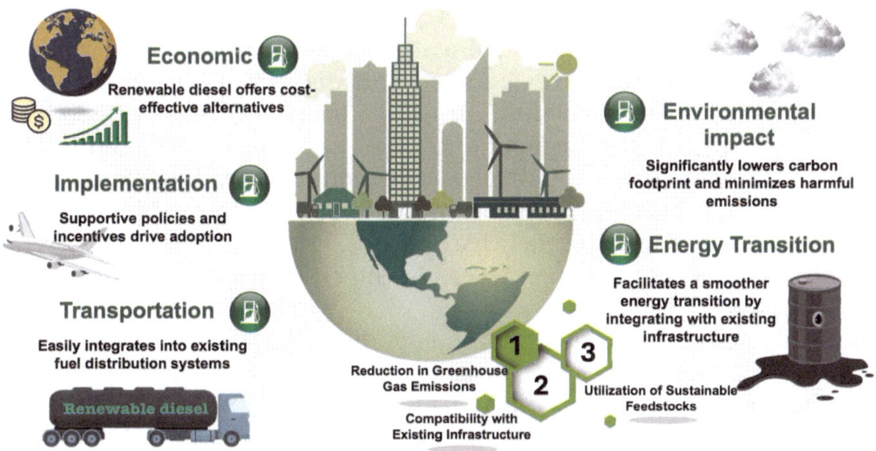

Figure 13.1 Some advantages of renewable diesel and its implementation.

13.2 Case Studies of Industrial Production Plants

13.2.1 Industrial-scale Production: Diamond Green Diesel (DGD)

Diamond Green Diesel (DGD) is a leading facility in the USA that produces renewable diesel. It is located in Norco, Louisiana.[8] This plant uses HEFA technology (hydrotreatment of esters and fatty acids), which makes it a recognized efficient synthesis for large-scale production using raw materials such as waste oils and fats.[8]

DGD currently operates with an installed production capacity of 1.1 billion gallons per year and can be listed among the most prominent renewable-diesel facilities globally.[8] The HEFA technology applied at these facilities consists of converting oils and fats into a fuel like fossil diesel, making it possible to apply it to the distribution chains already existing for conventional diesel.[9] In addition, DGD has a favourable geolocation, which allows easy integration between its facilities and refineries in the region. Thus, logistical and operational advantages and the ability to optimise and reduce transport costs and time are highlighted.

In DGD's production of renewable diesel, different raw materials are used to diversify its sources of resources and ensure a lower environmental impact.[10] Among the materials used are waste cooking oils, commonly collected from restaurants and even households, which leads to reduced waste and irregular disposal of these oils.[11] Animal fats that are a by-product of the meat industry, such as beef tallow or pork fat, are also used in this plant.[12] In addition, vegetable oils are also used, which can vary according to their availability, which corroborates the need to maintain diversified sources of raw materials.[13]

Among the consumers of the renewable diesel produced by DGD is the European Union, which is motivated by environmental legislation such as the Renewable Energy Directive II (RED II), which requires member countries to reduce emissions from the transportation sector. As mentioned, renewable diesel can be used and distributed using the same infrastructure as conventional diesel, so achieving gas-emission reduction targets becomes cheaper and faster with this new biofuel.[14] This case of use and distribution attracts other countries such as Canada, the United Kingdom, Ireland and Latin American countries, which are also growing markets due to local climate targets. Overall, renewable diesel is a highly competitive biofuel even in markets with high regulatory stringency.[11,13,15]

Regarding environmental and economic impacts, the technology used by DGD reduces greenhouse-gas emissions by 50–80% compared to diesel, a petroleum by-product.[8] In addition, using sustainable raw materials reduces improper disposal and transforms discarded materials into materials with high added value.[16] The DGD plant has the potential for positive economic impact, as it generates jobs, promotes the circular economy by reusing materials that would otherwise be discarded, and is internationally competitive due to its numerous consumer partners outside the US.[17]

13.2.2 Brazilian Renewable-diesel Project: RenovaBio

RenovaBio, established by the National Biofuels Policy (Law No. 13.576/2017), is a project aimed at increasing the production and use of biofuels in Brazil to reduce the emission of gases that aggravate the greenhouse effect.[18] Within its scope, this project promotes sustainability in the Brazilian energy sector and ensures Brazil's growth with a focus on global leadership in biofuel production. All these goals are based on integrating various public energy-development policies.[18]

These policies aim to reduce the use of conventional diesel and replace it with renewable diesel, so the RenovaBio project aligns with the Brazilian goals set out in the Paris Agreement.[19] In addition, this project is responsible for implementing decarbonization credits in the country, where producers or distributors are rewarded for using, producing or distributing renewable diesel as an alternative to petroleum diesel. In addition, this project strengthens the National Program for the Production and Use of Biodiesel (PNPB), as it reinforces the production chain for renewable diesel and allows for an increase in the percentage of biofuel mixed with conventional fuel sold to the end consumer.[20]

To ensure an increase in renewable-diesel production, numerous raw materials must be considered.[21] In this way, RenovaBio reinforces and encourages production from all available biomass and oils or animal fat that might otherwise be discarded, thus guaranteeing reuse and increasing the added value of this material.[18] This approach promotes the circular economy, reduces waste and diversifies Brazil's energy matrix, reducing the pressure on specific crops such as sugar cane.[22]

Brazil already dominates the biofuel industry and processing, as it excels in producing ethanol and biodiesel, so renewable diesel is emerging as a promising alternative for the country. Cargo in Brazil is mainly distributed by road, so heavy road transport, where the transition to electrification still faces technical and economic challenges, has great potential to absorb local biofuel production. The introduction of RenovaBio reduces dependence on fossil fuels, improves air quality and increases energy security.[18] Therefore, the expansion of renewable diesel in the domestic market strengthens Brazil's competitiveness in the renewable energy sector and contributes to cleaner and more efficient transportation.

13.3 Real Applications of Renewable Diesel

13.3.1 Road Transport

Due to its chemical similarities with traditional diesel, renewable diesel can be applied directly to existing engines without requiring any changes, which allows its use on a large scale.[23] Its use in the road transport sector focuses on fleets of urban buses and heavy-load trucks.[24] Concerning urban buses, diesel has been used to reduce local emissions of pollutants such as particulate matter (PM) and nitrogen oxides (NO_x), helping to improve air quality in regions with high population density.[25] In cargo transport, the fuel appears as an efficient option to reduce greenhouse gas emissions, preserving the performance and efficiency of large vehicles.[26]

In São Paulo, Brazil, a case study was conducted to partially replace fossil diesel with renewable diesel in urban public transport. Throughout testing, bus fleets used renewable-diesel blends that contained 30% hydrogenated vegetable oil.[27] The results indicated a decrease of around 70% in greenhouse-gas emissions, considering the complete fuel cycle and possible improvements in air quality due to the reduction of pollutants.[25]

Although renewable fuel has a higher initial cost, the environmental and social advantages, such as reducing respiratory diseases caused by urban pollutants, justified the project continuity and expansion of the project.[28] The action aligns with the National Biofuels Policy's (RenovaBio) objectives: to encourage a more ecological and sustainable means of transport in the national metropolis.[25]

13.3.2 Aviation Transport

Sustainable aviation fuel (SAF) has proven to be a promising option for decarbonizing the airline sector, contributing a considerable portion of global GHG emissions.[29] SAF, produced from renewable resources such as vegetable oils, animal fats, and waste through processes such as hydrogenation or the Fischer–Tropsch process, has chemical properties similar to traditional aviation kerosene (JetA-1).[30,31] Therefore, it is directly applicable to aircraft without requiring changes to engines or fuel infrastructure, providing an effective and practical solution to reduce emissions in the sector.[32]

In several countries, using renewable diesel in aviation is already a reality, with airlines carrying out commercial and demonstration flights using a combination of SAF and conventional kerosene.[33] A notable case was the United Airlines flight in 2021, in which a commercial plane flew with 100% SAF in one of the engines, highlighting remarkable technical progress.[34,35] Air France and Lufthansa have used up to 50% SAF daily, demonstrating its technical and ecological viability.[36,37] The results of these flights indicate that the partial exchange of fossil kerosene can reduce carbon emissions by up to 50%, depending on the origin and manufacturing process of the fuel, without compromising the safety or performance of the planes.[36,38]

The use of SAF provides significant advantages for reducing emissions in the airline sector, which currently contributes around 2% to 3% of global CO_2 emissions.[39] Among the beneficial consequences, the reduction in net carbon emissions by up to 80% during the complete cycle, from manufacturing to final use, is notable.[40] Additionally, SAF helps reduce regional pollutants, such as fine particles of sulfur oxides (SO_x), improving air quality in regions close to airports.[41] Another advantage is encouraging a circular economy that uses agricultural waste and by-products as inputs, promoting economic and ecological sustainability.[40]

13.3.3 Maritime Transport

Converting engines to renewable diesel in the maritime sector requires technical modifications to improve fuel combustion and increase efficiency.[42] Modern marine engines, particularly those produced to operate on low-sulfur fuels such as marine diesel, are already compatible with renewable diesel without significant changes.[43] Furthermore, renewable diesel presents technical benefits, such as excellent thermal stability, reduced waste generation, and reduced wear on engine components.[44]

A study was conducted in Sweden, where medium-sized vessels underwent operational tests with renewable-diesel blends.[45] The initiative covered cargo and passenger ships on coastal routes, areas where environmental sustainability is highly appreciated.[45] In the tests, combinations of up to 50% renewable diesel with fossil marine fuels were used.[45] The results indicated that the switch to diesel from renewable sources reduced carbon dioxide emissions by up to 80%, in addition to considerably reducing the release of local pollutants such as sulfur oxides and solid particles.[45–47]

The example of Sweden demonstrates how renewable diesel can be effectively incorporated into maritime transport, helping to decarbonize the sector and achieve global sustainability objectives, such as those defined by the International Maritime Organization (IMO).[48,49] This sustainable fuel reduces maritime transport's environmental impact while maintaining vessels' competitiveness and operational safety.[50]

13.4 Challenges Faced in Adopting Renewable Diesel

The shift to renewable energy, often known as the "decarbonization of the energy sector", remains a contentious topic among policymakers and economic stakeholders.[51] It represents a significant opportunity to achieve sustainable development goals. Renewable diesel is among the candidates with a high potential for replacing fossil diesel due to its relevant role in the global energy transition.[52–54] On the other hand, adopting its use on a large scale has faced barriers due to economic, technical, political and regulatory challenges.[55–57] Because of this, it is necessary to explain some of the difficulties encountered in each aspect mentioned above (Table 13.1).

13.4.1 Economic

The high cost of producing renewable diesel is one of the main bottlenecks for the large-scale implementation of its use and commercialization.[58–60] This is because this cost is related to the complexity of the technological matrix required to convert renewable raw materials (vegetable oils and animal fats) into

Table 13.1 Key barriers to the implementation of renewable diesel.

Challenges in renewable diesel adoption: economic, technical, political, and regulatory perspectives			
Economic	Technical	Political	Regulatory
High production costs compared to fossil diesel.	Need for adaptation of engines and distribution infrastructure.	Low prioritization in public policies for energy transition in some countries.	Lack of standardization in regulations for renewable diesel.
Dependency on government subsidies for economic viability.	Chemical instability under certain storage conditions.	Resistance from sectors linked to oil and gas.	Slow bureaucratic processes for approving and licensing new production plants.
Competition for raw materials with other sectors (food, cosmetics, *etc.*).	Lower efficiency in some cases due to properties differing from fossil diesel.	Absence of a national or regional strategic plan for energy transition.	Trade barriers and restrictions on importing/exporting raw materials and technologies.
Lack of tax incentives for producers and distributors.	Limited availability of optimized technologies for large-scale production.	Lack of coordination between states and municipalities in implementing renewable diesel policies.	Insufficient technical standards to ensure the quality and reliability of the final product.
	Complexity in obtaining suitable and consistent-quality raw materials.		

fuels with the desired properties equivalent to or better than those of the diesel commonly sold.[61–63] It is important to highlight relevant obstacles regarding seasonality and the quantity and availability of raw materials throughout the year, as well as competition between sectors.[64–66]

Another challenge to be overcome lies in the latent competitiveness in the fuel market. Diesel from non-renewable sources remains largely subsidized in several countries, making introducing alternative and renewable energy sources more difficult.[67–69] Furthermore, the high price of renewable diesel discourages its expansion in cost-sensitive sectors (freight transportation and agriculture).[70] The predicament is exacerbated by the reality that numerous economies lack robust carbon pricing, which does not consider the ecological benefits of renewable diesel at the comprehensive expense of fossil fuels.[71–73]

Ultimately, expanding renewable-diesel production poses considerable economic difficulty. The need for renewable feedstocks may compete with other sectors, such as the food industry, exerting pressure on costs.[74–76] In this framework, allocating resources to research and development is crucial to broaden feedstock options, including utilizing agro-industrial waste and algae.[77–79] Addressing these economic obstacles necessitates a comprehensive strategy, merging governmental, academic, and private sector initiatives to facilitate the shift to renewable fuels in a sustainable and economically advantageous way.

13.4.2 Technical

Technical challenges limit the widespread adoption and production of renewable diesel.[80] A significant issue is sourcing suitable raw materials.[81] While oily byproducts, animal fats, and plant-based oils are available, their quality and composition can vary significantly.[82] This variation demands the implementation of advanced technologies for purification and refinement, as it directly influences the efficiency of the conversion processes and the calibre of the product.[83]

Manufacturing processes face technical challenges that must follow specific protocols to maximize yields and reduce unwanted byproducts.[84,85] For example, hydrogenation is used to produce renewable diesel but requires high pressures, temperatures, and expensive noble-metal catalysts sensitive to contaminants.[86–89] Moreover, a significant concern that elevates operational expenses and complicates management is the degradation of catalysts throughout the process.[90,91]

Another essential technical consideration is whether renewable fuel can be utilized with existing engines. Although this fuel and fossil diesel share numerous chemical similarities, slight compositional differences may influence the durability and efficiency of conventional engines, especially in older vehicles.[92–94] Establishing specific standards and rigorous quality assessment is required to tackle this issue and ensure the interchangeability and comparable performance of the two fuel types.[95–97]

Therefore, the scalability of frameworks for producing and distributing renewable fuels is a significant concern. Modifying refineries, pipelines, and storage systems is necessary since the current fossil-diesel infrastructure cannot be entirely compatible with renewable fuels.[98–100] Managing the storage and transportation of renewable diesel is essential to prevent oxidation and the formation of deposits, which can compromise its stability.[101–103] Industry, academia, and public officials must collaborate to develop reliable, effective, and cost-efficient solutions to address these technical challenges.[104–106]

13.4.3 Political and Regulatory

Complex political and regulatory barriers to the use of renewable diesel make it challenging to create industry-wide rules that are both logical and efficient.[107–110] The lack of transparent laws governing the production and use of renewable fuels limits the ability to make strategic investments and plans in many countries.[111–113] Frequent changes to laws and government directives can create regulatory uncertainty, which can hinder the development of long-term projects and discourage the investments needed to expand the market for renewable diesel.[114–116]

Moreover, sustainable options encounter more significant challenges in the marketplace owing to the existing subsidies for fossil diesel and other traditional fuels.[117–119] Notwithstanding its ecological benefits, renewable diesel is pricier than fossil diesel due to production and infrastructure costs.[120,121] These subsidies often keep fossil-fuel prices at unnaturally low levels.[122,123] As the external costs linked to greenhouse-gas emissions are not reflected in the market, renewable fuels cannot achieve economic feasibility due to the lack of a substantial carbon price that accounts for the environmental impact of fossil diesel.[124–127]

Achieving political unity among various levels of government is essential for implementing public initiatives that promote renewable fuels such as renewable diesel.[128–130] In many countries, energy and environmental policies are fragmented, with differing federal, state, and municipal rules and incentives.[131–133] Because of this fragmentation, implementing renewable diesel might become even more complex, as effective policies in one region may be jeopardized by conflicting or out-of-date rules in other regions.[134–136]

The absence of a cohesive strategy to promote research and innovation in the biofuel domain presents a considerable challenge.[137–139] Although specific initiatives exist to stimulate biofuel production, most are sporadic and narrow in focus, lacking continuous backing for research into novel conversion technologies and feedstock options. To address these obstacles, legislators must implement a holistic approach that incorporates financial incentives, specific subsidies, and research funding to guarantee a stable regulatory framework that fosters the growth of the renewable-diesel industry.

13.5 Strategies to Overcome Challenges

13.5.1 Economic Incentives

Subsidies play a crucial role in reducing the cost of renewable diesel and stimulating its production and adoption.[140] For producers, countries like the United States lead with initiatives such as the Renewable Fuel Standard (RFS), which provides financial credits (RINs, or Renewable Identification Numbers) to encourage renewable fuel production.[141] In the European Union, direct subsidies and tax exemptions for biofuel producers are part of the Renewable Energy Directive II (RED II), establishing mandatory targets for renewable fuel use in transportation.[142] For consumers, programs such as California's Low Carbon Fuel Standard (LCFS) offer financial benefits, lowering the retail price of renewable diesel through incentives based on fuel carbon intensity.[143]

Carbon credits have also gained relevance as a tool to promote sustainable fuels. These credits, generated by reducing emissions, can be traded in markets, creating a financial reward for adopting low-carbon technologies.[144] The European Union Emissions Trading System (EU ETS) and voluntary markets, such as those promoted by the Climate Action Reserve in the US, provide opportunities for producers and users of renewable diesel to monetize its environmental benefits, further incentivizing investment in the sector.[145]

13.5.2 Technological Innovation

Technological efforts to reduce the costs of renewable-diesel production are advancing globally. Large-scale refineries, such as Neste's, continuously invest in optimizing hydrotreatment processes to maximize the use of waste materials as feedstocks, thereby reducing associated costs. Emerging technologies, such as more efficient bioreactors and advanced conversion methods like pyrolysis and gasification, are being developed to diversify feedstocks and scale production.[146]

A promising approach is the integration of renewable diesel with green hydrogen, creating hybrid fuels. This combination leverages the high energy density of renewable diesel and the carbon neutrality of green hydrogen.[147] Pilot projects in countries such as Germany and Japan have tested this technology for heavy transport, yielding promising emission reductions and technical performance results. Furthermore, public–private partnerships are establishing research centres dedicated to integrating renewable energy into industrial clusters, enhancing energy efficiency and lowering production costs.[148]

13.5.3 Public Policies

Government mandates have been fundamental in driving demand for renewable diesel. In the United States, the Renewable Fuel Standard requires an increasing percentage of renewable fuels to be blended with fossil fuels. At the same time, Brazil's RenovaBio program sets decarbonization targets for fuel distributors, incentivizing the domestic use of renewable diesel.[149,150] In the European Union, RED II includes similar obligations, such as achieving 14% of transportation energy consumption from renewable sources by 2030.[151,152]

Internationally, efforts to standardize regulations include initiatives like the International Sustainability and Carbon Certification (ISCC), which provides global criteria for producing and trading renewable fuels. Partnerships among OECD countries and organizations such as the International Renewable Energy Agency (IRENA) promote harmonization of technical and environmental standards, facilitating the trade and adoption of renewable diesel in global markets.[153] These examples underscore the importance of consistent and well-structured public policies in accelerating the energy transition.[154]

13.6 Conclusion

The renewable-diesel industry has made significant strides in recent years, driven by advances in feedstock diversification, refining technologies, and policies promoting sustainable-energy solutions. These developments underscore the growing feasibility of large-scale production and integration of renewable diesel into existing energy markets. As a low-carbon alternative to fossil fuels, renewable diesel is critical in the global transition toward decarbonization, particularly in hard-to-abate sectors such as transportation and heavy industry.

However, realizing the full potential of renewable diesel requires addressing key challenges, including feedstock availability, cost reduction, and the development of robust supply chains. Achieving these goals will depend on enhanced collaboration among governments, industries, and academic institutions. Governments must establish supportive regulatory frameworks and financial incentives, industries need to accelerate innovation and investment in sustainable practices, and academia should continue to advance research on alternative feedstocks and production efficiency. By fostering such synergies, the renewable-diesel industry can become a cornerstone of global efforts to combat climate change, contributing to a cleaner, more-sustainable energy future.

Acknowledgements

This study is supported in part by financial support from the following Brazilian agencies for scientific and technological development: Fundação Cearense de Apoio ao Desenvolvimento Científico e Tecnológico (FUNCAP) (PNE-0112-00048.01.00/16, PS1-0186-00216.01.00/21, PS1-00186-00255.01.00/21), Conselho Nacional de Desenvolvimento Científico e Tecnológico (CNPq) (311062/2019-9, 308452/2022-4), and Coordenação de Aperfeiçoamento de Ensino Superior (CAPES) (Finance Code 001, PROEX 23038.000509/2020-82).

References

1. H. Pitsch, D. Goeb, L. Cai and W. Willems, Potential of Oxymethylene Ethers as Renewable Diesel Substitute, *Prog. Energy Combust. Sci.*, 2024, **104**, 101173.
2. Y. Li, H. Xu, D. Northrup and M. Wang, Effects of Soybean Varieties on Life-cycle Greenhouse Gas Emissions of Biodiesel and Renewable Diesel, *Biofuels, Bioprod. Biorefin.*, 2023, **17**(3), 449–462.
3. A. G. M. B. Mustayen, M. G. Rasul, X. Wang, M. Negnevitsky and J. M. Hamilton, Remote Areas and Islands Power Generation: A Review on Diesel Engine Performance and Emission Improvement Techniques, *Energy Convers. Manage.*, 2022, **260**, 115614.
4. R. Koul, N. Kumar and R. C. Singh, A Review on the Production and Physicochemical Properties of Renewable Diesel and Its Comparison with Biodiesel, *Energy Sources, Part A*, 2021, **43**(18), 2235–2255.

5. A. A. V. Julio, T. S. Milessi, E. A. Ocampo Batlle, E. E. Silva Lora, D. M. Yepes Maya and J. C. Escobar Palacio, Techno-Economic and Environmental Potential of Renewable Diesel as Complementation for Diesel and Biodiesel in Brazil: A Comprehensive Review and Perspectives, *J. Cleaner Prod.*, 2022, **371**, 133431.
6. E. Kordouli, S. Lycourghiotis, K. Bourikas, A. Lycourghiotis and C. Kordulis, Renewable Diesel Synthesis by Hydro-Processing in Green Solvents, *Curr. Opin. Green Sustainable Chem.*, 2024, **48**, 100936.
7. H. Xu, U. Lee and M. Wang, Life-Cycle Energy Use and Greenhouse Gas Emissions of Palm Fatty Acid Distillate Derived Renewable Diesel, *Renewable Sustainable Energy Rev.*, 2020, **134**, 110144.
8. Diamond Green Diesel, *Renewable fuel for a low-carbon world.* https://www.diamondgreendiesel.com/ (accessed 2024-11-30).
9. K. Lahti-Leikas, E. Niemistö, H. Talvenmäki, N. Saartama, Y. Sun, L. Mercier and M. Romantschuk, Hydrotreated Vegetable Oil Migrates through Soil and Degrades Faster than Fossil Diesel and Hydrotreated Vegetable Oil-Fossil Diesel Blend, *Environ. Sci. Pollut. Res. Int.*, 2024, **31**(40), 53177–53192.
10. A. Giuliano, E. Catizzone and C. Freda, Process Simulation and Environmental Aspects of Dimethyl Ether Production from Digestate-Derived Syngas, *Int. J. Environ. Res. Public Health*, 2021, **18**(2), 1–21.
11. J. Feng, Y. Li, T. J. Strathmann and J. S. Guest, Characterizing the Opportunity Space for Sustainable Hydrothermal Valorization of Wet Organic Wastes, *Environ. Sci. Technol.*, 2024, **58**(5), 2528–2541.
12. K. B. Bhumula and G. N. Kumar, Using CRITIC-TOPSIS and Python to Examine the Effect of 1-Hepatnol on the Performance and Emission Characteristics of CRDI CI Engine with Split Injection, *Heliyon*, 2024, **10**(11), e31484.
13. H. H. Fayyaz, M. A. Mujtaba, S. Jahangir, S. Imran, M. A. Ijaz Malik, H. Fayaz, C. A. Saleel, U. Hassan, S. Quershi and H. farooq, A Comprehensive Study on the Performance and Emission Analysis in Diesel Engine *via* Optimization of Novel Ternary Fuel Blends: Diesel, Manganese, and Diethyl Ether, *Heliyon*, 2023, **9**(10), e21133.
14. S. Eswaran, S. Subramaniam, S. Geleynse, K. Brandt, M. Wolcott and X. Zhang, Dataset for Techno-Economic Analysis of Catalytic Hydrothermolysis Pathway for Jet Fuel Production, *Data Brief*, 2021, **39**, 107513.
15. K. R. Landwehr, R. Mead-Hunter, R. A. O'Leary, A. Kicic, B. J. Mullins and A. N. Larcombe, Respiratory Health Effects of In Vivo Sub-Chronic Diesel and Biodiesel Exhaust Exposure, *Int. J. Mol. Sci.*, 2023, **24**(6), 5130.
16. M. Dhanarasu, K. A. RameshKumar and P. Maadeswaran, Effect of Acetone as an Oxygenated Additive with Used Sunflower Oil Biodiesel on Performance, Combustion and Emission in Diesel Engine, *Environ. Technol.*, 2022, **43**(24), 3682–3692.

17. B. Das, P. V. Bhave, S. P. Puppala, S. Adhikari, S. Sainju, E. Mool and R. M. Byanju, Emission Factors and Emission Inventory of Diesel Vehicles in Nepal, *Sci. Total Environ.*, 2022, **812**, 152539.
18. RenovaBio, *National Agency for Petroleum, Natural Gas and Biofuels*. https://www.gov.br/anp/pt-br/assuntos/renovabio (accessed 2024-11-30).
19. Q. Liu, G. Xu, X. Wang, X. Liu and X. Mu, Highly Selective Upgrading of Biomass-Derived Alcohol Mixtures for Jet/Diesel-Fuel Components, *ChemSusChem*, 2016, **9**(24), 3465–3472.
20. M. M. Khan, S. M. M. Hasnain, A. K. Kadian, R. P. Sharma, M. Abbas and S. Pandey, Reduction of Fossil Fuel Usage and Emissions of Diesel Engines by the Application of Quaternary Blends and Varied Fuel Injection Pressures, *ACS Omega*, 2024, **9**(24), 26388–26399.
21. U. Rajak, P. K. Chaurasiya, T. N. Verma, A. Dasore, U. Ağbulut, K. Meshram, C. Saleel, S. Saboor, E. Cuce and Z. Mian, Optimizing Soybean Biofuel Blends for Sustainable Urban Medium-Duty Commercial Vehicles in India: An AI-Driven Approach, *Environ. Sci. Pollut. Res. Int.*, 2024, **31**(22), 32449–32463.
22. A. Anderson, V. Kumar M, B. Gavurová, H. Li, C. Xia and X. Zhang, Optimizing Engine Performance and Reducing Emissions of Greenhouse Gases through Spirulina Microalgae and Nano-Additive Blends, *Environ. Res.*, 2023, **231**(Pt 1), 115958.
23. A. Dimitriadis, I. Natsios, A. Dimaratos, D. Katsaounis, Z. Samaras, S. Bezergianni and K. Lehto, Evaluation of a Hydrotreated Vegetable Oil (HVO) and Effects on Emissions of a Passenger Car Diesel Engine, *Front. Mech. Eng.*, 2018, **4**, 7.
24. H. Xu, L. Ou, Y. Li, T. R. Hawkins and M. Wang, Life Cycle Greenhouse Gas Emissions of Biodiesel and Renewable Diesel Production in the United States, *Environ. Sci. Technol.*, 2022, **56**(12), 7512–7521.
25. P. Gerber Machado, A. C. Rodrigues Teixeira, F. Mendes de Almeida Collaço, A. Hawkes and D. Mouette, Assessment of Greenhouse Gases and Pollutant Emissions in the Road Freight Transport Sector: A Case Study for São Paulo State, Brazil, *Energies*, 2020, **13**(20), 5433.
26. A. Bradshaw and G. de Martino Jannuzzi, Governing Energy Transitions and Regional Economic Development: Evidence from Three Brazilian States, *Energy Policy*, 2019, **126**, 1–11.
27. F. M. de A. Collaço, L. P. Dias, S. G. Simoes, T. Pukšec, J. Seixas and C. Bermann, What If São Paulo (Brazil) Would like to Become a Renewable and Endogenous Energy-Based Megacity?, *Renewable Energy*, 2019, **138**, 416–433.
28. B. F. Honorato de Oliveira, L. F. de França, N. C. Fernandes Corrêa, N. F. da P. Ribeiro and M. Velasquez, Renewable Diesel Production from Palm Fatty Acids Distillate (PFAD) *via* Deoxygenation Reactions, *Catalysts*, 2021, **11**(9), 1088.
29. J. P. Ahire, R. Bergman, T. Runge, S. H. Mousavi-Avval, D. Bhattacharyya, T. Brown and J. Wang, Techno-Economic and

Environmental Impacts Assessments of Sustainable Aviation Fuel Production from Forest Residues, *Sustainable Energy Fuels*, 2024, **8**(19), 4602–4616.
30. P. H. Chen, U. Lee, X. Liu, H. Cai and M. Wang, Life-cycle Analysis of Sustainable Aviation Fuel Production through Catalytic Hydrothermolysis, *Biofuels, Bioprod. Biorefin.*, 2024, **18**(1), 42–54.
31. A. Midilli, H. Kucuk, M. E. Topal, U. Akbulut and I. Dincer, A Comprehensive Review on Hydrogen Production from Coal Gasification: Challenges and Opportunities, *Int. J. Hydrogen Energy*, 2021, **46**(50), 25385–25412.
32. D. Chiaramonti, Sustainable Aviation Fuels: The Challenge of Decarbonization, *Energy Procedia*, 2019, **158**, 1202–1207.
33. N. A. Huq, G. R. Hafenstine, X. Huo, H. Nguyen, S. M. Tifft, D. R. Conklin, D. Stück, J. Stunkel, Z. Yang, J. S. Heyne, M. R. Wiatrowski, Y. Zhang, L. Tao, J. Zhu, C. S. McEnally, E. D. Christensen, C. Hays, K. M. Van Allsburg, K. A. Unocic, H. M. Meyer, Z. Abdullah and D. R. Vardon, Toward Net-Zero Sustainable Aviation Fuel with Wet Waste–Derived Volatile Fatty Acids, *Proc. Natl. Acad. Sci. U. S. A.*, 2021, **118**(13), e2023008118.
34. L. N. Lino and J. A. Henkes, Impactos do transporte aéreo de passageiros uma análise da aplicação dos ods: um estudo de caso da united airlines: impacts of passenger air transport an analysis of the application of the sdgs: a case study from united airlines, *Revista Brasileira De Aviação Civil & Ciências Aeronáuticas*, 2023, **1**, 258–313.
35. T. M. Varma, Responsible Leadership and Reputation Management During a Crisis: The Cases of Delta and United Airlines, *J. Bus. Ethics*, 2021, **173**(1), 29–45.
36. Lufthansa Group, Demand for Sustainable Aviation Fuel on the rise. https://www.lufthansagroup.com/media/newsroom/group/2021/q4/20211118_PM_SAF_EN.pdf.
37. AIRFRANCE, *Sustainable Aviation Fuels (SAF).* https://corporate.airfrance.com/en/sustainable-aviation-fuel.
38. AIRFRANCE, *More sustainable aviation fuel.* https://wwws.airfrance.us/information/developpement-durable/carburants-aviation-durables.
39. K. S. Ng, D. Farooq and A. Yang, Global Biorenewable Development Strategies for Sustainable Aviation Fuel Production, *Renewable Sustainable Energy Rev.*, 2021, **150**, 111502.
40. L. Martinez-Valencia, M. Garcia-Perez and M. P. Wolcott, Supply Chain Configuration of Sustainable Aviation Fuel: Review, Challenges, and Pathways for Including Environmental and Social Benefits, *Renewable Sustainable Energy Rev.*, 2021, **152**, 111680.
41. M. F. Shahriar and A. Khanal, The Current Techno-Economic, Environmental, Policy Status and Perspectives of Sustainable Aviation Fuel (SAF), *Fuel*, 2022, **325**, 124905.
42. J. Herdzik, Decarbonization of Marine Fuels—The Future of Shipping, *Energies*, 2021, **14**(14), 4311.

43. L. A. Díaz-Secades, R. González, N. Rivera, E. Montañés and J. R. Quevedo, Waste Heat Recovery System for Marine Engines Optimized through a Preference Learning Rank Function Embedded into a Bayesian Optimizer, *Ocean Eng.*, 2023, **281**, 114747.
44. S. R. Chia, S. Nomanbhay, M. Y. Ong, A. H. B. Shamsuddin, K. W. Chew and P. L. Show, Renewable Diesel as Fossil Fuel Substitution in Malaysia: A Review, *Fuel*, 2022, **314**, 123137.
45. D. M. M. Yacout, M. Tysklind and V. K. K. Upadhyayula, Socio-Economic Implications of Forest-Based Biofuels for Marine Transportation in the Arctic: Sweden as a Case Study, *Front. Clim.*, 2024, **6**, 1414813.
46. C. Ghenai, M. Bettayeb, B. Brdjanin and A. K. Hamid, Hybrid Solar PV/PEM Fuel Cell/Diesel Generator Power System for Cruise Ship: A Case Study in Stockholm, Sweden, *Case Stud. Therm. Eng.*, 2019, **14**, 100497.
47. V. A. dos Santos, P. P. da Silva and L. M. Ventura Serrano, An Economic and CO_2 Assessment of Using Fischer-Tropsch Diesel in the European Maritime Sector, *Int. J. Environ. Stud.*, 2024, **81**(4), 1529–1562.
48. International Renewable Energy Agency, *A Pathway to Decarbonise the Shipping Sector*, International Renewable Energy Agency: Abu Dhabi, United Arab Emirates, 2021, ISBN 978-92-9260-330-4.
49. H. Xing, C. Stuart, S. Spence and H. Chen, Alternative Fuel Options for Low Carbon Maritime Transportation: Pathways to 2050, *J. Cleaner Prod.*, 2021, **297**, 126651.
50. S. Curran, A. Onorati, R. Payri, A. K. Agarwal, C. Arcoumanis, C. Bae, K. Boulouchos, F. Dal Forno Chuahy, M. Gavaises, G. J. Hampson, C. Hasse, B. Kaul, S.-C. Kong, D. Kumar, R. Novella, A. Pesyridis, R. Reitz, B. M. Vaglieco and N. Wermuth, The Future of Ship Engines: Renewable Fuels and Enabling Technologies for Decarbonization, *Int. J. Engine Res.*, 2024, **25**(1), 85–110.
51. H. Saadaoui and N. Chtourou, Do Institutional Quality, Financial Development, and Economic Growth Improve Renewable Energy Transition? Some Evidence from Tunisia, *J. Knowl. Econ.*, 2023, **14**(3), 2927–2958.
52. E. Kordouli, S. Lycourghiotis, K. Bourikas, A. Lycourghiotis and C. Kordulis, Renewable Diesel Synthesis by Hydro-Processing in Green Solvents, *Curr. Opin. Green Sustainable Chem.*, 2024, **48**, 100936.
53. P. J. Megía, A. J. Vizcaíno, J. A. Calles and A. Carrero, Hydrogen Production Technologies: From Fossil Fuels toward Renewable Sources. A Mini Review, *Energy Fuels*, 2021, **35**(20), 16403–16415.
54. S. A. A. Shah and Y. A. Solangi, A Sustainable Solution for Electricity Crisis in Pakistan: Opportunities, Barriers, and Policy Implications for 100% Renewable Energy, *Environ. Sci. Pollut. Res.*, 2019, **26**(29), 29687–29703.
55. W. L. Mahene, T. Kivevele and R. Machunda, The Role of Textural Properties and Surface Chemistry of Activated Carbon Support in Catalytic Deoxygenation of Triglycerides into Renewable Diesel, *Catal. Commun.*, 2023, **181**, 106737.

56. H. Pitsch, D. Goeb, L. Cai and W. Willems, Potential of Oxymethylene Ethers as Renewable Diesel Substitute, *Prog. Energy Combust. Sci.*, 2024, **104**, 101173.
57. N. Pérez-Rangel, C. Coronado and J. Ancheyta, Approaches to Conditioning of Vegetable Oil Feedstock for Hydrotreating to Produce Renewable Diesel, *Fuel*, 2025, **383**, 133897.
58. A. Arias, C.-E. Nika, V. Vasilaki, G. Feijoo, M. T. Moreira and E. Katsou, Assessing the Future Prospects of Emerging Technologies for Shipping and Aviation Biofuels: A Critical Review, *Renewable Sustainable Energy Rev.*, 2024, **197**, 114427.
59. T.-Y. Zhang, H.-Y. Hu, Y.-H. Wu, L.-L. Zhuang, X.-Q. Xu, X.-X. Wang and G.-H. Dao, Promising Solutions to Solve the Bottlenecks in the Large-Scale Cultivation of Microalgae for Biomass/Bioenergy Production, *Renewable Sustainable Energy Rev.*, 2016, **60**, 1602–1614.
60. V. Ram and S. R. Salkuti, An Overview of Major Synthetic Fuels, *Energies*, 2023, **16**(6), 2834.
61. P. Maheshwari, M. B. Haider, M. Yusuf, J. J. Klemeš, A. Bokhari, M. Beg, A. Al-Othman, R. Kumar and A. K. Jaiswal, A Review on Latest Trends in Cleaner Biodiesel Production: Role of Feedstock, Production Methods, and Catalysts, *J. Cleaner Prod.*, 2022, **355**, 131588.
62. N. S. Mat Aron, K. S. Khoo, K. W. Chew, P. L. Show, W. Chen and T. H. P. Nguyen, Sustainability of the Four Generations of Biofuels – A Review, *Int. J. Energy Res.*, 2020, **44**(12), 9266–9282.
63. Priya, P. S. Deora, Y. Verma, R. A. Muhal, C. Goswami and T. Singh, Biofuels: An Alternative to Conventional Fuel and Energy Source, *Mater. Today: Proc.*, 2022, **48**, 1178–1184.
64. S. M. Farouk, A. M. Tayeb, S. M. S. Abdel-Hamid and R. M. Osman, Recent Advances in Transesterification for Sustainable Biodiesel Production, Challenges, and Prospects: A Comprehensive Review, *Environ. Sci. Pollut. Res.*, 2024, **31**(9), 12722–12747.
65. E. G. Infante, F. T. F. Gomide, A. R. Secchi, L. F. Leite, A. M. Antunes, S. de and A. R. G. Caranton, Diesel Production from Lignocellulosic Residues: Trends, Challenges and Opportunities, *Biofuels, Bioprod. Biorefin.*, 2024, **18**(5), 1711–1738.
66. D. Neupane, Biofuels from Renewable Sources, a Potential Option for Biodiesel Production, *Bioengineering*, 2022, **10**(1), 29.
67. M. Zastempowski, Analysis and Modeling of Innovation Factors to Replace Fossil Fuels with Renewable Energy Sources – Evidence from European Union Enterprises, *Renewable Sustainable Energy Rev.*, 2023, **178**, 113262.
68. A. Antimiani, V. Costantini and E. Paglialunga, Fossil Fuels Subsidy Removal and the EU Carbon Neutrality Policy, *Energy Econ.*, 2023, **119**, 106524.
69. S. Matallah, S. Boudaoud, A. Matallah and M. Ferhaoui, The Role of Fossil Fuel Subsidies in Preventing a Jump-Start on the Transition to Renewable Energy: Empirical Evidence from Algeria, *Resour. Policy*, 2023, **86**, 104276.

70. F. Jelti, A. Allouhi and K. A. Tabet Aoul, Transition Paths towards a Sustainable Transportation System: A Literature Review, *Sustainability*, 2023, **15**(21), 15457.
71. H. Beath, S. Mittal, S. Few, B. Winchester, P. Sandwell, C. N. Markides, J. Nelson and A. Gambhir, Carbon Pricing and System Reliability Impacts on Pathways to Universal Electricity Access in Africa, *Nat. Commun.*, 2024, **15**(1), 4172.
72. K. W. Cheah, S. Yusup, H. K. Gurdeep Singh, Y. Uemura and H. L. Lam, Process Simulation and Techno Economic Analysis of Renewable Diesel Production via Catalytic Decarboxylation of Rubber Seed Oil – A Case Study in Malaysia, *J. Environ. Manage.*, 2017, **203**, 950–961.
73. J. Lilliestam, A. Patt and G. Bersalli, The Effect of Carbon Pricing on Technological Change for Full Energy Decarbonization: A Review of Empirical Ex-post Evidence, *WIREs Clim. Change*, 2021, **12**(1), e681.
74. C. B. Igbeghe, A. Nagy, Z. Gabnai and A. Bai, Exploring Biomass Linkages in the Food and Energy Market—A Systematic Review, *Energies*, 2024, **17**(3), 563.
75. W. N. A. Wan Osman, M. H. Rosli, W. N. A. Mazli and S. Samsuri, Comparative Review of Biodiesel Production and Purification, *Carbon Capture Sci. Technol.*, 2024, **13**, 100264.
76. S. K. Nair, R. M. Emerson and J. R. Solomon, Biomass Supply Chain Risk: Towards a Better Understanding of Feedstock Availability, Cost, Variability, and Uncertainty to Catalyze and De-Risk Biobased Investment, in *Handbook of Biorefinery Research and Technology: Biomass Logistics to Saccharification*, Springer Netherlands, Dordrecht, 2024, pp. 285–312.
77. C. I. Cano-Gómez, J. I. Hinojosa-López, D. B. Muñiz-Marquez and J. E. Wong-Paz, Opportunities of the Bioeconomy in Mexico: Valorization of Agro-Industrial Waste, *Environ. Qual. Manage.*, 2024, **34**(2), e22326.
78. S. Yadav, A. K. Choudhary, P. Yadav and D. B. Pal, Lignocellulosic Biomass Gasification for Bio-Circular Economy Sustainability: A Multiple Criteria Analysis Framework, *Biomass Convers. Biorefin.*, 2024, DOI: 10.1007/s13399-024-06387-3.
79. S. Kammler, A. Malvis Romero, C. Burkhardt, L. Baruth, G. Antranikian, A. Liese and M. Kaltschmitt, Macroalgae Valorization for the Production of Polymers, Chemicals, and Energy, *Biomass Bioenergy*, 2024, **183**, 107105.
80. M. Ali Ijaz Malik, S. Zeeshan, M. Khubaib, A. Ikram, F. Hussain, H. Yassin and A. Qazi, A Review of Major Trends, Opportunities, and Technical Challenges in Biodiesel Production from Waste Sources, *Energy Convers. Manage.: X*, 2024, **23**, 100675.
81. A. I. Osman, M. Nasr, M. Farghali, A. K. Rashwan, A. Abdelkader, A. H. Al-Muhtaseb, I. Ihara and D. W. Rooney, Optimizing Biodiesel Production from Waste with Computational Chemistry, Machine Learning

and Policy Insights: A Review, *Environ. Chem. Lett.*, 2024, **22**(3), 1005–1071.
82. V. Mittal and E. Lim, Aligning Advances in Biodiesel Technology with the Needs of the Defense Community, *Eng*, 2024, **5**(4), 2709–2727.
83. A. A. Babadi, S. Rahmati, R. Fakhlaei, B. Barati, S. Wang, W. Doherty and K. Ostrikov, Emerging Technologies for Biodiesel Production: Processes, Challenges, and Opportunities, *Biomass Bioenergy*, 2022, **163**, 106521.
84. K. Rambabu, G. Bharath, A. Hai, P. L. Show, F. Banat and S. Manickam, Enhancing Biodiesel Production from Waste Date Seed Oil through Ultrasonic-Assisted Optimization: A Sustainable Approach Using Non-Edible Feedstocks, *Chem. Eng. Process. – Process Intensif.*, 2023, **194**, 109601.
85. A. O. Esan, O. A. Olalere, C.-Y. Gan, S. M. Smith and S. Ganesan, Synthesis of Biodiesel from Waste Palm Fatty Acid Distillate (PFAD) and Dimethyl Carbonate (DMC) via Taguchi Optimisation Method, *Biomass Bioenergy*, 2021, **154**, 106262.
86. N. H. Abou-Taleb, D. R. El-Wasseef, D. T. El-Sherbiny and S. M. El-Ashry, Optimizing the Spectrofluorimetric Determination of Cefdinir through a Taguchi Experimental Design Approach, *Luminescence*, 2016, **31**(3), 856–864.
87. O. Trott and A. J. Olson, AutoDock Vina: Improving the Speed and Accuracy of Docking with a New Scoring Function, Efficient Optimization, and Multithreading, *J. Comput. Chem.*, 2010, **31**(2), 455–461.
88. Y. Sugami, E. Minami and S. Saka, Renewable Diesel Production from Rapeseed Oil with Hydrothermal Hydrogenation and Subsequent Decarboxylation, *Fuel*, 2016, **166**, 376–381.
89. N. Klinkaew, A. Wiangkham, A. Ariyarit, P. Aengchuan, A. Pumpuang, S. Sripratum, A. Maneedaeng, J. Srisertpol and E. Sukjit, Strategic Optimization of Engine Performance and Emissions with Bio-Hydrogenated Diesel and Biodiesel: A RVEA-GRNNs Framework, *Results Eng.*, 2024, **24**, 103072.
90. N. Pérez-Rangel, C. Coronado and J. Ancheyta, Approaches to Conditioning of Vegetable Oil Feedstock for Hydrotreating to Produce Renewable Diesel, *Fuel*, 2025, **383**, 133897.
91. E. Kordouli, S. Lycourghiotis, K. Bourikas, A. Lycourghiotis and C. Kordulis, Renewable Diesel Synthesis by Hydro-Processing in Green Solvents, *Curr. Opin. Green Sustainable Chem.*, 2024, **48**, 100936.
92. S. Curran, A. Onorati, R. Payri, A. K. Agarwal, C. Arcoumanis, C. Bae, K. Boulouchos, F. Dal Forno Chuahy, M. Gavaises, G. J. Hampson, C. Hasse, B. Kaul, S.-C. Kong, D. Kumar, R. Novella, A. Pesyridis, R. Reitz, B. M. Vaglieco and N. Wermuth, The Future of Ship Engines: Renewable Fuels and Enabling Technologies for Decarbonization, *Int. J. Engine Res.*, 2024, **25**(1), 85–110.
93. V. Patil, P. Singh and S. Kumar, Reliability of Renewable Oxygenated Fuels in Engines: A Comprehensive Review of Long-Term Testing, *Fuel*, 2024, **367**, 131451.

94. S. A. Shahir, H. H. Masjuki, M. A. Kalam, A. Imran, I. M. R. Fattah and A. Sanjid, Feasibility of Diesel–Biodiesel–Ethanol/Bioethanol Blend as Existing CI Engine Fuel: An Assessment of Properties, Material Compatibility, Safety and Combustion, *Renewable Sustainable Energy Rev.*, 2014, **32**, 379–395.
95. M. B. Oliveira, A. J. Queimada and J. A. P. Coutinho, Modeling of Biodiesel Multicomponent Systems with the Cubic-Plus-Association (CPA) Equation of State, *Ind. Eng. Chem. Res.*, 2010, **49**(3), 1419–1427.
96. U. Rashid, C. L. Lee, B. Hazmi, S. Gamal and M. Beygisangchin, Standard Specifications for Renewable Diesel, in *Renewable Diesel*, Elsevier, 2024, pp. 33–63.
97. H. Stančin, H. Mikulčić, X. Wang and N. Duić, A Review on Alternative Fuels in Future Energy System, *Renewable Sustainable Energy Rev.*, 2020, **128**, 109927.
98. J. Sun, L. Zhang and K.-C. Loh, Enhancing Scalability and Economic Viability of Lignocellulose-Derived Biofuels Production through Integrated Pretreatment and Methanogenesis Arrest, *Bioresour. Technol.*, 2023, **389**, 129790.
99. A. W. Bartling, P. T. Benavides, S. D. Phillips, T. Hawkins, A. Singh, M. Wiatrowski, E. C. D. Tan, C. Kinchin, L. Ou, H. Cai, M. Biddy, L. Tao, A. Young, K. Brown, S. Li, Y. Zhu, L. J. Snowden-Swan, C. R. Mevawala and D. J. Gaspar, Environmental, Economic, and Scalability Considerations of Selected Bio-Derived Blendstocks for Mixing-Controlled Compression Ignition Engines, *ACS Sustainable Chem. Eng.*, 2022, **10**(20), 6699–6712.
100. M. D. Kass, C. J. Janke, R. M. Connatser, S. A. Lewis, J. R. Keiser and K. Gaston, Compatibility Assessment of Fuel System Infrastructure Plastics with Bio-Oil and Diesel Fuel, *Energy Fuels*, 2018, **32**(1), 542–553.
101. N. Pérez-Rangel, C. Coronado and J. Ancheyta, Approaches to Conditioning of Vegetable Oil Feedstock for Hydrotreating to Produce Renewable Diesel, *Fuel*, 2025, **383**, 133897.
102. F. R. da Silva, F. J. L. Baumgardt da Silva, F. Vandresen, F. da Silva Lisboa and R. Sequinel, Oxidative Stability of Biodiesel: Challenges and Perspectives for the Sustainability of a Large-Scale Program in Brazil, *Biofuels*, 2024, 1–8.
103. J. Pullen and K. Saeed, An Overview of Biodiesel Oxidation Stability, *Renewable Sustainable Energy Rev.*, 2012, **16**(8), 5924–5950.
104. S. O. Jeje, T. Marazani, J. O. Obiko and M. B. Shongwe, Advancing the Hydrogen Production Economy: A Comprehensive Review of Technologies, Sustainability, and Future Prospects, *Int. J. Hydrogen Energy*, 2024, **78**, 642–661.
105. S. Shahzad, M. A. Abbasi, M. B. Shahid and J. M. Guerrero, Unlocking the Potential of Long-Duration Energy Storage: Pathways to Net-Zero Emissions through Global Innovation and Collaboration, *J. Energy Storage*, 2024, **97**, 112904.
106. M. N. A. M. Yusoff, N. W. M. Zulkifli, N. L. Sukiman, O. H. Chyuan, M. H. Hassan, M. H. Hasnul, M. S. A. Zulkifli, M. M. Abbas and

M. Z. Zakaria, Sustainability of Palm Biodiesel in Transportation: A Review on Biofuel Standard, Policy and International Collaboration Between Malaysia and Colombia, *Bioenergy Res.*, 2021, **14**(1), 43–60.
107. M. Boucher and M.-A. Pigeon, Scaling Renewable Energy Cooperatives for a Net-Zero Canada: Challenges and Opportunities for Accelerating the Energy Transition, *Energy Res. Soc. Sci.*, 2024, **115**, 103618.
108. A. Rahman, P. Dargusch and D. Wadley, The Political Economy of Oil Supply in Indonesia and the Implications for Renewable Energy Development, *Renewable Sustainable Energy Rev.*, 2021, **144**, 111027.
109. R. L. Naylor and M. M. Higgins, The Political Economy of Biodiesel in an Era of Low Oil Prices, *Renewable Sustainable Energy Rev.*, 2017, **77**, 695–705.
110. S. Sen and S. Ganguly, Opportunities, Barriers and Issues with Renewable Energy Development – A Discussion, *Renewable Sustainable Energy Rev.*, 2017, **69**, 1170–1181.
111. L. Langer, M. Brander, S. M. Lloyd, D. Keles, H. D. Matthews and A. Bjørn, Does the Purchase of Voluntary Renewable Energy Certificates Lead to Emission Reductions? A Review of Studies Quantifying the Impact, *J. Cleaner Prod.*, 2024, **478**, 143791.
112. S. O. Bade and O. S. Tomomewo, A Review of Governance Strategies, Policy Measures, and Regulatory Framework for Hydrogen Energy in the United States, *Int. J. Hydrogen Energy*, 2024, **78**, 1363–1381.
113. B. Wang, Z. J. Ting and M. Zhao, Sustainable Aviation Fuels: Key Opportunities and Challenges in Lowering Carbon Emissions for Aviation Industry, *Carbon Capture Sci. Technol.*, 2024, **13**, 100263.
114. À. Alonso-Travesset, D. Coppitters, H. Martín and J. de la Hoz, Economic and Regulatory Uncertainty in Renewable Energy System Design: A Review, *Energies*, 2023, **16**(2), 882.
115. F. Polzin, M. Migendt, F. A. Täube and P. von Flotow, Public Policy Influence on Renewable Energy Investments—A Panel Data Study across OECD Countries, *Energy Policy*, 2015, **80**, 98–111.
116. I. López, N. Goitia-Zabaleta, A. Milo, J. Gómez-Cornejo, I. Aranzabal, H. Gaztañaga and E. Fernandez, European Energy Communities: Characteristics, Trends, Business Models and Legal Framework, *Renewable Sustainable Energy Rev.*, 2024, **197**, 114403.
117. P. M. Falcone, Sustainable Energy Policies in Developing Countries: A Review of Challenges and Opportunities, *Energies*, 2023, **16**(18), 6682.
118. D. Song, B. Jia and H. Jiao, Review of Renewable Energy Subsidy System in China, *Energies*, 2022, **15**(19), 7429.
119. S. Matallah, S. Boudaoud, A. Matallah and M. Ferhaoui, The Role of Fossil Fuel Subsidies in Preventing a Jump-Start on the Transition to Renewable Energy: Empirical Evidence from Algeria, *Resour. Policy*, 2023, **86**, 104276.
120. V. Kumar, N. Sharma, A. S. Abdelaal, P. Chakraborty, J. Thomas, L. Duhan, R. Pasrija, S. Dogra and I. Jayaraj, Bio-Oil Production and

Catalytic Upgrade to Value Added Product: A Review on Recent Technologies, *J. Energy Inst.*, 2025, **118**, 101880.
121. J. Rentschler, M. Kornejew and M. Bazilian, Fossil Fuel Subsidy Reforms and Their Impacts on Firms, *Energy Policy*, 2017, **108**, 617–623.
122. W. Ginn, The Paradox of Fossil Fuel Subsidies, *Econ. Anal. Policy*, 2024, **83**, 333–358.
123. A. Wattanakuljarus, Diverse Effects of Fossil Fuel Subsidy Reform on Industrial Competitiveness in Thailand, *Eurasian Econ. Rev.*, 2021, **11**(3), 489–517.
124. Y. Guo, X. Gou, Z. Xu and M. Skare, Carbon Pricing Mechanism for the Energy Industry: A Bibliometric Study of Optimal Pricing Policies, *Acta Montan. Slovaca*, 2022, **27**, 49–69.
125. L. Xu and J. Yang, Carbon Pricing Policies and Renewable Energy Development: Analysis Based on Cross-Country Panel Data, *J. Environ. Manage.*, 2024, **366**, 121784.
126. A. O. Adelaja, Barriers to National Renewable Energy Policy Adoption: Insights from a Case Study of Nigeria, *Energy Strategy Rev.*, 2020, **30**, 100519.
127. J. H. Pratama, Z. Rahmawati, A. R. Widyanto, T. Gunawan, W. N. Wan Abdullah, N. L. Azua Jamari, A. Hamzah and H. Fansuri, Advancements in Green Diesel Production for Energy Sustainability: A Comprehensive Bibliometric Analysis, *RSC Adv.*, 2024, **14**(48), 36040–36062.
128. M. J. Burke and J. C. Stephens, Political Power and Renewable Energy Futures: A Critical Review, *Energy Res. Soc. Sci.*, 2018, **35**, 78–93.
129. M. I. Miskat, P. Sarker, H. Chowdhury, T. Chowdhury, M. S. Rahman, N. Hossain, P. Chowdhury and S. M. Sait, Current Scenario of Solar Energy Applications in Bangladesh: Techno-Economic Perspective, Policy Implementation, and Possibility of the Integration of Artificial Intelligence, *Energies*, 2023, **16**(3), 1494.
130. U. Mehmood, Assessing the Impacts of Eco-Innovations, Economic Growth, Urbanization on Ecological Footprints in G-11: Exploring the Sustainable Development Policy Options, *J. Knowl. Econ.*, 2024, **15**, 16849–16867.
131. G. Kissinger, M. Brockhaus and S. R. Bush, Policy Integration as a Means to Address Policy Fragmentation: Assessing the Role of Vietnam's National REDD+ Action Plan in the Central Highlands, *Environ. Sci. Policy*, 2021, **119**, 85–92.
132. L. Hofbauer, W. McDowall and S. Pye, Challenges and Opportunities for Energy System Modelling to Foster Multi-Level Governance of Energy Transitions, *Renewable Sustainable Energy Rev.*, 2022, **161**, 112330.
133. H. Busch, S. Ruggiero, A. Isakovic and T. Hansen, Policy Challenges to Community Energy in the EU: A Systematic Review of the Scientific Literature, *Renewable Sustainable Energy Rev.*, 2021, **151**, 111535.
134. O. Ogunkunle and N. A. Ahmed, A Review of Global Current Scenario of Biodiesel Adoption and Combustion in Vehicular Diesel Engines, *Energy Rep.*, 2019, **5**, 1560–1579.

135. A. Juaidi, F. Anayah, R. Assaf, A. A. Hasan, S. Monna, L. Herzallah, R. Abdallah, P. Dutournié and M. Jeguirim, An Overview of Renewable Energy Strategies and Policies in Palestine: Strengths and Challenges, *Energy Sustainable Dev.*, 2022, **68**, 258–272.
136. I. Lupu, A. Criste, T. Ciumara, C. Milea and R. Lupu, Addressing the Renewable Energy Challenges through the Lens of Monetary Policy—Insights from the Literature, *Energies*, 2024, **17**(19), 4820.
137. B. Sandeep, U. Baranwal and A. Mandpe, Biofuel Production from Microbial Sources: Advances and Challenges, in *Energy, Environment, and Sustainability*, 2024, pp. 323–357.
138. R. C. Rial, Biofuels *versus* Climate Change: Exploring Potentials and Challenges in the Energy Transition, *Renewable Sustainable Energy Rev.*, 2024, **196**, 114369.
139. D. Kour, K. L. Rana, N. Yadav, A. N. Yadav, A. A. Rastegari, C. Singh, P. Negi, K. Singh and A. K. Saxena, Technologies for Biofuel Production: Current Development, Challenges, and Future Prospects, in *Biofuel and Biorefinery Technologies*, 2019, pp. 1–50.
140. R. Repetto, Economic Incentives for Sustainable Production, *Ann. Reg. Sci.*, 1987, **21**(3), 44–59.
141. A. Jafari, T. Khalili, H. G. Ganjehlou and A. Bidram, Optimal Integration of Renewable Energy Sources, Diesel Generators, and Demand Response Program from Pollution, Financial, and Reliability Viewpoints: A Multi-Objective Approach, *J. Cleaner Prod.*, 2020, **247**, 119100.
142. J. Aburto and M. A. Amezcua-Allieri, Biodiesel and Green Diesel Fuels: A Techno-Economic Analysis, in *Advances in Sustainability Science and Technology*, 2022, pp. 309–324.
143. K. B. Ansari, S. Z. Hassan; S. A. Farooqui, R. Hasib, P. Khan, A. R. S. Rahman, M. S. Khan and Q. T. Trinh, Policies, Techno-Economic Analysis and Future Perspective of Green Diesel, in *Advances in Sustainability Science and Technology*, 2022, pp. 351–375.
144. A. L. Schmid and C. A. A. Hoffmann, Replacing Diesel by Solar in the Amazon: Short-Term Economic Feasibility of PV-Diesel Hybrid Systems, *Energy Policy*, 2004, **32**(7), 881–898.
145. S. B. Glisic, J. M. Pajnik and A. M. Orlović, Process and Techno-Economic Analysis of Green Diesel Production from Waste Vegetable Oil and the Comparison with Ester Type Biodiesel Production, *Appl. Energy*, 2016, **170**, 176–185.
146. M. Aslam, S. S. Maktedar and A. K. Sarma, *Green Diesel: An Alternative to Biodiesel and Petrodiesel*, ed. M. Aslam, S. Shivaji Maktedar and A. K. Sarma, Springer Nature Singapore, Singapore, 2022, DOI: 10.1007/978-981-19-2235-0.
147. T. N. Kalnes, K. P. Koers, T. Marker and D. R. Shonnard, A Techno-economic and Environmental Life Cycle Comparison of Green Diesel to Biodiesel and Syndiesel, *Environ. Prog. Sustainable Energy*, 2009, **28**(1), 111–120.

148. G. Rispoli, Biorefineries and Green Diesel: Process and Product Innovation, in *Biorefineries and green diesel: process and product innovation*, 2020, pp. 65–76.
149. F. H. Salina, I. A. de Almeida and F. R. Bittencourt, RenovaBio Opportunities and Biofuels Outlook in Brazil, *Innovative Renewable Energy*, 2020, 391–399.
150. R. S. Tiburcio, T. R. de Macêdo and A. M. P. Neto, Brazilian Biofuels Policy (RenovaBio): Overview and Generation of Decarbonization Credits by Biodiesel Production Facilities, *Energy Sustainable Dev.*, 2023, 77, 101334.
151. T. Heimann, R. Argueyrolles, M. Reinhardt, F. Schuenemann, M. Söder and R. Delzeit, Phasing out Palm and Soy Oil Biodiesel in the EU: What Is the Benefit?, *GCB Bioenergy*, 2024, **16**(1), e13115.
152. P. Dusser, The European Energy Policy for 2020–2030. RED II: What Future for Vegetable Oil as a Source of Bioenergy?, *OCL*, 2019, **26**, 51.
153. T. Bock, Sustainability Aspects of Powerfuels and Their Certification, in *Sustainability Aspects of Powerfuels and Their Certification*, 2025, pp. 977–1023.
154. C. Moser, T. Hildebrandt and R. Bailis, International Sustainability Standards and Certification, in *Sustainable Development of Biofuels in Latin America and the Caribbean*, Springer New York, New York, NY, 2014, pp. 27–69.

CHAPTER 14

Life-cycle and Techno-economic Assessment of Renewable Diesel and Biorefinery

MYRIAM ADELA AMEZCUA-ALLIERI,[a] JORGE ABURTO[b] AND ELIAS MARTINEZ-HERNANDEZ*[a]

[a] Energy Efficiency and Biofuels Division, Instituto Mexicano del Petróleo, Eje Central Lázaro Cárdenas Norte 152, Col. San Bartolo Atepehuacan, 07730, Alcaldía Gustavo A. Madero, Mexico; [b] Independent Researcher, Alc. Coyoacán, 04360, Mexico City, Mexico, Email: jaaburto@yahoo.com
*Email: emartinez@imp.mx

14.1 Introduction

The world faces complex challenges related to climate change and other environmental issues, mainly due to the overuse of fossil fuels in industrial activities releasing excess CO_2 emissions into the atmosphere. As part of the goal of decarbonizing the global economy, countries whose main source of wealth relies on fossil fuels plan an energy transition by incorporating alternative cleaner forms of energy. It is well known that transportation is a major contributor to pollution and has been the focus for several research and policy interventions. Although electric cars are increasingly becoming part of the solution, there are still countries that may need more options for the transition, such as renewable forms of fuels for the existing car fleets. When such alternatives can be used with existing infrastructure and car-engine designs they are called drop-in fuels due to their similar chemical nature to petroleum-derived fuels. Drop-in fuels can be used in mixtures

Green Chemistry Series No. 89
Biorefining of Renewable Diesel: From Production to Market
Edited by Mohammad Aslam, Sanjeev Mishra and Jorge Aburto

Published by the Royal Society of Chemistry, www.rsc.org

with fossil fuels. In this chapter, we focus on the production of renewable diesel. It is noteworthy that Rudolf Diesel's compression ignition engine originally operated using renewable vegetable oil, which was then replaced by the availability of cheap petroleum-based fuels. Nowadays, the search is for how to go back from fossil fuels to renewable fuels for diesel engines. In this chapter, we argue that renewable diesel, a drop-in fuel, can play a significant role in the energy transition. Renewable diesel can displace part of the diesel demands either as a stand-alone fuel or in diesel blends to decrease the carbon intensity of transportation systems.

Renewable biodiesel (RD) can be processed with petroleum diesel, seamlessly blended with petroleum diesel and/or biodiesel in various amounts, transported in petroleum pipelines, and sold at retail stations with or without blending. Renewable diesel can be made from multiple feedstocks and at plants that also produce sustainable aviation fuel and biodiesel. The California Air Resources Board reported that renewable diesel reduces carbon intensity on average by 65% when compared with petroleum diesel, in addition to reducing nitrogen oxide emissions when compared with petroleum fuel.[1] Renewable diesel is also able to meet the conventional petroleum ASTM D975 specification, allowing it to be used in existing infrastructure and diesel engines.

In this chapter, we focus first on the different biomass sources and process technologies and the fuel specifications and standards. We briefly review the techno-economic and lifetime environmental aspects according to results presented in the literature. We then focus on the hydrotreatment of vegetable oils and fats, which produces a hydrocarbon mixture usually called "green diesel", and propose an integrated biorefinery scheme for its production from utilization of the whole oily crop to supply the required energy and hydrogen for the process.

14.2 Processes for Renewable-diesel Production

Renewable diesel can be obtained through different synthetic pathways: (1) gas-to-liquids; (2) hydrogenation of oils and fats; (3) sugars to hydrocarbons; (4) upgrading of pyrolysis oil. A mixture of renewable and petrodiesel can also be produced by co-processing of biomass with petroleum fractions. The pathways are described as follows:

- *Gas-to-liquids.* In this route a synthesis gas ($CO + H_2$) is generated *via* gasification of lignocellulosic biomass. The gasification occurs by partial oxidation in the presence of a limited quantity of oxygen or steam to produce syngas ($CO + H_2$) as well as CO_2, CH_4 and water. Then syngas is transformed *via* the Fischer–Tropsch reactions into saturated or unsaturated hydrocarbons (C_1–C_{40}) with catalysts such as Fe and Co. The resulting product is generally named Fischer–Tropsch diesel.[2]
- *Hydrogenation of oils and fats.* The hydrogenation of triacyl glycerides present in vegetable oils and animal fats *via* the hydrodeoxygenation (HDO) reaction allows the production of propane from the glycerol

moiety and linear paraffins from the fatty acid part, with water as by-product. Decarbonylation and/or decarboxylation reactions can take place, and they generate linear paraffins with release of CO and/or CO_2, respectively. A general reaction scheme is presented in eqn (14.1)–(14.4).

Depropanation reaction:

$$C_3H_5(C_nH_{2n+1-2m}COO)_3 + (m+3)H_2 \rightarrow 3C_nH_{2n+1}COOH + C_3H_8 \tag{14.1}$$

Hydrodeoxygenation reaction:

$$C_nH_{2n+1-2m}COOH + (m+3)H_2 \rightarrow C_{n+1}H_{2n+4} + 2H_2O \tag{14.2}$$

Decarbonylation reaction:

$$C_nH_{2n+1-2m}COOH + (m+1)H_2 \rightarrow C_nH_{2n+2} + CO + H_2O \tag{14.3}$$

Decarboxylation reaction:

$$C_nH_{2n+1-2m}COOH + mH_2 \rightarrow C_nH_{2n+2} + CO_2 \tag{14.4}$$

In these equations, n is the number of carbon atoms in the aliphatic chain and m is the number of double bonds in the fatty acid chain. Hence, these reaction equations consider hydrogen saturation.[3]

- *Aqueous phase reforming.* Starches and sugars from biomass are especially suited for conversion through the BioForming® process that involves several steps: hydrolysis of biomass, hydrogenation, dehydration, and isomerization into hydrocarbons.[4] Nevertheless, the technology readiness level (TRL) of this approach was set at 3, which means that the proof of concept through analytical and experimental function was reached.
- *Upgrading of pyrolysis oil.* Another thermochemical approach is *via* the pyrolysis of biomass to produce bio-oil, followed by hydrotreating and hydrocracking to produce renewable diesel and gasoline.[5]

14.3 Renewable Diesel: Properties, Specifications, Blending and Storage

14.3.1 Fuel Properties and Standards

Renewable diesel has similar physical properties to, and can be used for the same purposes as, petroleum diesel and hence it is considered a drop-in fuel.[6] Depending on the technology used for its production, the composition and properties of the renewable-diesel product can vary. Fossil diesel must comply with international and national standards such as ASTM D975,[7] EN 590 [8] and NOM-016-CRE 2016.[9] The standards are summarized and compared in Table 14.1. Since RD is mainly composed of linear paraffins,

Table 14.1 International and national standards for diesel fuel and properties of renewable diesel.

Property	Unit	No. 1 D-S15[7]	No. 1 D-S500[7]	EN 590[8]	NOM-016-CRE[9]	EN 15940[10]	Renewable diesel[2]	FT diesel[2]
Flash point (min.)	°C	38	38	55	45	55	>59	*ca.* 71
Cetane number (min.)		40	40	51	45	70/51	>70	74–80
Cetane index (min.)		40	40	46	45	—	—	—
Aromatic hydrocarbons (max.)	vol%	35	35	11	35	—	<0.1	<0.2
Kinematic viscosity at 40 °C	$mm^2 s^{-1}$	Min. 1.3, max. 2.4	Min. 1.3, max. 2.4	Min. 2.0, max. 4.5	Min. 1.9, max. 4.1	Min. 2.0, max. 4.5	Min. 2.0, max. 4.0	Min. 2.0, max. 4.5
Lubricity, HFRR at 60 °C (max.)	microns	520	520	460	520	460	>700	420–570
Conductivity (min.)	$pS m^{-1}$	25	25	—	25	25	≥50	—
Ramsbotton carbon residue[a] (max.)	mass%	0.15	0.15	0.3	0.35	0.3	<0.1	—
Sulfur	$mg kg^{-1}$	15	500	Max. 50	Max. 15[b], max. 500[c]	Max. 5	<5	<5
Ash (max.)	mass%	0.01	0.01	0.01	0.01	0.01	<0.001	—

[a] On 10% distillation residue.
[b] In metropolitan areas of Mexico City, Guadalajara and Monterrey.
[c] Other regions of Mexico.

which is a fraction present in fossil diesel, blending recommendations or co-processed RD/diesel must follow specifications described in the mentioned standards. Pure RD is called R100 and blends must show the percentage by volume of RD/fossil diesel, *i.e.* 20% RD with 80% fossil diesel must be labelled as R20 and so on. The standard EN 15940 [10] defines specifications for a paraffinic diesel fuel that is synthetically produced from biomass, natural gas or coal, and blends of RD with fossil diesel must comply with these specifications.

14.3.2 Flash Point and Cetane Number

The flash point of RD is higher compared to that of fossil diesel by about 17–33 °C (Table 14.1), which means that RD requires a higher temperature to provide enough vapor to generate a flammable blend with air. This means RD handling, transport and storage is safer. Other studies reported that flash point may vary between 55 °C and 58 °C depending on hydrotreatment temperature.[11] The cetane number (CN) indicates the ignition delay from the injection of the fuel into the combustion chamber to the beginning of the combustion process. RD presents a higher CN, which indicates a shorter ignition delay and is associated with a higher thermal performance and a lower fuel consumption. This is attributed to the lower combustion activation energy of linear hydrocarbons when compared to unsaturated or aromatic hydrocarbons.

14.3.3 Aromatic Content and Emissions

The high content of aromatic hydrocarbons in fossil fuel is related to lower CN, lower thermal performance, and higher fuel consumption, but with higher unburned hydrocarbons, soot formation and NO_x emissions as well as with smaller-sized particulate matter (PM), which are associated with health and environmental issues.[2,12] RD has a low aromatic-hydrocarbon content and its blending with fossil diesel may help to avoid such negative impacts.

14.3.4 Viscosity and Lubricity

Fossil-fuel specifications like kinematic viscosity, carbon residue, conductivity, sulfur and ash content are all met by RD and no issues are envisaged in blends. But RD has a lower lubricity property, which may constrain blending volume with fossil diesel or require additives such as biodiesel at a low volume of 0.5–2.0%.[13] International fossil-fuel standards regulate the content of biodiesel to a maximum of 5% and 7% in ASTM D975 [7] and EN 590,[8] respectively. In Mexico, the use of biodiesel is not currently regulated by NOM-016-CRE.[9]

14.3.5 Storage and Stability

In terms of storage, RD, fossil diesel and their blends must meet oxidative and thermal stability requirements even if there is no minimum recommendation in ASTM D975 and NOM-016-CRE. Nevertheless, the specific standard ASTM D6468 for thermal stability was designed for diesel containing 5000 ppm and 500 ppm of sulfur, and it may be used only for qualitative information, where a reflectance level of 80% at 180 min was suggested for a premium diesel, and a lower reflectance indicates a decrease in fuel stability.[14] The European standard states that the oxidation stability of fossil fuel must be maximum 25 g cm^{-3} with a maximum of 20 h of ageing at 95 °C.[8]

Storage stability is approached by ASTM D2274 or D5304, but ASTM D7545 seems to be more appropriate for a diesel blend with RD since it works with biodiesel–diesel blends and fossil diesel with 15 ppm sulfur content. The last standard is an accelerated stability test to get insights on field-storage stability using an automated oxidation tester, but performance criteria have not been established to assure long-term storage. Here the results must be provided as the induction period in minutes to reach at 10% drop in pressure. A larger induction time indicates better storage stability.

Fuel-storage conditions must assure that tanks are kept free of water and oxygen to minimize corrosion, microorganism growth and formation of sediments. RD, fossil diesel and their blends must not exceed 0.5% v/v water, and draining sediments and water from the tank must be done on a regular basis. Moreover, it is recommended that fixed-roof tanks must be kept full of fuel to restrain oxygen supply.

14.3.6 Freezing Point

Cold properties are also an important aspect to consider for renewable-diesel utilization in diesel engines and this is highly dependent on the type of hydrocarbons and composition. A typical composition of green diesel obtained from HDO of palm oil is shown in Table 14.2. The freezing point of the mixture can be estimated from a linear mixing rule using molar composition and the freezing points of the components. It can be observed that the freezing point is 293.7 K, which corresponds to 20.62 °C. This represents a technical and operational challenge for cars in places with cold climate conditions. One way to improve cold properties is to add an isomerization

Table 14.2 Compositions of green diesel and estimated freezing points.

Component	Hydrocarbon name	Fraction (mol%)	Freezing point of component/K
C_{17}	Heptadecane	21.595	295.15
C_{16}	Hexadecane	26.970	291.35
C_{18}	Octadecane	32.532	301.35
C_{15}	Pentadecane	17.911	283.05
C_{14}	Tetradecane	0.596	279.05
C_{13}	Tridecane	0.395	267.75
Estimated freezing point of mixture			20.62 °C

stage, but this requires more hydrogen in the production process. The cost of production will likely increase when an isomerization stage is included. Another option is to use a green-diesel mixture in a blend with petroleum diesel up to a level where the cold properties are not affected for the proper operation of a diesel engine.

14.4 Techno-economic Challenges of Renewable Diesel

14.4.1 Commercial-scale Renewable Diesel

The commercial-scale renewable-diesel technologies are mainly based on the hydroprocessing route with vegetable oils, fats and waste cooking oil. In these technologies, there is a step for pretreatment of the oils to remove impurities such as colorants, gums, and bleaching. An optional pre-hydrogenation step is included when processing oils rich in unsaturated fatty acids. The hydrodeoxygenation step removes oxygen to generate hydrocarbons but is a highly exothermic reaction so that 2–4 catalytic beds are used. An additional hydroisomerization step may be needed to meet standards and specifications, especially cold-flow properties such as freezing and cloud points. A summary of technologies is presented in Table 14.3.

Table 14.3 Major technologies for renewable-diesel production from hydroprocessing of oils and fats.

Technology	Description	Location	Green-diesel production	Source
Bio-Synfining (Renewable Energy Group)	Hydroprocessing of vegetable oils, fats and waste cooking oil	USA	233 kt per year	15
Ecofining (ENI/ Honeywell-UOP)	Hydroprocessing of vegetable oils, fats and waste cooking oil	Italy, USA, Malaysia, South Korea	400 kt per year	16
HydroFlex (Haldor Topsoe)	Hydroprocessing of vegetable oils, fats and waste cooking oil	USA (California)	2.2 kt per year (730 million gallons per year)	17 and 18
		USA (North Dakota)	12 000 barrels per day (180 million gallons per year)	
		Sweden	16 000 barrels per day	
NEXBTL (Neste)	Hydroprocessing of palm oil, tallow and corn oil	Netherlands	800 kt per year	19

14.4.2 Effect of Vegetable Oil Price and Plant Scale

Previous studies agree with the fact that feedstock availability and costs are a major obstacle for large-scale production and commercialization of renewable diesel. Figure 14.1 shows the prices and availabilities of major vegetable oils and their main producer countries in 2020. The prices ranged from as cheap as 146 USD t^{-1} for US corn oil up to 1527 USD t^{-1} for peanut oil from China. This range of prices creates a wide range of possible costs of production (COP) for renewable diesel from vegetable oils.

An analysis can be conducted to determine the sensitivity of COP to varying production capacities and oil prices to account for both the oil availability and price challenges. This was carried out based on previous simulation works and using palm oil as feedstock for a range of capacities between 37 000 and 200 000 barrels per year, and for an oil price range between 289 USD t^{-1} and 500 USD t^{-1}.[3] The results are presented in Figure 14.2. It can be observed that for a low capacity of about 50 000 barrels per year, the COP can vary widely in the range 0.566–0.77 USD L^{-1}, while for the larger capacity of 200 000 barrels per year, the COP range is 0.46–0.67 USD L^{-1}. Although larger capacities lower the cost, the COP is more sensitive to oil price. In fact, oil feedstock contributes to about 68% of the total annual costs of the green-diesel process. A comparison of these costs with renewable diesel produced from pyrolysis bio-oil shows that this is within a similar range for the case of bio-oil from corn stover, for which the COP range is 0.56–0.82 USD L^{-1}.[21] The same technological route using agricultural wastes and woody biomass has a cost range of 0.98–1.27 USD L^{-1}.[22] These data could be a more realistic range for renewable diesel from bio-oil given the higher complexity of the process technologies.

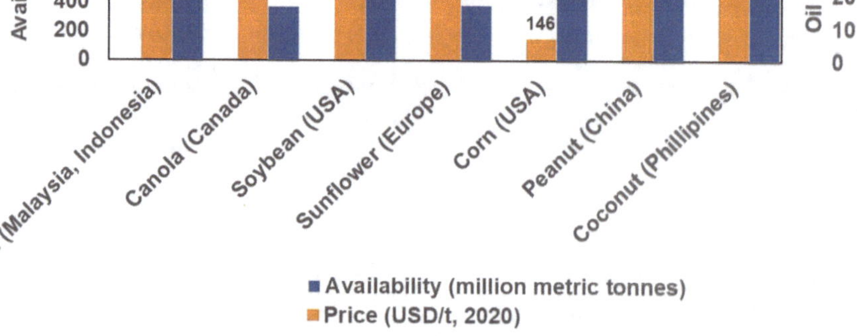

Figure 14.1 Main vegetable oil feedstocks with their main producer countries and their respective availabilities and prices. Data from ref. 20.

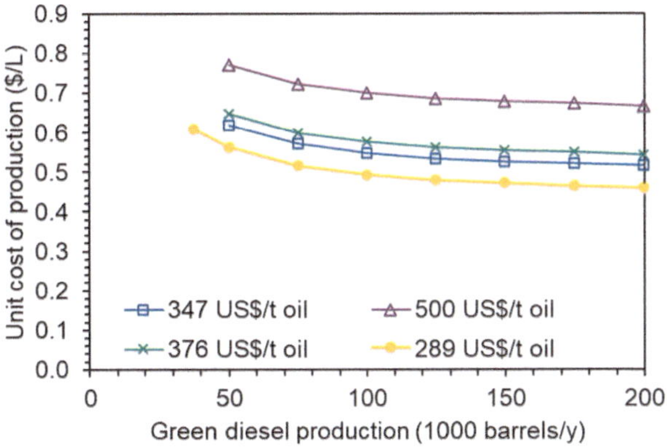

Figure 14.2 Sensitivity analysis of green-diesel cost of production *versus* production capacity at different oil prices.

Table 14.4 Comparison of scenarios for hydrodeoxygenation *vs.* decarboxylation reactions.

Indicator	HDO scenario	DCx scenario
Hydrodeoxygenation reaction (%)	96.2	0
Decarboxylation reaction (%)	0	96.2
Decarbonylation reaction (%)	3.8	3.8
Hydrogen consumption/kg h^{-1}	40.75	14.77
Green diesel produced/L h^{-1}	1827.7	1724.5
Reactor cooling duty/kW	263.1	104.7
Cost of production (COP)/USD kg^{-1}	1.08	1.03

14.4.3 Effect of Reaction Pathway and Hydrogen Consumption

Another economic challenge is the use of hydrogen, which needs to be imported to a renewable-diesel refinery. The consumption of hydrogen depends on the reactions favored in the HDO reactor. As seen in eqn (14.1), the HDO reactions consume more hydrogen than decarbonylation and decarboxylation reactions. Together with differences in the heats of reaction and product distributions, this will also have an impact on the economic results. Table 14.4 summarizes the hypothetical impacts of favoring the reaction routes of HDO (HDO scenario) or decarboxylation (DCx scenario). When decarboxylation is favored and no HDO reactions occur, the hydrogen consumption can be decreased by up to 63%. However, note that more carbon would be lost to CO_2 emissions and a lower amount of green diesel is produced. This in turn makes the cost of gas separations to recover propane higher due to higher CO_2 capture requirements. However, the lower reaction heat and lower hydrogen demand makes the COP slightly lower in the full DCx scenario. Thus, a balance between both reactions is desirable and this can guide further research in new catalysts for hydroprocessing vegetable oils.

14.5 Life-cycle Assessment of Renewable Diesel

Life-cycle assessment (LCA) emerges as a crucial tool for analyzing the environmental impacts of renewable-diesel production. This section reviews the literature on production of renewable diesel from different studies related to environmental impacts, arable land requirements, and the importance of using low-quality waste or fats as feedstocks to reduce costs. In addition, it highlights the importance of systematically assessing the whole life cycle of the fuel. A summary of the main data reported on life-cycle carbon emissions for renewable diesel is presented in Table 14.5, considering feedstock and allocation.

LCA of renewable-diesel production involves evaluating its environmental impact across various stages. The stages can vary according to the system boundaries and the feedstock used by the authors. Most of the papers used palm oil; however, other feedstocks including lignocellulosic biomass and miscanthus *via* anaerobic digestion can be used.[29] In the case of renewable-diesel production from palm oil, the main stages include:

1. Plantation–culture–harvest: this stage encompasses the cultivation of palm trees, including planting, maintenance, and eventual harvesting. The environmental impact here is influenced by factors such as water usage, land management practices, and application of fertilizers and agrochemicals. The use of agrochemicals during palm culture contributes significantly to human toxicity impacts. While plants mitigate CO_2 emissions, it is essential to address the environmental impact of agrochemical use during palm culture.[25]
2. Palm-oil extraction: the process of extracting palm oil from harvested fruits. Environmental considerations include energy consumption, water usage, and waste generation during extraction, including palm-oil mill effluent (POME) and the extraction cake, which can be used for feed products.
3. Palm-oil refining: refining palm oil to produce a cleaner feedstock for renewable diesel, including removal of salts, pigments, gums, and other substances that can affect the performance of the renewable-diesel production. Energy consumption, emissions, and waste management are key factors in this stage.
4. Renewable-diesel production: the actual conversion of palm oil (or other feedstocks) into renewable diesel. Technologies like hydroprocessed esters and fatty acids (HEFA) are used. Compared to conventional-diesel production, renewable-diesel production significantly reduces CO_2 emissions. Key emissions come from the reactions carried out in the process and the generation of the hydrogen required for the hydrotreatment.
5. Renewable-diesel use: the final stage involves using the produced biofuel as a fuel. Here, the environmental impact depends on combustion efficiency, emissions, and overall energy performance.

Table 14.5 Carbon emissions of renewable diesel from different technologies and feedstocks.[23–29]

		Hydrogenation of oils and fats		Upgrading of pyrolysis oil	Fischer–Tropsch	Sugar to hydrocarbons	Anaerobic digestion	
Author	Allocation method	Palm oil ($g_{CO_2 eq.}$ MJ^{-1})	Cooking oil ($g_{CO_2 eq.}$ MJ^{-1})	Oil ($g_{CO_2 eq.}$ MJ^{-1})	Sorghum biomass ($g_{CO_2 eq.}$ MJ^{-1})	Jatropha ($g_{CO_2 eq.}$ MJ^{-1})	Miscanthus ($g_{CO_2 eq.}$ MJ^{-1})	Sorghum ($g_{CO_2 e}$ MJ^{-1})
Larnaudie *et al.*, 2020[23]	Mass allocation				17.7			17.7
Barla *et al.*, 2024[29]	Energy allocation						22.5–38.1	
Arguelles-Arguelles *et al.*, 2021[25]	Mass allocation	11.2						
Staples *et al.*, 2014[28]	System expansion				62.6			62.6
	Market allocation				4.13			4.13
Ou *et al.*, 2022[27]	Mass allocation	21.9–31.5						
	Energy allocation	5.0–15.0						
Bailis *et al.*, 2010[26]	Mass allocation					4.9		
	Energy allocation					65.6		
Omidkar *et al.*, 2023[24]	Mass allocation		15.88	35.68	12.7			12.7
	Energy allocation		10.49	75.26				

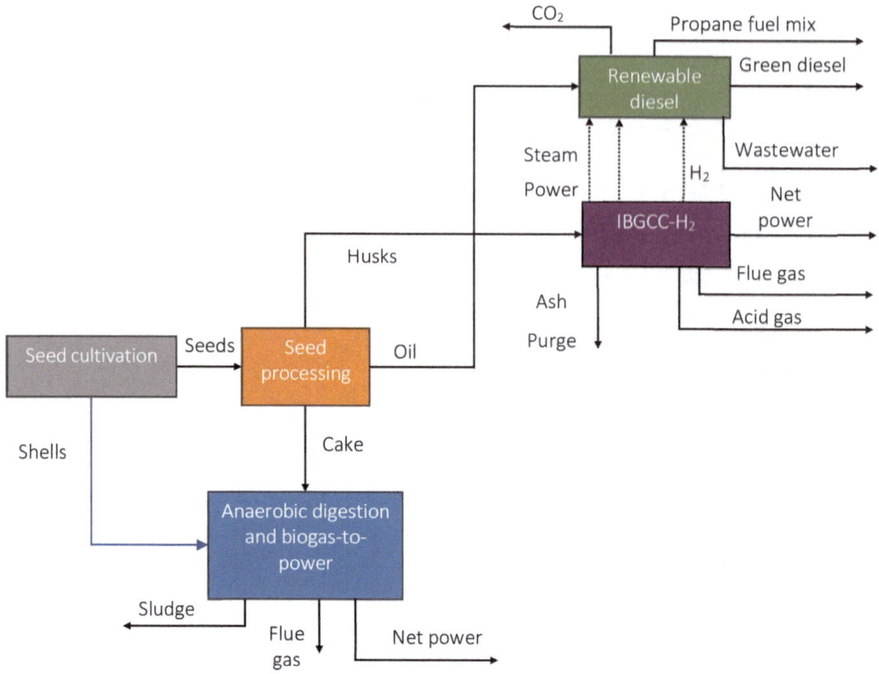

Figure 14.3 Integrated biorefinery system for renewable-diesel production using the HEFA process and on-site hydrogen and energy generation.

14.6 Towards an Integrated Renewable-diesel Biorefinery

Here a biorefinery for production of renewable diesel from oily crops is conceptualized in which the whole crop fractions are used, including extraction cake, oil and shells and kernels, for on-site energy and hydrogen generation by integration of several process technologies. Four main sections of the biorefinery are presented in Figure 14.3 and include the following:

1. Biorefinery Feedstock Preparation
    - Reception of fresh seeds and empty fruit hulls
    - Seed drying and conditioning
    - Husk removal
    - Oil extraction from kernels and refining
    - Crushing of fruit hulls for processing
2. Renewable-diesel Process
    - Hydroprocessing
    - Water and off-gas separation
    - Green-diesel distillation

3. Hydrogen and utility generation in an integrated biomass gasification and combined cycle and H_2 generation (IBGCC-H_2)
   - Seed husk gasification. In the renewable-diesel biorefinery the seed husk is gasified into syngas, which is a mixture of hydrogen and CO. This gas can be used as fuel to supply heat and power or can be used as a hydrogen source for the oil hydroprocessing. The gasifier comprises a fluidized-bed steam gasification reactor and a char combustor for indirect heating. This section includes separation using a cyclone separator equipment to remove any solids and tars.
   - Hydrogen generation. The syngas is enriched with hydrogen by passing through a water-gas shift reactor (WGSR) where CO reacts with water to form more hydrogen and CO_2. This is a reversible reaction limited by equilibrium and depends on the temperature and the amount of steam. The water can come from the initial gasification step, or as an additional input. From a sensitivity study of the equilibrium reactions, a steam-to-biomass mass ratio of 0.7 was determined so that no additional steam was required for the WGSR. The effluent goes into a CO_2 capture unit using absorption with amines. Any H_2S also needs to be removed. Then the hydrogen is purified to >99.9% purity in a pressure-swing adsorption (PSA) unit.
   - Heat and power generation. The remaining syngas that is not used to satisfy hydrogen demand is used as fuel to generate heat and power for the process. Any excess power is then exported to the electricity grid. In this way, fossil energy use and environmental impact can be reduced. Thus, remaining syngas as well as the off-gas streams from hydrogen separation processes are sent to the combustor. Part of the combustion heat is used in the gasifier and the rest is recovered into heat and power generation *via* steam. Any excess heat is extracted into steam generation. The superheated steam generation must be maximized at the highest pressure level, such as very high pressure (VHP) steam. Thereafter, lower pressure superheated steam can be generated if heat is available at lower pressure levels. The steam can be used to cover heating demands in the process.
4. Biogas production from fruit shells and extraction cake. Anaerobic digestion of fruit shells and oil extraction cake is used to produce biogas. Biogas is then used as fuel to generate more electricity.

With such an integrated system, the biorefinery will not only provide green diesel but also electricity to the grid as well as a green propane stream. All crop fractions are used and maintained within the system, thus contributing to resource conservation and a circular economy while decarbonizing transport and energy sectors.

14.7 Conclusions

Renewable diesel is an attractive option for decarbonization of transport fuels as it is an infrastructure-ready alternative. The main process technology that has been commercialized is the hydroprocessing of vegetable oils and fats, mainly due to the technological legacy from crude oil refining where hydrotreatment is extensively used. Thus, large capacities are in place and more are planned. The environmental savings of carbon emissions will depend on the type of feedstock and its cultivation management with low fertilizer and pesticide use as well as higher yields to lower eutrophication and land-use impacts. The challenges that remain are feedstock availability, quality and cost, followed by the search for new catalysts that favor milder process conditions and less hydrogen consumption. Other technologies such as production from syngas and cheaper feedstocks such as bio-oil are also pending further development to increase technological maturity and commercialization.

References

1. K. J. Kelly and A. C. Ragatz, Economy and Emissions Impacts from Solazyme Fuel in UPS Delivery Vehicles. Golden, CO (United States), 2018. Available from: https://research-hub.nrel.gov/en/publications/economy-and-emissions-impacts-from-solazyme-fuel-in-ups-delivery-.
2. S. L. Douvartzides, N. D. Charisiou, K. N. Papageridis and M. A. Goula, Green Diesel: Biomass Feedstocks, Production Technologies, Catalytic Research, Fuel Properties and Performance in Compression Ignition Internal Combustion Engines, *Energies*, 2019, **12**, 809. Available from: https://www.mdpi.com/1996-1073/12/5/809/htm.
3. E. Martinez-Hernandez, L. F. Ramírez-Verduzco, M. A. Amezcua-Allieri and J. Aburto, Process simulation and techno-economic analysis of bio-jet fuel and green diesel production — Minimum selling prices, *Chem. Eng. Res. Des.*, 2019, **146**, 60–70.
4. Virent, Fractional Multistage Hydrothermal Liquefaction of Biomass and Catalytic Conversion into Hydrocarbons Virent Final Report for, 2017. Available from: https://www.osti.gov/servlets/purl/1392550.
5. S. Jones, D. Elliott, C. Kinchin, C. Valkenburg, J. Holladay, S. Czernik, *et al.* Production of Gasoline and Diesel from Biomass via Fast Pyrolysis, Hydrotreating, and Hydrocracking, 2009. Available from: https://www.energy.gov/eere/bioenergy/articles/design-case-summary-production-gasoline-and-diesel-biomass-fast-pyrolysis.
6. V. Piemonte, Green Diesel-. Oil and Gas portal, 2015. Available from: https://www.oil-gasportal.com/green-diesel/.
7. ASTM, D975 Standard Specification for Diesel Fuel, 2022. Available from: https://www.astm.org/d0975-20a.html.
8. European Standard, EN 590:2009 – Automotive fuels – Diesel – Requirements and test methods, 2009. Available from: https://standards.iteh.ai/catalog/standards/cen/c3d8c75c-ea89-4b8a-afc3-0fb533afbc14/en-590-2009.
9. Gobierno de Mexico, DOF – Diario Oficial de la Federación. 2016. Available from: https://www.dof.gob.mx/nota_detalle.php?codigo=5450011&fecha=29/08/2016#gsc.tab=0.

10. European Standard, EN 15940:2023 – Automotive fuels - Paraffinic diesel fuel from synthesis or hydrotreatment-, 2023. Available from: https://standards.iteh.ai/catalog/standards/cen/81f825ab-600a-4696-802c-34829c64a642/en-15940-2023.
11. A. Zikri, I. Puspita, P. L. Erlinawati, M. S. Agus, P. E. Zalita and K. Andre, Production of Green Diesel From Crude Palm Oil (CPO) Through Hydrotreating Process by Using Zeolite Catalyst, Proceedings of the 4th Forum in Research, Science, and Technology (FIRST-T1-T2-2020), Atlantis Press, 2021, vol. 7, pp. 67–74. Available from: https://www.atlantis-press.com/proceedings/first-t1-t2-20/125952486.
12. Y. Qian, Y. Qiu, Y. Zhang and X. Lu, Effects of different aromatics blended with diesel on combustion and emission characteristics with a common rail diesel engine, *Appl. Therm. Eng.*, 2017, **125**, 1530–8.
13. J. Hu, Z. Du, C. Li and E. Min, Study on the lubrication properties of biodiesel as fuel lubricity enhancers, *Fuel*, 2005, **84**, 12–131601–6.
14. ASTM. D6468 Standard Test Method for High Temperature Stability of Middle Distillate Fuels, 2022. Available from: https://www.astm.org/d6468-22.html.
15. B. Zhang, J. Wu, C. Yang, Q. Qiu, Q. Yan and R. Li, *et al.*, Recent Developments in Commercial Processes for Refining Bio-Feedstocks to Renewable Diesel, *Bioenergy Res.*, 2018, **11**(3), 689–702. Available from: https://link.springer.com/article/10.1007/s12155-018-9927-y.
16. ENI, Ecofining™: technology to produce biofuels | Eni, 2023. Available from: https://www.eni.com/en-IT/actions/energy-transition-technologies/biofuels/biomass-ecofining.html.
17. M. Burgess, Topsoe to supply HydroFlexTM technology to Californian renewable diesel plant | Decarbonisation | gasworld, Gas World Magazine, 2023. Available from: https://www.gasworld.com/story/topsoe-to-supply-hydroflex-technology-to-californian-renewable-diesel-plant/.
18. E. Voegele, HydroFlex technology exceeds process guarantees at ND plant | Biomass Magazine, Biomass Magazine. 2021. Available from: https://biomassmagazine.com/articles/hydroflex-technology-exceeds-process-guarantees-at-nd-plant-18244.
19. NESTE OIL, Neste Oil to build a NExBTL renewable diesel plant in Rotterdam, 2008. Available from: https://www.neste.com/news/neste-oil-to-build-a-n-ex-btl-renewable-diesel-plant-in-rotterdam.
20. N. Asikin-Mijan, J. C. Juan, Y. H. Taufiq-Yap, H. C. Ong, Y. C. Lin and G. AbdulKareem-Alsultan, *et al.*, Towards sustainable green diesel fuel production: Advancements and opportunities in acid-base catalyzed H2-free deoxygenation process, *Catal. Commun.*, 2023, **182**, 106741.
21. M. M. Wright, D. E. Daugaard, J. A. Satrio and R. C. Brown, Techno-economic analysis of biomass fast pyrolysis to transportation fuels, *Fuel*, 2010, **89**(SUPPL. 1), S2–S10.
22. M. Patel, A. O. Oyedun, A. Kumar and R. Gupta, What is the production cost of renewable diesel from woody biomass and agricultural residue

based on experimentation? A comparative assessment, *Fuel Process. Technol.*, 2019, **191**, 79–92.
23. V. Larnaudie, M. Bule, K. Y. San, P. V. Vadlani, J. Mosby and S. Elangovan, *et al.*, Life cycle environmental and cost evaluation of renewable diesel production, *Fuel*, 2020, **279**, 118429.
24. A. Omidkar, H. Xu, Z. Li, K. Haddadian and H. Song, Techno-economic and life cycle assessment of renewable diesel production *via* methane-assisted catalytic waste cooking oil upgrading, *J. Cleaner Prod.*, 2023, **414**, 137512.
25. A. Arguelles-Arguelles, M. A. Amezcua-Allieri and L. F. Ramírez-Verduzco, Life Cycle Assessment of Green Diesel Production by Hydrodeoxygenation of Palm Oil, *Front. Energy Res.*, 2021, **9**, 690725.
26. R. E. Bailis and J. E. Baka, Greenhouse gas emissions and land use change from Jatropha curcas -based jet fuel in brazil, *Environ. Sci. Technol.*, 2010, **44**(22), 8684–8691. Available from: https://pubs.acs.org/doi/abs/10.1021/es1019178.
27. L. Ou, S. Li, L. Tao, S. Phillips, T. Hawkins and A. Singh, *et al.*, Techno-economic Analysis and Life-Cycle Analysis of Renewable Diesel Fuels Produced with Waste Feedstocks, *ACS Sustainable Chem. Eng.*, 2022, **10**(1), 382–393. Available from: https://pubs.acs.org/doi/full/10.1021/acssuschemeng.1c06561.
28. M. D. Staples, R. Malina, H. Olcay, M. N. Pearlson, J. I. Hileman and A. Boies, *et al.*, Lifecycle greenhouse gas footprint and minimum selling price of renewable diesel and jet fuel from fermentation and advanced fermentation production technologies, *Energy Environ. Sci.*, 2014, 7(5), 1545–1554. Available from: https://pubs.rsc.org/en/content/articlehtml/2014/ee/c3ee43655a.
29. R. J. Barla, A. Anand, S. Raghuvanshi and S. Gupta, Life cycle assessment of renewable diesel production, in *Renewable Diesel: Value Chain, Sustainability, and Challenges*, Elsevier, 2024, pp. 65–86.

CHAPTER 15

Integrating Renewable-diesel Biorefineries with Existing Petroleum Refineries: Scope and Challenges

A. DIMITRIADIS, L. C. CHRYSIKOU AND S. BEZERGIANNI*

Centre for Research & Technology Hellas (CERTH), Chemical Process & Energy Resources Institute (CPERI), India
*Email: sbezrg@certh.gr

15.1 Introduction

Oil depletion and air pollution caused by the extensive use of conventional fossil fuels have triggered the research of more environmentally friendly and renewable fuels. Although renewable forms of energy continue to increase, fossil fuels still constitute about 80% of the world's energy.[1,2] Over 96% of the world's energy demand for transport is obtained and produced from non-renewable sources,[1] with road vehicles (*e.g.* cars, trucks, buses) accounting for three-quarters of the CO_2 emissions derived from transport.[3]

The first-generation biofuels were based on energy crops (biodiesel and bioethanol), which limit their environmental benefits while leading to incompatibility problems with their fossil-based counterparts and combustion systems. FAME (fatty acid methyl ester) biodiesel is the most common biodiesel and is produced *via* transesterification of vegetable oils. Transesterification targets chemically breaking the fatty acids from vegetable oils with the use of alcohol. The resulting products are alcohol esters and glycerol.

Green Chemistry Series No. 89
Biorefining of Renewable Diesel: From Production to Market
Edited by Mohammad Aslam, Sanjeev Mishra and Jorge Aburto

Published by the Royal Society of Chemistry, www.rsc.org

The most common alcohol that is used for the transesterification process is methanol due to its price. However, ethanol is the preferred alcohol for transesterification due to its renewable-biomass origin and lower toxicity level. Other alcohols such as isopropanol may also be used.

FAME biodiesel has many advantages, such as high cetane number, good lubricity, high flash point, and absence of sulfur or aromatics. However, it also presents many disadvantages. More specifically, FAME's high viscosity, low energy content, high oxygen and water content, as well as high cloud and pour points, render technical limitations regarding its end-use (*e.g.* injector coking in diesel engines, microbial contamination of fuel tanks, degradation of quality during long-term storage). In addition to that, the initial feedstock strongly influences the quality of the produced biodiesel. In order to increase the quality, fresh vegetable oils are preferable, however biodiesel produced *via* the transesterification of fresh vegetable oils (sunflower, rape, soy, jatropha, *etc.*) has raised economic, environmental and social implications due to the associated "food *versus* fuel" debate. Moreover, good flow properties, thermal stability and low freezing point are other important criteria that have to be met by alternative fuels, especially in the aviation sector. Consequently, FAME biodiesel is not acceptable as an aviation fuel due to its low energy density and high freezing point in comparison with jet fuels.

As a result, despite the several advantages, biodiesel has not met the expectations for a green and environmentally friendly substitute for conventional diesel fuel and it is inappropriate for diesel engines at high blends (more than 10% v/v). Therefore, the transformation of triglyceride-containing feedstocks by other thermal and/or catalytic processes has been widely explored towards the production of synthetic paraffinic hydrocarbons. The most promising lipid-based alternative is the so-called HVO (hydroprocessing vegetable oil), which has significant advantages compared to FAME in terms of fuel properties as well as sustainability.

Catalytic hydroprocessing of vegetable oils is focused on producing a high-quality biofuel that is compatible with existing diesel-fuel infrastructure. The HVO technology consists of two steps: one catalytic hydrotreatment step which produces normal paraffins, and one catalytic isomerization step which leads to a mixture of *n*- and iso-paraffins. HVO fuels are also referred to as "renewable diesel" instead of "biodiesel", which is a term usually reserved for FAME. Since 2009, HVO production in Europe has been developing rapidly. More and more oil majors, such as ENI or TOTAL, or dedicated renewable-fuel producers, *e.g.* UPM or Preem, started investing in HVO-production units. The growth has been especially visible in Europe and the US. There is also growing interest coming from the Asian market that might soon change the map of the HVO sector. If, in the future, all the announced HVO projects come into life, the world production capacity will grow by more than 40%. It is more than obvious that HVO will play a significant role in the fuel market in the coming decades.[4]

Although HVO production has many advantages, standalone biofuel production requires high investment costs and energy consumption. As a result, biomass integration in an existing petroleum refinery has received growing interest due to environmental concerns and increasing energy demands. Petroleum refineries already have a well-developed infrastructure to produce fuels and base chemicals and, consequently, would not require additional intensive investments for the processing of alternative feedstocks. From this point of view, co-processing of biomass-derived feedstocks with petroleum fractions for production of hybrid fuels is an attractive option, which in some cases has found its practical realization already.[5–10] There are two main technologies that could be used for the co-processing of biomass feedstock with petroleum fractions; the first one is catalytic hydroprocessing and the second one is fluid catalytic cracking (FCC). Both technologies are found in virtually any conventional refinery. Figure 15.1 presents the units of a typical refinery where the FCC and hydrotreating (HDT)/hydrocracking (HDC) units can be easily noticed.

Catalytic hydroprocessing is a well-established and commonly-used refining technology generally employed to upgrade petroleum distillate fractions. There are two types of hydroprocessing technologies: catalytic hydrotreating and catalytic hydrocracking, both of which are normally expected in a typical refinery (see Figure 15.1). Both HDT and HDC units can be used for co-hydroprocessing. Co-hydroprocessing technology has many advantages, including: (a) it can be applied in an existing or slightly modified

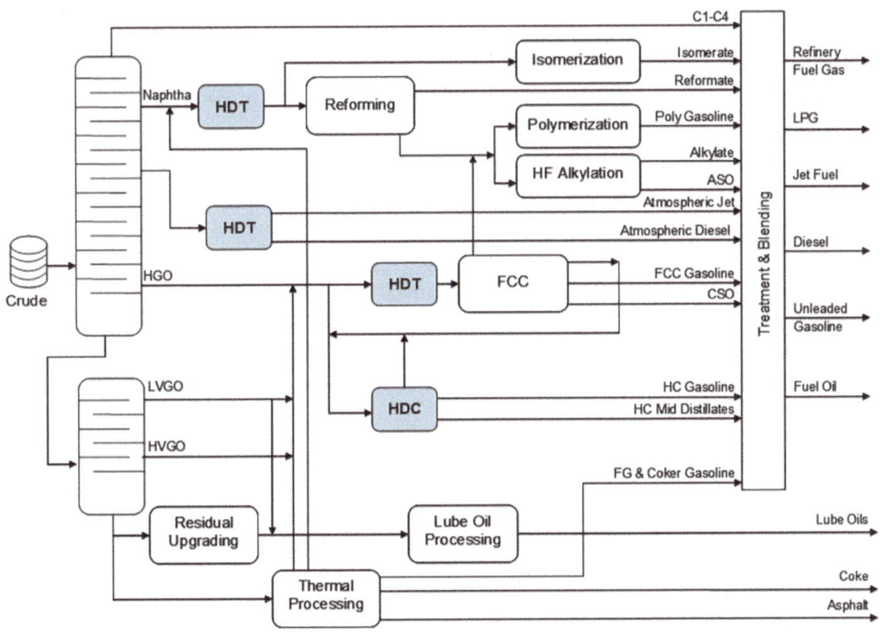

Figure 15.1 Various hydroprocessing units of a typical refinery.

gas oil desulfurization plant, minimizing investment costs; (b) it requires only one catalytic step; (c) it can achieve full conversion of the triglycerides contained in lipid feedstocks; and (d) it can upgrade lower quality fossil fractions (low cetane number; high aromatic, sulfur and nitrogen content) by co-hydroprocessing with triglycerides.

Fluid catalytic cracking is the most widely used process in petrochemical refineries for conversion of the heavy fractions of crude oil (*e.g.* vacuum gas oil or long residue (LR)) to gasoline and propylene, as the key products, and other hydrocarbon fractions such as C_4 olefins or a middle distillate, *i.e.* light cycle oil. Several reviews considering the FCC co-processing of biomass-derived and petroleum-derived feedstocks have been published to date.[11–14] While valuable information and conclusions related to co-processing of bio-based feedstocks have been provided, the information and relevant conclusions on the implementation of co-processing of bio-based feedstocks in industrial FCC technology is scarce.

The current chapter will focus on co-processing of bio-based feedstocks with petroleum-derived fractions. First, the most common petroleum-derived fractions will be presented, followed by the most common bio-based feeds. Then co-hydroprocessing technology will be explored, aiming to summarize all the advancements of this technology. Fluid catalytic cracking as a possible co-processing technology will be described in the next section, presenting all the key aspects of the current alternative option. The evaluation of the quality and sustainability of the hybrid fuel will be the main topics of the following parts, while finally the challenges and prospects of co-processing will be analysed in the last section. At the end of the current chapter, the reader will be able to identify the most promising bio-based feedstocks and petroleum fractions for hybrid-fuel production. In addition, significant knowledge will be gained regarding the integration of various bio-based feedstocks in hydrotreating and catalytic cracking units within a refinery, evaluating as well the key properties of the hybrid fuels.

15.2 Feedstocks

One of the main advantages of co-processing is the ability to combine a variety of bio- and fossil-based feedstocks/fractions without compromising product quality. The aim of this section is to present the most important fossil- and bio-based feedstocks that are using in co-processing for hybrid-fuel production.

15.2.1 Petroleum Fractions

The most commonly used fossil-based feeds for co-processing according to literature are mentioned below while their properties are presented on Table 15.1:

- straight run gas oil (SRGO)[15–18]
- straight run diesel, also called light atmospheric gas oil (LAGO)[19–22]

Table 15.1 Properties of fossil-based feedstocks.[53] Reproduced from ref. 53 with permission from Elsevier, Copyright 2018.

Properties	Units	SRGO	LAGO	HAGO	HVGO	VGO	LCO
Density	g ml^{-1}	0.83–0.84	0.85–0.86	0.85–0.86	0.91–0.97	0.84–0.92	0.95–0.96
Sulfur content	wt%	0.72–1.35	1.76	1.20–1.30	0.35–3.59	0.19–2.02	3.50
Nitrogen content	wppm	127–570	140	195–473	1600–1800	29–1900	1000
Oxygen content	wt%	—	—	0–0.50	0.37	0–0.28	—
Cetane index	—	54.4	—	—	—	—	26.2
Aromatic content	wt%	27–33	27–43	30	59	23	76–87
Viscosity	mm^2 s^{-1}	2.7–3.0	2.4–5.0	—	—	8.9–10.8	2.5–2.7
Distillation temperature							
10 wt%	°C	234–277	215–2.66	254	337–400	253–400	227–257
30 wt%	°C	296	280	308	390–430	329–437	276
50 wt%	°C	276	306	334	427–455	373–460	262–295
70 wt%	°C	326	331	359	468–490	422–495	327
90 wt%	°C	310–350	346–371	392	530–550	492–550	313–368

- heavy atmospheric gas oil (HAGO)[23]
- vacuum gas oil (VGO)[24]
- heavy vacuum gas oil (HVGO)[25]
- light cycle oil (LCO)[26–29]

15.2.1.1 *Straight Run Gas Oil*[30–32]

In general, SRGO has a diesel-fuel odour and relatively low vapour pressure. It is obtained by crude-oil distillation and is used as an intermediate feedstock for further processing within the refining system to increase the yields of lighter products. It is characterized by a low aromatic content, a high cetane index and a boiling-point range of 234–350 °C. SRGO is a good candidate for co-processing due to its low sulfur content and its suitable boiling-point range.

15.2.1.2 *Light Atmospheric Gas Oil*

LAGO is a cheap liquid petroleum distillate that is heavier than naphtha. LAGO is primarily made up of molecules with 14–20 carbon atoms. The LAGO distillation has a boiling range of 215–343 °C, a low total acid number (TAN) of 0.45 mg KOH per g) and a pour point of 12 °C. The primary use of light atmospheric gas oil is as a blendstock to make diesel or heating oil. However, it must typically go through a distillate hydrotreater first to remove sulfur. Diesel is typically blended from LAGO. LAGO has been tested in different blends with waste soya oil,[33] sunflower oil,[15,18] rapeseed oil,[21] tallow oil[21] and tall oil.[21]

15.2.1.3 *Heavy Atmospheric Gas Oil*

HAGO is a heavier petroleum distillate than LAGO, with a high sulfur content and is also used for diesel production. HAGO must also go through a distillate hydrotreater to remove the high sulfur content and break the large molecules, rendering high-quality blending components for diesel production.[34–36] Furthermore, HAGO can also be fed into the FCC unit when a refinery is trying to maximize the yield of gasoline over diesel.

15.2.1.4 Vacuum Gas Oil

VGO is produced from the vacuum distillation of the atmospheric residue. In general, the boiling point of VGO is 280–580 °C and it consists mainly of paraffins, naphthenes and aromatics. In addition, it is characterized by high sulfur and nitrogen content, which can poison the catalysts during the upgrading process in FCC units.[37] Due to the high sulfur and nitrogen content, VGO sometimes goes through hydrotreaters in order to remove the sulfur and nitrogen heteroatoms prior its co-processing in FCC units.[38–41] It is worth noting that the properties of VGO (density, viscosity, boiling-point range, *etc.*) may vary in different studies. For example, it is suggested that the usual feedstocks used in Brazil are heavier than those used in the USA or Europe.[42] In recent years, VGO has been investigated as a potential feedstock for co-processing in blends with different lignocellulosic[43–49] and triglyceride-based feedstocks.[50,51]

15.2.1.5 Heavy Vacuum Gas Oil

HVGO is part of the petroleum hydrocarbon heavy distillate family. HVGO is one of the two outputs of the vacuum distillation tower and is produced through the separation of the heaviest fraction from the atmospheric distillation column. The primary use of HVGO is to feed the cracking units such as the FCC or the hydrocracker. These units upgrade HVGO into more valuable products, namely gasoline, diesel and many other by-products. The fluidized catalytic cracker is a refining unit that aims to crack the HVGO in the presence of catalyst to produce gasoline rich mix products. On the other hand, hydrocracker, which is also a refining unit, aims to crack the HVGO utilizing hydrogen to produce a mixture of products rich in middle distillates such as diesel and kerosene. In general, HVGO is a promising feedstock for co-processing with various lipids as it consists of similar-range heavy hydrocarbons. Both technologies, catalytic cracking[16,52] and hydrocracking,[25] have been investigated for co-processing of HVGO in blends with various vegetable oils such as canola oil, sunflower oil *etc.*

15.2.1.6 Light Cycle Oil[29]

Cycle oil is the diesel-range product from the FCC unit. Cycle oil is not very highly valued as a diesel blendstock because of its high aromatic content, which gives it a low cetane number. In addition to the high aromatic content (40–60%), it also contains tetralins (20–30%) and it is characterized by high sulfur and nitrogen content. The boiling-point range of the hydrocarbons varies, while the side chains are very short paraffins.[18] Thus, LCO requires upgrading in order to improve its quality. A common practice is to feed LCO in a mixture with SRGO to the hydrotreater. However, the co-hydroprocessing of blends of SRGO and LCO and bio-based feeds is preferred, leading to hybrid-fuel production.[26] Several studies have examined co-hydroprocessing blends of SRGO and LCO with different types of lipids. For instance, the co-hydroprocessing of different prepared mixtures from straight run gas oil (70–100%), light cycle oil (0–10%) and waste animal fat (0–20%) was carried out to yield fuels that would satisfy the requirements of the EN 590:2013 standard.[26,27] A similar study of co-hydroprocessing was also carried out for a mixture of straight run gas oil (70–85%), light cycle oil (15%) and waste cooking oil (0–15%).[28]

15.2.2 Bio-based Feedstocks

15.2.2.1 Lipids

Triglycerides are characterized as a suitable raw material for high-quality fuel production. However, depending on the choice of the triacylglyceride source, the implementation of biofuel technology can have different economic and social outcomes.[54] First-generation biofuels were produced *via* transesterification of fresh vegetable oils such as soybean, peanut, sunflower, palm and coconut oils. However, the cost of these types of lipids is high and their availability is unstable due to market competition. Furthermore, the increase in demand for fresh oil for biofuel production has led to an increase in food prices, leading to the well-known "food *versus* fuel" debate. The use of non-edible oils was promoted in order to mitigate the drawbacks of fresh edible oils. Non-edible oils include rapeseed oil, jatropha oil, jojoba oil, waste or used cooking oils from homes and restaurants, and animal fats from slaughterhouses and industries. Thus, the development of alternative feedstocks and processing technologies has been gaining interest to further augment the feasibility of biofuel or hybrid-fuel production.[55] However, non-edible oils such as residual lipids are limited in quantity; therefore, alternative lipid-based feedstocks like microalgae oils have been studied. Microalgae have gained considerable attention as an alternative feedstock for biofuel production, particularly in combination with genetic-modification strategies to enhance lipid productivity. Another promising lipid-based feedstock is microbial oils. Microbial oils can be produced by many types of microorganisms including microalgae, fungi and bacteria, with the advantages of a small footprint, high lipid content and efficient uptake of carbon dioxide. The different types of lipids are presented in more detail below. Table 15.2 presents the fatty acids composition of various lipids.

15.2.2.1.1 Edible Oils. Today, the most common edible oils that are used for biofuel production are palm, soybean, sunflower, rapeseed and

Table 15.2 Fatty acid composition of different types of oils.

Fatty-acid composition (wt%)												
Edible oils												
	C8:0	C10:0	C12:0	C14:0	C16:0	C16:1	C18:0	C18:1	C18:2	C18:3	C20:0–C22:0	C20:1–C22:1
Palm	0	0	0	1–6	32–47	0	1–6	40–52	2–11	0	0	0
Soybean	0	0	0	0.3	7–11	0–1	3–6	22–34	50–60	2–10	5–10	0
Sunflower	0	0	0	0	6	0	4.2	18.7	69.3	0.3	1.4	0
Rapeseed	0	0	0	0	2–5	0.2	1–2	10–15	10–20	5–10	9	50–60
Peanut	0	0	0	0.5	6–11	1–2	3–6	39–66	17–38	0	5–10	0
Coconut	5–9	4–10	44–51	13–18	7–10	0	1–4	5–8	1–3	0	0	0
Safflower	0	0	0	0	5.2	0	2.2	76.3	16.2	0	0	0
Corn	0	0	0	0–2	8–10	1–2	1–4	30–50	34–56	0	0	0–2
Linseed	0	0	0	0.2	5–9	0	0–1	9–29	8–29	45–67	0	0
Non-edible oils												
Jatropha	0	0	0	0.1	14.2	0.7	7.0	44.7	32.8	0.2	0.2	0
Castor[a]	0	0	0	0	1–2	0	1–2	3–4	4–5	0.5–0.7	0	0
Jojoba	0	0	0	0	0.3	0.3	0.2	9.3	0	0	0	88.8
Cottonseed	0	0	0	0–3	17–23	0	1–3	23–41	34–55	0	0	2–3
Residual lipids												
Used cooking oil	0.007	0	0	0.05	7.4	0.09	2.9	33.5	54.8	0.3	0.2	0
Lard	0	0	0	1	25–30	2–5	12–16	41–51	4–232	0	0	2–3
Tallow	0	0	0.2	2–3	25–30	2–3	21–26	39–42	2	0	0.4–1	0.3

[a] Castor oil consists of 85–90 wt% ricinoleic acids.

peanut oils.[56] Other types of edible oils are linseed, corn and safflower. Traditionally, the edible oils are used for first-generation biodiesel production, the well-known FAME biodiesel, but they can also be used for co-processing with petroleum fractions for hybrid-fuel production. The edible oils are produced from vegetable resources and they are used mainly for direct human consumption as food intake. The biggest disadvantage of edible oils for biofuel production is that they compete with food production. In the literature the described drawback is called the "food *versus* fuel debate", and has led to increase of the food prices. In addition, edible oils have a high production cost compared to non-edible oils and, as they are mainly used for food production, their supply is limited.

15.2.2.1.2 Non-edible Oils. The use of non-edible oils is preferable for biofuel production as non-edible oils do not compete with food production. The most common non-edible oil feedstocks that are used for biofuel production are castor, jatropha, jojoba and cottonseed. However, non-edible oils are characterized by a high amount of free fatty acids (FFAs) that react with basic catalysts producing emulsions and soap during the production of first-generation biodiesel. Therefore, the use of this type of oil for FAME biodiesel production requires an initial mild treatment step before the transesterification process. However, non-edible oils are very promising feedstocks for hybrid-fuel production *via* co-processing with fossil-based feedstocks as they do not raise the cost of food production. Studies on co-processing of jatropha oil with SRGO at various blending ratios (10–50% jatropha in SRGO) have shown that the final product consists of hydrocarbons in the diesel-fuel range.[19]

15.2.2.1.3 Residual Lipids. Due to the limitations of first-generation biofuels, second-generation biofuel technologies have been developed utilizing residual biomass feedstocks, increasing the sustainability of the process. Residual oils are considered to be waste or used cooking oils from homes and restaurants as well as animal fats from slaughterhouses and industries. The animal fats can be either from lard or from tallow, depending on the industry. These feedstock sources have, however, higher amounts of so-called free fatty acids and water in comparison with edible and non-edible oils and may thus need pre-treatment before processing.

In general, waste cooking oils are characterized by their low cost, ready availability and ease of collection from households and restaurants. In many countries, used cooking oils are thrown away unsafely causing environmental pollution, despite their importance in biodiesel production. Today, in most developed countries there are laws that forbid the disposal of used cooking oils from restaurants. There are collecting companies whose main task is to collect used cooking oils from restaurants and homes and sell them to biofuel refineries. In general, waste cooking oils are a very promising feedstock for biofuel production and there are many publications in the literature that study the production of biofuels and hybrid fuels from used cooking oils utilizing various technologies. Although waste oils are residual feedstocks with a very low cost, one of their main disadvantages is large-scale collection, which has not been solved yet.

Another promising feedstock for biofuel production or co-processing is animal fat from either tallow or lard. In general, animal fats are very viscous or solid at room temperature and have poor vaporization characteristics because of their high saturated fatty acid content. In most developed countries the disposal of the animal fats on the environment is forbidden, and thus the use of animal fats for biofuel or hybrid-fuel production is a very promising feedstock that does not compete with the food industry. The main drawback of animal fats is their limited availability.

15.2.2.1.4 Algae Oils. Algae is the biomass used for third- and fourth-generation biofuels. The main difference between third- and fourth-generation biofuel is the implementation of genetic modification/engineering of algae for algal-based biofuel production.[57] In recent years, microalgae have gained importance and are popular candidates as they have a high production rate and high lipid content. In general, algae show rapid growth as they can double every few hours and be harvested daily. It is believed that each area unit of land used for a microalgae plant produces 10 times more oil compared to that of a typical terrestrial oleaginous crop.[58] One of the main advantages is that they can be cultivated in a saline water medium and thus do not compete with other terrestrial crops. In addition, they can grow in and purify wastewater and can tolerate extreme weather conditions. Finally, algae, as a photosynthetic microorganism, use sunlight, capture carbon dioxide and release oxygen during their growth.

Although the process of biofuel production from microalgae is well studied, the economic feasibility of the overall biorefinery process remains a challenge for its commercialization. A very important drawback of microalgae is the high cost of the oil extraction. Generally, the oil production from microalgae includes three steps: (1) microalgae cultivation (in open ponds or photobioreactors; (2) harvesting/concentration; and (3) oil extraction. Identifying an optimal balance among these stages is essential to reduce the environmental impacts and costs of the overall process. There are many studies in the literature that investigate the optimization of biofuel production *via* algae cultivation.[59]

15.2.2.1.5 Microbial Oils. As well as vegetable oils, used oils and animal fats, microbial oils are another important source for renewable-fuel production. Microbial oil can be produced *via* fermentation of the carbohydrate and protein fractions after enzymatic hydrolysis of food wastes from restaurants that are rich in oils/fats, starch and proteins. Microbial systems are more controllable, economic, and scalable for biorefining to green fuels and chemicals.[60,61] Microbial lipids are rich in storage lipids (mainly triacylglycerides, TAGs), making them an excellent feedstock for biofuel production. Due to the wide availability of this feedstock, local or centralized biorefineries could be developed in all countries for biodiesel production.

15.2.2.2 *Pyrolysis Bio-oil*

Further to the traditional lipid-based feedstocks, catalytic pyrolysis bio-oil (CPO) is an attractive alternative feedstock that can be co-processed with petroleum fractions to produce hybrid fuels. Pyrolysis is an attractive process to thermally crack the initial biomass at 500 °C in the absence of oxygen to produce a liquid product. Besides the liquid pyrolysis bio-oil product, gases as well as char are also produced by the process. Typical feedstocks for pyrolysis bio-oil production includes lignocellulosic biomass such as wood and forestry residues, crops and crop residues, marine products and bio-wastes. Such biomass utilization can also offer an optimum solution to waste management. It is estimated that the biomass waste generated from agricultural and forestry industries is in the order of 140 Gt annually. Traditional handling of biomass residues such as burning in the open air causes severe environmental pollution. These agricultural and forestry wastes can be used as low-cost biomass sources for energy generation while effectively mitigating biomass waste-management issues.[24] As bio-oils are carbon-neutral, the application of bio-oils can effectively mitigate the burden of greenhouse gas (GHG) emission arising from fossil fuel combustion.

Pyrolysis bio-oil, in general, is a free-flowing liquid at room temperature which is green, brown or black in color, depending upon the chemical composition and the amount of micro-carbon content. There are three types of pyrolysis, slow, fast and flash pyrolysis, all requiring an initial drying step. In slow pyrolysis, the heating rate is low and the target is mostly biochar production. Fast pyrolysis is characterized by fast heating rates (10–200 $^{\circ}C\,s^{-1}$) and short residence times (0.5–1.0 s), with the operation is taking place at 850–1250 °C. Last but not least is flash pyrolysis, which is characterized by very high heating rates (>1000 $^{\circ}C\,s^{-1}$) and high reaction temperatures, approximately 850–1300 °C, rendering a better-quality product compared to fast pyrolysis,[62] as the fast pyrolysis bio-oil (FPO) has several undesirable properties such as instability, low heating value and high acidity (see Table 15.3). In addition, the viscosity of bio-oil increases rapidly with heating, which leads to bio-oil polymerization, phase separation, and coke formation. The low heating value is a result of its high-water content, which comes from free water in the biomass and from condensation reactions during production. These properties are related to its content of solids, ash, alkali metals, water, carbonyls, and acids, as well as polarity and inhomogeneity. Ash refers to the inorganic part of solids such as metals, ranging from 0.1 wt% to 0.4 wt% depending on biomass feedstock. Solid particles have a negative influence on the storage stability and the combustion performance of crude bio-oils. The pH level of crude bio-oils is low due to the presence of an appreciable amount of organic acids such as acetic and formic acids, which are responsible for the corrosiveness of crude bio-oils.

Table 15.3 Physical properties of differently prepared bio-oils (PO: pyrolysis oil; HPO: hydrotreated pyrolysis bio-oil).[53] Reproduced from ref. 53 with permission from Elsevier, Copyright 2018.

Properties	Units	PO	HPO
Yield of bio-oil from biomass	wt%	46.4–69.0	24.0
Water content	wt%	15.0–61.7	2.1–15.9
Density	$g\,ml^{-1}$	1.05–1.25	0.93
Carbon content	wt%	29.4–65.0	63.0–84.4
Hydrogen content	wt%	5.0–8.5	8.0–10.6
Oxygen content	wt%	25.0–61.7	4.9–28.0
Nitrogen content	wt%	0.2–4.3	0.0–4.9
Sulfur content	wt%	≤0.4	0.0–0.11
Heating value	$MJ\,kg^{-1}$	16.0–27.9	25.2–46.0
Viscosity at 40 °C	cP	11.2–100.0	~45

Co-processing bio-oils with petroleum fractions provides a promising solution owing to the hydrogen transfer from hydrogen-rich petroleum hydrocarbons to bio-oils. However, due to its composition, pyrolysis bio-oil, in general, is not miscible with the fossil fuel fractions and affects the combustion of the final blend.

To allow blending, the high polarity of the bio-oil and its high content of water need to be reduced. Thus, an upgrade step for pyrolysis oil is considered to be important prior to its use as a liquid fuel or chemical feedstock. Two different upgrade routes are generally considered to improve bio-oil properties: hydrotreatment and fluid catalytic cracking.

Hydrotreatment is a promising technology for mild upgrading of pyrolysis bio-oil. There are many studies in the literature that investigate the upgrading of the pyrolysis bio-oil.[53] However, the upgrading of bio-oil is not an easy step, and coke formation and catalyst deactivation during hydrotreatment have been reported by many researchers. In order to assess these issues, a two- or multi-stage system is needed to hydrotreat and stabilize the bio-oil at lower temperatures (<280 °C) and hydrodeoxygenate the oil at higher temperatures (370–400 °C). Even if the carbon yield from hydrodeoxygenation (HDO) is high, the process has not yet been commercialized as a result of the high process and equipment costs and the technical challenges related to reactor plugging and catalyst deactivation.[63] Typical catalysts are sulfided $CoMo/Al_2O_3$ or $NiMo/Al_2O_3$.

Fluid catalytic cracking of raw pyrolysis bio-oil is hardly possible as a result of extensive coking. Therefore, research has been conducted on catalytic pyrolysis, where typically a zeolite-based catalyst has been used as bed material to produce a bio-oil with a lower oxygen content. The challenge with catalytic pyrolysis of biomass is how to improve catalyst properties and process conditions to obtain high yields of hydrocarbons while minimizing coke formation and deactivation of the catalyst by the metals in the feedstock.

The partially deoxygenated bio-oils are more stable compared to the crude bio-oils and chemically more resemble petroleum intermediates, and therefore are deemed potential bio-feedstocks for coprocessing in conventional refineries. Considering multistep hydroprocessing as the most promising technology to upgrade raw pyrolysis bio-oil, the upgraded bio-oil can be called hydrotreated pyrolysis bio-oil (HPO), and it can be easily co-processed with petroleum-refinery streams. A major advantage of co-processing partially stabilized pyrolysis oil in existing petroleum-refining infrastructure is to leverage existing capital instead of constructing new parallel infrastructure for standalone hydroprocessing of pyrolysis oil. Typical properties of fast pyrolysis bio-oil and the hydrotreated pyrolysis bio-oil are presented in Table 15.3.

Today, wood-based fast pyrolysis is a maturing technology (technology readiness level (TRL) of 7–9) with near-commercial production facilities based on the technology of BTG Bioliquids in Finland (Green Fuel Nordic, 24 000 tonnes per year), the Netherlands (Twence, 24 000 tonnes per year), and Sweden (Pyrocell, 24 000 tonnes per year), and based on the technology of Ensyn in Canada (Renfrew and Côte Nord, 40 ML per year).[63]

15.2.2.3 Hydrothermal Liquefaction to Biocrude

Another promising thermochemical technology for converting lignocellulosic biomass to alternative fuels is hydrothermal liquefaction (HTL). In contrast to pyrolysis, feedstock drying is not necessary for HTL, significantly reducing the biomass conversion cost. In HTL, wet feedstock is liquefied in a hot (250–400 °C) and pressurized (40–220 bar) water environment into biocrude. Typically, alkaline conditions are used to modify the ionic medium to favor certain base-catalyzed condensation reactions, which can lead to aromatic oil formation. In comparison to pyrolysis bio-oil, the biocrude from HTL is more deoxygenated and more hydrophobic with less dissolved water. The biocrude is more viscous but less dense than pyrolysis bio-oil. The major compound groups identified in biocrude include acids, alcohols, cyclic ketones, phenols, methoxyphenols, and more condensed structures like naphthols and benzofuran.[64]

Solvents can be employed in the HTL process, rendering higher-quality products compared to those obtained from the pyrolysis process due to oxygen content around 5–10 wt%, lower content of corrosive carboxylic acids, kinematic viscosities as low as 10–20 cSt, and lower water content of around 0–5 wt% depending on the separation method. On the other hand, the need for high-pressure operation increases the operating costs. HTL to biocrude is also known as hydrous pyrolysis. In general, HTL can be characterized as a flexible technology in terms of feedstock type because a wide variety of bio-based and waste feedstocks can be utilized, such as lignocellulosic biomass, industrial wastes, food wastes, swine manure, algae, arborous crops, wastes from forest industry, *etc.* The HTL process directly converts the initial biomass to the biocrude product utilizing a solvent and in some cases a catalyst.

Although HTL biocrude has better properties compared to pyrolysis bio-oil, an upgrading step is also necessary prior to its use as a blending component with fossil-based feedstocks for hybrid-fuel production. The reason is the oxygenate compounds in the biocrude that have to be removed *via* hydrodeoxygenation. As a result, similar to pyrolysis bio-oil, hydrotreatment technology can also be applied in order to remove the high oxygen content from the initial HTL biocrude prior its use in a petroleum refinery. The situation is different with nitrogen, which cannot be completely removed during hydrotreatment. Other challenges with hydrotreatment of HTL oil have been the high viscosity of the liquid, which requires replacement of small-diameter piping and heat tracing, and the minerals in the oil, which might deactivate the catalysts.[63] The advantage of this approach is the use of decentralized pyrolysis and HTL plants, which can be located near the biomass production site. Only the biocrude is transported to a centralized refinery, which reduces transportation costs as a result of the increase in the volumetric energy of the oil compared to the original biomass. Many oil refineries have different configurations and product profiles, including fuels, chemicals, and asphalt. There are therefore many possible locations for inserting the biocrude in a refinery, depending upon the nature of the biocrude and target product.

Inserting the upgraded biocrude into petroleum fuels as finished fuels represents the lowest risk for the refinery, while inserting it at the pre-distillation phase poses the highest risk. The greatest benefit for the refinery is achieved if the biocrude is introduced in some of the processing units after mild upgrading. Both cracking and the capability to remove oxygen from the biocrude are required from these processing units, and therefore co-processing could potentially take place in the fluid catalytic cracker, hydrotreater, or hydrocracker.[65–67]

In 2014 an extensive technoeconomic and life-cycle assessment made for the US Department of Energy modelled the large-scale production cost of liquid fuels from wood by pyrolysis + hydrotreating and HTL + hydrotreating, respectively. HTL was concluded to be both more energy efficient and cheaper compared to pyrolysis and the production cost of a completely deoxygenated HTL fuel was estimated to be 0.53 USD per litre gasoline equivalent.[68]

HTL is being commercialized by a number of stakeholders, with the prominent players being Canadian/Danish Steeper Energy, Australia-based Licella and Southern Oil Refining Pty Ltd, ENI from Italy, and US company Genifuel. No large-scale or commercial production of HTL biocrude yet exists. The largest HTL plant existing is the Silva Green Fuel demo plant located in Tofte, Norway, with a production capacity of 4000 L per day.[63]

15.3 Hydroprocessing Technology

In the previous sections, various bio-based as well as fossil-based feedstocks for co-processing were presented. The current section will present the hydroprocessing technology as well as a promising candidate for hybrid-fuel production. Hydroprocessing is not a new technology: it is a widespread technology that is used for the upgrading of fossil-based streams in a typical petroleum refinery. Hydroprocessing aims to first remove the sulfur, nitrogen and oxygen heteroatoms, and in some cases also targets the reduction of olefins and aromatics. Secondly, it aims to saturate the unsaturated hydrocarbons, increasing the hydrogen-to-oil ratio. The main reactions of hydroprocessing are hydrodesulfurization (HDS), hydrodenitrogenation (HDN), hydrodeoxygenation, cracking and saturation.

Hydrodesulfurization removes the sulfur content by forming hydrogen sulfide. During the hydrodesulfurization process, hydrogen breaks carbon–sulfur bonds and saturates the remaining hydrocarbon chains. The most basic sulfur types are mercaptans, sulfides, disulfides, thiophenes, benzothiophenes, and dibenzothiophenes. In general, sulfur removal from mercaptans, sulfides and disulfides is an easy process. On the other hand, sulfur removal from the heteroatomic aromatic compounds is more difficult as it proceeds by ring-opening followed by sulfur removal and saturation of the olefin group. In general, heavy hydrocarbon fractions contain more difficult species compared to lighter hydrocarbon feeds.

The order of difficulty for removal of sulfur compounds, from easiest to hardest, is: mercaptans ⇒ sulfides ⇒ disulfides ⇒ thiophenes ⇒ benzothiophenes ⇒ dibenzothiophenes. The reactions involved are:

- mercaptans: $RSH + H_2 \rightarrow RH + H_2S$
- sulfides: $R_2S + 2H_2 \rightarrow 2RH + H_2S$
- disulfides: $(RS)_2 + 3H_2 \rightarrow 2RH + 2H_2S$
- thiophenes: $C_4H_4S + 4H_2 \rightleftharpoons C_4H_{10} + H_2S$

Hydrodenitrogenation reactionsremove the nitrogen heteroatoms in the form of ammonia (NH_3). Nitrogen is mostly found in the heavier petroleum fractions in the form of five- and six-membered aromatic ring structures. The complexity and quantity of nitrogen-containing molecules increases with a rise in boiling range and make them more difficult to convert. In general, denitrogenation reactions require more severe conditions compared to desulfurization reactions as they are more difficult reactions with a different path. The path of desulfurization reactions is to remove the sulfur leading to the formation of olefins as intermediate products, which are then saturated. In contrast, during denitrogenation reactions the aromatic ring is saturated and then the nitrogen is removed. Below is presented the denitrogenation path for pyridine (C_5H_5N):

$$\begin{aligned} C_5H_5N\,(\text{pyridine}) + 5H_2 &\rightarrow C_5H_{11}N\,(\text{piperidine}) + 2H_2 \\ &\rightarrow C_5H_{11}NH_2\,(\text{amylamine}) + H_2 \\ &\rightarrow C_5H_{12}\,(\text{pentane}) + NH_3\,(\text{ammonia}) \end{aligned}$$

The overall reaction for denitrogenation of pyridine is as follows:

$$C_5H_5N + 5H_2 \rightarrow C_5H_{12} + NH_3$$

Hydrodeoxygenation reactions remove the oxygenated compounds in the form of water and the corresponding hydrocarbon by the reaction of hydroxyl groups with the hydrogen. In general, the petroleum crudes contain low levels of oxygen. The lower-molecular-weight oxygen compounds can easily be removed, while the higher-molecular-weight compounds like furans are more difficult to convert.

Another type of reaction during hydrotreatment is the saturation reaction that saturates the unsaturated hydrocarbons by breaking the double bonds. During saturation of olefins, organic compounds containing double bonds are converted to their saturated homologues. Olefin saturation reactions are very rapid and highly exothermic. When the concentration of olefins is high, it can cause excessive coking which can lead to pressure drop and poor liquid flow distribution over the catalyst bed. Olefins are not found in crude oils but are formed in thermal or catalytic units. Fractions containing olefins are unstable and must be protected from contact with oxygen to prevent formation of polymer gums.

Aromatic saturation is also referred to as hydrodearomatization. During these reactions, some of the aromatic compounds are converted to naphthenes. Aromatic saturation is necessary for the improvement of petroleum product quality, *e.g.* smoke point, diesel index, *etc.* Aromatic saturation is the most difficult reaction. The aromatics are present as one-, two-, and three-ring structures called mono-, di-, and tri-aromatics. Naphthenes are the products from the saturation of aromatics. The aromatic saturation reaction requires high hydrogen partial pressure.

Demetallation or hydrodemetalation is the reaction where metals (organometallics) are removed and converted to the respective metal sulfides. Most metallic impurities are present in distillates as organometallic compounds and occur at ppm or ppb levels, but their effect is significant. One of the reasons for catalyst deactivation is the deposition of metals in the catalyst's pores, which cannot be removed by regeneration. The most common metals are arsenic, mercury, nickel, vanadium, silicon, iron, Ca, Mg, Na, Pb, phosphorus, *etc.* Na, Ca and Mg are present in the form of inorganic compounds.

Hydrodehalogenation reactions remove organic halides which are converted to hydrogen halides. Organic halides, such as chlorides or bromides, are present in petroleum fractions in trace amounts. During hydrotreatment, the organic halides react with hydrogen and are converted to hydrogen halides. The hydrogen halides react with ammonia and produce ammonium chloride salts, which are then removed with addition of wash water in the downstream sections.

During hydroprocessing gaseous products are also formed. The gaseous products may consist of some gaseous hydrocarbons such as methane, ethane, propane, *etc.* In addition, some quantity of CO and CO_2 may be observed as gaseous products, which are produced by decarbonylation and decarboxylation reactions.

Examples of reactions are given in below (R represents a hydrocarbon group):

$$RSH + H_2 \rightarrow RH + H_2S \text{ (hydrodesulfurizaion reaction)}$$

$$RCH = CH_2 + H_2 \rightarrow RCH_2CH_3 \text{ (olefin saturation)}$$

$$RNH_2 + H_2 \rightarrow RH + NH_3 \text{ (hydrodenitrogenation)}$$

$$ROH + H_2 \rightarrow RH + H_2O \text{ (hydrodeoxygenation)}$$

$$RCH_2COOH + H_2 \rightarrow RCH_3 + CO + H_2O \text{ (decarbonylation)}$$

$$RCH_2COOH \rightarrow RCH_3 + CO_2 \text{ (decarboxylation)}$$

$$RCH_2COOH + 3H_2 \rightarrow RCH_2CH_3 + 2H_2O \text{ (hydrodeoxygenation)}$$

$$CO + 3H_2 \rightleftharpoons CH_4 + H_2O \text{ (methanation 1)}$$

$$2CO + 2H_2 \rightleftharpoons CH_4 + CO_2 \text{ (methanation 2)}$$

$$CO_2 + 4H_2 \rightleftharpoons CH_4 + 2H_2O \text{ (methanation 3)}$$

$$CO_2 + H_2 \rightleftharpoons CO + H_2O \text{ (water-gas shift)}$$

Depending on the targeted product the choice of the optimum catalyst is crucial and strongly influences the rate of reactions. The main components of catalysts for hydrotreating naphtha and diesel are active metals dispersed on high-surface-area alumina. The most-known categories of industrial catalysts are those based on cobalt–molybdenum (CoMo) and nickel–molybdenum (NiMo). Activities of these catalysts vary over a wide range. It is generally accepted that CoMo catalysts are the most appropriate for desulfurization, while they exhibit low activity for the removal of nitrogen and the saturation of aromatic compounds. NiMo catalysts exhibit higher activity in nitrogen reduction and saturation of aromatics, while they are less efficient for sulfur removal.

As already mentioned, catalytic hydroprocessing is widely used by all refineries worldwide to upgrade fossil-based intermediate streams. In most refineries more than one hydroprocessing unit can be found so more reactions can be targeted (see Figure 15.1). As a result, due to the availability and flexibility of the underlying refinery hydroprocessing plants, the described process is a promising candidate technology for co-processing of bio-based feedstocks with fossil streams in order to convert the bio-based feedstocks/intermediates in parallel. In many cases, some refinery streams are of such low quality (*e.g.* having high polyaromatic and sulfur content) that their individual upgrading to fuels is not profitable. Therefore, co-hydroprocessing of these low-value fossil feedstocks with liquid bio-based feedstocks may better justify their upgrading, while in some cases even improve the exploitation potential of such low-quality refinery streams. The final products can be called hybrid fuels, as they contain a small percentage of biomass. In that case, the produced hybrid fuels fulfill the high-bio-content fuel requirements of the relevant directives of the European Union.

The efficiency of the process is strongly influenced by the operating parameters. The most important parameters of hydroprocessing are the reaction temperature, the reaction pressure, the space velocity (SV) and the hydrogen-to-oil ratio. Most hydrotreatment units (gasoline, diesel and gas oil) operate at a 290–400 °C temperature range. Hydrogen or hydrogen sulfide partial pressures significantly affect desulfurization reactions. The latter are first-order reactions concerning hydrogen partial pressure and are hindered by hydrogen sulfide partial pressure. Catalyst deactivation is also significantly affected by hydrogen partial pressure. Deactivation rate is inversely proportional to hydrogen partial pressure and hydrogen flow rate. Special attention should be given in the case of feed that has a high olefin concentration as saturation reactions consume large amounts of hydrogen. In the case of hydrogen-to-oil ratio, increasing the hydrogen flow rate increases feed desulfurization. This is attributed to the fact that desulfurization is proportional to the partial pressure of hydrogen. In addition, when having a constant flow rate of feed, increasing the hydrogen flow rate contributes to the vaporization of a large amount of the feed. Finally, the SV (space velocity) expresses the retention time of the feed inside the catalyst. Thus, higher SV values result in lower retention times and thus smaller reactors. On the other hand, for the same capacity when the SV is low, reactor size increases proportionally resulting in high investment and operation costs.

15.3.1 Co-hydroprocessing

Catalytic hydroprocessing of bio-based feedstocks has been extensively applied in pilot- and industrial-scale for the production of renewable-diesel fuels. Companies like Neste Oil and UOP are perfect examples of the commercialization of this technology. However, the investment cost of a standalone hydroprocessing unit for bio-based feedstocks is very high, which has driven research activity to find alternative solutions such as the co-hydroprocessing of bio-based feedstocks with petroleum-derived raw materials. This solution allows utilization of existing refinery technology and equipment, rendering it a more economically attractive solution.

When co-hydroprocessing a bio-based feedstock with a fossil-based feedstock, the process has two targets: first, to perform all reactions for the upgrading of the fossil-based feedstock and, second, to reduce the oxygen content and improve the oxidation stability of the bio-based fraction. Therefore, the catalyst choice as well as the optimum operating conditions are crucial factors during co-hydroprocessing, as they can influence the product yields and quality and the associated production cost.

15.3.1.1 Co-hydroprocessing of lipids with fossil fractions

The co-hydroprocessing of various lipid-based feedstocks with petroleum fractions has been widely explored by many researchers in recent years. This approach produces hybrid fuels with reduced carbon emissions enhancing sustainability while enabling the gradual decarbonization of the transportation and aviation sector *via* a biomass infusion in refineries. Lipid-based feedstocks can be classified into previously mentioned categories, such as edible oils, non-edible oils, residual lipids, algal oils, and microbial oils all of which primarily consist of triacylglycerides.

As already mentioned, during co-hydroprocessing, the catalysts have a dual role: first, to hydroprocess the fossil-based fraction and, second, to process the triacylglyceride feedstock. In the case of triacylglycerides, the role of hydrotreating is first to deoxygenate and saturate the large heavy molecules to the corresponding paraffins and then to hydrocrack and isomerize the paraffins yielding straight and branched hydrocarbons (iso-paraffins). Triacylglycerides are characterized by a glycerol backbone connected by ester bonds (oxygen atoms) to three carbon chains that are usually straight chains (normal paraffins) with an even number of carbon atoms with some unsaturated (double) carbon bonds in the chain. Figure 15.2 presents the major hydrotreating reactions of triacylglycerides feedstocks. The carbon-chain length of the fatty acids in the TAGs determines the main properties of the fuels that can be produced by hydroprocessing. The chain

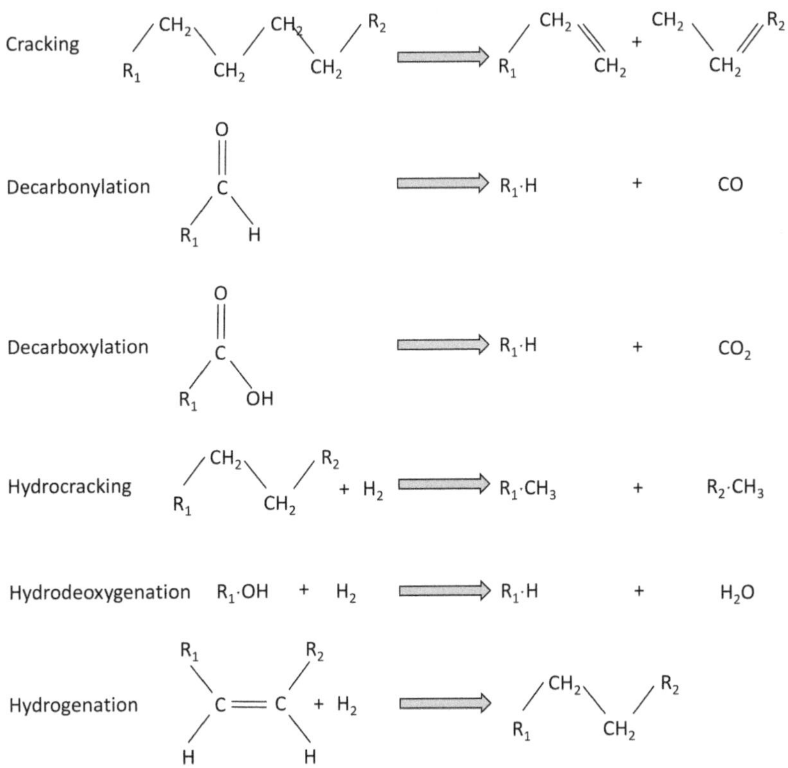

Figure 15.2 Hydroprocessing reactions for triacylglycerides.

lengths and numbers of unsaturated bonds in the fatty acids of some representative triacylglycerides are presented in Table 15.2. Thus, the choice of the catalyst, the hydrogen supply, and the optimum operating conditions for co-hydroprocessing is crucial for the efficiency of the process. When there is not sufficient hydrogen, the catalysts will start to convert the TAGs into coke. The coke will accumulate on the surface and deactivate the catalyst.

15.3.1.1.1 Catalysts for Co-hydroprocessing of Lipids. The choice of the optimum catalyst for co-hydroprocessing of petroleum-derived feedstocks with lipid-based feedstocks is crucial as it determines not only the product yield and quality, but also the duration of the co-processing operation, *i.e.* the maximum catalyst life expectancy. As a result, catalyst evaluation for co-hydroprocessing is a very important research area, as there are two major challenges for the proper selection: (a) catalyst activity varies significantly, as commercial catalysts are optimized for different feedstocks, *e.g.* feedstocks with high sulfur concentration or heavy feedstocks (containing large molecules) or feedstocks with high oxygen concentration, *etc.*; (b) currently there are only a few commercial hydrotreatment catalysts available for bio-based feedstock/intermediate processing.[69]

At the moment, the most common catalysts used in co-hydroprocessing are commercial hydrotreating catalysts based on cobalt–molybdenum (CoMo) and nickel–molybdenum (NiMo) sulfides. However, the use of common catalysts for co-hydroprocessing raises two important issues. The first one is that the triacylglycerides have high oxygen content and the deoxygenation reactions take place on the active sites of the catalyst that in other cases would be used for desulfurization reactions, resulting in a product with higher sulfur content. Secondly, the triacylglyceride molecules are bulky and, to overcome diffusion limitations, large pore-size catalysts, *i.e.* catalysts with larger mean pore size than are normally used in common hydrotreating catalysts, are required. Experiments reported in the literature on the co-hydroprocessing of lipid-based feedstocks with petroleum-derived feedstocks utilizing commercial CoMo and NiMo catalysts have shown that although the activity of NiMo catalyst was not strongly influenced by the addition of the lipids to the petroleum feedstock, the efficiency of CoMo catalyst was severely hindered with the addition of lipids.[53] For example, it has been found that the deactivation rate of the NiMo/γ-Al_2O_3 catalyst was approximately 20% while in case of CoMo/γ-Al_2O_3 catalyst it was more than 30%, rendering NiMo a better choice for lipid-containing feedstock.

Another important issue that is raised during co-hydroprocessing is that the hydrotreatment of triacylglycerides leads to the production of H_2O, CO and CO_2 as by-products. Several studies have examined the effect of these by-products on the catalyst activity and have found that although the presence of H_2O does not influence the hydrodesulfurization and hydrodenitrogenation reactions, the presence of CO and CO_2 limits both heteroatom removal (sulfur and nitrogen) and saturation reactions, reducing in that way the life of the catalyst when NiMo and CoMo catalysts are employed. Therefore, other types of catalysts have also been explored in the literature such as NiW catalysts. In general, NiW catalysts favor decarboxylation and decarbonylation reactions during co-hydroprocessing and, thus, are more appropriate when high triglyceride content is integrated in the petroleum feed.

Further to the main catalyst, the supporting material is also important when lipid-based feedstocks are processed. Such materials include Al_2O_3, TiO_2, NaY, ZrO_2 and zeolite beta (BEA). For instance, the addition of BEA in co-hydroprocessing of straight run gas oil with sunflower oil helped in cracking the triglyceride molecules and enhanced the hydrodesulfurization activity of the NiMo catalyst.[17]

In order to overcome the negative influence of triacylglycerides on catalyst efficiency, a combined approach has also been suggested by various researchers. In that scenario, different catalysts are loaded on the different layers of the reactor. The aim of this approach is to take advantage of the different activities of each catalyst loaded in each layer. Another suggested approach to improve the CFPP (cold filter plugging point) and increase the product yields is the two-stage process, where the catalyst in the first step performs the hydrogenation reactions and the catalyst in the second step performs the isomerization reactions.

Today, co-hydroprocessing of lipid-based feedstocks has already reached commercial status. For example, ConocoPhillips have produced renewable-diesel fuel in its refinery in Ireland since 2006; however, information about the plant capacity, the lipid content, *etc.*, is limited. Another example is Petrobras, which has developed a co-process for vegetable oils and animal fats with fossil-based feedstocks at a lipid content of up to 20%. Moreover, Topsoe, a catalyst company, supplies catalysts to three refineries in Australia, Europe and USA for co-processing bio-feedstocks such as animal fats, chicken fats, *etc.*, with petroleum-derived feeds.

15.3.1.1.2 Operating Conditions for Co-hydroprocessing of Lipids. The rate of reactions in hydroprocessing is strongly influenced by the operating conditions of the process. The operating conditions include four main parameters: reactor temperature, process pressure, space velocity (SV) and hydrogen-to-liquid feed ratio (H_2/oil ratio).

15.3.1.1.2.1 Reaction Temperature. In general, the temperature range of a hydrotreater unit is 290–400 °C for petroleum feedstocks. Lower temperatures favor conversion and saturation reactions while higher temperatures favor hydrocracking reactions. In the case of light petroleum feedstocks when the target is heteroatom removal and conversion reactions, temperatures lower than 350 °C are preferable. On the other hand, when the target is hydrocracking, reaction temperatures higher than 350 °C are more optimum. However, with heavier petroleum feedstocks such as heavy vacuum gas oil, higher temperatures are necessary for saturation and conversion. When a lipid-based feedstock is integrated into a light petroleum stream, slightly higher temperatures are required compared to the neat light petroleum stream. This is due to the fact that the lipid feedstocks consist of heavier molecules than the light petroleum feed and thus higher temperatures are required for hydrocracking reactions. However, when a lipid-based feedstock is integrated into a heavy petroleum feedstock, then slightly lower temperatures can be applied compared to neat heavy petroleum feedstocks as the lipids consists of lighter molecules compared to the heavy petroleum streams. Thus, it is obvious that the type of petroleum feedstock and the targeted reactions define the selection of the reactor temperature during co-hydroprocessing.

15.3.1.1.2.2 Reactor Pressure. A typical pressure range for the hydrotreating units in a refinery range from 20 bar to 110 bar. The choice of reactor pressure can influence the heteroatom removal reactions and the product yields. High hydrogen pressures are necessary for the hydrogen-demanding reactions and to increase the availability of hydrogen in the liquid phase. The integration of lipid-based feedstocks in a petroleum stream demands higher hydrogen pressure, especially when the lipids have high concentration of olefinic bonds. In general, 50–80 bar is a common pressure range for co-processing.

15.3.1.1.2.3 Space Velocity (SV). There are two types of space velocities: the liquid hourly space velocity (LHSV) and the weight hourly space velocity (WHSV). The LHSV is defined as the feed volume flow rate per catalyst volume loaded in the reactor. The WHSV is defined as the mass feed flow rate per dry catalyst mass loaded in the reactor. In both cases, lower values are used when the target is to increase the retention time of the feedstock in the reactor. For the same unit capacity, lower space velocity values required higher reactor sizes, increasing the investment cost. Lower space velocities increase the reaction time between liquid feed, hydrogen and catalyst, leading to lighter hydrocarbons. The typical range of space velocity in a hydrotreater is 0.5–4 h^{-1}, with the most common value close to 1 h^{-1}. In the case of co-hydroprocessing, a preferable range of space velocity is 1–2 h^{-1}. According to the literature, lower space velocities favor saturation reactions while higher space velocities lead to higher middle distillate hydrocarbon yields.

15.3.1.1.2.4 Hydrogen-to-oil Ratio. In the hydrotreating process, excess of hydrogen is very important for the effectiveness of the reactions. A typical hydrogen-to-oil ratio for petroleum streams is 400–800 nL per L of feed. However, higher H_2/oil ratio is preferable when a lipid-based feedstock is integrated in a petroleum refinery compared to neat petroleum stream. This is due to the fact that the addition of a lipid-based feedstock increases the hydrogen consumption of the overall process rather substantially. The amount of additional hydrogen consumed depends on the type of feedstock (saturated or unsaturated) and the reaction path, which is related to the mechanism and kinetics of the process. For example, animal fats are highly saturated feeds and offer the advantage of lower hydrogen consumption when the target is renewable diesel. To conclude, for co-processing of lipid-based feedstocks, a typical hydrogen-to-oil ratio is 420–590 nL per L of feed.

15.3.1.2 Co-hydroprocessing of Pyrolysis Bio-oil

Lipids are the major biomass feedstocks that are currently co-processed in existing refineries. However, as already discussed, lipids from fresh vegetable oils face the "food *versus* fuel" debate; on the other hand, residual lipids from animal fats and waste cooking oils are limited. Therefore, biogenic liquid intermediates generated from lignocellulosic biomass (such as forest residues, agricultural wastes, and algae) and wastes (such as sewage sludge and organic municipal solid waste) from thermochemical technologies (such as fast pyrolysis and HTL) will have great potential to provide the volumes required for co-processing to produce renewable biofuels.[70]

Pyrolysis bio-oil, as already mentioned in previous sections, can be produced *via* various processes. The simplest way is fast pyrolysis oil which is the most cost-effective technique to produce pyrolysis oil or bio-oil from lignocellulosic biomass. However, pyrolysis oil faces many drawbacks, limiting its usability as a direct fuel.[71] In addition, pyrolysis bio-oil is not miscible with the nonpolar hydrocarbons of petroleum-based feedstocks (paraffin, naphtha and aromatic). Thus, an initial upgrading step is necessary prior to use as an intermediate refinery feedstock. Catalytic fast pyrolysis (CFP) and hydrogenation are the two promising approaches. Catalytic fast pyrolysis oil uses catalysts during the fast pyrolysis process in order to reduce the oxygen content and the associated reactive functional groups in the pyrolysis vapor and to produce a water-lean upgraded bio-oil. Hydrogenation allows the nearly full deoxygenation of the pyrolysis oil by a combination of hydrodeoxygenation and decarboxylation reactions leading to a two-phase liquid product consisting of an organic phase and an aqueous phase. In most cases the aqueous phase is up to 40%. Once stabilized pyrolysis oil is produced, it can be further processed in a conventional refinery as an intermediate feed for co-processing with petroleum streams.

During hydrogenation, the pyrolysis bio-oil is processed under a high-pressure hydrogen atmosphere (~140 bar) and at a temperature range of 300–450 °C in the presence of heterogeneous catalysts. However, direct hydrogenation of pyrolysis bio-oil at higher temperatures leads to coke formation on the surface of the catalysts and thus leads to DP build up and product deterioration. In order to solve this issue, a two-step approach was suggested, with a low- to medium-temperature stabilization stage to deoxygenate the most reactive species (carbonyls and sugars) at a temperature range of 250–280 °C followed by a more severe (higher temperature) deoxygenation stage. The rate of pyrolysis bio-oil upgrading depends on the severity of the process. However, more severe conditions require higher processing costs due to the high temperature and operating pressure requirements. Therefore, co-processing of partially stabilized pyrolysis bio-oil into the existing refinery infrastructures can reduce both the capital investment cost and the operation cost. In that case, the most important issue is to investigate the minimum required pre-treatment to stabilize the raw pyrolysis bio-oil to a possible intermediate stream that a refinery can successfully co-process with petroleum fractions.

An important drawback of co-processing is that the integration of a stabilized pyrolysis bio-oil into an existing refinery will negatively affected the hydrogen consumption during co-hydroprocessing. Thus, the cost associated with the use of H_2 should be balanced by improving H_2 efficiency and product quality. This is a hot topic today that needs further investigation. Furthermore, it has been observed that even when much of the oxygen is removed from raw pyrolysis oil *via* the upgrading processes, the resulting hydrocarbon product still demonstrates differences to petroleum-derived fuels. Thus, if the target is a compatible blend feedstock, further downstream processing of HDO in a refinery is necessary to meet the qualifying qualities of existing fuel grades. From all the above, it is important for the upgraded pyrolysis oil to be miscible with the petroleum candidate streams for co-hydroprocessing in an existing refinery infrastructure. According to recent studies,[71] the most common petroleum candidate streams are straight run gas–oil, light cycle oil[29] and heavy vacuum gas oil.[72]

Integration of a partially stabilized pyrolysis bio-oil in a petroleum-based feedstock leads also to the production of carbon monoxide and carbon dioxide due to decarbonylation and decarboxylation reactions. The production of CO and CO_2 during the reactions inhibits the hydrodesulfurization, hydrodenitrogenation and aromatic hydrogenation (HDA) reactions of the petroleum-based feedstock.

15.3.1.2.1 Catalysts for Co-hydroprocessing of Pyrolysis Bio-oil. Co-hydroprocessing of both raw and stabilized pyrolysis bio-oil is still at the very early stages of investigation and there is a lack of literature and data. Most published studies use model compounds instead of real stabilized pyrolysis bio-oil. Some common catalysts that have been investigated for co-hydroprocessing are commercial $CoMo/\gamma\text{-}Al_2O_3$, $NiMo/\gamma\text{-}Al_2O_3$ and ReS_2 supported on SiO_2 and $\gamma\text{-}Al_2O_3$; however, further studies with different types of catalysts should take place in the future to better understand the effect of the catalyst type on the extent of the hydrodeoxygenation and hydrodesulfurization reactions and to learn about the stability of these catalysts under various co-hydroprocessing conditions. At the moment, from the limited current data with the model compounds, it is generally accepted that the addition of model compounds to fossil-based streams inhibits the hydrodesulfurization reactions, as hydrodeoxygenation reactions take place on the same active sites of the catalyst that are used for hydrodesulfurization reactions.

15.3.1.3 Co-hydroprocessing of HTL Biocrude

Hydrothermal liquefaction is a thermochemical process that can lead to the production of a dark-brown liquid called biocrude. Biocrude can be produced from different types of biomass feedstocks, such as forest and agricultural biomass, sewage sludge, algae and solid wastes. Due to the different available types of biomass, the biocrude properties vary depending on the biomass feedstock. In general, biocrudes have high oxygen content, which however may vary depending on the initial feed; a typical oxygen range is 5–50%. The oxygen is removed *via* hydrotreatment through hydrodeoxygenation reactions, increasing in that way the hydrogen demand. The oxygen in biocrude is present in the form of carboxylic acids, alcohols, aldehydes, esters, sugars, furans, ethers and hydroxyls, depending on the source of the biomass feedstock. These oxygenate compounds are very reactive and they can promote unwanted chemical reactions during co-hydroprocessing. In order to remove the oxygenate compounds, similar catalysts to those used for nitrogen and sulfur removal can be employed.

Hydrodeoxygenation, decarbonylation and decarboxylation are the main reactions for oxygen removal. Hydrodeoxygenation removes the oxygen in the form of water (H_2O), decarbonylation in the form of carbon monoxide and water (CO and H_2O), and decarboxylation in the form of carbon dioxide and water (CO_2 and H_2O). The most preferable reaction is hydrodeoxygenation, as during decarbonylation and decarboxylation some carbon in the feedstock is lost as low-value gases (CO and CO_2), leading in that way to lower liquid hydrocarbon yields. During co-hydroprocessing, the above reactions take place at the same time, and only the ratio of the reactions is changed with the operating conditions of the process. Refineries may prefer the hydrodeoxygenation reactions to limit the amount of CO and CO_2 formed, as it may have an impact on catalyst deactivation. Finally, it should be noted here that additional side reactions also occur, such as the water–gas shift and the methanation reactions.

Another important issue that occurs during co-hydroprocessing of biocrude with fossil streams is the generation of heat. The hydrodeoxygenation and decarboxylation reactions are exothermic, decarbonylation is slightly endothermic, the reverse water–gas shift reaction is endothermic, and CO methanation is strongly exothermic. As a result, the total heat contribution of all of these reactions can vary depending on processing conditions and different feedstocks. The heat generation is very important as it can cause significant problems during co-hydroprocessing, such as the overheating of the reactor which could damage the catalyst or causing polymerization of the biocrude and eventually coke formation. From all the described reactions, the methanation reaction, which uses high amounts of hydrogen, is the most undesirable reaction from the perspective of heat generation.

In general, most refineries can accept feedstocks with low oxygen levels. When the blending levels of biocrude are low (5–10%), they can be acceptable for co-hydroprocessing. However, at higher blending rations (>10%), extensive refinery modifications will be required to accommodate the high oxygen levels, as these levels will likely result in high levels of CO, CO_2 and H_2O production and increase reactor temperatures. In this case, a first mild hydrodeoxygenation step for biocrude is preferable before its integration within a refinery for co-hydroprocessing with petroleum-based feedstocks. A first mild hydrotreatment upgrading step can remove or partially remove the oxygen species such as carboxylic acids and alcohols that cause polymerization reactions and eventually coking. In that way, the initial biocrude can be stabilized and easily used as an intermediate refinery stream.

In addition to containing water and oxygen, biogenic feedstocks may also contain alkali metals such as Mg, Na, K, Ca and other contaminants such as chlorides and phosphorus. As these contaminants can cause permanent deactivation of catalysts, they need to be removed prior to co-processing. Furthermore, another important property is the total acid number that strongly influences the storage stability of the biocrude. TAN is also affected by the type of biomass feedstock. Biocrudes produced from lignocellulosic biomass are characterized by high aromatic content due to the lignin. Most of the time, this high aromatic content affects the co-hydroprocessing. In order to solve this issue, more severe hydrotreatment is required. Another important property of biocrudes is the nitrogen content, which also varies depending on the initial biomass; for example, sewage sludge and algae contain high levels of nitrogen. These high-nitrogen feedstocks require more severe hydrotreating conditions in order to force hydrodenitrogenation reactions, which significantly increases the hydrogen consumption of the process. Furthermore, the biocrude consists of large compounds that are water-soluble and large polyaromatic compounds that are insoluble in water. As a result, biocrude miscibility with fossil feeds during co-hydroprocessing is strongly influenced.

To conclude, biocrudes derived from hydrothermal liquefaction are, potentially, easier feedstocks to co-process in the hydrotreater than pyrolysis bio-oil. The team of Sauvanaud *et al.* (2018)[73] studied the co-hydroprocessing of biocrude with straight run gas oil and they did not experience catalyst deactivation during the 50 hour run. On the other hand, according to the researchers at NRCan, Devon,[74] the efficiency of hydrodesulfurization and hydrodenitrogenation reactions was reduced after a long-term experiment with 10% biocrude. However, larger-scale studies involving co-processing of biocrude in the hydrotreater have not been reported, indicating its lower technology readiness level.

15.4 Catalytic Cracking Technology

In general, FCC is used by refineries to convert the high boiling-point, high molecular-weight hydrocarbon fractions of petroleum into lighter hydrocarbons such as gasoline, alkene gases, *etc.* The most common petroleum feedstocks for FCC are HAGO, HVGO, *etc.* During the process, a fluidized bed of catalyst particles is brought into contact with the gas–oil feed along with injected steam at the riser, which is the entrance of the reactor. The hot catalyst particles that come from the regenerator unit evaporate the feed gas oil upon contact in the riser, and the cracking starts as the gas–oil vapors and the catalyst particles move upward in the reactor. As the evaporation of the gas oil and endothermic cracking reactions proceed during the upward movement, the temperature of the catalyst particles drops. During this procedure, coke is formed on the catalyst leading to its deactivation. After removing the absorbed hydrocarbons by steam stripping, the coked catalyst is moved to the regeneration unit in order to burn the coke with air. The heat that is released during the burning process of the coke increases the temperature of the catalyst particles that are returned to the riser to complete the cycle. Burning off the rejected coke in the regenerator provides the energy necessary for the cracking without much loss, which increases the thermal efficiency of the process. Finally, the products from cracking are sent to the fractionator for recovery, after they are separated from the catalyst particles in the upper section of the reactor.

15.4.1 Co-processing in Catalytic Cracking Units

The fluid catalytic cracker is another potential insertion point for co-processing. The advantages of a FCC unit compared to a hydrotreater is that in a FCC unit there is no need for external hydrogen, which may make this process economically more attractive. In addition, the FCC unit provides a lower-risk option as the FCC catalysts are more tolerant than hydroprocessing catalysts to higher oxygen levels in the bio-feeds. Furthermore, the FCC catalysts are continuously regenerated on-site by burning off any coke deposits in the regenerator attached to the FCC unit before recirculating the catalyst.

15.4.1.1 Co-processing of Lipids

Fluid catalytic cracking is a promising technology for the production of hybrid fuels *via* co-processing of lipids with petroleum streams. Lipids are characterized by high oxygen content; however, the oxygen can be removed as CO, CO_2 and water during co-processing in a FCC unit. At the same time, the coke that is formed during the process can be successfully burned during the regeneration of the catalyst and the generated heat can be used to provide heat for the unit. In general, the catalyst carries the heat that supplies the energy needed for the endothermic reactions to take place. The advantage of this process is that all the energy of the bio-feed is utilized in the process.

During co-processing the catalyst has a double role: to perform all reactions needed for the petroleum hydrocarbons, and also to perform all reactions needed for the bio-feed. In the case of petroleum-derived feedstocks, the most common reactions include cracking reactions (cracking of alkanes, alkenes, naphthenes and alkyl aromatics to lighter products), hydrogen transfer, isomerization and coking reactions. On the other hand, in case of the bio-feed, the reactions include dehydration, cracking of large oxygenated molecules to smaller molecules, hydrogen-producing reactions, hydrogen-consuming reactions and production of larger molecules by C–C bond-forming reactions (aldol condensation or Diels–Alder reactions). During co-processing, all the described reactions take place simultaneously and their combination determines the chemical composition and characteristics of the final product.

Lipid-based feedstock is an easy bio-based feed for co-processing in FCC units compared to pyrolysis bio-oil or HTL biocrude. This is due to the fact that the lipid-based feedstocks are homogeneous mixtures and mix well with petroleum fractions. In addition, seed oils possess high crackability derived from their low thermal stability. For example, neat crude palm oil can be successfully exposed to thermal cracking by means of free radicals in the preheating zone of the FCC reactor. Furthermore, the usual operating conditions of the FCC unit are severe enough for the successful catalytic decomposition of the triglycerides to different hydrocarbons *via* decarboxylation, decarbonylation and dehydration reactions.

15.4.1.1.1 Operating Parameters

15.4.1.1.1.1 Reaction Temperature. Reaction temperature strongly influences the yields of the process. More specifically, in blends of canola oil with heavy gas oil, a temperature of 530 °C can achieve higher conversions compared to lower temperatures such as 490 °C. The higher temperature can promote cracking reactions leading to more gaseous products and lighter liquid products, while at the same time it can reduce catalyst poisoning which is more pronounced at lower temperatures. The product yields are higher in the case of blends of canola oil and gas oil compared to neat gas oil as the canola oil has a higher content of aliphatic chains. Thus, high temperatures favor liquid product yields such as gasoline and diesel, as higher temperatures enhance cracking reactions and reduce nitrogen poisoning of the catalyst. However, if lighter fossil-based feedstocks are used such as LCO, then lower reaction temperatures are favorable. This is due to the fact that lighter fossil-based feeds (*e.g.* LCO) are more crackable compared to their heavier counterparts (*e.g.* HAGO). Thus, it is easily noticed that the choice of reaction temperature depends from the fossil-based feedstock. Furthermore, at higher temperatures the coke content on the catalysts increased, as it is formed through the polymerization of aromatic hydrocarbons in the cracked liquid product together with a condensation reaction involving the lipids.

15.4.1.1.1.2 Catalyst-to-oil Mass Ratio. The severity of the catalytic cracking process can be also modified by the catalyst-to-oil ratio (CTO). Increasing the CTO leads to more severe cracking reactions and thus higher conversion. Because higher CTOs (as high as 9–11) provide proportionally more active acid sites of the catalyst to crack a given amount of feed molecules. However, the cracking rate is reduced at high CTOs once the crackable components in the feed are gradually exhausted.

15.4.1.1.1.3 Bio-content Ratio. The bio-content in the feedstock is very crucial in FCC co-processing. A small amount of lipid content (up to 25 wt%) promotes the activation of paraffinic and naphthenic hydrocarbons in the fossil-based feedstock by hydrogen-transfer reactions. During the process, hydrocarbons of the fossil-based part serve as hydrogen donors, while the primary semi-products of lipid thermal cracking (*i.e.* C_{15}–C_{17} olefins) are hydrogen acceptors. Thus, carbocations formed by the interaction of the olefins with the catalyst active site react with paraffinic and naphthenic hydrocarbons in the fossil-based feed. As a result, hydrogen-transfer reactions result in the formation of aromatic hydrocarbons and isoparaffins and, thus, in an increased fossil-based feed conversion. However, higher amounts of lipids (>25 wt%) can lead to catalyst deactivation as large amounts of mono- and polyaromatic hydrocarbons are formed and are adsorbed on the acid sites of the catalysts.

To conclude, lipids contain large amounts of oxygenate compounds while the end product should be an oxygen-free liquid product. The FCC process should promote the effective deoxygenation of the lipid-derived molecules. In general, the operating conditions of the FCC process are severe enough to decompose the heavy oxygenated hydrocarbons *via* decarboxylation, decarbonylation and dehydration reactions, leading to the production of oxygen-free hydrocarbons and also to the production of CO, CO_2 and water. At the moment the current technology is in pilot-scale and further research is necessary prior its commercialization.

15.4.1.2 Co-processing of Bio-oils

Pyrolysis bio-oil can also be co-processed in Fluid Catalytic Cracking (FCC) units, which are considered key conversion units in most heavy-oil refineries due to their high profit margins and relatively low operational risk. Furthermore, the versatility and operative flexibility of FCC support the process as a candidate for bio-oil co-processing. In a FCC unit, heavy gas oils are converted on fluidized catalyst particles by cracking the carbon–carbon bonds to lower the average molecular weight and produce a high yield of gasoline and diesel precursors. The produced coke is deposited on the catalyst in the reactor and burned off in the regenerator to provide heat for the process.[24]

The co-processing of bio-oil with petroleum feeds under FCC conditions causes fewer technological challenges compared to cracking neat bio-oil. Problems such as plugging can be avoided due to the ratio of the feeds used when blended. However, special attention should be given to the choice of the feedstocks. The thermal and oxidative stability of bio-oils is very important because it may cause storage problems due to changes in density, viscosity, acidity and an increase in polymer formation. The instability of the pyrolysis oil is mainly caused by lignin oligomers formed during pyrolysis.

In addition, the presence of oxygenated compounds such as phenols and carboxylic acids contribute to the corrosiveness of bio-oils. The instability of bio-oil can be overcome through flash pyrolysis, hydrotreating or zeolite upgrading.

The oxygenated compounds of bio-oil create two key challenges. The first challenge is the lack of miscibility of the bio-oil with the petroleum-based feedstocks. In contrast to lipids, the use of bio-oil creates miscibility problems when blended with fossil feed. The second challenge is the insertion of bio-oil at high temperatures that results in extensive coking. In order to overcome these two key challenges, a separate injection nozzle can be used. This nozzle allows the bio-oil to be kept at temperature below 50 °C, thus avoiding temperature-related polymerization during feeding while also helping to avoid the problems of miscibility. The described method is the most feasible option by which bio-oil with an oxygen content of 40 wt% can be co-processed in an FCC unit.

At the moment there are plenty of published studies in co-processing of bio-oils with petroleum-based feeds in different types of FCC units. All these set-ups are intended for an effective and operational evaluation of the performance of the modified FCC catalysts and different FCC feedstocks. However, it can be questioned whether the results obtained by catalyst testing in laboratory units simulate the real catalyst performance during commercial operation in some way. For instance, the coke results obtained in circulating FCC pilot risers were frequently lower than those observed in lab-scale experiments. Hence, knowledge about the experimental unit set-up and operating conditions is essential to draw correct conclusions, particularly when predicting future performance in an industrial-scale unit.

Another important issue that has been observed from lab-scale experiments in co-processing raw bio-oils with fossil streams is that reactor plugging might be a crucial issue because char and coke formation was substantial even at low temperatures. Therefore, upgrading the raw bio-oil mostly *via* hydrotreatment is often performed prior to its co-processing with petroleum feedstocks. This is because the raw bio-oil has high oxygen content in the oxygenate compounds that react during the process forming water that poisons and deactivates the catalysts. The co-processing of hydrogenated pyrolysis oils with fossil feeds in refineries constitutes a promising route toward the production of greener transportation fuels. Nevertheless, an obvious disadvantage of the hydrotreatment of bio-oil with oxygen content close to 5 wt% is the necessity to use significant quantities of hydrogen, which would make the technology economically unviable. In order to decrease the cost of bio-oil hydrotreatment, milder conditions for bio-oil upgrading can be used. This is because bio-oils do not require complete hydrodeoxygenation but only elimination of the highly reactive components and functional groups, minimizing the hydrogen consumption. Nonetheless, such upgrading technology is still under development and requires the use of an expensive noble-metal catalyst.

In addition to hydrotreatment technology, catalytic pyrolysis oil and flash pyrolysis oil may be considered as promising bio-additives to petroleum-derived feedstock for FCC co-processing. However, it is difficult to evaluate the efficiency of the different bio-oils as intermediate refinery streams for co-processing because there are many differences in the properties between the bio-oils studied in the literature as well as the corresponding operating conditions.

15.4.1.2.1 Operating Conditions

15.4.1.2.1.1 Reaction Temperature. Reaction temperature is very important for co-processing bio-oil with petroleum-derived feedstocks. Typical operating temperatures range from 500 °C to 550 °C. Higher temperatures close to 550 °C lead to higher conversions; however, temperatures that are too high may have adverse effects such as the formation of coke.

15.4.1.2.1.2 Catalyst-to-oil Mass Ratio. The catalyst-to-oil mass ratio can vary over a wide range of 1–9, thus allowing for significant modification of the conversion and yields. According to a published study,[75] at a catalyst-to-oil ratio of 6, the CPO/gas oil blends can be used to produce gasoline with similar yields to petroleum gas oil, thus constituting a true replacement for gasoline produced entirely from petroleum.

15.4.1.3 Co-processing of HTL Biocrude

In order to produce on-specification products, a potential bio-refinery that produces final fuel products from biocrude only requires complex and thus expensive processing units that can handle a varying biocrude composition. This is due to the heterogeneous nature of the biomass that the biocrude is made from. In addition, market penetration will become difficult by competing rather than cooperating with the petroleum industry. By coprocessing the biocrude at conventional refineries the point of market introduction is moved upstream, which makes it an attractive market proposition compared to producing transport-grade finished fuels.

Co-processing of HTL biocrude in FCC units has been less studied than pyrolysis bio-oil. Even if HTL biocrudes have a lower oxygen content and are more stable compared to pyrolysis bio-oil, the co-processing of HTL biocrude by FCC is still challenging in terms of processability, *e.g.* as a result of high viscosity. Recent studies have shown that small amounts (<10 wt%) of untreated HTL biocrude can be co-processed with VGO in a modified micro-activity test unit, leading to lower LPG and gasoline yields but higher diesel and coke yields. A proposed path is to perform a first mild upgrading of HTL biocrude *via* hydrotreating before its insertion for co-processing. In that case, the hydrotreatment will remove the sulfur, nitrogen and oxygen content from the HTL biocrude and then it can be used further for co-processing with petroleum-based feedstocks at higher amounts. However, as the technology is still at the very early stages, further optimization of the HTL and

HDO steps is needed to find the most feasible route and allow for the co-processing of larger fractions of biocrude than 10 wt%.

15.5 Evaluation of Hybrid-fuel Quality

Standalone processing of bio-based feedstocks leads to the production of drop-in biofuels; however, co-processing of bio-based feedstocks with petroleum fractions leads to hybrid fuels. Since hybrid fuels contain a small percent of bio-based feedstock, they present slightly different properties compared to conventional fuels. Before evaluating the quality of hybrid fuels, a short description of the most important fuel properties is presented below.

15.5.1 Density

Density is a measure of a fuel's mass per unit volume. Density depends on temperature and for fuels is normally determined at 15 °C. The overall density of a fuel depends on the composition of the fuel and, more specifically, on the hydrocarbon compounds. Density is strongly correlated with other fuel parameters, particularly the cetane number, aromatic content, viscosity and distillation characteristics. The energy content of the fuel is approximately proportional to the mass of the fuel injected in the combustion engine. Thus, for a constant volume injection system, variations in the fuel density can lead to variations in the energy content of the fuel injected. As a result, density may affect the engine power, emissions and fuel consumption, especially in diesel car engines. In order to optimize the engine performance and exhaust emissions, the fuel density must be controlled within a fairly narrow range.

15.5.2 Heating Value

Heating value is the heat released when a fuel undergoes complete combustion with oxygen under standard conditions. For the gross heating value, the water produced by the combustion is assumed to be re-condensed to liquid. For the net heating value, water remains in the gas phase. Since engine-exhaust water is in the gas phase, the net heating value is the appropriate value for comparing fuels.

15.5.3 Distillation Characteristics

Mass-recovery curves give the amount of a fuel sample that is evaporated at atmospheric pressure at each temperature point when the temperature is increased gradually. Distillation characteristics illustrate how the fuel is evaporated when it is sprayed into the combustion chamber of an engine. Some fractions that boil at low temperatures are needed for engine start-up, while the fractions that boil at temperatures that are too high may not combust completely and form engine deposits and increase tailpipe emissions. The typical ranges of the most common fuels are 60–200 °C for the gasoline fraction, 200–300 °C for the jet-fuel fraction, and 300–360 °C for the diesel fraction. In some cases, when the refinery target is solely diesel fuel production, the diesel range can be extended to 220–360 °C, including the jet-fuel fraction.

15.5.4 Cold Flow Properties

Cold filter plugging point, cloud point, pour point and freezing point are called the cold flow properties of a fuel. CFPP is the lowest temperature at which the fuel can pass through a standard test filter under standard conditions. Cloud point is the temperature at which wax crystals start to precipitate out and the fuel becomes cloudy. Pour point is the lowest temperature at which oil will flow. This property is crucial for fuels that must flow at low temperatures. Freezing point is a property especially for the jet-fuel fraction, for which it should be approximately −40 °C.

15.5.5 Oxidation Stability

Oxidation stability or induction time is an indication of the aging rate of a fuel. It is estimated *via* the Rancimat test, which measures the time required for the formation of gums and sediments (causing plugging of filters and engine deposits) after accelerated oxidation. The oxidation stability shows the threshold for storage and utilization period after the production of a fuel.

15.5.6 Cetane Number

The cetane number (CN) of a fuel is a measure of its propensity for autoignition. In practical terms, the CN has a strong influence on the time interval between the fuel injection and the combustion in a diesel engine. The higher the CN, the shorter this ignition delay period is; therefore, the CN is preferably high. The CN affects the ease of starting, the noise generated by combustion, and the exhaust emissions of diesel engines.

15.5.7 Octane Number

An octane rating or octane number is a standard measure of a fuel's ability to withstand compression in an internal combustion engine without detonating. The higher the octane number, the more compression the fuel can withstand before detonating. Octane number is a measure especially for gasoline-type fuels.

15.5.8 Sulfur Content

Sulfur is naturally present in crude oils and must be removed to an acceptable level during the refining process. Sulfur in diesel fuel contributes to the formation of particulate matter (PM) in the engine's exhaust and affects the performance of vehicle emissions control equipment. It has therefore an indirect effect on emissions of CO, hydrocarbons and NO_x. In general, sulfur specifications are very strict, especially for diesel (<10 wppm), paraffinic fuels (<5 wppm) and gasoline (<8 wppm).

15.5.9 Carbon Residue

Carbon residue is a measure of the tendency of the fuel to form carbonaceous deposits in engines, which can result in hot spots leading to stress, corrosion or cracking of components. The deposits of most concern are those which build up in the nozzles of fuel injectors. The amount of carbon in fuel can be correlated with a tendency to form deposits, hence the use of a carbon residue test. The test is performed on the residual volume after 90% of the fuel has been boiled off (10% residual).

15.5.10 Copper-strip Corrosion

The copper-strip corrosion test is a measure of the corrosiveness of the fuel to metals. Corrosion can affect metallic components in vehicle fuel systems, dispenser pumps and fuel-storage systems, and is measured in order to provide protection for fuel tanks, dispenser pumps and vehicle-engine components.

15.5.11 Color

The general appearance and color as well as the diesel clarity are useful indicators of contamination.

15.5.12 Viscosity

Viscosity is a measure of the resistance of the fuel to flow at a stated temperature. According to ISO, the kinematic viscosity of heavy fuels should be specified in centistokes ($1\ cSt = 10^{-6}\ m^2\ s^{-1}$) at 100 °C. Viscosity is generally an important parameter for diesel fuel. Fuel which is too viscous can cause damage in the fuel pump (*e.g.* cam and follower wear) due to higher pressure. Viscosity that is too low may lead to a lack of lubrication. Viscosity also influences the fuel delivery rate and the atomization of the fuel during injection.

15.5.13 Flash Point

The flash point of a liquid is defined as the lowest temperature at which a substance generates a sufficient amount of vapor to form a vapor/air mixture that can be ignited (piloted ignition). At that temperature, the vapor pressure of the liquid provides a vapor concentration that is equal to the lower flammability limit. If ignition is attempted when the liquid reaches its flash point, a flash flame will occur but the flame will not sustain.

It should be noted that the product yield and quality of the fuels is not compromised by the refineries that have to meet the fuel specifications. Co-processing of bio-based feedstocks with petroleum fractions may improve the characteristics of the fuels due to the synergy between fossil feeds and the biofeeds.[76] Co-hydroprocessing of lipids with fossil feeds, for example, can lead to hybrid fuels with higher cetane levels due to the paraffins that are produced from the hydrotreatment of the lipid fraction.[11] Furthermore, co-processing of biocrudes and lipids in the FCC can, under certain circumstances, produce gasoline with a higher octane rating due to the increased formation of aromatics resulting from hydrogen transfer reactions.[77,78] An important drawback is the reduced yields that may be observed due to the carbon losses by decarbonylation and decarboxylation reactions that lead to production of CO and CO_2.[79] Some studies have shown that when co-processing of lipids with fossil feeds in the FCC unit, although it can lead to the reduction of the overall yield due to the carbon loss as CO and CO_2, gasoline production remains constant with an increase in residue and gases. However, this result can be influenced by several factors such as the type of lipids (saturated or unsaturated), the blending level and the operating conditions of the process (temperature and catalyst-to-oil ratio).[11] In general, lipids are more susceptible to cracking, leading to the production of more gases.[80] Some related studies have suggested that gasoline production *via* co-processing of lipids with fossil feeds in FCC units can lead to gasoline fuel with higher research octane number (RON) and motor octane number (MON) due to the formation of higher aromatics.[11] Some other studies that investigated the co-processing of different types of pyrolysis biocrudes, such as fast pyrolysis, catalytic pyrolysis and hydrotreated fast pyrolysis bio-oil, in FCC units have noticed that similar yields of gasoline can be produced compared to neat fossil feed (VGO), with only a slight increase in coke formation.[81] In general, it should be mentioned that all the above results are from different research studies that have been performed with small-scale lab units and thus the results do not fully represent the real-life FCC conditions.

In the case of hydrotreaters, during co-hydroprocessing of lipid-based feeds with fossil feeds, the paraffinic content of the produced fuels increased. Due to this increase of paraffinic content, the cloud point, melting point, pour point and cold filter plugging point are increased.[11] In general, the quality of the produced hybrid fuel is influenced by the type of lipids, the catalyst and the operating conditions of the hydroprocessing. For instance, when vegetable oils are co-hydroprocessed in a hydrotreater, the preferred temperature range is between 360 °C and 390 °C; on the other hand, when animal fats are co-hydroprocessed in a hydrotreater, the preferred temperature range is between 320 °C and 340 °C, as co-processing animal fats above 340 °C reduces the yields due to cracking reactions.[53] One potential challenge with co-processing in the hydrotreater is the competition between the hydrodeoxygenation and hydrodesulfurization reactions, as this can impact the diesel specifications.[82] Some studies[53] claim that the sulfur content of the organic product decreased with the addition of a bio-based feedstock. As lipid-based feeds are virtually sulfur-free, their addition results in lower sulfur content of the blend which may be responsible for lowering the sulfur content in the liquid products. On the other hand, other studies[15] mention that oxygen-containing compounds (triglycerides or other oxygen-containing products) bind strongly to the active sites of the desulfurization catalyst and hence limit the desulfurization reactions. Consequently, when a higher content of vegetable oil is used the sulfur content in the products increased. Hence, the deoxygenation reactions restricted the desulfurization reactions because of the intake of large quantities of oxygen with the triglyceride molecules of the vegetable oil.[83] The total aromatic content of the final products decreases compared to that of the petroleum feedstocks because the lipid-based feeds are free of aromatics. As far as the kinematic viscosity is concerned, it decreased with increasing temperature and pressure due to triglyceride conversion, saturation of aromatics and cracking reactions. Furthermore, an increase in the products' viscosity was also observed with an increase in the concentration of triglycerides in the feedstock. This happens because a feedstock with higher triglyceride content creates more viscous *n*-paraffins.

Studies have shown that the addition of lipid-based feeds into an existing refinery improves the product quality.[34] Co-hydroprocessing is a key technology to produce high-quality hybrid fuels that meet fuel specifications. Hydroprocessing of lipid feedstocks converts triglycerides into normal paraffins. As a result, in the case of co-hydroprocessing, the increase in lipid content results in higher paraffinic content in the final product. This leads to a final product with a higher cetane number and heating value, which improves the combustion efficiency of the engines. However, if the content of lipids is too high, then problems with cold flow properties will occur.

15.6 Sustainability

Considering that the EU aims to be carbon-neutral by 2050 – *i.e.* an economy with net-zero GHG emissions – the transition to a carbon-neutral society is an opportunity for sustainable transportation. The EU can lead the way by investing in and promoting technological solutions that ensure actions to support a smooth transition in key areas such as industrial policy and research.

Co-processing is the simultaneous processing of bio-based material, such as bio-oils, residual lipids and other feedstocks, with fossil-based feeds in refinery infrastructure. By using the existing refinery infrastructure, major benefits are offered in terms of GHG emissions and cost savings since new dedicated processing units are not required. So, co-processing that can utilize the existing refineries without modification or with minor modifications provides a more convenient, faster and ready-to-use pathway towards renewable fuel production. Compared to a standalone biorefinery, upgrading biomass-derived oils using co-processing of bio-oils with petroleum intermediates in the existing refining infrastructure can save ~20% cost.[24]

As sustainable technological schemes are continuously evolving under the context of the circular economy for production of renewable fuels, their environmental assessment is also substantially validating the process. Particularly, the sustainability assessment of refineries quantifies their environmental profile based on a multitude of individual results. The life-cycle assessment (LCA) methodology is a broadly recognized support tool to assess and reduce environmental impacts considering the entire life cycle of a system.[84] LCA evaluates the environmental impacts associated with fuels based on a compiled inventory of inputs and outputs and interprets the results to support the decision-making process.[85]

LCA studies provide valuable information embracing environmental aspects that are fundamental to ensure that sustainability is achieved. Furthermore, the integrated environmental assessment approach has emerged as a robust methodology to guide and assist developers and policymakers in developing more environmentally sustainable technologies.[86] In general, in an LCA study the system boundaries commonly define the feedstock production chain and its subsequent conversion to the envisioned products and efficiently valorizing any byproducts (Figure 15.3). The LCA data are collected from all the processes in the examined system, whereas the functional unit provides a quantitative illustration related to the targeted products. The functional unit of these LCA studies commonly refers to mass (kg) or energy (MJ) when considering hybrid fuels.[87]

LCA examines the environmental impacts and burdens of production systems from cradle-to-grave (*i.e.* from feedstock/biomass to industrial operations, usage stage and end-of-life management such as waste disposal) or cradle-to-gate (from feedstock/biomass provision to the usage stage). In this context, LCA is applied to examine how the whole value chain for producing hybrid fuels *via* co-hydroprocessing would influence the environment. Moreover, the goal of an LCA-based refinery study might be to compare the

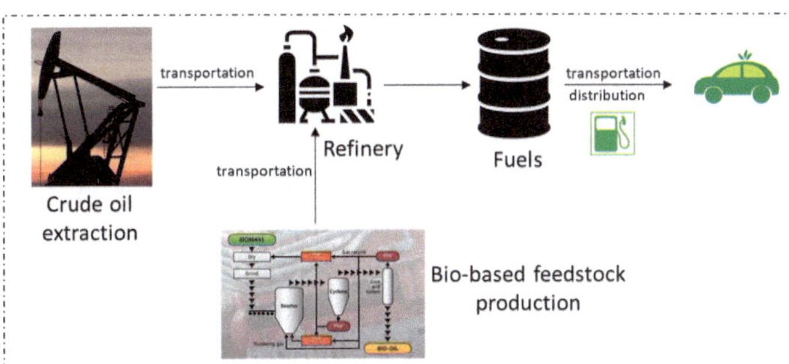

Figure 15.3 Generalized system boundaries of co-hydroprocessing biomass with petroleum fractions.

environmental effects of different process alternatives (technologies or designs) or to decide on the best uses of biomass-feedstock valorization *via* co-processing.[88] The boundaries of each technological system depend on the LCA study, but cradle-to-grave, starting from the biomass/feedstock and involving the products/fuels final use, is usually preferred over cradle-to-gate.

LCA studies employ software tools such as SimaPro, GaBi and Open LCA and databases such as Ecoinvent in order to assist the modeling of the examined system, complete the inventory and obtain the impact assessment results. The environmental performance is commonly focused on climate change-related indicators (*e.g.* global warming potential), while other impacts are also considered, such as eutrophication, land-use change, or biodiversity, to avoid shifting burdens.[89] The environmental assessment of an examined technological approach constitutes a powerful tool that could potentially guide a wide range of stakeholders towards informed decisions, ensuring sustainability.[90]

One apparent issue when analyzing such technological systems is the uncertainty and difficulty regarding the energy systems that will undergo a significant transformation in the upcoming years to achieve targeted climate goals and future (uncertain) developments. Previous studies have considered the feasibility, kinetics, modeling, optimization and economics of co-processing bio-oil and petroleum fractions, and co-processing research is still active due to its complexity. The core advantage of the co-processing technique is to decrease the environmental impacts by adding renewable energy to a fossil-fuel refinery.[91]

A few LCA studies have assessed the environmental sustainability of co-processing in existing refineries, considering mainly global-warming impacts and energy balances as impact categories, whereas other environmental burdens such as acidification and eutrophication are also quantified, though characterized as site-oriented. In general, it appears that the co-processing technique is an environmentally-friendly technology rendering renewable fuels. Nonetheless, the LCA approach to a co-processing system depends on the life-cycle stages and the inputs and outputs considered, generating discrepancies between results in several cases; however, some representative results can be presented.

The LCA for co-processing bio-oil in a FCC unit was evaluated for both gasoline and diesel production, employing the GREET model (GA-GREET 2.0). The results showed ∼70% decrease of CO_2 emissions compared to those from an equivalent fossil fuel.[93]

In another LCA study, it was found that the co-processing of petroleum fractions with residual lipids (heavy atmospheric gas oil and waste cooking oils at 95%/5% v/v) in the refinery renders environmentally friendly fuels (*i.e.* hybrid fuels) as compared to the market diesel production.[94] These results highlight that the co-processing of petroleum fractions with residual biomass in a refinery constitutes a beneficial production process exploiting existing infrastructure, thus enforcing future research studies in this direction.

It should be noted that co-processing ~10 vol% bio-based feedstock in the processing unit does not significantly vary the operation of the modeled refinery in terms of energy efficiency or product slate at the facility level.[92]

The environmental impacts of the co-processing scheme should be considered when designing the scheme. As the impacts of the non-renewable feedstock are the highest, the future direction of the co-processing approach may be to increase the bio-oil quality so that more bio-oil can be added into the FCC without decreasing the gasoline yield. However, it should be noted that the absence of real experimental data on long-term impacts will likely delay their utilization in an attempt to decrease the carbon intensity of refineries.

Nevertheless, in order to evaluate the emissions reduction of co-processing, it is important to quantify (estimate/measure) the biogenic portion of the final fuel products. The results show that the refinery carbon intensities for drop-in fuels do not present notable variations from co-processing, whereas significant reductions in life-cycle GHG emissions originate from biogenic carbon content in the final fuel products. Particularly, co-processing using hydrotreating and hydrocracking lessens the carbon intensities of jet fuel by 3.7–7.5% compared to petroleum jet fuels. Likewise, diesel production through hydrotreating and hydrocracking of bio-based feedstocks presumably reduces the GHG emissions by 1.1–4.5% compared to the fossil-based diesel fuels. Considering that co-processed jet fuels contain a higher biogenic portion than diesel fuels, jet fuel presents a higher reduction in carbon intensity per megajoule (MJ) of energy production. Co-processing of pyrolysis oil in an FCC does not present substantial reductions in GHG emissions due to the higher content of oxygenates and water in the bio-based feedstock that are converted into emissions (*i.e.* 29% of bio-based feedstock is lost to emissions). Therefore, as refinery emissions do not fluctuate with 10 vol% bio-based feedstock inputs at the co-processing unit level, it is estimated that the biogenic carbon intensities values are ~72–90% lower than the petroleum baseline on a MJ basis.[92]

As far as co-processing for jet-biofuel production is concerned, the utilization of feedstocks with high lipid content (waste cooking oils) and renewable energy could reduce GHG emissions. Specifically, co-processing of jet fuels takes advantage of the reduction of energy consumption in the fuel stage, while GHG reductions of ~8% are achieved by employing used cooking oils.

According to process simulation model analysis, the carbon footprint of a refinery can be significantly decreased by co-processing HDO oil and biochar.[95–97] Thorough economic analysis was conducted to complement the above environmental assessment by means of life-cycle costing (LCC), and net present value (NPV) was selected as the life-cycle economic indicator. Co-processing in a FCC showed decreased operating-unit costs and increased product revenues compared to processing crude oil in a conventional refinery, due to the higher production rate. On the contrary, the operational costs associated with the addition of a pyrolysis plant as well as FCC-catalyst replacement and services were increased, leading to a lower NPV. As a result, despite an economically feasible process, co-processing in a FCC performs slightly less eco-efficiently compared to processing crude oil in a conventional refinery. On the other hand, co-processing in a catalytic hydrocracking unit achieves improved eco-efficiency performance.

The high prices of biodiesel and hydrogen make directly blending bio-oils into refinery final products economically less favorable than co-processing HDO oils with petroleum intermediates using refinery facilities. Techno-economic analysis of co-processing showed economic attractiveness when the price of bio-oil is above 60 USD per bbl for crude bio-oils or 120 USD per bbl for HDO bio-oils. The LCA results clearly presented a significant reduction in GHG emissions for the production of crude bio-oils (−0.30 kg CO_2-eq per MJ) and upgraded bio-oils (−0.14 kg CO_2-eq per MJ) compared to those for the production of VGO (0.0052 kg CO_2-eq per MJ) and petroleum-derived diesel (0.013 kg CO_2-eq per MJ).[98]

15.7 Challenges and Prospects

In general, the properties of crude pyrolysis bio-oil are different than the fossil feeds. This is due to the high oxygen content of bio-oil that makes it immiscible with the conventional fossil streams. As a result, there is a big interest in investigating other more suitable bio-oils as potential feedstocks for co-refining. At the moment, there is a significant gap on the literature as far as the impact on co-processing of biomass type and conversion process. Thus, the influence of bio-oil production and properties on bio-oil co-processing is considered in great need of future research.

Catalyst performance is of great importance in refineries. However, processing of crude bio-oils in refineries negatively influences catalyst life and performance. In addition, the alkaline metals in bio-oils may neutralize the acidity of catalysts and cause loss of activity. As a result, commercial catalysts may not work well in co-processing bio-oils with fossil streams. In order to overcome this drawback, advancing refinery-compatible catalysts is required to increase incorporation of biogenic carbon in fuels while maintaining consistent catalyst performance in refineries.

An important challenge is the corrosion risk faced by refiners, particularly in atmospheric distillation units where biocrudes are introduced into the feedstock stream.[99] Bio-oils contains acids and metals that may cause corrosion to the unit equipment during co-processing. Although this is an important concern, research on this field is rather limited. Considering the reactor material (carbon steel), how much bio-oil carbon steel can endure from the corrosion point of view is another research interest.

Furthermore, the large amount of phenolics contained in bio-oils cause coke formation, especially when the percentage of bio-oil is >20 wt%. The coke formation accelerates the deactivation of FCC catalysts and limits the possibility of processing more bio-oils in the feedstock. Leveraging coke formation and feed conversion is needed to optimize the process performance.

Another interesting research area is the need to lower the oxygen level and remove heteroatoms from blends of bio-oils and petroleum fractions prior their co-hydrocracking or co-FCC. However, the literature on co-processing HTL biocrude with petroleum-based feeds is still limited. Further research in this field is very important prior the commercialization of the technology. In addition, the higher requirement for hydrogen with integration of HTL and bio-oil in an existing refinery unit needs to be overcome. Therefore, seeking fungible hydrogen sources such as hydrogen-donor solvents or solar hydrogen may be a solution.

Biogenic carbon is an important indicator to measure the efficiency of co-processing. Robust methods are therefore required to quantify the biogenic (renewable) fractions in co-processing products. Today, the most common methods are ^{14}C isotopic analysis, mass balance approaches (based on carbon mass balance and product yields), stable carbon isotope ratios ($^{13}C/^{12}C$) and the energy content method. However, these may not be reliable for determination of the renewable carbon when the bio-feedstock fraction is under 10 wt%.[99] The wide application of this method is limited because of high costs and lack of onsite availability. Suitable quantitative methods for measuring renewable carbon in finished products of co-processing need to be determined.

15.8 Conclusions

The EU considers that by far the most serious challenge of the transportation sector is to decrease the associated GHG emissions, and road transport constitutes the highest proportion of these emissions. However, the largest increases of GHG emissions up to 2030 are projected to come from the aviation sector, followed by maritime transport, as they are not prioritized by national policies. Therefore, these sub-sectors are expected to constitute a higher proportion of transportation-sector emissions in the coming years. Currently, less than 1% of all fuels used in aviation and shipping are sustainable. The need for drop-in sustainable biofuels *via* targeted decarbonization pathways is apparent. Although, there are many benefits of drop-in biofuels, including the domestic production of fuels, decreased GHG emissions, decreased dependence on fossil fuels, improvement of rural economies and increased national security, standalone biofuel production requires high investment costs and energy consumption. Therefore, biomass integration in an existing petroleum refinery has received growing interest due to environmental concerns and increasing energy demands.

Petroleum refineries have well-developed infrastructure to produce fuels and therefore do not require additional intensive investments for the processing of alternative feedstocks. The co-processing approach enables utilization of the existing refinery infrastructure while maintaining the nominal operating conditions, producing high-quality products and avoiding the establishment of parallel systems. Today, there are two main technologies that could be used for co-processing of biomass feedstock with petroleum fractions: catalytic hydroprocessing and FCC. One of the main advantages of co-processing is the ability to combine a variety of bio- and fossil-based feedstocks/fractions without compromising product quality. The most common bio-based feedstocks for co-processing include edible oils (such as soybean, peanut, sunflower, palm and coconut oils), non-edible oils (such as the rapeseed oil, jatropha oil, jojoba oil, used cooking oils and animal fats), algae oils, microbial oils (produced *via* fermentation of the carbohydrate and protein fractions after enzymatic hydrolysis of food wastes from restaurants), pyrolysis bio-oil and hydrothermal liquefaction biocrude.

The co-processing of various lipid-based feedstocks with petroleum fractions has been widely explored by many researchers in recent years. This approach renders low-carbon hybrid fuels (of partially bio-origin and partially fossil-origin) of increased sustainability while enabling the gradual decarbonization of the transportation and aviation sectors *via* integration of biomass in refineries. As far as pyrolysis bio-oil and HTL biocrude is concerned, fast pyrolysis and HTL technologies are almost mature technologies with commercial and demonstration plants in the Netherlands, Sweden, Finland, Canada, and Norway. FPO has thus far been used mainly as heating fuel, but a goal is to use both FPO and HTL biocrude for the production of renewable marine and aviation fuels in the future. The fuel production from biocrudes could be carried out competitively by co-processing them in an oil refinery. Both cracking and the ability to remove oxygen from the biocrude are needed, and therefore co-processing could potentially take place in a fluid catalytic cracker, hydrotreater, or hydrocracker. However, the research on co-processing of pyrolysis bio-oil and HTL biocrude within a refinery is quite limited and the data are mostly from small-scale pilot plants. As a result, there is still a lot of room for improvement and further research is required prior the commercialization of the technology.

LCA studies of the co-processing pathway in refineries have recently been conducted, incorporating biomass provision and its co-processing with petroleum fractions as well as the production and usage of the final fuels. These results have identified the potential environmental benefits of co-processing, but future comprehensive LCA studies on industrial-scale implementation are required, incorporating techno-economic approaches to support informed decision-making.

Abbreviations

CFPP	Cold filter plugging point
CN	Cetane number
CPO	Catalytic pyrolysis oil
CTO	Catalyst-to-oil ratio
DP	Drop pressure
FAME	Fatty acids methyl ester
FCC	Fluid catalytic cracking
FFA	Free fatty acid

FPO	Fast pyrolysis oil
GHG	Greenhouse gas
HAGO	Heavy atmospheric gas oil
HDA	Aromatic hydrogenation
HDC	Hydrocracking
HDN	Hydrodenitrogenation
HDO	Hydrodeoxygenation
HDS	Hydrodesulfurization
HDT	Hydrotreating or hydrotreatment
HPO	Hydrotreated pyrolysis oil
HTL	Hydrothermal liquefaction
HVGO	Heavy vacuum gas oil
HVO	Hydroprocessing vegetable oil
LAGO	Light atmospheric gas oil
LCA	Life-cycle assessment
LCC	Life-cycle costing
LCO	Light cycle oil
LHSV	Liquid hourly space velocity
LR	Long residue
MON	Motor octane number
NPV	Net present value
PM	Particulate matter/mass
PO	Pyrolysis oil
RON	Research octane number
SRGO	Straight run gas oil
SV	Space velocity
TAG	Triacylglyceride
TAN	Total acid number
TRL	Technology readiness level
VGO	Vacuum gas oil
WHSV	Weight hourly space velocity

References

1. REN21, Renewables 2020 Global Status Report, 2020.
2. IEA, Renewables 2020 Analysis and forecast to 2025, 2020.
3. IEA, Tracking Transport 2020 – Analysis 2021.
4. A. Dimitriadis, PhD thesis, Aristotle University of Thessaloniki, Faculty of Engineering, School of Mechanical Engineering, 2021.
5. A. Primo and H. Garcia, *Chem. Soc. Rev.*, 2014, **43**, 7548–7561.
6. H. Aatola, M. Larmi, T. Sarjovaara and S. Mikkonen, *SAE Int. J. Eng.*, 2009, **1**, 1251–1262.
7. ConocoPhillips begins production of renewable diesel, Biodiesel Magazine, 2007.
8. J. S. D. Azevedo, Presentation in biofuels Houston summit III, Houston, 2008.

9. T. Kalnes, T. Marker and D. R. Shonnard, *Int. J. Chem. React. Eng.*, 2007, **5**, A48.
10. J. Rekoske, Farnborough Air Show, 2010.
11. M. Al-Sabawi and J. W. Chen, *Energy Fuels*, 2012, **26**, 5373–5399.
12. B. M. Weckhuysen, *Chem. Soc. Rev.*, 2015, **44**, 7342–7370.
13. I. Graca, J. M. Lopes, H. S. Cerqueira and M. F. Ribeiro, *Ind. Eng. Chem. Res.*, 2013, **52**, 275–287.
14. D. Kubicka and O. Kikhtyanin, *Catal. Today*, 2015, **243**, 10–22.
15. C. Toth, P. Baladincz and J. Hancsok, Proceedings of the 15th international conference on process integration, modelling and optimisation for energy saving and pollution reduction, 2012, p. 613–618.
16. G. W. Huber, P. O'Connor and A. Corma, *Appl. Catal., A*, 2007, **329**, 120–129.
17. T. M. Sankaranarayanan, M. Banu, A. Pandurangan and S. Sivasanker, *Bioresour. Technol.*, 2011, **102**(22), 10717–10723.
18. C. Toth, P. Baladincz and J. Hancsok, *Top. Catal.*, 2011, **54**, 1084–1093.
19. J. K. Satyarthi, T. Chiranjeevi, D. T. Gokak and P. S. Viswanathan, *Fuel Process. Technol.*, 2014, **118**, 180–186.
20. J. Walendziewski, M. Stolarski, R. Luzny and B. Klimek, *Fuel Process. Technol.*, 2009, **90**(5), 686–691.
21. J. Mikulec, J. Cvengros, L. Jorikova, M. Banic and A. Kleinova, *J. Cleaner Prod.*, 2010, **18**(9), 917–926.
22. B. Watkins, C. Olsen, K. Sutovich and N. Petti, *Catalagram*, 2008, **103**, 1–13.
23. P. Baladincz and J. Hancsok, *Chem. Eng. J.*, 2015, **282**, 152–160.
24. X. Han, H. Wang, Y. Zeng and J. Liu, *Energy Convers. Manage.: X*, 2021, **10**, 100069.
25. J. W. Chen, H. Farooqi and C. Fairbridge, *Energy Fuels*, 2013, **27**(6), 3306–3315.
26. D. Sagi, P. Baladincz, Z. Varga and J. Hancsok, *J. Cleaner Prod.*, 2016, **111**, 34–41.
27. P. Baladincz, M. Szepesi and J. Hancsok, *Chem. Eng. Trans.*, 2014, **39**, 1159–1164.
28. C. Toth, D. Sagi and J. Hancsok, *Top. Catal.*, 2015, **58**(14–17), 948–960.
29. A. Dimitriadis, G. Meletidis, U. Pfisterer, M. Auersvald, D. Kubicka and S. Bezergianni, *Fuel Process. Technol.*, 2022, **30**, 107220.
30. S. Bezergianni and V. D. Can, *J. Chem. Eng.*, 2015, **93**(6), 1017–1023.
31. S. Bezergianni, V. Dagonikou and S. Sklari, *Fuel Process. Technol.*, 2016, **144**, 20–26.
32. Z. Đukanović, S. B. Glišić, V. J. Čobanin, M. Nićiforović, C. A. Georgiou and A. M. Orlović, *Fuel Process. Technol.*, 2013, **106**, 160–165.
33. B. S. Rana, R. Kumar, R. Tiwari, R. Kumar, R. K. Joshi and M. O. Garg, *et al.*, *Biomass Bioenergy*, 2013, **56**, 43–52.
34. S. Bezergianni, A. Dimitriadis and D. Karonis, *Fuel*, 2014, **136**, 366–373.
35. S. Bezergianni, A. Dimitriadis and G. Meletidis, *Fuel*, 2014, **125**, 129–136.
36. A. Dimitriadis and S. Bezergianni, *J. Cleaner Prod.*, 2016, **131**, 321–326.

37. P. Bielansky, A. Reichhold and C. Schonberger, *Chem. Eng. Process.*, 2010, **49**(8), 873–880.
38. V. Abbasov, T. Mammadova, N. Aliyeva, M. Abbasov, N. Movsumov and A. Joshi, *et al.*, *Fuel*, 2016, **181**, 55–63.
39. P. Bielansky, A. Weinert, C. Schonberger and A. Reichhold, *Fuel Process. Technol.*, 2011, **92**(12), 2305–2311.
40. P. Lovas, P. Hudec, M. Hadvinova and A. Haz, *Fuel Process. Technol.*, 2015, **134**, 223–230.
41. V. P. Doronin, O. V. Potapenko, P. V. Lipin and T. P. Sorokina, *Fuel*, 2013, **106**, 757–765.
42. A. D. Pinho, M. B. B. de Almeida, F. L. Mendes, V. L. Ximenes and L. C. Casavechia, *Fuel Process. Technol.*, 2015, **131**, 159–166.
43. Y. X. Wang, T. He, K. T. Liu, J. H. Wu and Y. M. Fang, *Bioresour. Technol.*, 2012, **108**, 280–284.
44. G. Fogassy, N. Thegarid, Y. Schuurman and C. Mirodatos, *Energy Environ. Sci.*, 2011, **4**(12), 5068–5076.
45. G. Fogassy, N. Thegarid, Y. Schuurman and C. Mirodatos, *Green Chem.*, 2012, **14**(5), 1367–1371.
46. L. Gueudre, N. Thegarid, L. Burel, B. Jouguet, F. Meunier and Y. Schuurman, *et al.*, *Catal. Today*, 2015, **257**, 200–212.
47. A. Ibarra, E. Rodriguez, U. Sedran, J. M. Arandes and J. Bilbao, *Ind. Eng. Chem. Res.*, 2016, **55**(7), 1872–1880.
48. A. Ibarra, A. Veloso, J. Bilbao, J. M. Arandes and P. Castano, *Appl. Catal., B*, 2016, **182**, 336–346.
49. N. Thegarid, G. Fogassy, Y. Schuurman, C. Mirodatos, S. Stefanidis and E. F. Iliopoulou, *et al.*, *Appl. Catal., B*, 2014, **145**, 161–166.
50. J. A. Melero, M. M. Clavero, G. Calleja, A. Garcia, R. Miravalles and T. Galindo, *Energy Fuels*, 2010, **24**, 707–717.
51. H. Tian, C. Y. Li, C. H. Yang and H. H. Shan, *Chin. J. Chem. Eng.*, 2008, **16**(3), 394–400.
52. S. H. Ng, M. Al-Sabawi, J. S. Wang, H. Ling, Y. Zheng and Q. Wei, *et al.*, *Fuel*, 2015, **156**, 163–176.
53. S. Bezergianni, A. Dimitriadis, O. Kikhtyanin and D. Kubicka, *Prog. Energy Combust. Sci.*, 2018, **68**, 29–64.
54. X. Li, E. Mupondwa and L. Tabil, *Bioresour. Technol.*, 2018, **249**, 196–205.
55. G. Perin and P. Jones, *Curr. Opin. Biotechnol.*, 2019, **57**, 175–182.
56. A. Elgharbawy, W. Sadik, O. Sadek and M. Kasaby, *J. Chil. Chem.*, 2021, **66**, 1.
57. K. Dutta, A. Daverey and J.-G. Lin, *Renewable Energy*, 2014, **69**, 114–122.
58. B. Sajjadi, W.-Y. Chen, A. Aziz, A. Raman and S. Ibrahim, *Renewable Sustainable Energy Rev.*, 2018, **97**, 200–232.
59. S. Kumar, A. Tirlangi, A. Kumar and M. Imran, *et al.*, *Sustainable Energy Technol. Assess.*, 2023, **60**, 103498.
60. H. Liu, J. Han, Q. Huang, H. Shen, L. Lei, Z. Huang, Z. Zhang, Z. Zhao and F. Wang, *Ind. Eng. Chem. Res.*, 2020, **59**, 17440–17450.

61. M. Carmona-Cabello, L. García, A. Papadaki, E. Tsouko, A. Koutinas and M. P. Dorado, *Bioresour. Technol.*, 2021, **323**, 124597.
62. M. Zhang, Y. Hu, H. Wang, H. Li, X. Han, Y. Zeng and C. C. Xu, *Mol. Catal.*, 2011, **504**, 1418–1423.
63. C. Lindfors, D. C. Ellito, W. Prins, A. Oasmaa and J. Lehtonen, *Energy Fuels*, 2023, **37**, 799–804.
64. A. Dimitriadis and S. Bezergianni, *Renewable Sustainable Energy Rev.*, 2017, **68**, 113–125.
65. S. van Dyk, J. Su, J. D. Mcmillan and J. J. Saddler, *Biofuels, Bioprod. Biorefin.*, 2019, **13**, 760.
66. National Advanced Biofuels Consortium (NABC), Biofuels for Advancing America, NABC, 2010.
67. J. Holladay, Commercial Aviation Alternative Fuels Initiative (CAAFI), 2014.
68. I. Tews, Y. Zhu, C. Drennan, D. Elliott, L. Snowden-Swan, K. Onarheim, *et al.* Tech rep. Prepared for U.S. Department of Energy, 2014.
69. D. Kubicka and V. Tukac, *Adv. Chem. Eng.*, 2013, **42**, 141–194.
70. D. M. Santosa, I. Kutnyakov, M. Flake and H. Wang, *Energy Fuels*, 2022, **36**, 12641–12650.
71. A. Dimitriadis, D. Liakos, U. Pfisterer, M. Moustaka-GOuni, D. Karonis and S. Bezergianni, *Biomass Bioenergy*, 2021, **151**, 106171.
72. N. Bergvall, L. Sandstorm, F. Weiland and O. G. W. Ohrman, *Energy Fuel*, 2020, **34**, 8452–8465.
73. L. Sauvanaud, Y. Mathieu, A. Corma, L. Humphreys, W. Rowlands and T. Maschmeyer, *Appl. Catal., A*, 2018, **551**, 139–145.
74. T. Xing, A. Alvarez-majmutov, R. Gieleciak and J. Chen, *Energy Fuels*, 2019, **33**, 11135–11144.
75. F. A. Agblevor, O. Mante, R. McClung and S. T. Oyama, *Biomass Bioenergy*, 2012, **45**, 130–137.
76. G. W. Huber and A. Corma, *Angew. Chem., Int. Ed.*, 2007, **46**, 7184–7201.
77. A. D. R. Pinho, M. B. B. De Almeida, F. L. Mendes, V. L. Ximenes and L. C. Casavechia, *Fuel Process. Technol.*, 2015, **131**, 159–166.
78. J. A. Melero, J. Iglesias and A. Garcia, *Energy. Environ Sci*, 2012, **5**, 7393.
79. S. Arbogast, D. Bellman, D. Paynter and J. Wykowski, *Hydrocarbon Process.*, 2017, 189–207.
80. A. D. R. Pinho, M. Silva, A. P. Da Silva Neto and J. A. R. Cabral, *US Pat.*, US7540952B2, 2009.
81. D. Castello, T. H. Pedersen and L. A. Rosendahl, *Energies*, 2018, **11**, 3165.
82. S. van Dyk, J. Su, M. Ebadian and J. Saddler, *Fuel*, 2022, **324**, 124636.
83. C. Toth, P. Baladincz, S. Kovacs and L. Hancsok, *Clean Technol. Environ. Policy*, 2011, **13**, 581–585.
84. D. Pérez-Almada, A. Galan-Martın, M. del Mar Contreras and E. Castro, *Sustainable Energy Fuels*, 2023, 7, 4031.
85. R. Mahmud, S. M. Moni, K. High and M. Carbajales-Dale, *J. Cleaner Prod.*, 2021, **317**, 128247.

86. Ó. Ogmundarson, S. Sukumara, M. J. Herrgard and P. Fantke, *Trends Biotechnol.*, 2020, **38**, 1203–1214.
87. T. Bhaskar, A. Pandey, E. R. Rene and D. C. W. Tsang, *Waste Biorefinery Integrating Biorefineries for Waste Valorisation*, Elsevier, Amsterdam, Netherlands, 2020.
88. S. Ahlgren, A. Björklund, A. Ekman, H. Karlsson, J. Berlin, P. Börjesson, T. Ekvall, G. Finnveden, M. Janssen and I. Strid, *Biofuels Bioproducts and Biorefinery*, 2015, **9**, 606–619.
89. E. Zuiderveen, K. Kuipers, C. Caldeira, S. Hanssen, M. van der Hulst, M. de Jonge, R. van Zelm and M. Huijbregts, Preprint (Version 1), 2022, available at Research Square, DOI: 10.21203/rs.3.rs-1816061/v1.
90. R. Mahmud, S. M. Moni, K. High and M. Carbajales-Dale, *J. Cleaner Prod.*, 2021, **317**, 128247.
91. M. Shi, X. Zhao, Q. Wang and L. Wu, *Processes*, 2021, **9**, 187.
92. Argonne National Laboratory, ANL/ESD-21/20, 2022.
93. Biofuels digest, Available at: http://www.biofuelsdigest.com/bdigest/2018/ 03/19/the-digests-2018-multi-slide-guide-to-honeywells-uop-and-renewable-jetand-diesel-technology/11/, 2018. Access: August, 2021.
94. L. P. Chrysikou, V. Dagonikou, A. Dimitriadis and S. Bezergianni, *J. Cleaner Prod.*, 2019, **219**, 566.
95. P. L. Cruz, E. Montero and J. Dufour, *Fuel*, 2017, **196**, 362–370.
96. P. L. Cruz, D. Iribarren and L. Dufour, *Biofuels, Bioprod. Biorefin.*, 2020, **14**, 30–42.
97. P. L. Cruz, D. Iribarren and J. Dufour, *Energies*, 2019, **12**, 4664.
98. D. L. van Schalkwyk, M. Mandegari, S. Farzad and J. F. Gorgens, *Energy Convers. Manage.*, 2020, **213**, 112815.
99. M. Talmadge, H. Chum, E. Christensen, C. Kinchin, Y. Zhang, M. Biddy *et al.*, *Fuel*, 2016, **188**, 462–473.

CHAPTER 16

Current Status and Future Outlook for Renewable-diesel Biorefineries

PRAKASH PARTHASARATHY,[a] SNIGDHENDUBALA PRADHAN,[a] ANABEL FERNANDEZ,[b] ROSA RODRIGUEZ,[b] GERMÁN MAZZA,[c] TAREQ AL-ANSARI[a] AND GORDON MCKAY*[a]

[a] Division of Sustainable Development, College of Science and Engineering, Hamad Bin Khalifa University, Qatar Foundation, Doha, Qatar; [b] Instituto de Ingeniería Química – Grupo Vinculado al PROBIEN (CONICET-UNCo), Facultad de Ingeniería, Universidad Nacional de San Juan, Av. Libertador Gral. San Martín (Oeste) 1109, J5400ARL, San Juan, Argentina; [c] Instituto de Investigación y Desarrollo en Ingeniería de Procesos, Biotecnología y Energías Alternativas, PROBIEN (CONICET-Universidad Nacional del Comahue), Calle Buenos Aires, 1400, 8300, Neuquén, Argentina
*Email: gmckay@hbku.edu.qa

16.1 Introduction

16.1.1 The Necessity for Biofuels in Energy Consumption

The global need for primary energy is steadily rising and primarily met by fossil fuel sources, including coal, natural gas, and petroleum crude.[1] Given the limited availability and impending depletion of fossil fuels, there is a growing interest in exploring alternative energy fuels, such as biofuels.[2] Biofuels are liquid or gaseous fuels that are derived from organic materials specifically upgraded for transportation purposes. They provide numerous

Green Chemistry Series No. 89
Biorefining of Renewable Diesel: From Production to Market
Edited by Mohammad Aslam, Sanjeev Mishra and Jorge Aburto

Published by the Royal Society of Chemistry, www.rsc.org

advantages, including enhanced supply security, sustainability, promotion of circular economy, and employment opportunities.[3,4]

Biofuels are considered a viable solution for reducing global warming. This is due to the absorption of carbon during the growth of biomass balancing out the greenhouse gas (GHG) emissions produced by vehicles. Additionally, biofuels have better fuel characteristics than fuels derived from petroleum and can provide advantages such as improved fuel efficiency. Many developed and developing countries are increasingly adopting modern technology and efficient bioenergy conversion techniques that utilize various biofuels. These biofuels are becoming economically competitive with fossil fuels.[5]

In general, biofuels include ethanol, biodiesel, and renewable diesel. Bioethanol is derived from starch- or sugar-rich raw materials like corn, wheat, sugar cane and sugar beet by fermentation processes. Other sources of raw materials include trees, wood debris, and the organic component of municipal waste. Currently, most ethanol is derived from corn and sugarcane.[6] Ethanol can replace traditional gasoline/petrol because of its high-octane number, low cetane number, and high vaporisation heat. The worldwide biofuel production since 2019 and projection up to 2025 is presented in Figure 16.1.[7]

The projected volumes of ethanol, biodiesel, and renewable diesel for the years 2023–2025 are estimated to reach 119 billion litres, 46 billion litres, and 17 billion litres, respectively. By 2025, it is projected that biofuels will

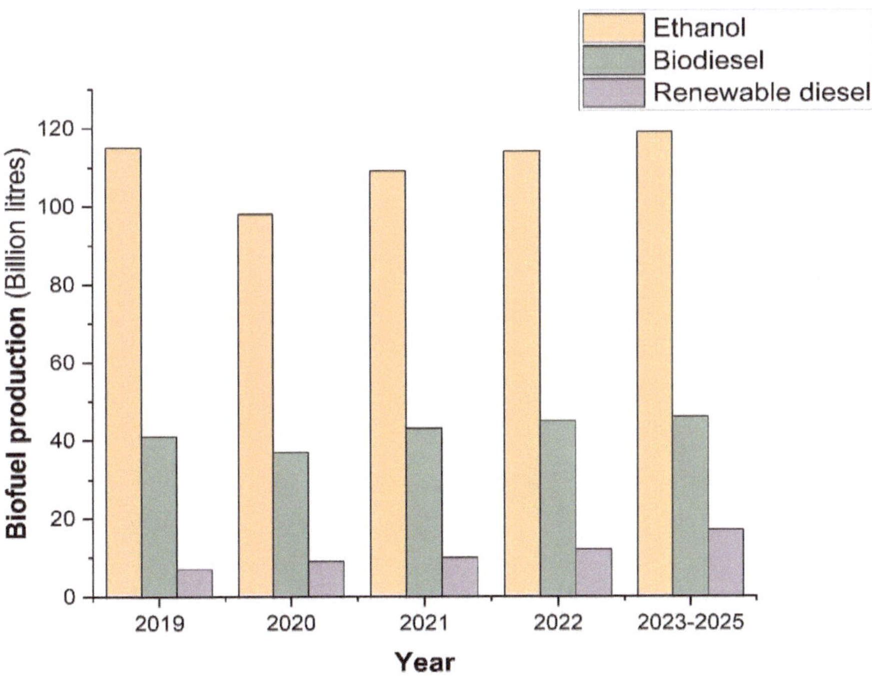

Figure 16.1 Worldwide biofuel production since 2019 and its projection up to 2025.

account for around 5.4% of the total energy demand for road transport, which is an increase from 4.8% in 2019.[8] Global biofuel production for transportation has significantly increased, with ethanol production playing an important role in the transportation sector. Moreover, the utilization of renewable diesel and biodiesel is also experiencing a significant surge. The recent increase in biofuel production and usage can be attributed, in part, to government support in the form of tax credits.[9] The provision of this support is mostly attributed to the acknowledgment of the effectiveness of biofuels in supplementing conventional fossil fuels.

16.1.2 Biodiesel and Its Advantages and Disadvantages

Biodiesel is a biofuel produced through the transesterification process, which entails reacting a feedstock (such as waste cooking oil, vegetable oil, non-edible plant oil, and animal fat) with an alcohol and a catalyst. Biodiesel, usually recognised as a fatty acid methyl ester (FAME), is the outcome of the transesterification process.[10] Biodiesel is a substitute fuel that closely resembles traditional diesel fuel. The biodegradability, non-toxicity, and absence of sulfur and aromatics give it an edge over traditional petroleum diesel. Figure 16.2 presents a schematic of biodiesel production.

The advantages of biodiesel are:

i. Biodiesel possesses the following qualities: it is renewable, portable, readily produceable, emits no visible smoke, toxic vapours, or odours, and is devoid of aromatic and sulfur compounds.

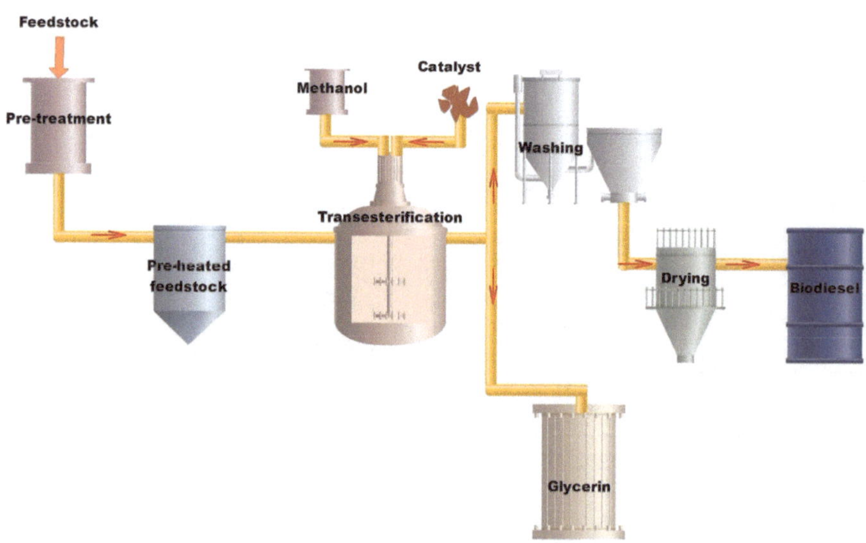

Figure 16.2 Schematic of biodiesel production.

ii. Biodiesel emits fewer GHGs, including CO_2, SO_2, particulate matter, and hydrocarbons, in comparison to diesel.
iii. Biodiesel exhibits increased combustion efficiency due to the presence of 10–11% oxygen.
iv. Biodiesel, because of its elevated flash point (100–170 °C), is relative safer for distribution, transportation, handling, and use.[11]
v. Due to its higher cetane number in comparison to diesel, biodiesel exhibits a reduced ignition delay.
vi. Biodiesel properties such as fuel penetration, atomization, and droplet size significantly impact fuel combustion and, consequently, emissions.
vii. In contrast to diesel engines, the performance of biodiesel does not require the addition of lubricant due to its exceptional clarity and purity.
viii. Biodiesel's superior lubricating qualities enhance lubrication in injector units and fuel pumps, thereby reducing engine wear and tear and increasing engine efficiency while reducing engine maintenance requirements.
ix. The fact that biodiesel is biodegradable and safe to handle reduces the cost and difficulty of cleanup in the event of any spillage.
x. The feedstock for the biodiesel can be easily sourced and maintained through systematic plantation schemes and replanting.
xi. Producing and consuming biodiesel requires a shorter time than diesel.
xii. The process of excavating, transporting, and refining is excluded, as the biodiesel can be synthesised locally from a wide range of feedstock materials and can be synthesised in any nation; therefore, it does not incur importation tariffs or taxes.
xiii. Production and application of biodiesel potentially augments the gross domestic product (GDP) of a nation.
xiv. Production and application of biodiesel also promotes rural employment prospects by means of agricultural development.

Some disadvantages of biodiesel include:

i. One of the primary impediments pertaining to biodiesel as a fuel is its comparatively increased production cost. This is because the transesterification process is expensive (biodiesel fuels necessitate costly fatty acid separation or usage of high-priced acid catalysts).
ii. There are environmental impacts associated with the transesterification process, including water consumption for cleansing, waste disposal, soap formation, *etc.*
iii. Compared to diesel, the 12% reduced energy content of biodiesel leads to a 2–10% raise in fuel consumption.[12]
iv. Due to its lower volatility and relatively high viscosity (11–18 times that of diesel), it necessitates a greater injector pressure.[13]
v. In comparison to diesel, biodiesel exhibits a reduced oxidation stability. In the presence of oxygen, it oxidises into fatty acids and induces corrosion in injectors, fuel tanks, and pipes.

vi. In engines, biodiesel induces excessive carbon deposition and gel formation (polymerization), which contaminates biodiesel and disturbs its flow.
vii. Due to the elevated oxygen content, biodiesel generates NO_x levels in the range of 10–14% during combustion, which are comparatively higher than conventional diesel.[14]
viii. Utilising biodiesel in an internal combustion (IC) engine results in injector coking, piston ring sticking, filter clogging, and other durability issues.
ix. Biodiesel has the potential to induce corrosion in vehicle components made of copper and brass, leading to issues such as blockage in the fuel system, failure of seals, clogging of filters, and the formation of deposits in injection pumps.
x. Given that over 95% of biodiesel is derived from edible oil, numerous allegations have surfaced suggesting that this may stimulate additional economic challenges. The conversion of edible lipids into biodiesel results in the utilisation of food resources as fuels for automobiles. The potential consequence of producing biodiesel on a large scale from edible oils is an imbalance in global food demand and supply.[15]

16.1.3 Renewable Diesel and Its Advantages and Disadvantages

Renewable diesel is synthesised through the process of hydrotreating, which entails the hydrogenation of triglycerides (fats) to eliminate metals and molecules that include nitrogen and oxygen. It is also referred to as 'hydrotreated vegetable oil (HVO)'.[16] It differs significantly from FAME biodiesel as it exclusively consists of hydrogen and carbon, making it a hydrocarbon fuel like diesel obtained from crude oil (petroleum). Often, renewable diesel is regarded as a direct substitute for petroleum fuel. Figure 16.3 presents a schematic of renewable-diesel production.

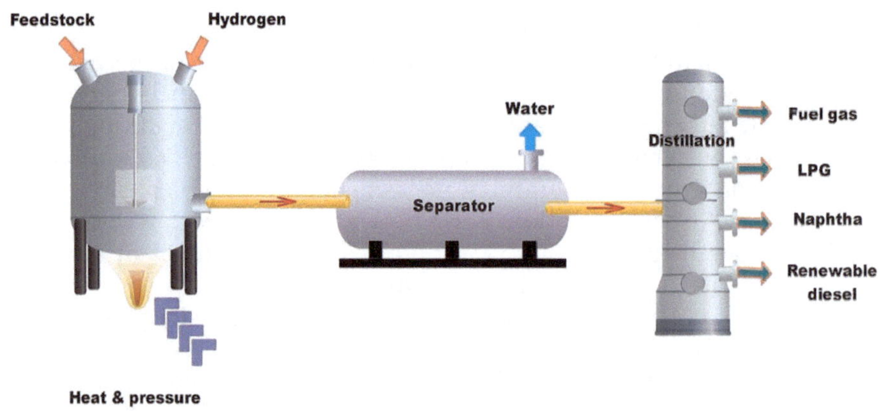

Figure 16.3 Schematic of renewable-diesel synthesis.

The advantages of renewable diesel are:

i. Renewable diesel has a high energy content due to the absence of double bonds and oxygen.
ii. It typically has a lower viscosity compared to biodiesel.
iii. It has a higher cetane number, indicating that it burns more easily.
iv. The fuel does not include sulfur, which can decrease the performance of exhaust catalysts.
v. Additionally, renewable diesel can be easily blended with other fuels.
vi. Renewable diesel has excellent storage stability.
vii. It is also less corrosive compared to biodiesel.
viii. Renewable diesel is regarded as a more advanced substitute for biodiesel, overcoming its limitations and enabling cleaner combustion relative to normal diesel fuel.
ix. Furthermore, the manufacture of renewable diesel does not necessitate the setting up of a separate facility for treating raw materials, as it can be readily handled in existing petrochemical plant facilities.

16.2 General Properties of a Liquid Fuel

16.2.1 Common Properties of Fuels

Fuel quality is an essential determinant of sustained success for fuel performance. Various fuel characteristics are crucial for preserving the effectiveness and integrity of fuel in cold weather conditions, throughout engine activities, transit, and storage. The features of a fuel are defined by its physicochemical attributes. These properties encompass various characteristics such as density, calorific value, viscosity, cloud and pour points, flash point, cetane number, oxidation stability, moisture content, ash content, acid value, titer, free glycerine, total glycerol, water content and sediment, sulfonated ash, copper-strip corrosion, presence of phosphorus/calcium/magnesium, carbon residue, distillation range, and sulfur content. Table 16.1 provides key properties of fuels. All of these properties are defined in the biodiesel standard ASTM D 6751. This standard specifies the criteria that pure biodiesel (B100) must satisfy to be utilised as a standalone fuel or mixed with diesel fuel derived from petroleum.

16.2.2 Comparison of Diesel, Biodiesel, and Renewable-diesel Properties

Table 16.2 compares the characteristics of diesel, biodiesel, and renewable diesel.[20,21]

Biodiesel has a higher density compared to renewable diesel. Renewable diesel exhibits superior viscosity compared to biodiesel, resulting in improved atomisation and increased power generation. Renewable diesel exhibits superior combustion compared to biodiesel due to its lower flash point, resulting in reduced ignition delay between fuel injection and the

Table 16.1 Critical properties of fuels.

Property	Short description
Calorific value ($MJ\,kg^{-1}$)	The calorific value of a fuel refers to the quantity of thermal energy that is released when a specific amount of fuel undergoes combustion.[17] The moisture level of the fuel is a crucial factor in determining its heating value.
Moisture content (% m/m)	Moisture refers to the quantity of water that cannot be transformed into biodiesel. Water content is a significant determinant of fuel quality, as it reduces the calorific value and can increase corrosion within engine compartments.
Density ($kg\,m^{-3}$) and specific gravity	Density is the measure of mass per unit volume. Denser oils contain more energy. Relative density is the ratio of a density of a component to that of water. The specific gravity of a fuel is critical for converting mass to volume, determining viscosity, flow, and the design of fuel tanks.
Kinematic viscosity ($mm^2\,s^{-1}$)	Viscosity refers to the measure of the resistance of a fuel to flow under the influence of gravity, considering its density. It is a crucial characteristic of fuel since it signifies its flowability. Consequently, it impacts the functioning of the fuel injection system and the dispersion of fuel spray, especially in cold weather conditions when the rise in viscosity hampers the flow of fuel.
Flash point (°C)	Flash point refers to the minimum temperature at which the fuel produces a momentary flash upon exposure to an open spark or flame. It is inversely proportional to the fuel's volatility. Biodiesel usually has a higher flash point than normal diesel fuel. Hence, it is safer to store, transport, and handle than normal diesel.
Cloud point (°C), pour point (°C), and cold filter plugging point (°C)	Cloud point refers to the temperature at which, upon cooling the fuel, the formation of wax crystals turns the transparent liquid cloudy. The pour point is the minimum temperature at which a fuel can still flow. This is the temperature at which the amount of wax in the fuel is high enough to solidify it. If the temperature falls too low, fuel components solidify and block the test filter. This temperature is called the 'cold filter plugging point'.
Cetane number	The cetane number quantifies the ignition characteristics of a fuel, and specifically its capacity for rapid self-ignition after injection. The cetane number is a key property of diesel fuels that has a substantial impact on both engine performance and emissions. The cetane number in diesel fuels varies depending on the types of hydrocarbons present. A higher cetane value is consistently linked to improved ignition quality of the gasoline.
Oxidation stability (h at a certain °C)	The assessment of biodiesel quality is significantly influenced by the process of biodiesel fuel oxidation. Oxidation stability refers to the extent of oxidation, *i.e.* the likelihood of reacting with air, and can help identify the necessity of antioxidants.
Acid value ($g\,kg^{-1}$ of KOH)	The acid number quantifies the concentration of unbound fatty acids present in a fuel sample. Saturated/unsaturated monocarboxylic acids that are found naturally in oils, greases and fats make up free fatty acids that are not chemically bound to the glycerol.

Table 16.1 (*Continued*)

Property	Short description
Titer (°C)	Titer refers to the specific temperature at which the state of oil transitions from a solid to a liquid. It is an essential variable because the transesterification process mostly involves liquids. Biodiesel manufacturing facilities may incur higher energy demands and production expenses when dealing with oils that have a high titer, as heating may be required.
Free glycerin (% m/m)	Free glycerin refers to the residual glycerol content in the final biodiesel product. The concentration of unbound glycerol is based on the method of synthesis. Inadequate separation during the ester product's washing process may be the cause of excessive glycerol generation. Glycerol is predominantly insoluble in biodiesel, thus allowing for efficient removal through centrifugation or settling processes. Unbound glycerol can exist in either the form of tiny droplets that are suspended in the biodiesel or as a minor quantity that is dissolved in the biodiesel. Elevated levels of free (unbound) glycerol can lead to the formation of deposits and harm the fuel injection system.
Total glycerol (% m/m)	Total glycerol is a quantitative assessment of the amount of triglyceride that has not been transformed into methyl esters. The calculation is based on the quantity of free glycerin and total mono-, di-, and triglycerides.
Water content and sediment (% m/m)	Presence of water and sediments are primarily maintenance concerns for biodiesel production. Water can exist in two states: either as water that is dissolved or as water droplets that are suspended. Though biodiesel is insoluble in water, it absorbs a significantly larger quantity of water compared to diesel fuel. Biodiesel can have a maximum water content of 1500 ppm,[18] but diesel fuel typically has a water content of just approximately 50 ppm.[19] Sediment may come from the fuel itself as insoluble compounds generated during fuel oxidation, or it may be composed of dirt and suspended rust.
Sulfated ash (% m/m)	Sulfated ash refers to the quantity of inorganic impurities, such as coarse particles, residual catalysts, and a certain amount of metal soaps in soluble form, present in a fuel sample.
Copper-strip corrosion (h at a certain °C)	Copper-strip corrosion assesses the nature of a fuel to corrode copper, bronze, or brass components. Corrosion caused by biodiesel could be due to acidic and sulfur compounds. Therefore, this characteristic is associated with the acid number.
Phosphorus, calcium, and magnesium (ppm)	Biodiesel contains many metallic elements, including calcium, magnesium, and phosphorus, which require purification. According to ASTM D6751 regulations, in biodiesel the phosphorus composition must be below 10 ppm while the combined magnesium and calcium content must be below 5 ppm.[12]
Carbon residue (% m/m)	Carbon residue indicates the fuel's tendency to deposit carbon after combustion. The Conradson carbon residue in biodiesel is more significant than that in normal diesel due to its strong association with the existence of free fatty acids, polymers, soaps, glycerides, inorganic contaminants, and highly unsaturated fatty acids. The maximum permissible limit according to ASTM D4530 is 0.050% (m/m).[20]

Table 16.2 A comparative analysis of the properties of diesel, biodiesel, and renewable diesel.

Properties	Units	Diesel	Biodiesel	Renewable diesel
Chemical structure		C_8–C_{25}	Methyl esters of C_{12}–C_{22} fatty acids	C_{15}–C_{18}
Heating value	$MJ\,kg^{-1}$	43.0–46.0	37.0–38.0	44.0–47.0
Density	$g\,cm^{-3}$	0.82–0.85	0.87	0.77
Viscosity	$mm^2\,s^{-1}$ at 40 °C	3.6	4.6	3.0–4.0
Flash point	°C	73.8	100.0–170.0	125.0
Cloud point	°C	−5.0	−5.0 to +15.0	−20.0 to +20.0
Cetane number		40.0–55.0	48.0–65.0	70.0–90.0
Oxygen content	%	0.0	11.0	0.0
Sulfur content	ppm	Less than 10.0	Less than 10.0	Less than 1.0
Stability		Good	Marginal	Good
Percentage change in NO_x emissions	ppm		+10.0	0.0–10.0
Cold flow properties		Good	Poor	Good
Lubricity		Additives required	Excellent	Additives required

start of combustion. Renewable diesel with long H–C chains exhibits a high cetane value, while short or isomerized H–C chains shows good cold flowing qualities. Mixing fossil fuel as high as 20% enhances the cold flowing qualities of renewable diesel.[22] Biodiesel undergoes rapid decomposition because of microbial activity, primarily due to its high oxygen content within its molecular structure. In contrast, renewable diesel exhibits greater stability compared to biodiesel, owing to the existence of a saturated HC chain and the absence of oxygen in its chemical structure.

Desirable characteristics for high-quality fuel include a higher heating value (HHV) and cetane value, strong stability, and low oxygen content. Renewable diesel, consisting solely of HC with a structure comparable to traditional diesel fuels, does not necessitate any alterations prior to being blended with normal diesel fuel. Furthermore, its performance is more on par with biodiesel. Nevertheless, biodiesel exhibits superior lubricity in comparison to diesel fuel and renewable diesel mostly because of its glyceride presence.

16.3 Life-cycle Assessment (LCA) Process of Renewable Diesel

LCA is a commonly used method for evaluating the environmental impact of a product/process across its entire lifespan. The ISO 14040 standard delineates the four primary phases of any LCA: objective and definition of scope, life-cycle inventory investigation, life-cycle impact evaluation, and outcome interpretation.[23] The LCA of a process needs the accurate computation of material flows, pollutant emissions, and the inclusion of all fluxes (mass and energy) in both preceding and subsequent processes. The objective of the LCA is to assess and optimise the utilisation and recuperation of energy and minimise its effect on the environment. The LCA's distinctive attributes render it a potent technique for assessing the ecological ramifications of energy generation and aiding decision-makers in attaining a comprehensive understanding.

The LCA comprises a sequence of consecutive steps: the initial phase of the technique involves determining the purpose and scope of the research. Subsequently, a comprehensive inventory analysis is conducted to ascertain and quantify the inputs and outputs of the system under investigation. The next phase entails performing an impact assessment to comprehensively examine the possible environmental and socio-economic outcomes of the system. In the last phase, the study's findings are thoroughly analysed to draw conclusions and provide recommendations. A thorough comprehension of the extensive consequences of this process is attained by conducting a rigorous examination of the energy and material inputs and outputs of the system.

LCA offers a comprehensive data set on a range of environmental factors, such as eutrophication potential, climate change impact, photochemical ozone formation potential (POFP), ozone depletion, ionising radiation, fossil depletion potential, human toxicity potential, terrestrial acidification potential, ecotoxicity potential, particulate matter formation potential, and metal exhaustion.

16.3.1 LCA of Renewable-diesel Synthesis from Sunflower Plant Waste

Spinelli *et al.* performed a LCA on the generation of renewable fuel from sunflower plant waste.[24] The objective of the analysis was to determine the extent to which the production of biodiesel from sunflower is decreasing the usage of fossil fuels. The LCA encompassed the agricultural phase, the extraction of sunflower oil, the process of transesterification, and logistics activities within the defined system boundary. The findings of the study can be interpreted as follows. The total raw material utilisation was 1.25×10^4 g kg^{-1}. The synthesis of 1 kg of biodiesel necessitated an energy input of 2.22×10^2 g of crude oil. The combustion stoichiometry indicates that the CO_2 release for 0.7 kg of biodiesel is 3.2 kg CO_2 per kg oil fuel. The specific energy required for biodiesel production was 7.24×10^{12} seJ kg^{-1}. The environmental loading ratio revealed that the fossil fuel energy used for biodiesel production was 1.2-fold greater than the renewable energy. The uncertainty estimate for carcinogens was highest at 2.74×10^{-9} DALY. The uncertainty range for other environmental impact categories was between 5% and 17%.

The implications on human health, resources, and ecosystem quality caused by biodiesel were quantified at 7.09×10^{-3} Pt, 7.26×10^{-2} Pt, and 2.99×10^{-1} Pt respectively. The findings indicate that the release of CO_2 from the burning of 1 kg of biodiesel is 0.7 kg, equivalent to 19.4 g of CO_2 per MJ of energy produced.

16.3.2 LCA of Renewable-diesel Synthesis from Palm Oil

Yee *et al.* carried out a LCA on the manufacture of biofuel from palm oil.[25] The study examined the GHG emissions associated with the manufacture of renewable fuel from the feedstock. The study encompassed the initial step of establishing plantations, the stage of palm-oil milling, and transesterification for the generation of biodiesel from palm. The annual energy consumption for the transesterification stage was 1.8 GJ per metric tonne of crude palm oil. The annual energy release resulting from the burning of palm diesel was 39.2 GJ per metric tonne of crude palm oil. The annual energy content of the biomass, glycerol, and biodiesel was 60.7 GJ per metric tonne of crude palm oil. The mean CO_2 emission was 42.5 metric tonnes of CO_2 per metric tonne of crude palm oil, equivalent to 211 997 kg of CO_2 emitted per metric tonne of biodiesel generated. The combustion of fibre and shell resulted in the emission of 117 234 kg of CO_2 per metric tonne of biodiesel. A total of 2087 kg of CO_2 was emitted per metric tonne of biodiesel throughout the transesterification process. The agricultural phase yielded a total of 19 metric tonnes of fresh fruit bunches per hectare of planted area. The processing of each tonne of fresh fruit bunch resulted in an output of 0.20 tonnes of crude palm oil. The co-products generated from the process, including fibre and shell, palm-oil mill effluent, and empty fruit bunches, had yields of 190 kg, 600–700 kg, and 230 kg per metric tonne of fresh fruit bunch, respectively. The conversion efficiency of crude palm oil to biodiesel was 99%. The yearly output amounted to 126 900 metric tonnes, equivalent to 1.0 metric tonnes of biodiesel per metric tonne of crude palm oil feedstock.

The results of the aforementioned LCA demonstrate that renewable diesel has minimal environmental consequences compared to fossil fuels. The findings emphasise environmental sustainability and potential areas for enhancement in the production of renewable diesel.

16.4 Renewable-diesel Production Potential

16.4.1 Commercial Production

Renewable-diesel production facilities have been constructed globally in commercial industries. A total of 5.5 billion litres of renewable diesel are now being generated, with an intended increase to 13 billion litres by the year 2024.[8] Multiple prominent companies are manufacturing renewable

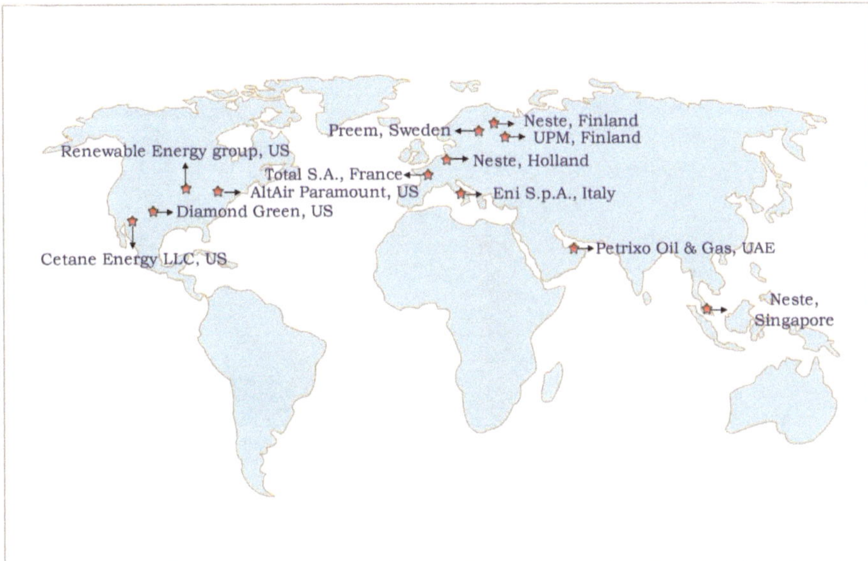

Figure 16.4 Geographical location of renewable-diesel producers worldwide.

diesel on a higher magnitude. The geographical distribution of certain companies is depicted in Figure 16.4.[26,27]

NExBTL, the first renewable diesel, was unveiled in 2005 by Neste Oil, a Finnish corporation. Over recent years, NESTE has manufactured almost 50% of renewable diesel, establishing itself as the foremost producer of renewable diesel in the world market.[26] Apart from Neste, other companies such as Eni S.p.A., Total S.A., Petrixo Oil & Gas, Diamond Green, Renewable Energy Group, AltAir Paramount, UPM, Preem, and Cetane Energy have constructed multiple commercial plants. The proposed expansions towards the establishment of renewable-diesel production facilities have significant potential for the expansion of renewable diesel in the worldwide market.

16.4.2 Worldwide Production Potential

Figure 16.5 illustrates the trend of renewable-diesel generation capacity worldwide between 2019 and 2022.[7] Renewable-diesel production has consistently risen in all regions in recent years and experienced a significant surge in 2022 following the COVID-19 epidemic. The upward trend is anticipated to persist in the forthcoming years.

North America is the primary producer of renewable diesel among the regions, with Europe being the second largest producer. The consistent expansion of renewable-diesel production capacity in various regions provides a positive indicator for replacing petroleum diesel and conventional biodiesel, thanks to its advantageous attributes. Markets have room to expand due to rising gasoline consumption and the backing of biofuel regulations in

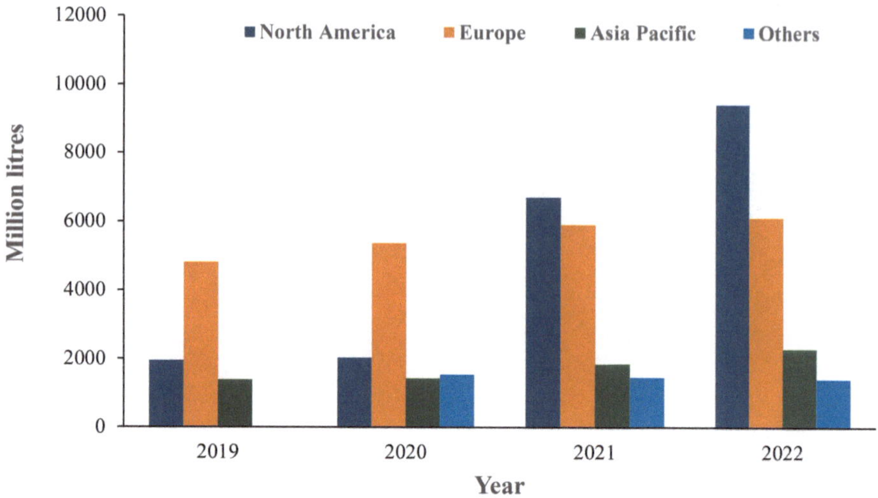

Figure 16.5 Global renewable-diesel production capacity.

Asian and American nations. Worldwide biofuel synthesis is projected to reach 182 billion litres in 2023–2025,[8] with ethanol production seeing the greatest growth in China and Brazil.[28,29] The American and Asian regions are anticipated to have the greatest biodiesel and renewable-diesel production levels, respectively.

Biofuels have gained global attention as a potential replacement for fossil fuels over the last five years. Key factors driving their market development are their good effects on rural development, lower carbon emissions compared to traditional fuels, and the current high oil costs. The percentage share of biofuel production through the years 2019–2022 is depicted in Figure 16.6.[7] It can be observed that the share of biofuels is steadily growing.

In 2022, the share of the biofuel production has increased from 5.1% in 2019 to 11.1%. At the very least, throughout the next few decades, it is anticipated that this upward tendency will continue. On the foreseeable future, biofuel will undoubtedly have a substantial impact on meeting the global energy requirements. To achieve self-sufficiency in biofuel production, a country must farm the primary biofuel crops domestically without compromising food needs.

16.4.3 Promising Prospects

The viability of renewable-diesel synthesis relies on the development of innovative technologies and achieving cost efficiency. Furthermore, the sustainability of energy consumption can be enhanced and preserved by ongoing research that investigates the potential of renewable diesel. The production of biofuels through the hydrolysis of lignocellulosic waste presents a favourable opportunity due to its environmentally friendly and sustainable nature. The advancement of renewable-diesel synthesis has been

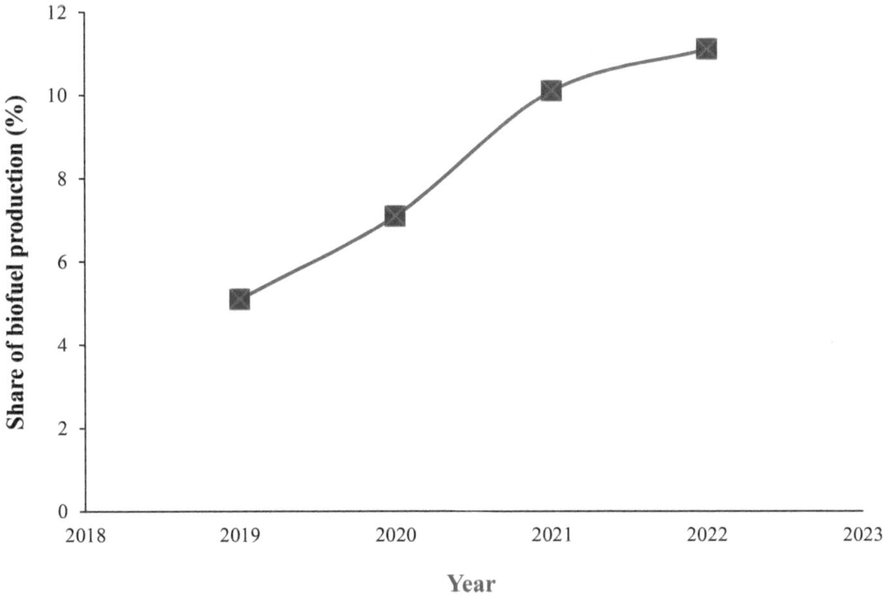

Figure 16.6 Percentage share of biofuel production.

significantly impacted by the successful implementation of commercial production of renewable fuel, along with the decreasing price of enzymes.[30]

The outlook for exploring new sources of renewable diesel, optimising the production and degradation process, affordability, and formulation is highly promising. It is imperative for both government and corporate sectors to seize this opportunity and adopt renewable diesel. It is crucial for the government to allocate funds for ongoing research to assist researchers in uncovering the most environmentally friendly and cost-effective technology for producing renewable diesel. Meanwhile, private entities should promote the widespread use of renewable diesel.

Due to the escalating energy fluctuations and the growth in the global economic framework, it is certain that there will be a rise in both average energy consumption and end-use demands. Hence, it is advantageous for countries that have significant agricultural exports and generate substantial waste to capitalise on the increasing energy demand by allocating more resources to renewable-diesel investments. Greater emphasis should be placed on allocating resources towards the expansion of large-scale production of renewable diesel and establishing a conducive framework for adopting policies that promote the commercialisation of renewable fuel.

16.4.4 Policy and Government Incentives

A government energy policy typically encompasses international agreements, laws governing commercial energy activities (storage, transport, and trading), incentives for investment, regulations for energy production and conversion, emissions standard framework, research and development, taxation and other public strategies, and promotion of energy diversity. Energy policy has a direct impact on the growth of the biodiesel business. As key stakeholders, all governments have a vital role in deciding the direction and magnitude of biodiesel development. This can be achieved by framing of biodiesel policy and implementation of appropriate incentives, such as tax exemptions, price restrictions, and direct subsidies.

While governments prioritise enhancing biodiesel production and usage, it is crucial for them to adequately address outstanding concerns such as rainforest depletion and the rise in food prices. Globally, deforestation is responsible for around 20% of GHG emissions.[31] Given the circumstances, it is advisable for the government to promote the use of feedstocks sourced from crops cultivated by biofuel companies on degraded lands. Furthermore, at present, there is a lack of public awareness of the biodiesel industry. Most of the general population is either uninformed or possesses inadequate understanding of biodiesel subjects. Hence, it is imperative for governments to bolster the promotion of the biodiesel industry and raise public consciousness about it. This would facilitate the government's biodiesel policies in gaining ample popular backing.

A comprehensive awareness of the economic and technical intricacies is still necessary for the commercialisation of renewable diesel. In addition to formulating guidelines that support the commercialisation of renewable diesel, it is crucial to ensure the continued existence and effective execution of these policies. In nations primarily dependent on fossil fuels, the government should offer subsidies to domestic investors to facilitate their transition from non-renewable energy sources to alternative energy sources including renewable diesel. The primary considerations for the commercialisation of renewable diesel include capital, operational, maintenance, and feedstock expenses. Despite the availability of government intervention funds, the maintenance cost could pose a hurdle until profit accumulates. Therefore, it is crucial to conduct a comprehensive evaluation of the production process, encompassing the advancement of cost-effective technology, the pursuit of affordable raw materials, and the implementation of supportive policies. These measures are necessary to enhance the synthesis of renewable diesel and overcome the current obstacle of high costs, which impedes biodiesel business.

16.5 Future Demands

16.5.1 Climate Change Perspective

Given the climate change policy targets for 2030 or 2050, there is a significant urgency to meet the goals for reducing GHG emissions.[32] There are sustainable and low-GHG biofuels made from biomass materials that are accessible worldwide. This represents the start of a development plan, spanning a medium- to long-term period, aimed at substituting fossil fuels and biofuels derived from waste materials. For these initiatives to evolve, there needs to be a timeframe that is difficult to forecast in terms of how important it will be in the future market. Some of these options have not been competitive for years, and they need a lot of public money to develop an appropriate infrastructure (like charging stations and distribution networks). These initiatives should also be backed with government investment promotion and incentives.

The countries that signed the Paris climate agreement committed to a globally enforceable goal of restricting the increase in global warming to a maximum of 2 °C by 2050.[33] The promise is to mitigate the observable impacts of climate change. Hence, it is the duty of all countries to expedite this commitment by implementing ambitious policies and adopting cutting-edge technologies.

Given that the global population is expected to reach over 9 billion by 2050, rising global traffic is also a major concern.[34] Climate experts are raising the alarm and advocating for an immediate transportation transformation. Currently, transport is responsible for over 14% of global GHG release.[35] Hence, policymakers and automobile industries are under enormous pressure to respond to the growing demand for renewable fuels.

16.5.2 Biofuel Research

Globally, a great deal of research is being done to enhance the production of biofuels. Without a doubt, the field of biofuel research is expanding quickly, as is the biofuel industry. Significant technological advancements have been made, and a thorough understanding of the processes involved in biofuel production has also been gained. However, biofuels cannot yet fully replace fossil fuels, and several integrated approaches combining biology, biotechnology, process engineering, bioengineering, automobile engineering, and artificial intelligence are still needed to optimise the production of biofuels on a large scale.

Developed nations are aiming to develop the biofuel business for the transportation sector through research and development (R & D). Research in the future needs to support the use of already available raw materials, such as vegetable oil fuels from decentralised processing and biodiesel made from vegetable oils. The current technologies for producing ethanol are not very cost-effective; instead, research is focused on finding ways to lower the total cost of production, which calls for low energy consumption during pretreatment, more effective separation techniques, improved and sophisticated methods for converting lignocellulosic biomass waste to sugars, and adding value to lignin.[36]

The fuel characteristics of biofuels are determined by the fatty acid content of the feedstock used. Biodiesel must possess similar fuel characteristics to diesel fuel to operate the engine proficiently. While the direct practical implementation of biofuel is still in its early stages, engine modification and fuel modification alone require a substantial quantity of research. It is necessary to conduct research on the optimisation of engine and exhaust gas treatment for renewable diesel and their mixtures to meet the escalating emission standards of the future.

16.6 Conclusion

Renewable diesel is a highly promising substitute for conventional diesel fuel. This chapter examines the crucial necessity of biofuels in energy consumption, with a specific emphasis on biodiesel and renewable diesel, as well as their advantages and disadvantages. Biodiesel and renewable diesel, apart from being renewable, also exhibit significantly reduced greenhouse gas emissions. As a result, the use of renewable diesel and biodiesel is also witnessing a substantial increase.

Renewable diesel possesses several positive characteristics compared to biodiesel and diesel. These include improved viscosity, combustion, stability, blending flexibility, and lubricity. Nevertheless, the production of renewable diesel incurs substantial costs, which may deter many businesses and government agencies from adopting it. Consequently, there has been a strong push to quickly develop renewable diesel at a reasonably cheap cost to eventually replace non-renewable fuels. The expected expansions aimed at establishing renewable-diesel manufacturing facilities holds significant potential for expanding the renewable-diesel market worldwide. The prospects for investigating novel sources of renewable diesel, enhancing the efficiency of the production and degrading process, ensuring cost-effectiveness, and refining the process formulation are quite encouraging. Both the government and corporate sectors must urgently take this opportunity and embrace the use of renewable diesel. A thorough understanding of the complex economic and technical aspects is still required for the commercialisation of renewable diesel.

Abbreviations

AI	Artificial intelligence
ASTM	American society for testing and materials
COVID	Coronavirus disease
DALY	Disability adjusted life years
FAME	Fatty acid methyl ester
GDP	Gross domestic product
GHG	Greenhouse gas
HHV	Higher heating value
HVO	Hydrotreated vegetable oil
IC	Internal combustion
ISO	International Organization for Standardization
LCA	Life-cycle analysis
LPG	Liquefied petroleum gas

POFP	Photochemical ozone formation potential
R & D	Research and development

Glossary

- **Aromatics:** Benzene-ringed hydrocarbons, *e.g.* toluene.
- **Biodiesel:** A form of diesel fuel, derived from biological sources like vegetable oils, animal fats, *etc.*
- **Biofuels:** Fuels derived from organic substances, including plants and animals.
- **Carbon deposition:** Byproduct of combustion in an engine produced due to incomplete fuel combustion.
- **Circular economy:** An endeavour to make the most use of resources and to reduce waste to a least minimum.
- **Greenhouse gases (GHG):** Atmospheric gases that capture heat.
- **Gross domestic product:** Total market value of the goods and services produced by a country during a certain time period.
- **Hydrotreatment:** Treatment process for removing sulfur and nitrogen impurities.
- **IC engine:** Heat engine in which the combustion of a fuel occurs in the presence of air in a combustion chamber.
- **Life-cycle analysis (LCA):** A widely employed technique for evaluating the environmental implications of a process/product across its complete life cycle.
- **Renewable diesel:** A liquid fuel derived from organic materials that is chemically the same as normal diesel fuel.
- **Tax credit:** Amount of money that taxpayers can deduct from the income taxes that they owe.
- **Transesterification:** The reaction between an ester and an alcohol in the presence of an acid catalyst.
- **Vegetable oil:** Oil that is extracted from plant parts such as seeds.

Acknowledgement

The authors would like to thank Qatar National Research Fund (QNRF) for their support of this research through NPRP-11S-0117-180328, the Supreme Committee for Delivery and Legacy (SCDL) and Hamad Bin Khalifa University (HBKU) and Qatar Foundation (QF).

References

1. P. Parthasarathy, M. Alherbawi, M. Shahbaz, H. R. Mackey, G. McKay and T. Al-Ansari, Conversion of oil palm waste into value-added products through pyrolysis: a sensitivity and techno-economic investigation, *Biomass Convers. Biorefin.*, 2022, **1**, 1–21.
2. T. Al-Ansari, A. AlNouss, N. Al-Thani, P. Parthasarathy, S. ElKhalifa and G. Mckay, *et al.*, Optimising multi biomass feedstock utilisation

considering a multi technology approach, *Comput.-Aided Chem. Eng.*, 2020, **48**, 1633–1638.

3. P. Parthasarathy, N. K. Gupta and K. S. Narayanan, Effect of composting on products of slow pyrolysis, *Biofuels*, 2015, **6**(5–6), 313–321.
4. S. Pradhan, A. H. Abdelaal, K. Mroue, T. Al-Ansari, H. R. Mackey and G. McKay, Biochar from vegetable wastes: agro-environmental characterization, *Biochar*, 2020, **2**(5), 439–453.
5. H. I. Mahdi, A. Bazargan, G. McKay, N. I. W. Azelee and L. Meili, Catalytic deoxygenation of palm oil and its residue in green diesel production: A current technological review, *Chem. Eng. Res. Des.*, 2021, **174**, 158–187.
6. G. Knothe, Biodiesel and renewable diesel: A comparison, *Prog. Energy Combust. Sci.*, 2010, **36**(3), 364–373.
7. S. R. Chia, S. Nomanbhay, M. Y. Ong, A. H. B. Shamsuddin, K. W. Chew and P. L. Show, Renewable diesel as fossil fuel substitution in Malaysia: A review, *Fuel*, 2022, **314**, 123137.
8. D. Kumar, B. Singh and S. K. Gupta, *Renewable Diesel: Value Chain, Sustainability, and Challenges*, Elsevier, Amsterdam, Netherlands, 2023, pp. 1–195.
9. D. Ogunkoya, W. L. Roberts, T. Fang and N. Thapaliya, Investigation of the effects of renewable diesel fuels on engine performance, combustion, and emissions, *Fuel*, 2015, **140**, 541–554.
10. R. Koul, N. Kumar and R. C. Singh, A review on the production and physicochemical properties of renewable diesel and its comparison with biodiesel, *Energy Sources, Part A*, 2021, **43**(18), 2235–2255.
11. M. H. M. Ashnani, A. Johari, H. Hashim and E. Hasani, A source of renewable energy in Malaysia, why biodiesel?, *Renewable Sustainable Energy Rev.*, 2014, **35**, 244–257.
12. A. E. Atabani, A. S. Silitonga, I. A. Badruddin, T. M. I. Mahlia, H. H. Masjuki and S. Mekhilef, A comprehensive review on biodiesel as an alternative energy resource and its characteristics, *Renewable Sustainable Energy Rev.*, 2012, **16**(4), 2070–2093.
13. S. Li, Y. Wang, S. Dong, Y. Chen, F. Cao and F. Chai, *et al.*, Biodiesel production from Eruca Sativa Gars vegetable oil and motor, emissions properties, *Renewable Energy*, 2009, **34**(7), 1871–1876.
14. P. Nakpong and S. Wootthikanokkhan, Roselle (Hibiscus sabdariffa L.) oil as an alternative feedstock for biodiesel production in Thailand, *Fuel*, 2010, **89**(8), 1806–1811.
15. M. M. Gui, K. T. Lee and S. Bhatia, Feasibility of edible oil *vs.* non-edible oil *vs.* waste edible oil as biodiesel feedstock, *Energy*, 2008, **33**(11), 1646–1653.
16. D. S. Kim, M. Hanifzadeh and A. Kumar, Trend of biodiesel feedstock and its impact on biodiesel emission characteristics, *Environ. Prog. Sustainable Energy*, 2018, **37**(1), 7–19.
17. P. Prakash and K. N. Sheeba, Prediction of pyrolysis and gasification characteristics of different biomass from their physico-chemical properties, *Energy Sources, Part A*, 2016, **38**(11), 1530–1536.

18. S. Fernando, P. Karra, R. Hernandez and S. K. Jha, Effect of incompletely converted soybean oil on biodiesel quality, *Energy*, 2007, **32**(5), 844–851.
19. M. S. Shehata, Emissions, performance and cylinder pressure of diesel engine fuelled by biodiesel fuel, *Fuel*, 2013, **112**, 513–522.
20. L. C. Meher, S. D. Vidya and S. N. Naik, Technical aspects of biodiesel production by transesterification—a review, *Renewable Sustainable Energy Rev.*, 2006, **10**(3), 248–268.
21. A. G. Alsultan, N. Asikin Mijan, N. Mansir, S. Z. Razali, R. Yunus and Y. H. Taufiq-Yap, Combustion and Emission Performance of CO/NOx/SOx for Green Diesel Blends in a Swirl Burner, *ACS Omega*, 2021, **6**(1), 408–415.
22. M. Unglert, D. Bockey, C. Bofinger, B. Buchholz, G. Fisch and R. Luther, *et al.*, Action areas and the need for research in biofuels, *Fuel*, 2020, **268**, 117227.
23. P. Parthasarathy, F. Tahir, S. Pradhan, T. Al-Ansari and G. McKay, Life cycle assessment of biofuel production from waste date stones using conventional and microwave pyrolysis, *Energy Convers. Manage.: X*, 2024, **21**, 100510.
24. D. Spinelli, S. Jez, R. Pogni and R. Basosi, Environmental and life cycle analysis of a biodiesel production line from sunflower in the Province of Siena (Italy), *Energy Policy*, 2013, **59**, 492–506.
25. K. F. Yee, K. T. Tan, A. Z. Abdullah and K. T. Lee, Life cycle assessment of palm biodiesel: Revealing facts and benefits for sustainability, *Appl. Energy*, 2009, **86**(Suppl. 1), S189–S196.
26. B. Zhang, J. Wu, C. Yang, Q. Qiu, Q. Yan and R. Li, *et al.*, Recent Developments in Commercial Processes for Refining Bio-Feedstocks to Renewable Diesel, *BioEnergy Res.*, 2018, **11**(3), 689–702.
27. H. Bateni, A. Saraeian and C. Able, A comprehensive review on biodiesel purification and upgrading, *Biofuel Res. J.*, 2017, **4**(3), 668–690.
28. B. Wu, Y. W. Wang, Y. H. Dai, C. Song, Q. L. Zhu and H. Qin, *et al.*, Current status and future prospective of bio-ethanol industry in China, *Renewable Sustainable Energy Rev.*, 2021, **145**, 111079.
29. A. S. Duden, P. A. Verweij, A. P. C. Faaij, D. Baisero, C. Rondinini and F. van der Hilst, Biodiversity impacts of increased ethanol production in Brazil, *Land*, 2020, **9**(1), 12.
30. E. F. Aransiola, T. V. Ojumu, O. O. Oyekola, T. F. Madzimbamuto and D. I. O. Ikhu-Omoregbe, A review of current technology for biodiesel production: State of the art, *Biomass Bioenergy*, 2014, **61**, 276–297.
31. Asner G. P., Environmental defense fund, 2010. Measuring carbon emissions from tropical deforestation: An overview. Available from: http://unfccc.int/methods_science/redd/items/4531.php.
32. T. H. Joung, S. G. Kang, J. K. Lee and J. Ahn, The IMO initial strategy for reducing greenhouse gas (GHG) emissions, and its follow-up actions towards 2050, *J. Int. Marit. Saf., Environ. Aff., Shipp.*, 2020, **4**(1), 1–7.
33. N. Glanemann, S. N. Willner and A. Levermann, Paris Climate Agreement passes the cost-benefit test, *Nat. Commun.*, 2020, **11**(1), 1–11.

34. C. Béné, M. Barange, R. Subasinghe, P. Pinstrup-Andersen, G. Merino and G. I. Hemre, *et al.*, Feeding 9 billion by 2050 – Putting fish back on the menu, *Food Secur.*, 2015, 7(2), 261–274.
35. WHO, World Health Organisation, 2024. Environment, Climate Change and Health. Available from: https://www.who.int/teams/environment-climate-change-and-health/healthy-urban-environments/transport/health-risks#:~:text=Climate impacts,greenhouse gases (GHG) budget.
36. V. Larnaudie, M. Bule, K. Y. San, P. V. Vadlani, J. Mosby and S. Elangovan, *et al.*, Life cycle environmental and cost evaluation of renewable diesel production, *Fuel*, 2020, **279**, 118429.

Subject Index